Connections in Classical and Quantum Field Theory

L. Mangiarotti
University of Camerino, Italy

G. Sardanashvily
Moscow State University, Russia

Connections in Classical and Quantum Field Theory

World Scientific
Singapore • New Jersey • London • Hong Kong

Published by

World Scientific Publishing Co. Pte. Ltd.

P O Box 128, Farrer Road, Singapore 912805

USA office: Suite 1B, 1060 Main Street, River Edge, NJ 07661

UK office: 57 Shelton Street, Covent Garden, London WC2H 9HE

British Library Cataloguing-in-Publication Data
A catalogue record for this book is available from the British Library.

CONNECTIONS IN CLASSICAL AND QUANTUM FIELD THEORY

ISBN 981-02-2013-8

Printed in Singapore by Uto-Print

Preface

The present book is based on the graduate and post graduate courses of lectures given at the Department of Theoretical Physics of Moscow State University and the Department of Mathematics and Physics of Camerino University. It is addressed to a wide audience of physicists and mathematicians, and aims at showing in a unified way the role that the concept of a connection plays both in classical and quantum field theory. To our knowledge, this is the first attempt to present connections as a main geometrical object which underlies many relevant physical ideas.

The concept of a connection in quantum field theory is rather new. It is phrased in algebraic terms, in comparison with the purely geometric one used in classical field theory. This concept is based on modern development of quantum mechanics, SUSY models, BRST formalism, and non-commutative field theory. In our opinion, connections provide a new link between classical and quantum physics. In classical field theory, we follow the general notion of a connection on fibre bundles, formulated in terms of jet manifolds. We emphasize the role of connections as the main ingredient in dynamic equation theory, and Lagrangian and Hamiltonian formalisms. As is well-known, connections on principal bundles have a role of gauge potentials of fundamental interactions while, in gravitation theory, connections characterize the space-time geometry. Classical mechanics is also presented as a particular field theory, involving the concept of a connection in many aspects. The reader will find connections in quantum mechanics, too. In fact, we collect together the basic methods concerning different types of connections in quantum field theory. These are superconnections, connections in BRST formalism, topological field theory, theory of anomalies, and non-commutative geometry.

Mathematics is not the primary scope of our book, but provides the powerful methods of studying contemporary field models.

Contents

Introduction

The main reasons why connections play a prominent role in contemporary field theory lie in the fact that they enable us to deal with invariantly defined objects. Gauge theory shows clearly that this is a basic physical principle.

In gauge theory, connections on principal bundles are well known to provide the mathematical description of gauge potentials of the fundamental interactions. Furthermore, since the characteristic classes of principal bundles are expressed in terms of gauge strength, one also meets the topological phenomena in classical and quantum gauge models, e.g., anomalies. All gauge gravitation models belong to the class of metric-affine theories where a pseudo-Riemannian metric and a connection on a world manifold are considered on the same footing as dynamic variables.

Though gauge theory has made great progress in describing fundamental interactions, it is a particular case of field models on fibre bundles. Differential geometry of fibre bundles and formalism of jet manifolds give the adequate mathematical formulation of classical field theory, where fields are represented by sections of fibre bundles. This formulation is also applied to classical mechanics seen as a particular field theory on fibre bundles over a time axis. In summary, connections are the main ingredient in describing dynamic systems on fibre bundles, and in Lagrangian and Hamiltonian machineries.

Jet manifolds provide the appropriate language for theory of (non-linear) differential operators and equations, the calculus of variations, and Lagrangian and Hamiltonian formalisms. For this reason, we follow the general definition of connections as sections of jet bundles. This enables us to deal with non-linear connections and to include connections in a natural way in describing field dynamics.

In quantum field theory requiring the theory of Hilbert and other linear spaces, one needs another concept of a connection, phrased in algebraic terms as a connection on modules and sheaves. This notion is equivalent to the above-mentioned geometric one in the case of structure modules of smooth vector bundles. Extended to the case of modules over graded and non-commutative rings, it provides the geometric language of supersymmetry and non-commutative geometry.

Our book is not a book on differential geometry, but it collects together the basic mathematical facts about various types of connections in Lagrangian and Hamiltonian formalisms, gauge and gravitation theory, classical and quantum mechanics, topological field theory and anomalies, BRST formalism, and non-commutative geometry. Additionally, it provides the detailed exposition of relevant theoretical methods both in classical and quantum field theory.

We have tried throughout to give the necessary mathematical background, thus making the book self-contained. We hope that our book will attract new interest, from theoretical and mathematical physicists, in modern geometrical methods in field theory.

Chapter 1

Geometric interlude

This Chapter summarizes the basic notions on fibre bundles and jet manifolds which find an application in the sequel [123, 179, 265, 274].

Unless otherwise stated, all maps are smooth, while manifolds are real, finite-dimensional, Hausdorff, second-countable (hence, paracompact) and connected.

We use the standard symbols $\otimes$, $\vee$, and $\wedge$ for the tensor, symmetric, and exterior products, respectively. By ∂_A^B are meant the partial derivatives with respect to the coordinates with indices ${}^A_B$. The symbol $\circ$ stands for a composition of maps.

1.1 Fibre bundles

Subsections: *Fibre bundles*, 3; *Vector bundles*, 6; *Affine bundles*, 8; *Tangent and cotangent bundles*, 9; *Tangent and cotangent bundles of fibre bundles*, 10.

Fibre bundles

Let M and N be manifolds of dimensions m and n, respectively. Recall that by the *rank of a morphism* $f : M \to N$ at a point $p \in M$ is meant the rank of the linear map of the tangent spaces

$$T_p f : T_p M \to T_{f(p)} N,$$

i.e., the rank of the Jacobian matrix of f at p. Suppose that f is of maximal rank at $p \in M$. Then f at the point p is said to be a local diffeomorphism if $m = n$, an *immersion*, if $m < n$, and a *submersion* if $m > n$.

A manifold Y is called a *fibred manifold* over a base X if there is a surjective submersion (a *projection*)

$$\pi : Y \to X. \tag{1.1.1}$$

A fibred manifold $Y \to X$ is provided with an atlas of fibred coordinates (x^λ, y^i), where x^λ are coordinates on the base X with transition functions $x^\lambda \to x'^\lambda$ independent of the coordinates y^i.

A fibred manifold $Y \to X$ is called a *fibre bundle* if it is locally trivial. It means that the base X admits an open covering $\{U_\xi\}$ so that Y is locally equivalent to the splittings

$$\psi_\xi : \pi^{-1}(U_\xi) \to U_\xi \times V, \tag{1.1.2}$$

together with the *transition functions*

$$\begin{aligned} &\rho_{\xi\zeta} : (U_\xi \cap U_\zeta) \times V \to (U_\xi \cap U_\zeta) \times V, \\ &\rho_{\xi\zeta} : (x, v) \mapsto (x, \rho_{\xi\zeta}(x, v)), \\ &\psi_\xi(y) = (\rho_{\xi\zeta} \circ \psi_\zeta)(y), \qquad y \in \pi^{-1}(U_\xi \cap U_\zeta), \end{aligned} \tag{1.1.3}$$

which fulfill the *cocycle relations*

$$\rho_{\xi\zeta} \circ \rho_{\zeta\iota} \circ \rho_{\iota\xi} = \mathrm{Id}\,((U_\xi \cap U_\zeta \cap U_\iota) \times V). \tag{1.1.4}$$

The manifold V is one for all local splittings (1.1.2). It is called a *typical fibre.* Trivialization charts (U_ξ, ψ_ξ) constitute an *atlas*

$$\Psi = \{(U_\xi, \psi_\xi), \rho_{\xi\zeta}\}$$

of the fibre bundle Y. Given an atlas Ψ, a fibre bundle Y is provided with the associated fibred coordinates (x^λ, y^i), called *bundle coordinates,* where

$$y^i(y) = (y^i \circ \mathrm{pr}_2 \circ \psi_\xi)(y), \qquad y \in Y,$$

are coordinates on the typical fibre V. Note that a fibre bundle $Y \to X$ is uniquely defined by a bundle atlas Ψ. Two bundle structures on a manifold Y are said to be equivalent if the corresponding bundle atlases are equivalent, i.e., a union of these atlases is also a bundle atlas.

Hereafter, we will restrict our consideration to fibre bundles.

A fibre bundle $Y \to X$ is called trivial if Y is diffeomorphic to the product $X \times V$. Different trivializations of a fibre bundle $Y \to X$ differ from each other in projections $Y \to V$.

THEOREM 1.1.1. Each fibre bundle over a contractible base is trivial ([286], p.53). □

By a *section* (or a *global* section) of a fibre bundle (1.1.1) is meant a morphism $s : X \to Y$ such that $\pi \circ s = \mathrm{Id}\, X$. A section s is an *imbedding*, i.e., $s(X) \subset Y$ is both a (closed) submanifold and a topological subspace of Y. Similarly, a section s of a fibre bundle $Y \to X$ over a submanifold $N \subset X$ is said to be a morphism $s : N \to Y$ such that

$$\pi \circ s = i_N : N \hookrightarrow X$$

is a natural injection. A section of a fibre bundle over an open subset $U \subset X$ is called simply a *local* section. A fibre bundle, by definition, admits a local section around each point of its base.

THEOREM 1.1.2. Each fibre bundle $Y \to X$ whose typical fibre is diffeomorphic to $\mathbb{R}^m$ has a global section. Its section over a closed subset of $N \subset X$ can always be extended to a global section ([157]; [286], p.55). □

A *fibred* (or *bundle*) *morphism* of two bundles $\pi : Y \to X$ and $\pi' : Y' \to X'$ is a pair of maps $\Phi : Y \to Y'$ and $f : X \to X'$ such that the diagram

$$\begin{array}{ccc} Y & \xrightarrow{\Phi} & Y' \\ {\scriptstyle\pi}\downarrow & & \downarrow{\scriptstyle\pi'} \\ X & \xrightarrow[f]{} & X' \end{array} \qquad (1.1.5)$$

is commutative, i.e., Φ sends fibres to fibres. In brief, we will say that (1.1.5) is a fibred morphism $\Phi : Y \underset{f}{\to} Y'$ over f. If $f = \mathrm{Id}\, X$, then $\Phi : Y \underset{X}{\to} Y'$ is called a fibred morphism over X.

If a fibred morphism Φ (1.1.5) is a diffeomorphism, it is called an *isomorphism* of fibre bundles. Two fibre bundles over the same base X are said to be *equivalent* if there exists their isomorphism over X.

A fibred morphism Φ (1.1.5) over X (or its image $\Phi(Y)$) is called a *subbundle* of the fibre bundle $Y' \to X$ if $\Phi(Y)$ is a submanifold of Y'. There are the following useful criteria for an image and an inverse image of a fibred morphism to be subbundles ([247], p.19; [302]).

PROPOSITION 1.1.3. Let $\Phi : Y \to Y'$ be a fibred morphism over X. Given a global section s of the fibre bundle $Y' \to X$ such that $s(X) \subset \Phi(Y)$, by the *kernel of the fibred morphism* Φ *with respect to the section* s is meant the inverse image

$$\mathrm{Ker}_s\Phi = \Phi^{-1}(s(X))$$

of $s(X)$ by Φ. If $\Phi : Y \to Y'$ is a fibred morphism of constant rank over X, then $\Phi(Y)$ and $\mathrm{Ker}_s\Phi$ are subbundles of Y' and Y, respectively. □

Given a fibre bundle $\pi : Y \to X$ and a morphism $f : X' \to X$, the *pull-back* of Y by f is called the manifold

$$f^*Y = \{(x', y) \in X' \times Y \;:\; \pi(y) = f(x')\} \tag{1.1.6}$$

together with the natural projection $(x', y) \mapsto x'$. It is a fibre bundle over X' such that the fibre of f^*Y over a point $x' \in X'$ is that of Y over the point $f(x') \in X$. There is the canonical fibred morphism

$$f_Y : f^*Y \ni (x', y) \underset{f}{\mapsto} y \in Y. \tag{1.1.7}$$

In particular, if $X' \subset X$ is a submanifold of X and $i_{X'}$ is the corresponding natural injection, then the pull-back

$$i^*_{X'}Y = Y\mid_{X'}$$

is the *restriction* of a fibre bundle Y to the submanifold $X' \subset X$. In particular, let $\pi : Y \to X$ and $\pi' : Y' \to X$ be fibre bundles over the same base X. Their *fibred product* $Y \times Y'$ over X is defined as the pull-back

$$Y \underset{X}{\times} Y' = \pi^*Y' \quad \text{or} \quad Y \underset{X}{\times} Y' = \pi'^*Y.$$

Vector bundles

A *vector bundle* is a fibre bundle $Y \to X$ such that:

- its typical fibre V and all the fibres $Y_x = \pi^{-1}(x)$, $x \in X$, are real finite-dimensional vector spaces;
- there is a bundle atlas $\Psi = \{(U_\xi, \psi_\xi)\}$ of $Y \to X$ whose trivialization morphisms ψ_ξ restrict to linear isomorphisms

$$\psi_\xi(x) : Y_x \to V, \qquad \forall x \in U_\xi.$$

Dealing with a vector bundle Y, we will always use linear bundle coordinates (y^i) associated with a linear bundle atlas Ψ, i.e.,

$$(\mathrm{pr}_2 \circ \psi_\xi)(y) = y^i e_i,$$
$$y = y^i e_i(x) = y^i \psi_\xi(x)^{-1}(e_i),$$

where $\{e_i\}$ is a fixed basis for the typical fibre V of Y, while $\{e_i(x)\}$ are the associated fibre bases (or the *frames*) for the fibres Y_x of Y. By a morphism of vector bundles $\Phi : Y \to Y'$ is meant a fibred morphism whose restriction to each fibre of Y is a linear map. It is called a *linear bundle morphism*.

By virtue of Proposition 1.1.3, if $\Phi : Y \to Y'$ is a linear bundle morphism of vector bundles of constant rank, then $\Phi(Y)$ and $\mathrm{Ker}\,\Phi$ are vector subbundles of Y and Y', respectively (see also [159]). Injection and surjection of vector bundles exemplify such kind morphisms.

Recall the following constructions of new vector bundles from old.

- Let $Y \to X$ be a vector bundle with a typical fibre V. By $Y^* \to X$ is meant the *dual vector bundle* with the typical fibre V^* dual of V. The *interior product* of Y and Y^* is defined as a fibred morphism

$$\rfloor : Y \otimes Y^* \underset{X}{\longrightarrow} X \times \mathbb{R}.$$

- Let $Y \to X$ and $Y' \to X$ be vector bundles with typical fibres V and V', respectively. Their *Whitney sum* $Y \underset{X}{\oplus} Y'$ is a vector bundle over X with the typical fibre $V \oplus V'$.
- Let $Y \to X$ and $Y' \to X$ be vector bundles with typical fibres V and V', respectively. Their *tensor product* $Y \underset{X}{\otimes} Y'$ is a vector bundle over X with the typical fibre $V \otimes V'$. Similarly, the *exterior product* of vector bundles $Y \underset{X}{\wedge} Y$ is defined.

Remark 1.1.1. Let Y and Y' be vector bundles over the same base X. Every linear morphism

$$\Phi : Y \ni e_i \to \Phi_i^k(x)e'_k \in Y'$$

over X can be seen as a global section

$$\Phi : x \to \Phi_i^k(x)e'_k \otimes e^i$$

of the tensor bundle $Y' \otimes Y^* \to X$, and *vice versa*. •

By virtue of Theorem 1.1.2, a vector bundle has a global sections, e.g., the canonical zero-valued global section $\widehat{0}$.

Let us consider an *exact sequence* of vector bundles over the same base X

$$0 \to Y' \xrightarrow{i} Y \xrightarrow{j} Y'' \to 0, \tag{1.1.8}$$

where 0 denotes a 0-dimensional vector bundle, $Y' \xrightarrow{i} Y$ is an injection and $Y \xrightarrow{j} Y''$ is a surjection of vector bundles such that $\operatorname{Im} i = \operatorname{Ker} j$. This is equivalent to the fact that $Y'' = Y/Y'$ is the quotient bundle. One says that the exact sequence (1.1.8) of vector bundles admits a *splitting* if there exists a linear bundle monomorphism $\Gamma : Y'' \to Y$ over X such that $j \circ \Gamma = \operatorname{Id} Y''$. Then

$$Y = i(Y') \oplus \Gamma(Y'').$$

THEOREM 1.1.4. Every exact sequence of vector bundles admits a splitting ([157], p.56; [159]). □

Affine bundles

Let $\overline{\pi} : \overline{Y} \to X$ be a vector bundle with a typical fibre $\overline{V}$. An *affine bundle* modelled over the vector bundle $\overline{Y} \to X$ is a fibre bundle $\pi : Y \to X$ whose typical fibre V is an affine space modelled over $\overline{V}$, while the following conditions hold.

- All fibres Y_x of Y are affine spaces modelled over the corresponding fibres $\overline{Y}_x$ of the vector bundle $\overline{Y}$.

- There is a bundle atlas $\Psi = \{(U_\alpha, \psi_\alpha)\}$ of $Y \to X$ whose trivialization morphisms restrict to affine isomorphisms

$$\psi_\zeta(x) : Y_x \to V, \qquad \forall x \in U_\zeta.$$

In particular, every vector bundle has a natural structure of an affine bundle. Dealing with an affine bundle, we will use only *affine bundle coordinates* (x^λ, y^i) associated with an affine bundle atlas Ψ. There are the fibred morphisms

$$Y \underset{X}{\times} \overline{Y} \underset{X}{\longrightarrow} Y, \qquad (y^i, \overline{y}^i) \mapsto y^i + \overline{y}^i,$$
$$Y \underset{X}{\times} Y \underset{X}{\longrightarrow} \overline{Y}, \qquad (y^i, y'^i) \mapsto y^i - y'^i,$$

where $(\overline{y}^i)$ are linear coordinates on the vector bundle $\overline{Y}$.

By a morphism of affine bundles is meant a fibred morphism $\Phi : Y \to Y'$ whose restriction to each fibre of Y is an affine map. It is called an *affine bundle morphism*. Every affine bundle morphism $\Phi : Y \to Y'$ from an affine bundle Y modelled over a vector bundle $\overline{Y}$ to an affine bundle Y' modelled over a vector bundle $\overline{Y}'$ determines uniquely the linear bundle morphism

$$\overline{\Phi} : \overline{Y} \to \overline{Y}', \qquad \overline{y}'^i \circ \overline{\Phi} = \frac{\partial \Phi^i}{\partial y^j} \overline{y}^j, \tag{1.1.9}$$

called the *linear derivative* of Φ.

Similarly to vector bundles, if $\Phi : Y \to Y'$ is an affine bundle morphism of affine bundles of constant rank, then $\Phi(Y)$ and $\operatorname{Ker} \Phi$ are affine subbundles of Y and Y', respectively.

By virtue of Theorem 1.1.2, an affine bundle has a global section, but there is no canonical global section of an affine bundle.

Let $\pi : Y \to X$ be an affine bundle modelled over a vector bundle $\overline{Y} \to X$. Every global section s of an affine bundle $Y \to X$ yields the fibred morphism

$$D_s : Y \ni y \to y - s(\pi(y)) \in \overline{Y}. \tag{1.1.10}$$

Tangent and cotangent bundles

Tangent and cotangent bundles exemplify vector bundles. The fibres of the *tangent bundle*

$$\pi_Z : TZ \to Z$$

of a manifold Z are tangent spaces to Z. Given an atlas $\Psi_Z = \{(U_\xi, \phi_\xi)\}$ of a manifold Z, the tangent bundle is provided with the *holonomic atlas*

$$\Psi = \{U_\xi, \psi_\xi = T\phi_\xi)\}, \tag{1.1.11}$$

where by $T\phi_\xi$ is meant the tangent map to ϕ_ξ. The associated linear bundle coordinates with respect to the holonomic frames $\{\partial_\lambda\}$ in tangent spaces T_zZ. They are called *induced* or *holonomic* coordinates, and are denoted by $(\dot z^\lambda)$ on TZ. The transition functions of holonomic coordinates read

$$\dot z'^\lambda = \frac{\partial z'^\lambda}{\partial z^\mu}\dot z^\mu.$$

Every manifold map $f : Z \to Z'$ generates the linear bundle morphism of the tangent bundles

$$Tf : TZ \underset{f}{\longrightarrow} TZ', \qquad \dot z'^\lambda \circ Tf = \frac{\partial f^\lambda}{\partial z^\mu}\dot z^\mu,$$

called the *tangent map* to f.

The *cotangent bundle* of a manifold Z is the dual

$$\pi_{*Z} : T^*Z \to Z$$

of the tangent bundle $TZ \to Z$. It is equipped with the holonomic coordinates $(z^\lambda, \dot z_\lambda)$ with respect to the coframes $\{dz^\lambda\}$ in T^*Z dual of $\{\partial_\lambda\}$. Their transition functions read

$$\dot z'_\lambda = \frac{\partial z^\mu}{\partial z'^\lambda}\dot z_\mu.$$

Various tensor products

$$T = (\overset{m}{\otimes} TZ) \otimes (\overset{k}{\otimes} T^*Z) \tag{1.1.12}$$

over Z of tangent and cotangent bundles are called *tensor bundles*. Their sections are *(m,k)-tensor fields*.

Tangent and cotangent bundles of fibre bundles

Let $\pi_Y : TY \to Y$ be the tangent bundle of a fibre bundle $\pi : Y \to X$. Given fibred coordinates (x^λ, y^i) on Y, the tangent bundle TY is equipped with the holonomic coordinates $(x^\lambda, y^i, \dot x^\lambda, \dot y^i)$. The tangent bundle $TY \to Y$ has the *vertical tangent subbundle*

$$VY = \operatorname{Ker} T\pi,$$

given by the coordinate relations $\dot{x}^\lambda = 0$. This subbundle consists of the vectors tangent to fibres of Y. The vertical tangent bundle VY is provided with the holonomic coordinates $(x^\lambda, y^i, \dot{y}^i)$ with respect to the frames $\{\partial_i\}$.

Let $T\Phi$ be the tangent map to a fibred morphism $\Phi : Y \to Y'$. Its restriction

$$V\Phi = T\Phi|_{VY} : VY \to VY',$$
$$\dot{y}'^i \circ V\Phi = \partial_V \Phi^i = \dot{y}^j \partial_j \Phi^i,$$

to VY is a linear bundle morphism of the vertical tangent bundle VY to the vertical tangent bundle VY', called the *vertical tangent map* to Φ.

Vertical tangent bundles of many fibre bundles are equivalent to the pull-backs

$$VY \cong Y \underset{X}{\times} \overline{Y} \tag{1.1.13}$$

where $\overline{Y} \to X$ is some vector bundle. It means that VY can be provided with bundle coordinates $(x^\lambda, y^i, \overline{y}^i)$ such that the transition functions of coordinates $\overline{y}^i$ are independent of y^i. For instance, every affine bundle $Y \to X$ modelled over a vector bundle $\overline{Y} \to X$ admits the canonical *vertical splitting*

$$VY \cong Y \underset{X}{\times} \overline{Y} \tag{1.1.14}$$

because the holonomic coordinates $\dot{y}^i$ on VY have the same transformation law as the linear coordinates $\overline{y}^i$ on the vector bundle $\overline{Y}$. If Y is a vector bundle, the splitting (1.1.14) reads

$$VY \cong Y \underset{X}{\times} Y. \tag{1.1.15}$$

The *vertical cotangent bundle* $V^*Y \to Y$ of a fibre bundle $Y \to X$ is defined as the vector bundle dual of the vertical tangent bundle $VY \to Y$. There is the canonical projection

$$\zeta : T^*Y \underset{Y}{\to} V^*Y, \tag{1.1.16}$$
$$\zeta : \dot{x}_\lambda dx^\lambda + \dot{y}_i dy^i \mapsto \dot{y}_i \overline{d}y^i,$$

where $\{\overline{d}y^i\}$ are the bases for the fibres of V^*Y, which are dual of the holonomic frames $\{\partial_i\}$ for the vertical tangent bundle VY.

It should be emphasized that there is no canonical injection of V^*Y to T^*Y as it follows from the coordinate transformation laws

$$dy'^i = \frac{\partial y'^i}{\partial y^j} dy^j + \frac{\partial y'^i}{\partial x^\lambda} dx^\lambda, \qquad \overline{d}y'^i = \frac{\partial y'^i}{\partial y^j} \overline{d}y^j.$$

With VY and V^*Y, one has the following two exact sequences of vector bundles over Y:

$$0 \to VY \hookrightarrow TY \xrightarrow{\pi_T} Y \underset{X}{\times} TX \to 0, \tag{1.1.17a}$$

$$0 \to Y \underset{X}{\times} T^*X \hookrightarrow T^*Y \xrightarrow{\zeta} V^*Y \to 0. \tag{1.1.17b}$$

Every splitting

$$\Gamma : Y \underset{X}{\times} TX \underset{Y}{\hookrightarrow} TY, \qquad \partial_\lambda \mapsto \partial_\lambda + \Gamma^i_\lambda \partial_i, \tag{1.1.18}$$

of (1.1.17a) and the dual splitting

$$\Gamma : V^*Y \underset{Y}{\hookrightarrow} T^*Y, \qquad \overline{d}y^i \mapsto dy^i - \Gamma^i_\lambda dx^\lambda, \tag{1.1.19}$$

of (1.1.17b), by definition, are a connection Γ on the fibre bundle $Y \to X$ (see Section 2.1).

For the sake of simplicity, the pull-backs $Y \underset{X}{\times} TX$ and $Y \underset{X}{\times} T^*X$ will be denoted by TX and T^*X, respectively.

1.2 Differential forms and multivector fields

In this Section, we are concerned with vector and multivector fields, exterior and tangent-valued differential forms on manifolds and fibre bundles. Note that connections on fibre bundles are represented by tangent-valued forms.

Vector fields

A *vector field* on a manifold Z is defined as a global section of the tangent bundle $TZ \to Z$. The set $\mathcal{T}(Z)$ of vector fields on Z is both a locally free module over the ring $C^\infty(Z)$ of smooth functions on Z and a real Lie algebra with respect to the *Lie bracket*

$$[v, u] = (v^\lambda \partial_\lambda u^\mu - u^\lambda \partial_\lambda v^\mu)\partial_\mu, \quad u = u^\lambda \partial_\lambda, \quad v = v^\lambda \partial_\lambda.$$

A curve $c : () \to Z$, $() \subset \mathbb{R}$, in Z is said to be an *integral curve* of a vector field u on Z if

$$\dot{c} = u \circ c,$$

where by $\dot{c}$ is meant the morphism

$$\dot{c}(t) = Tc(t, \dot{t} = 1) : () \to TZ, \tag{1.2.1}$$

called the *tangent prolongation* of the curve c. For every point $z \in Z$, there exists a unique integral curve c of a vector field u through $z = c(0)$.

Recall the relationship between vector fields on a manifold Z and its diffeomorphisms. Given an open subset $U \subset Z$, by a *local 1-parameter group of local diffeomorphisms* of Z defined on $(-\epsilon, \epsilon) \times U$ is meant a mapping

$$G : (-\epsilon, \epsilon) \times U \ni (t, z) \mapsto G_t(z) \in Z, \qquad \epsilon > 0,$$

which possesses the following properties:

- for each $t \in (-\epsilon, \epsilon)$, the mapping G_t is a diffeomorphism of U onto the open subset $G_t(U) \subset Z$;
- $G_{t+t'}(z) = (G_t \circ G_{t'})(z)$ if $t + t' \in (-\epsilon, \epsilon)$.

If G is defined on $\mathbb{R} \times Z$, it is called a *1-parameter group of diffeomorphisms* of Z. Each local 1-parameter group of local diffeomorphisms G on $U \subset Z$ defines a local vector field u on U by setting $u(z)$ to be the tangent vector to the curve $c(t) = G_t(z)$ at $t = 0$. Conversely, let u be a vector field on a manifold Z. For each $z \in Z$, there exist a real number $\epsilon > 0$, a neighbourhood U of z and a unique local 1-parameter group of local diffeomorphisms on $(-\epsilon, \epsilon) \times U$, which determines u [177]. In brief, one can say that every vector field u on a manifold Z is the *generator* of a local 1-parameter group of local diffeomorphisms. A vector field u on a manifold Z is is called *complete*, if it is induced by a 1-parameter group of diffeomorphisms of Z.

A vector field u on a fibre bundle $Y \to X$ is said to be *projectable* if it projects over a vector field τ on X, i.e., if the diagram

$$\begin{array}{ccc} Y & \xrightarrow{u} & TY \\ {\scriptstyle\pi}\downarrow & & \downarrow{\scriptstyle T\pi} \\ X & \xrightarrow[\tau]{} & TX \end{array}, \qquad \tau \circ \pi = T\pi \circ u,$$

commutes. A projectable vector field has the coordinate expression

$$u = u^\lambda(x^\mu)\partial_\lambda + u^i(x^\mu, y^j)\partial_i, \qquad \tau = u^\lambda\partial_\lambda.$$

A projectable vector field $u = u^i\partial_i$ on a fibre bundle $Y \to X$ is said to be *vertical* if it projects over the zero vector field $\tau = 0$ on X.

A vector field $\tau = \tau^\lambda\partial_\lambda$ on a base X of a fibre bundle $Y \to X$ can give rise to a projectable vector field on the total space Y by means of some connection on this fibre bundle (see the expression (2.1.6) below). Nevertheless, every tensor bundle (1.1.12) admits the *canonical lift*

$$\widetilde{\tau} = \tau^\mu\partial_\mu + [\partial_\nu\tau^{\alpha_1}\dot{x}^{\nu\alpha_2\cdots\alpha_m}_{\beta_1\cdots\beta_k} + \ldots - \partial_{\beta_1}\tau^\nu\dot{x}^{\alpha_1\cdots\alpha_m}_{\nu\beta_2\cdots\beta_k} - \ldots]\frac{\partial}{\partial\dot{x}^{\alpha_1\cdots\alpha_m}_{\beta_1\cdots\beta_k}} \tag{1.2.2}$$

of any vector field τ on X. In particular, we have the canonical lift

$$\widetilde{\tau} = \tau^\mu\partial_\mu + \partial_\nu\tau^\alpha\dot{x}^\nu\frac{\partial}{\partial\dot{x}^\alpha} \tag{1.2.3}$$

of τ onto the tangent bundle TX and its canonical lift

$$\widetilde{\tau} = \tau^\mu\partial_\mu - \partial_\beta\tau^\nu\dot{x}_\nu\frac{\partial}{\partial\dot{x}_\beta} \tag{1.2.4}$$

onto the cotangent bundle T^*X. Hereafter, we will use the compact notation

$$\dot{\partial}_\lambda = \frac{\partial}{\partial\dot{x}^\lambda}. \tag{1.2.5}$$

Let $Y \to X$ be a vector bundle. Due to the canonical vertical splitting (1.1.15), there exists the canonical vertical vector field

$$u_Y = y^i\partial_i \tag{1.2.6}$$

on Y, called the *Liouville vector field*. For instance, the Liouville vector field on the tangent bundle TX reads

$$u_{TX} = \dot{x}^\lambda\dot{\partial}_\lambda. \tag{1.2.7}$$

Accordingly, any vector field $\tau = \tau^\lambda\partial_\lambda$ on a manifold X has the canonical *vertical lift*

$$\tau_V = \tau^\lambda\dot{\partial}_\lambda \tag{1.2.8}$$

onto the tangent bundle TX.

Exterior forms

An *exterior r-form* on a manifold Z is a section

$$\phi = \frac{1}{r!}\phi_{\lambda_1\dots\lambda_r} dz^{\lambda_1}\wedge\cdots\wedge dz^{\lambda_r}$$

of the exterior product $\overset{r}{\wedge} T^*Z \to Z$. We denote by $\mathfrak{O}^r(Z)$ the vector space of exterior r-forms on a manifold Z. This is also a locally free module over the ring $\mathfrak{O}^0(Z) = C^\infty(Z)$. All exterior forms on Z constitute the exterior $\mathbb{Z}$-graded algebra $\mathfrak{O}^*(Z)$ with respect to the exterior product $\wedge$. This algebra is provided with exterior differential

$$d : \mathfrak{O}^r(Z) \to \mathfrak{O}^{r+1}(Z),$$
$$d\phi = \frac{1}{r!}\partial_\mu\phi_{\lambda_1\dots\lambda_r} dz^\mu \wedge dz^{\lambda_1}\wedge\cdots dz^{\lambda_r},$$

which obeys the relations

$$d(\phi\wedge\sigma) = d(\phi)\wedge\sigma + (-1)^{|\phi|}\phi\wedge d(\sigma), \qquad d\circ d = 0.$$

The symbol $|\phi|$ stands for the form degree.

Given a morphism $f : Z \to Z'$, by $f^*\phi$ is meant the *pull-back* on Z of an r-form ϕ on Z' by f. It is defined by the condition

$$f^*\phi(v^1,\dots,v^r)(z) = \phi(Tf(v^1),\dots,Tf(v^r))(f(z)), \qquad \forall v^1,\cdots v^r \in T_zZ,$$

and obeys the relations

$$f^*(\phi\wedge\sigma) = f^*\phi\wedge f^*\sigma, \qquad df^*\phi = f^*(d\phi).$$

Let $\pi : Y \to X$ be a fibre bundle with fibred coordinates (x^λ, y^i). The pull-back onto Y of exterior forms on X by π provides the inclusion

$$\pi^* : \mathfrak{O}^*(X) \to \mathfrak{O}^*(Y).$$

Elements of its image are called *basic forms*. Exterior forms

$$\phi : Y \to \overset{r}{\wedge} T^*X,$$
$$\phi = \frac{1}{r!}\phi_{\lambda_1\dots\lambda_r} dx^{\lambda_1}\wedge\cdots\wedge dx^{\lambda_r},$$

on Y such that $\vartheta\rfloor\phi = 0$ for an arbitrary vertical vector field ϑ on Y are said to be *horizontal forms*. A horizontal n-form is called a *horizontal density*. We will use the notation

$$\omega = dx^1 \wedge \cdots \wedge dx^n, \qquad \omega_\lambda = \partial_\lambda\rfloor\omega, \qquad \omega_{\mu\lambda} = \partial_\mu\rfloor\partial_\lambda\rfloor\omega. \tag{1.2.9}$$

In the case of the tangent bundle $TX \to X$, there is a different way to lift onto TX the exterior forms on X, besides the pull-back by π_X. Let f be a function on X. Its *tangent lift* onto TX is defined as the function

$$\widetilde{f} = \dot{x}^\lambda \partial_\lambda f. \tag{1.2.10}$$

Let σ be an r-form on X. Its *tangent lift* onto TX is said to be the r-form $\widetilde{\sigma}$ given by the relation

$$\widetilde{\sigma}(\widetilde{\tau}_1, \ldots, \widetilde{\tau}_r) = \sigma(\widetilde{\tau_1, \ldots, \tau_r}), \tag{1.2.11}$$

where τ_i are arbitrary vector fields on X, and $\widetilde{\tau}_i$ are their canonical lifts (1.2.3) onto TX. We have the coordinate expression

$$\begin{aligned} \sigma &= \frac{1}{r!}\sigma_{\lambda_1\cdots\lambda_r} dx^{\lambda_1} \wedge \cdots \wedge dx^{\lambda_r}, \\ \widetilde{\sigma} &= \frac{1}{r!}[\dot{x}^\mu \partial_\mu \sigma_{\lambda_1\cdots\lambda_r} dx^{\lambda_1} \wedge \cdots \wedge dx^{\lambda_r} + \\ &\sum_{i=1}^r \sigma_{\lambda_1\cdots\lambda_r} dx^{\lambda_1} \wedge \cdots \wedge d\dot{x}^{\lambda_i} \wedge \cdots \wedge dx^{\lambda_r}]. \end{aligned} \tag{1.2.12}$$

It is easily verified that $d\widetilde{\sigma} = \widetilde{d\sigma}$.

The *interior product* of a vector field $u = u^\mu\partial_\mu$ and an exterior r-form ϕ is given by the coordinate expression

$$\begin{aligned} u\rfloor\phi = \sum_{k=1}^r \frac{(-1)^{k-1}}{r!} u^{\lambda_k} \phi_{\lambda_1\ldots\lambda_k\ldots\lambda_r} dz^{\lambda_1} \wedge \cdots \wedge \widehat{dz}^{\lambda_k} \wedge \cdots \wedge dz^{\lambda_r} = \\ \frac{1}{(r-1)!} u^\mu \phi_{\mu\alpha_2\ldots\alpha_r} dz^{\alpha_2} \wedge \cdots \wedge dz^{\alpha_r}, \end{aligned} \tag{1.2.13}$$

and satisfies the relations

$$\phi(u_1, \ldots, u_r) = u_r\rfloor \cdots u_1\rfloor\phi, \tag{1.2.14}$$

$$u\rfloor(\phi \wedge \sigma) = u\rfloor\phi \wedge \sigma + (-1)^{|\phi|}\phi \wedge u\rfloor\sigma, \tag{1.2.15}$$

$$[u, u']\rfloor\phi = u\rfloor d(u'\rfloor\phi) - u'\rfloor d(u\rfloor\phi) - u'\rfloor u\rfloor d\phi, \qquad \phi \in \mathfrak{O}^1(Z). \tag{1.2.16}$$

The *Lie derivative* of an exterior form ϕ along a vector field u is

$$\mathbf{L}_u\phi = u\rfloor d\phi + d(u\rfloor\phi),$$
$$\mathbf{L}_u(\phi\wedge\sigma) = \mathbf{L}_u\phi\wedge\sigma + \phi\wedge\mathbf{L}_u\sigma.$$

In particular, if f is a function, then

$$\mathbf{L}_u f = u(f) = u\rfloor df.$$

Given the tangent lift $\widetilde{\phi}$ (1.2.12) of an exterior form ϕ, we have

$$\mathbf{L}_u(\phi) = u^*\widetilde{\phi}.$$

Multivector fields

A *multivector field* ϑ of degree $\mid\vartheta\mid = r$ (or simply an r-vector field) on a manifold Z is a section

$$\vartheta = \frac{1}{r!}\vartheta^{\lambda_1\dots\lambda_r}\partial_{\lambda_1}\wedge\cdots\wedge\partial_{\lambda_r} \tag{1.2.17}$$

of the exterior product $\overset{r}{\wedge} TZ\to Z$. Let $\mathcal{T}_r(Z)$ denote the vector space of r-vector fields on Z. In particular, $\mathcal{T}_1(Z)$ is the space of vector fields on Z (denoted by $\mathcal{T}(Z)$ for the sake of simplicity), while $\mathcal{T}_0(Z)$ is the vector space $C^\infty(Z)$ of smooth functions on Z. All multivector fields on a manifold Z make up the exterior $\mathbb{Z}$-graded algebra $\mathcal{T}_*(Z)$ with respect to the exterior product of multivector fields.

Given a manifold Z, the *tangent lift* $\widetilde{\vartheta}$ onto TZ of an r-vector field ϑ (1.2.17) on Z is defined by the relation

$$\widetilde{\vartheta}(\widetilde{\sigma}^r,\dots,\widetilde{\sigma}^1) = \widetilde{\vartheta(\sigma^r,\dots,\sigma^1)} \tag{1.2.18}$$

where: (i) $\sigma^k = \sigma^k_\lambda dx^\lambda$ are arbitrary 1-forms on the manifold Z, (ii) by

$$\widetilde{\sigma}^k = \dot{x}^\mu\partial_\mu\sigma^k_\lambda dx^\lambda + \sigma^k_\lambda d\dot{x}^\lambda$$

are meant their tangent lifts (1.2.12) onto the tangent bundle TZ of Z, and (iii) the right-hand side of the equality (1.2.18) is the tangent lift (1.2.10) onto TZ of the function $\vartheta(\sigma^r,\dots,\sigma^1)$ on Z. We have the coordinate expression

$$\widetilde{\vartheta} = \frac{1}{r!}[\dot{z}^\mu\partial_\mu\vartheta^{\lambda_1\dots\lambda_r}\dot{\partial}_{\lambda_1}\wedge\cdots\wedge\dot{\partial}_{\lambda_r} + \vartheta^{\lambda_1\dots\lambda_r}\sum_{i=1}^r\dot{\partial}_{\lambda_1}\wedge\cdots\wedge\partial_{\lambda_i}\wedge\cdots\wedge\dot{\partial}_{\lambda_r}]. \tag{1.2.19}$$

In particular, if τ is a vector field on a manifold Z, its tangent lift (1.2.19) coincides with the canonical lift $\widetilde{\tau}$ (1.2.3). If an r-vector field ϑ is *simple*, i.e.,

$$\vartheta = \tau^1 \wedge \cdots \wedge \tau^r,$$

its tangent lift (1.2.19) reads

$$\widetilde{\vartheta} = \sum_{i=1}^{r} \tau_V^1 \wedge \cdots \wedge \widetilde{\tau}^i \cdots \wedge \tau_V^r,$$

where τ_V^k is the vertical lift (1.2.8) onto TZ of the vector field τ^k.

The exterior algebra of multivector fields on a manifold Z is provided with the *Schouten–Nijenhuis* bracket which generalizes the Lie bracket of vector fields as follows:

$$\begin{aligned}
&[.,.]_{\rm SN} : \mathcal{T}_r(M) \times \mathcal{T}_s(M) \to \mathcal{T}_{r+s-1}(M), \qquad (1.2.20)\\
&\vartheta = \frac{1}{r!}\vartheta^{\lambda_1\ldots\lambda_r}\partial_{\lambda_1}\wedge\cdots\wedge\partial_{\lambda_r}, \qquad \upsilon = \frac{1}{s!}\upsilon^{\alpha_1\ldots\alpha_s}\partial_{\alpha_1}\wedge\cdots\wedge\partial_{\alpha_s},\\
&[\vartheta,\upsilon]_{\rm SN} \overset{\rm def}{=} \vartheta\star\upsilon + (-1)^{rs}\upsilon\star\vartheta,\\
&\vartheta\star\upsilon = \frac{r}{r!s!}(\vartheta^{\mu\lambda_2\ldots\lambda_r}\partial_\mu\upsilon^{\alpha_1\ldots\alpha_s}\partial_{\lambda_2}\wedge\cdots\wedge\partial_{\lambda_r}\wedge\partial_{\alpha_1}\wedge\cdots\wedge\partial_{\alpha_s}).
\end{aligned}$$

The following relations hold:

$$[\vartheta,\upsilon]_{\rm SN} = (-1)^{|\vartheta||\upsilon|}[\upsilon,\vartheta]_{\rm SN}, \qquad (1.2.21)$$

$$[\nu,\vartheta\wedge\upsilon]_{\rm SN} = [\nu,\vartheta]_{\rm SN}\wedge\upsilon + (-1)^{(|\nu|-1)|\vartheta|}\vartheta\wedge[\nu,\upsilon]_{\rm SN}, \qquad (1.2.22)$$

$$\begin{aligned}
&(-1)^{|\nu|(|\upsilon|-1)}[\nu,[\vartheta,\upsilon]_{\rm SN}]_{\rm SN} + (-1)^{|\vartheta|(|\nu|-1)}[\vartheta,[\upsilon,\nu]_{\rm SN}]_{\rm SN} + \qquad (1.2.23)\\
&\qquad (-1)^{|\upsilon|(|\vartheta|-1)}[\upsilon,[\nu,\vartheta]_{\rm SN}]_{\rm SN} = 0.
\end{aligned}$$

In particular, the Lie derivative of a multivector field υ along a vector field u is

$$\begin{aligned}
&\mathbf{L}_u\upsilon = [u,\upsilon]_{\rm SN},\\
&\mathbf{L}_u(\vartheta\wedge\upsilon) = \mathbf{L}_u\vartheta\wedge\upsilon + \vartheta\wedge\mathbf{L}_u\upsilon.
\end{aligned}$$

The Schouten–Nijenhuis bracket commutes with the tangent lift (1.2.19) of multivector fields, i.e.,

$$[\widetilde{\vartheta},\widetilde{\upsilon}]_{\rm SN} = \widetilde{[\vartheta,\upsilon]}_{\rm SN}. \qquad (1.2.24)$$

The generalization of the interior product (1.2.13) is the *left interior product*

$$\vartheta\rfloor\phi = \phi(\vartheta), \qquad |\,\vartheta\,|\leq|\,\phi\,|, \qquad \phi\in\mathfrak{O}^*(Z), \qquad \vartheta\in\mathcal{T}_*(Z),$$

of multivector fields and exterior forms, which is derived from the equality

$$\phi(u_1 \wedge \cdots \wedge u_r) = \phi(u_1, \ldots, u_r), \qquad \phi \in \mathfrak{O}^*(Z), \qquad u_i \in \mathcal{T}(Z),$$

for simple multivector fields. There is the relation

$$\vartheta\rfloor\upsilon\rfloor\phi = (\upsilon \wedge \vartheta)\rfloor\phi = (-1)^{|\upsilon||\vartheta|}\upsilon\rfloor\vartheta\rfloor\phi, \qquad \phi \in \mathfrak{O}^*(Z), \qquad \vartheta, \upsilon \in \mathcal{T}_*(Z).$$

The *right interior product*

$$\vartheta\lfloor\phi = \vartheta(\phi), \qquad |\phi| \leq |\vartheta|, \qquad \phi \in \mathfrak{O}^*(Z), \qquad \vartheta \in \mathcal{T}_*(Z),$$

of exterior forms and multivector fields is given by the equalities

$$\vartheta(\phi_1, \ldots, \phi_r) = \vartheta\lfloor\phi_r \cdots \lfloor\phi_1, \qquad \phi_i \in \mathfrak{O}^1(Z), \qquad \vartheta \in \mathcal{T}_r(Z),$$
$$\vartheta\lfloor\phi = \frac{1}{(r-1)!}\vartheta^{\alpha_1\ldots\alpha_{r-1}\mu}\phi_\mu\partial_{\alpha_1} \wedge \cdots \wedge \partial_{\alpha_{r-1}}, \qquad \phi \in \mathfrak{O}^1(Z).$$

It satisfies the relations

$$(\vartheta \wedge \upsilon)\lfloor\phi = \vartheta \wedge (\upsilon\lfloor\phi) + (-1)^{|\upsilon|}(\vartheta\lfloor\phi) \wedge \upsilon, \qquad \phi \in \mathfrak{O}^1(Z),$$
$$\vartheta(\phi \wedge \sigma) = \vartheta\lfloor\sigma\lfloor\phi, \qquad \phi, \sigma \in \mathfrak{O}^*(Z).$$

In particular, if $|\vartheta| = |\phi|$, there is the natural contraction

$$\langle , \rangle : \mathcal{T}_r(Z) \times \mathfrak{O}^r(Z) \to C^\infty(Z),$$
$$\langle \vartheta, \phi \rangle = \vartheta\rfloor\phi = \vartheta\lfloor\phi = \vartheta(\phi) = \phi(\vartheta). \tag{1.2.25}$$

Tangent-valued forms

A *tangent-valued r-form* on a manifold Z is a section

$$\phi = \frac{1}{r!}\phi^\mu_{\lambda_1\ldots\lambda_r} dz^{\lambda_1} \wedge \cdots \wedge dz^{\lambda_r} \otimes \partial_\mu$$

of the tensor bundle $\overset{r}{\wedge} T^*Z \otimes TZ \to Z$.

Example 1.2.1. There is one-to-one correspondence between the tangent-valued 1-forms ϕ on a manifold Z and the linear bundle endomorphisms

$$\widehat{\phi} : TZ \to TZ, \qquad \widehat{\phi} : T_zZ \ni v \mapsto v\rfloor\phi(z) \in T_zZ, \tag{1.2.26}$$
$$\widehat{\phi}^* : T^*Z \to T^*Z, \qquad \widehat{\phi}^* : T^*_zZ \ni v^* \mapsto \phi(z)\rfloor v^* \in T^*_zZ, \tag{1.2.27}$$

(see Remark 1.1.1). In particular, the *canonical tangent-valued 1-form*

$$\theta_Z = dz^\lambda \otimes \partial_\lambda \tag{1.2.28}$$

on Z corresponds to the identity morphisms (1.2.26) and (1.2.27). •

Example 1.2.2. Let $Z = TX$. There is the fibred endomorphism J of the double tangent bundle TTX of TX such that, for every vector field τ on X, we have

$$J \circ \widetilde{\tau} = \tau_V, \qquad J \circ \tau_V = 0,$$

where $\widetilde{\tau}$ is the canonical lift (1.2.3) and τ_V is the vertical lift (1.2.8) onto TTX of a vector field τ on TX. This endomorphism reads

$$J(\partial_\lambda) = \dot{\partial}_\lambda, \qquad J(\dot{\partial}_\lambda) = 0. \tag{1.2.29}$$

It corresponds to the tangent-valued form

$$\theta_J = dx^\lambda \otimes \dot{\partial}_\lambda \tag{1.2.30}$$

on the tangent bundle TX. It is readily observed that $J \circ J = 0$. •

The space $\mathfrak{O}^*(M)\otimes\mathcal{T}(M)$ of tangent-valued forms is provided with the *Frölicher–Nijenhuis bracket* (*F-N bracket*) which generalizes the Lie bracket of vector fields as follows:

$$\begin{aligned}
&[\,,\,]_{\rm FN} : \mathfrak{O}^r(M)\otimes\mathcal{T}(M)\times\mathfrak{O}^s(M)\otimes\mathcal{T}(M)\to\mathfrak{O}^{r+s}(M)\otimes\mathcal{T}(M),\\
&[\alpha\otimes u,\, \beta\otimes v]_{\rm FN} = (\alpha\wedge\beta)\otimes[u,\,v] + (\alpha\wedge\mathbf{L}_u\,\beta)\otimes v - \\
&\qquad (\mathbf{L}_v\,\alpha\wedge\beta)\otimes u + (-1)^r(d\alpha\wedge u\rfloor\beta)\otimes v + (-1)^r(v\rfloor\alpha\wedge d\beta)\otimes u,\\
&\alpha\in\mathfrak{O}^r(M), \qquad \beta\in\mathfrak{O}^s(M), \qquad u,v\in\mathcal{T}(M).
\end{aligned} \tag{1.2.31}$$

Its coordinate expression is

$$\begin{aligned}
&[\phi,\sigma]_{\rm FN} = \frac{1}{r!s!}(\phi^\nu_{\lambda_1\ldots\lambda_r}\partial_\nu\sigma^\mu_{\lambda_{r+1}\ldots\lambda_{r+s}} - \sigma^\nu_{\lambda_{r+1}\ldots\lambda_{r+s}}\partial_\nu\phi^\mu_{\lambda_1\ldots\lambda_r} - \\
&r\phi^\mu_{\lambda_1\ldots\lambda_{r-1}\nu}\partial_{\lambda_r}\sigma^\nu_{\lambda_{r+1}\ldots\lambda_{r+s}} + s\sigma^\mu_{\nu\lambda_{r+2}\ldots\lambda_{r+s}}\partial_{\lambda_{r+1}}\phi^\nu_{\lambda_1\ldots\lambda_r})dz^{\lambda_1}\wedge\cdots\wedge dz^{\lambda_{r+s}}\otimes\partial_\mu,\\
&\phi\in\mathfrak{O}^r(M)\otimes\mathcal{T}(M), \qquad \sigma\in\mathfrak{O}^s(M)\otimes\mathcal{T}(M).
\end{aligned}$$

There are the relations

$$[\phi,\psi]_{\rm FN} = (-1)^{|\phi||\psi|+1}[\psi,\phi]_{\rm FN}, \tag{1.2.32}$$

$$[\phi,[\psi,\theta]_{\rm FN}]_{\rm FN} = [[\phi,\psi]_{\rm FN},\theta]_{\rm FN} + (-1)^{|\phi||\psi|}[\psi,[\phi,\theta]_{\rm FN}]_{\rm FN}, \tag{1.2.33}$$

$$\phi,\psi,\theta \in \mathfrak{O}^*(M)\otimes \mathcal{T}(M).$$

Given a tangent-valued form θ, the *Nijenhuis differential* on $\mathfrak{O}^*(M)\otimes\mathcal{T}(M)$ is defined as the morphism

$$d_\theta : \sigma \mapsto d_\theta\sigma = [\theta,\sigma]_{\rm FN}, \qquad \forall\sigma\in\mathfrak{O}^*(M)\otimes\mathcal{T}(M).$$

By virtue of the relation (1.2.33), it has the property

$$d_\phi[\psi,\theta]_{\rm FN} = [d_\phi\psi,\theta]_{\rm FN} + (-1)^{|\phi||\psi|}[\psi,d_\phi\theta]_{\rm FN}.$$

In particular, if $\theta = u$ is a vector field, the Nijenhuis differential is the *Lie derivative of tangent-valued forms*

$$\mathbf{L}_u\sigma = d_u\sigma = [u,\sigma]_{\rm FN} = (u^\nu\partial_\nu\sigma^\mu_{\lambda_1\ldots\lambda_s} - \sigma^\nu_{\lambda_1\ldots\lambda_s}\partial_\nu u^\mu +$$
$$s\sigma^\mu_{\nu\lambda_2\ldots\lambda_s}\partial_{\lambda_1}u^\nu)dx^{\lambda_1}\wedge\cdots\wedge dx^{\lambda_s}\otimes\partial_\mu, \qquad \sigma\in\mathfrak{O}^s(M)\otimes\mathcal{T}(M).$$

Let $Y\to X$ be a fibre bundle. We consider the following subspaces of the space $\mathfrak{O}^*(Y)\otimes\mathcal{T}(Y)$ of tangent-valued forms on Y:

- *tangent-valued horizontal forms*

$$\phi : Y \to \overset{r}{\wedge} T^*X \underset{Y}{\otimes} TY,$$
$$\phi = dx^{\lambda_1}\wedge\ldots\wedge dx^{\lambda_r}\otimes[\phi^\mu_{\lambda_1\ldots\lambda_r}(y)\partial_\mu + \phi^i_{\lambda_1\ldots\lambda_r}(y)\partial_i],$$

- *projectable tangent-valued horizontal forms*

$$\phi = dx^{\lambda_1}\wedge\ldots\wedge dx^{\lambda_r}\otimes[\phi^\mu_{\lambda_1\ldots\lambda_r}(x)\partial_\mu + \phi^i_{\lambda_1\ldots\lambda_r}(y)\partial_i],$$

- *vertical-valued horizontal forms*

$$\phi : Y \to \overset{r}{\wedge} T^*X \underset{Y}{\otimes} VY,$$
$$\phi = \phi^i_{\lambda_1\ldots\lambda_r}(y)dx^{\lambda_1}\wedge\ldots\wedge dx^{\lambda_r}\otimes\partial_i,$$

- vertical-valued horizontal 1-forms, called *soldering forms*,

$$\sigma = \sigma^i_\lambda(y)dx^\lambda \otimes \partial_i,$$

- basic soldering forms

$$\sigma = \sigma^i_\lambda(x)dx^\lambda \otimes \partial_i.$$

Remark 1.2.3. The tangent bundle TX is provided with the canonical soldering form θ_J (1.2.30). Due to the canonical vertical splitting

$$VTX = TX \underset{X}{\times} TX, \tag{1.2.34}$$

this soldering form defines the canonical tangent-valued form θ_X (1.2.28) on X. By this reason, tangent-valued 1-forms on a manifold X are also called soldering forms. •

Remark 1.2.4. Let $Y \to X$ be a fibre bundle, $f : X' \to X$ a morphism, $f^*Y \to X'$ the pull-back of Y by f and $f_Y : f^*Y \to Y$ the corresponding fibred morphism (1.1.7). Since

$$Vf^*Y = f^*VY = f^*_YVY, \qquad V_{y'}Y' = V_{f_Y(y')}Y,$$

one can define the pull-back $f^*\phi$ onto f^*Y of any vertical-valued form f on Y in accordance with the relation

$$f^*\phi(v^1, \ldots, v^r)(y') = \phi(Tf_Y(v^1), \ldots, Tf_Y(v^r))(f_Y(y')).$$

•

We also mention the TX-valued forms

$$\phi : Y \to \overset{r}{\wedge} T^*X \underset{Y}{\otimes} TX, \tag{1.2.35}$$

$$\phi = \phi^\mu_{\lambda_1\ldots\lambda_r} dx^{\lambda_1} \wedge \cdots \wedge dx^{\lambda_r} \otimes \partial_\mu,$$

and V^*Y-valued forms

$$\phi : Y \to \overset{r}{\wedge} T^*X \underset{Y}{\otimes} V^*Y, \tag{1.2.36}$$

$$\phi = \phi_{\lambda_1\ldots\lambda_r i} dx^{\lambda_1} \wedge \cdots \wedge dx^{\lambda_r} \otimes \overline{d}y^i.$$

It should be emphasized that (1.2.35) are not tangent-valued forms, while (1.2.36) are not exterior forms. They exemplify vector-valued forms.

Distributions

An r-dimensional smooth *distribution* on a k-dimensional manifold Z is an r-dimensional subbundle $\mathbf{T}$ of the tangent bundle TZ. We will say that a vector field v on Z is *subordinate* to a distribution $\mathbf{T}$ if it is a section of $\mathbf{T} \to Z$.

A distribution $\mathbf{T}$ is said to be *involutive* if the Lie bracket $[u, u']$ is subordinate to $\mathbf{T}$, whenever u and u' are subordinate to $\mathbf{T}$.

A connected submanifold N of a manifold Z is called an *integral manifold* of a distribution $\mathbf{T}$ on Z if the tangent spaces to N belong to the fibres of this distribution. Unless otherwise stated, by an integral manifold we mean an *integral manifold of maximal dimension*, equal to $\dim \mathbf{T}$. An integral manifold N is called *maximal* if there is no other integral manifold which contains N. There is the well-known Frobenius theorem ([302], p.75).

THEOREM 1.2.1. Let $\mathbf{T}$ be a smooth involutive distribution on a manifold Z. For any point $z \in Z$, there exists a unique maximal integral manifold of $\mathbf{T}$ passing through z. □

In view of this fact, involutive distributions are also called *completely integrable distributions*.

If a distribution $\mathbf{T}$ is not involutive, there are no integral submanifolds of dimension equal to $\dim \mathbf{T}$. However, integral submanifolds always exist, e.g., the integral curves of vector fields, subordinate to $\mathbf{T}$.

A *codistribution* $\mathbf{T}^*$ on a manifold Z is a subbundle of the cotangent bundle. For instance, the *annihilator* $\mathrm{Ann}\,\mathbf{T}$ of an r-dimensional distribution $\mathbf{T}$ is a $(k-r)$-dimensional codistribution. $\mathrm{Ann}\,\mathbf{T}_z$, $z \in Z$, consists of covectors $w \in T^*_z$ such that $v \rfloor w = 0$, $\forall v \in \mathbf{T}_z$.

THEOREM 1.2.2. Let $\mathbf{T}$ be a distribution and $\mathrm{Ann}\,\mathbf{T}$ its annihilator. Let $\wedge \mathrm{Ann}\,\mathbf{T}(Z)$ be the ideal of the exterior algebra $\mathfrak{O}^*(Z)$ which is generated by sections of $\mathrm{Ann}\,\mathbf{T} \to Z$. A distribution $\mathbf{T}$ is involutive if and only if the ideal $\wedge \mathrm{Ann}\,\mathbf{T}(Z)$ is a *differential ideal*, i.e., $d(\wedge \mathrm{Ann}\,\mathbf{T}(Z)) \subset \wedge \mathrm{Ann}\,\mathbf{T}(Z)$ ([302], p.74). □

COROLLARY 1.2.3. Let $\mathbf{T}$ be an involutive r-dimensional distribution on a k-dimensional manifold Z. Every point $z \in Z$ has an open neighbourhood $U \ni z$

which is a domain of a coordinate chart $(z^1,\ldots,z^k)$ such that the restrictions of the distribution $\mathbf{T}$ and its annihilator $\operatorname{Ann}\mathbf{T}$ to U are generated by the r vector fields $\partial/\partial z^1,\cdots,\partial/\partial z^r$ and the $(k-r)$ 1-forms $dz^{k-r+1},\ldots,dz^k$, respectively. It follows that integral manifolds of an involutive distribution make up a foliation. □

An r-dimensional (regular) *foliation* on a k-dimensional manifold Z is said to be a partition of Z into connected leaves F_ι with the following property. Every point of Z has an open neighbourhood U which is a domain of a coordinate chart (z^α) such that, for every leaf F_ι, the connected components $F_\iota\cap U$ are described by the equations

$$z^{r+1}=\text{const.},\qquad \cdots,\qquad z^k=\text{const.}$$

[166, 252]. Note that leaves of a foliation fail to be imbedded submanifolds, i.e., topological subspaces in general.

Example 1.2.5. Submersions $\pi : Y\to X$ and, in particular, fibre bundles are foliations with the leaves $\pi^{-1}(x)$, $x\in\pi(Y)\subset X$. A foliation is called *simple* if it is a fibre bundle. Any foliation is locally simple. •

Example 1.2.6. Every real function f on a manifold Z with nowhere vanishing differential df is a submersion $Z\to\mathbb{R}$. It defines a 1-codimensional foliation whose leaves are given by the equations

$$f(z)=c,\qquad c\in f(Z)\subset\mathbb{R}.$$

This is the *foliation of level surfaces* of the function f, called a *generating function*. Every 1-codimensional foliation is locally a foliation of level surfaces of some function on Z. •

The level surfaces of an arbitrary function $f\neq$ const. on a manifold Z define a *singular foliation* F on Z [166]. Its leaves are not submanifolds in general. Nevertheless if $df(z)\neq 0$, the restriction of F to some open neighbourhood U of z is a foliation with the generating function $f|_U$.

1.3 Jet manifolds

Subsections: *First order jet manifolds*, 25; *Second order jet manifolds*, 27; *Higher order jet manifolds*, 28; *Jets of submanifolds*, 30; *Differential equations and differential operators*, 32.

First order jet manifolds

Given a fibre bundle $Y \to X$ with bundle coordinates (x^λ, y^i), let us consider the equivalence classes $j^1_x s$, $x \in X$, of its sections s, which are identified by their values $s^i(x)$ and the values of their first order derivatives $\partial_\mu s^i(x)$ at points $x \in X$. The equivalence class $j^1_x s$ is called the *first order jet of sections* s at the point $x \in X$. The set J^1Y of first order jets is provided with a manifold structure with respect to the adapted coordinates $(x^\lambda, y^i, y^i_\lambda)$ such that

$$\begin{aligned}
&(x^\lambda, y^i, y^i_\lambda)(j^1_x s) = (x^\lambda, s^i(x), \partial_\lambda s^i(x)),\\
&y'^i_\lambda = \frac{\partial x^\mu}{\partial x'^\lambda}(\partial_\mu + y^j_\mu \partial_j) y'^i.
\end{aligned} \tag{1.3.1}$$

It is called the *jet manifold* of the fibre bundle $Y \to X$.

The jet manifold J^1Y admits the natural fibrations

$$\pi^1 : J^1Y \ni j^1_x s \mapsto x \in X, \tag{1.3.2}$$

$$\pi^1_0 : J^1Y \ni j^1_x s \mapsto s(x) \in Y, \tag{1.3.3}$$

where, by virtue of the transformation law (1.3.1), the latter is an affine bundle modelled over the vector bundle

$$T^*X \underset{Y}{\otimes} VY \to Y. \tag{1.3.4}$$

For the sake of convenience, the fibration (1.3.2) is called a *jet bundle*, while the fibration (1.3.3) is an *affine jet bundle*.

There are the following two canonical imbeddings of the jet manifold J^1Y:

$$\begin{aligned}
&\lambda_1 : J^1Y \underset{Y}{\hookrightarrow} T^*X \underset{Y}{\otimes} TY,\\
&\lambda_1 = dx^\lambda \otimes (\partial_\lambda + y^i_\lambda \partial_i) = dx^\lambda \otimes d_\lambda,
\end{aligned} \tag{1.3.5}$$

where d_λ are *total derivatives*, and

$$\begin{aligned}
&\theta_1 : J^1Y \hookrightarrow T^*Y \underset{Y}{\otimes} VY,\\
&\theta_1 = (dy^i - y^i_\lambda dx^\lambda) \otimes \partial_i = \theta^i \otimes \partial_i,
\end{aligned} \tag{1.3.6}$$

where

$$\theta^i = dy^i - y^i_\lambda dx^\lambda \tag{1.3.7}$$

are called *contact forms*. In accordance with Remark 1.1.1, the morphism θ_1 (1.3.6) can also be rewritten as

$$\theta_1 : J^1Y \underset{Y}{\times} TY \to VY, \qquad \theta_1 = (\dot{y}^i - \dot{x}^\lambda y^i_\lambda)\partial_i. \tag{1.3.8}$$

Remark 1.3.1. From now on, we will identify the jet manifold J^1Y with its image under the canonical morphisms (1.3.5) and (1.3.6), and represent the jet $(x^\lambda, y^i, y^i_\mu)$ by the tangent-valued forms

$$dx^\lambda \otimes (\partial_\lambda + y^i_\lambda \partial_i) \quad \text{and} \quad (dy^i - y^i_\lambda dx^\lambda) \otimes \partial_i. \tag{1.3.9}$$

•

Each fibred morphism $\Phi : Y \to Y'$ over a diffeomorphism f is extended to the fibred morphism of the corresponding affine jet bundles

$$J^1\Phi : J^1Y \underset{\Phi}{\to} J^1Y',$$
$$J^1\Phi : j^1_x s \mapsto j^1_{f(x)}(\Phi \circ s \circ f^{-1}),$$
$$y'^i_\lambda \circ J^1\Phi = \frac{\partial (f^{-1})^\mu}{\partial x'^\lambda} d_\mu \Phi^i,$$

called the *jet prolongation* of the morphism Φ.

Each section s of a fibre bundle $Y \to X$ has the *jet prolongation* to the section

$$(J^1s)(x) \overset{\text{def}}{=} j^1_x s,$$
$$(y^i, y^i_\lambda) \circ J^1 s = (s^i(x), \partial_\lambda s^i(x)),$$

of the jet bundle $J^1Y \to X$. A section $\overline{s}$ of the jet bundle $J^1Y \to X$ is said to be *holonomic* if it is the jet prolongation of some section of the fibre bundle $Y \to X$.

Any projectable vector field

$$u = u^\lambda(x^\mu)\partial_\lambda + u^i(x^\mu, y^j)\partial_i$$

on a fibre bundle $Y \to X$ admits the *jet prolongation* to the projectable vector field

$$J^1u = r_1 \circ J^1u : J^1Y \to J^1TY \to TJ^1Y,$$
$$J^1u = u^\lambda \partial_\lambda + u^i \partial_i + (d_\lambda u^i - y^i_\mu \partial_\lambda u^\mu)\partial^\lambda_i, \tag{1.3.10}$$

on the jet manifold J^1Y. In order to obtain (1.3.10), the canonical fibred morphism

$$r_1 : J^1TY \to TJ^1Y, \qquad \dot{y}^i_\lambda \circ r_1 = (\dot{y}^i)_\lambda - y^i_\mu \dot{x}^\mu_\lambda$$

is used. In particular, there is the canonical isomorphism

$$VJ^1Y = J^1VY, \qquad \dot y^i_\lambda = (\dot y^i)_\lambda. \tag{1.3.11}$$

Second order jet manifolds

Taking the first order jet manifold of the jet bundle $J^1Y \to X$, we obtain the *repeated jet manifold* J^1J^1Y provided with the adapted coordinates

$$\begin{aligned}
&(x^\lambda, y^i, y^i_\lambda, \widehat{y}^i_\mu, y^i_{\mu\lambda}),\\
&\widehat{y}'^i_\lambda = \frac{\partial x^\alpha}{\partial x'^\lambda} d_\alpha y'^i,\\
&y'^i_{\mu\lambda} = \frac{\partial x^\alpha}{\partial x'^\mu} d_\alpha y'^i_\lambda, \qquad d_\alpha = \partial_\alpha + \widehat{y}^j_\alpha \partial_j + y^j_{\nu\alpha}\partial^\nu_j.
\end{aligned}$$

There exist two different affine fibrations of J^1J^1Y over J^1Y:

- the familiar affine jet bundle (1.3.3)

$$\pi_{11} : J^1J^1Y \to J^1Y, \qquad y^i_\lambda \circ \pi_{11} = y^i_\lambda, \tag{1.3.12}$$

- and the affine bundle

$$J^1\pi^1_0 : J^1J^1Y \to J^1Y, \qquad y^i_\lambda \circ J^1\pi^1_0 = \widehat{y}^i_\lambda. \tag{1.3.13}$$

In general, there is no canonical identification of these fibrations. The points $q \in J^1J^1Y$, where $\pi_{11}(q) = J^1\pi^1_0(q)$, make up the affine subbundle $\widehat{J}^2Y \to J^1Y$ of J^1J^1Y, called the *sesquiholonomic jet manifold*. This is given by the coordinate conditions $\widehat{y}^i_\lambda = y^i_\lambda$, and is coordinated by $(x^\lambda, y^i, y^i_\lambda, y^i_{\mu\lambda})$.

The *second order jet manifold* J^2Y of a fibre bundle $Y \to X$ is the affine subbundle $\pi^2_1 : J^2Y \to J^1Y$ of the fibre bundle $\widehat{J}^2Y \to J^1Y$, given by the coordinate conditions $y^i_{\lambda\mu} = y^i_{\mu\lambda}$ and coordinated by $(x^\lambda, y^i, y^i_\lambda, y^i_{\lambda\mu} = y^i_{\mu\lambda})$. It is modelled over the vector bundle

$$\overset{2}{\vee} T^*X \underset{J^1Y}{\otimes} VY \to J^1Y.$$

The second order jet manifold J^2Y can also be seen as the set of the equivalence classes $j^2_x s$ of sections s of the fibre bundle $Y \to X$, which are identified by their values and the values of their first and second order partial derivatives at points $x \in X$, i.e.,

$$y^i_\lambda(j^2_x s) = \partial_\lambda s^i(x), \qquad y^i_{\lambda\mu}(j^2_x s) = \partial_\lambda \partial_\mu s^i(x).$$

Let s be a section of a fibre bundle $Y \to X$ and $J^1 s$ its jet prolongation to a section of the jet bundle $J^1Y \to X$. The latter gives rise to the section J^1J^1s of the repeated jet bundle $J^1J^1Y \to X$. This section takes its values into the second order jet manifold J^2Y. It is called the *second order jet prolongation* of the section s, and is denoted by J^2s.

PROPOSITION 1.3.1. Let $\overline{s}$ be a section of the jet bundle $J^1Y \to X$ and $J^1\overline{s}$ its jet prolongation to the section of the repeated jet bundle $J^1J^1Y \to X$. The following three facts are equivalent:

- $\overline{s} = J^1 s$ where s is a section of the fibre bundle $Y \to X$;
- $J^1\overline{s}$ takes its values into $\widehat{J}^2Y$;
- $J^1\overline{s}$ takes its values into J^2Y.

□

Higher order jet manifolds

The notion of first and second order jet manifolds is naturally extended to higher order jet manifolds. Here, we touch on only a few elements of the higher order jet technique, and refer the reader to Section 11.1 for a detailed exposition.

The *k-order jet manifold* J^kY of a fibre bundle $Y \to X$ comprises the equivalence classes $j^k_x s$, $x \in X$, of sections s of Y identified by the $k+1$ terms of their Taylor series at the points $x \in X$. The jet manifold J^kY is provided with the adapted coordinates

$$(x^\lambda, y^i, y^i_\lambda, \ldots, y^i_{\lambda_k \cdots \lambda_1}),$$
$$y^i_{\lambda_l \cdots \lambda_1}(j^k_x s) = \partial_{\lambda_l} \cdots \partial_{\lambda_1} s^i(x), \qquad 0 \leq l \leq k.$$

Every section s of a fibre bundle $Y \to X$ gives rise to the section $J^k s$ of the jet bundle $J^kY \to X$ such that

$$y^i_{\lambda_l \cdots \lambda_1} \circ J^k s = \partial_{\lambda_l} \cdots \partial_{\lambda_1} s^i, \qquad 0 \leq l \leq k.$$

We will use the following operators on exterior forms on jet manifolds (see Section 11.1 for their intrinsic definitions):

- the *total derivative*

$$d_\lambda = \partial_\lambda + y^i_\lambda \partial_i + y^i_{\lambda\mu}\partial^\mu_i + \cdots, \tag{1.3.14}$$

  given by the relations

$$d_\lambda(\phi\wedge\sigma) = d_\lambda(\phi)\wedge\sigma + \phi\wedge d_\lambda(\sigma),$$
$$d_\lambda(d\phi) = d(d_\lambda(\phi)),$$

  e.g.,

$$d_\lambda(f) = \partial_\lambda f + y^i_\lambda \partial_i f + y^i_{\lambda\mu}\partial^\mu_i f + \cdots, \qquad f\in C^\infty(J^kY),$$
$$d_\lambda(dx^\mu) = 0, \qquad d_\lambda(dy^i_{\lambda_l\cdots\lambda_1}) = dy^i_{\lambda\lambda_l\cdots\lambda_1};$$

- the *horizontal projection* h_0 given by the relations

$$h_0(dx^\lambda) = dx^\lambda, \qquad h_0(dy^i_{\lambda_k\cdots\lambda_1}) = y^i_{\mu\lambda_k\ldots\lambda_1}dx^\mu, \tag{1.3.15}$$

  e.g.,

$$h_0(dy^i) = y^i_\mu dx^\mu, \qquad h_0(dy^i_\lambda) = y^i_{\mu\lambda}dx^\mu;$$

- the *horizontal differential*

$$d_H(\phi) = dx^\lambda\wedge d_\lambda(\phi) \tag{1.3.16}$$

  possessing the properties

$$d_H\circ d_H = 0, \qquad h_0\circ d = d_H\circ h_0.$$

Jets of submanifolds

The notion of jets of sections of a fibre bundle is generalized to jets of submanifolds of a manifold Z which has no fibration [123, 185]. We will appeal to jets of submanifolds both in order to introduce the general notion of a differential equation and in relativistic mechanics.

Let Z be a manifold of dimension $m+n$. The *k-order jet of n-dimensional submanifolds* of Z at a point $z \in Z$ is defined as the equivalence class $[S]^k_z$ of n-dimensional imbedded submanifolds of Z which pass through z and which are tangent to each other at z with order $k \geq 0$. The disjoint union

$$J^k_n Z = \bigcup_{z\in Z} [S]^k_z, \qquad k > 0, \tag{1.3.17}$$

of jets $[S]^k_z$ is said to be the *k-order jet manifold of the n-dimensional submanifolds of* Z. Lowering the order of tangency, one obtains the natural surjections $J^k_n Z \to J^{k-i}_n Z$ and $J^k_n Z \to Z$. By definition, $J^0_n Z = Z$.

Remark 1.3.2. The above definition of jets of submanifolds does not provide for jets of n-dimensional submanifolds of an n-dimensional manifold Z. Jets of this type are widely known due to their application to the study of G-structures [123, 178, 252]. •

Hereafter, we will restrict our consideration to first order jets of submanifolds. The set of these jets $J^1_n Z$ is provided with a manifold structure as follows. Let $Y \to X$ be an $(m+n)$-dimensional fibre bundle over an n-dimensional base X, and let Φ be an imbedding of Y into Z. Then there is the natural injection

$$\begin{aligned} &J^1\Phi : J^1Y \to J^1_n Z, \\ &j^1_x s \mapsto [S]^1_{\Phi(s(x))}, \qquad S = \mathrm{Im}\,(\Phi\circ s), \end{aligned} \tag{1.3.18}$$

where s are sections of $Y \to X$. This injection defines a chart on $J^1_n Z$. Indeed, given a submanifold $S \subset Z$ which belongs to the jet $[S]^1_z$, there exist a neighbourhood U_z of the point z and the tubular neighbourhood U_S of $S\cap U_z$ so that the fibration $U_S \to S\cap U_z$ takes place. It means that every jet $[S]^1_z$ lives in a chart of the above-mentioned type. These charts cover the set $J^1_n Z$, and transition functions between them are differentiable.

It is convenient to use the following coordinate atlas of the jet manifold $J^1_n Z$ of n-dimensional submanifolds of Z. Let Z be equipped with a coordinate atlas

$$\{(U; z^A)\}, \qquad A = 1, \ldots, n+m. \tag{1.3.19}$$

Though $J^0_n Z = Z$, let us provide $J^0_n Z$ with the atlas where every chart $(U; z^A)$ on a domain $U \subset Z$ is replaced with the

$$\binom{n+m}{m} = \frac{(n+m)!}{n!m!}$$

charts on the same domain U which correspond to different partitions of the collection $(z^1 \cdots z^A)$ in the collections of n and m coordinates, denoted by

$$(x^\lambda, y^i), \qquad \lambda = 1, \ldots, n, \qquad i = 1, \ldots, m. \tag{1.3.20}$$

The transition functions between the coordinate charts (1.3.20) of $J^0_n Z$, associated with the same coordinate chart (1.3.19) of Z, reduce to an exchange between coordinates x^λ and y^i. Transition functions between arbitrary coordinate charts (1.3.20) of the manifold $J^0_n Z$ read

$$\begin{aligned} &\widetilde{x}^\lambda = \widetilde{g}^\lambda(x^\mu, y^j), \qquad &\widetilde{y}^i = \widetilde{f}^i(x^\mu, y^j), \\ &x^\lambda = g^\lambda(\widetilde{x}^\mu, \widetilde{y}^j), \qquad &y^i = f^i(\widetilde{x}^\mu, \widetilde{y}^j). \end{aligned} \tag{1.3.21}$$

Given the coordinate atlas (1.3.20) of the manifold $J^0_n Z$, the jet manifold $J^1_n Z$ is provided with the adapted coordinates

$$(x^\lambda, y^i, y^i_\lambda), \qquad \lambda = 1, \ldots, n, \qquad i = 1, \ldots, m. \tag{1.3.22}$$

The transition functions of the coordinates y^i_λ (1.3.22) under coordinate transformations (1.3.21) take the form

$$\begin{aligned} \widetilde{y}^i_\lambda = d_{\widetilde{\lambda}} \widetilde{f}^i = [d_{\widetilde{\lambda}} g^\alpha(\widetilde{x}^\mu, \widetilde{y}^i)] d_\alpha \widetilde{f}^i(x^\mu, y^j) = \\ \left[\left(\frac{\partial}{\partial \widetilde{x}^\lambda} + \widetilde{y}^p_\lambda \frac{\partial}{\partial \widetilde{y}^p}\right) g^\alpha(\widetilde{x}^\lambda, \widetilde{y}^i)\right] \left(\frac{\partial}{\partial x^\alpha} + y^j_\alpha \frac{\partial}{\partial y^j}\right) \widetilde{f}^i(x^\mu, y^j). \end{aligned} \tag{1.3.23}$$

It is readily observed that the transition functions (1.3.1) are a particular case of the coordinate transformations (1.3.23) when the transition functions g^α (1.3.21) are independent of coordinates $\widetilde{y}^i$. In contrast with (1.3.1), the coordinate transformations (1.3.23) are not affine. It follows that the fibration $J^1_n Z \to Z$ is not an affine bundle. Similarly to the morphism (1.3.5), there is one-to-one correspondence

$$\lambda_1 : [S]^1_z \mapsto \dot{x}^\lambda(\partial_\lambda + y^i_\lambda([S]^1_z)\partial_i) \tag{1.3.24}$$

between the jets $[S]^1_z$ at a point $z \in Z$ and n-dimensional vector subspaces of the tangent space T_zZ. It follows that the fibration $J^1_nZ \to Z$ is a fibre bundle with the structure group $GL(n,m;\mathbb{R})$ of linear transformations of the vector space $\mathbb{R}^{m+n}$ which preserve its subspace $\mathbb{R}^n$. The typical fibre of $J^1_nZ \to Z$ is the *Grassmann manifold*

$$\mathfrak{G}(n,m;\mathbb{R}) = GL(n+m;\mathbb{R})/GL(n,m;\mathbb{R}).$$

In particular, let $Y \to X$ be an $(m+n)$-dimensional fibre bundle over an n-dimensional base X, J^1Y the first order jet manifold of its sections, and J^1_nY the first order jet manifold of n-dimensional subbundles of Y. Then the injection $J^1Y \to J^1_nY$ (1.3.18) is an affine subbundle of the jet bundle $J^1_nY \to Y$. Its fibre at a point $y \in Y$ consists of the n-dimensional subspaces of the tangent space T_yY whose intersections with the vertical subspace V_yY of T_yY reduce to the zero vector.

Differential equations and differential operators

Jet manifolds provide the standard language for the theory of differential equations and differential operators [46, 123, 185, 247]. We will refer to the following general notion of a differential equation.

DEFINITION 1.3.2. Let Z be an $(m+n)$-dimensional manifold. A *system of k-order partial differential equations in n variables on Z* is defined as a closed submanifold $\mathfrak{E}$ of the k-order jet bundle J^k_nZ of n-dimensional submanifolds of Z. □

In brief, we will call $\mathfrak{E}$ simply a *differential equation*. By its *classical solution* is meant an n-dimensional submanifold S of Z whose k-order jets $[S]^k_z$, $z \in S$, belong to $\mathfrak{E}$.

DEFINITION 1.3.3. A k-order differential equation in n variables on a manifold Z is called a *dynamic equation* if it can be algebraically solved for the highest order derivatives, i.e., it is a section of the fibration $J^k_nZ \to J^{k-1}_nZ$. □

In particular, a *first order dynamic equation* in n variables on a manifold Z is a section of the jet bundle $J^1_nZ \to Z$. Its image in the tangent bundle $TZ \to Z$ by the correspondence λ_1 (1.3.24) is an n-dimensional vector subbundle of TZ. If $n = 1$, a dynamic equation is given by a vector field

$$\dot{z}^\lambda(t) = u^\lambda(z(t)) \tag{1.3.25}$$

on a manifold Z. Its classical solutions are integral curves $c(t)$ of the vector field u.

Let $Y \to X$ be a fibre bundle and J^kY its k-order jet manifold.

DEFINITION 1.3.4. In accordance with Definition 1.3.2, a k-order differential equation on $Y \to X$ is defined as a closed subbundle $\mathfrak{E}$ of the jet bundle $J^kY \to X$. Its classical solution is a (local) section s of the fibre bundle $Y \to X$ such that its k-order jet prolongation J^ks lives in $\mathfrak{E}$. □

Henceforth, we will consider differential equations associated with differential operators. Given a fibre bundle $Y \to X$, let $E \to X$ be a vector bundle coordinated by (x^λ, v^A).

DEFINITION 1.3.5. A fibred morphism

$$\begin{aligned} &\mathcal{E} : J^kY \underset{X}{\to} E, \qquad (1.3.26)\\ &v^A \circ \mathcal{E} = \mathcal{E}^A(x^\lambda, y^i, y^i_\lambda, \ldots, y^i_{\lambda_k\cdots\lambda_1}), \end{aligned}$$

is called a k-order *differential operator* on the fibre bundle $Y \to X$. It sends each section $s(x)$ of $Y \to X$ onto the section $(\mathcal{E} \circ J^ks)(x)$ of the vector bundle $E \to X$:

$$(\mathcal{E} \circ J^ks)^A(x) = \mathcal{E}^A(x^\lambda, s^i(x), \partial_\lambda s^i(x), \ldots, \partial_{\lambda_k} \cdots \partial_{\lambda_1} s^i(x)).$$

□

Let us assume that $\widehat{0}(X) \subset \mathcal{E}(J^kY)$, where $\widehat{0}$ is the zero section of the vector bundle $E \to X$. Then the *kernel* of a differential operator is the subset

$$\operatorname{Ker}\mathcal{E} = \mathcal{E}^{-1}(\widehat{0}(X)) \subset J^kY. \qquad (1.3.27)$$

If $\operatorname{Ker}\mathcal{E}$ (1.3.27) is a closed subbundle of the fibre bundle $J^kY \to X$, it defines a differential equation

$$\mathcal{E} \circ J^ks = 0$$

written in the coordinate form

$$\begin{cases} \qquad\qquad \cdots \\ \mathcal{E}^A(x^\lambda, y^i, y^i_\lambda, \ldots, y^i_{\lambda_k\cdots\lambda_1}) = 0, \\ \qquad\qquad \cdots \end{cases}$$

The following condition is sufficient for a kernel of a differential operator to be a differential equation.

PROPOSITION 1.3.6. Let the morphism (1.3.26) be of constant rank. By virtue of Proposition 1.1.3, its kernel (1.3.27) is a closed subbundle of the fibre bundle $J^kY \to X$ and, consequently, is a k-order differential equation. □

Remark 1.3.3. Linear differential operators are usually phrased in terms of jets of modules (see Section 8.1). •

Chapter 2

Connections

This Chapter is devoted to the general notion of a connection on a fibre bundle. We start from the traditional definition of connections as splittings of the exact sequences (1.1.17a) – (1.1.17b), but then follow their definition as global section of the affine jet bundle [123, 179, 212, 265, 274]. These connections are represented by tangent-valued forms. The algebraic definition of connections on modules and sheaves is given in Chapter 8. It is appropriate for quantum field theory, and is equivalent to the above mentioned ones in the case of vector bundles.

2.1 Connections as tangent-valued forms

A *connection* on a fibre bundle $Y \to X$ is defined traditionally as a linear bundle monomorphism

$$\Gamma : Y \underset{X}{\times} TX \to TY, \qquad (2.1.1)$$

$$\boxed{\Gamma : \dot{x}^\lambda \partial_\lambda \mapsto \dot{x}^\lambda (\partial_\lambda + \Gamma^i_\lambda \partial_i),}$$

over Y which splits the exact sequence (1.1.17a), i.e.,

$$\pi_T \circ \Gamma = \mathrm{Id}\,(Y \underset{X}{\times} TX).$$

By virtue of Theorem 1.1.4, a connection always exists. The local functions $\Gamma^i_\lambda(y)$ in (2.1.1) are said to be *components* of the connection Γ with respect to the fibred coordinates (x^λ, y^i) on $Y \to X$.

The image of $Y \times TX$ by the connection Γ defines the *horizontal distribution* $HY \subset TY$ which splits the tangent bundle TY as follows:

$$TY = HY \underset{Y}{\oplus} VY, \tag{2.1.2}$$
$$\dot{x}^\lambda \partial_\lambda + \dot{y}^i \partial_i = \dot{x}^\lambda(\partial_\lambda + \Gamma^i_\lambda \partial_i) + (\dot{y}^i - \dot{x}^\lambda \Gamma^i_\lambda)\partial_i.$$

Its annihilator is locally generated by the 1-forms $dy^i - \Gamma^i_\lambda dx^\lambda$.

The connection Γ (2.1.1) can be written in several equivalent forms (see (2.1.3), (2.1.4), (2.1.7), (2.1.8) and (2.1.10) below). We will use the same symbol for all of them.

Given the horizontal splitting (2.1.2), the projection

$$\Gamma : TY \underset{Y}{\to} VY, \tag{2.1.3}$$
$$\boxed{\dot{y}^i \circ \Gamma = \dot{y}^i - \Gamma^i_\lambda \dot{x}^\lambda,}$$

defines a connection on $Y \to X$ in an equivalent way.

The linear morphism Γ over Y (2.1.1) yields uniquely the horizontal tangent-valued 1-form

$$\boxed{\Gamma = dx^\lambda \otimes (\partial_\lambda + \Gamma^i_\lambda \partial_i)} \tag{2.1.4}$$

on Y which projects over the canonical tangent-valued form θ_X (1.2.28) on X. With this form, the morphism (2.1.1) reads

$$\Gamma : \partial_\lambda \mapsto \partial_\lambda \rfloor \Gamma = \partial_\lambda + \Gamma^i_\lambda \partial_i.$$

One can think of the tangent-valued form Γ (2.1.4) as being another definition of a connection on a fibre bundle $Y \to X$.

Given a connection Γ and the corresponding horizontal distribution (2.1.2), a vector field u on the fibre bundle $Y \to X$ is called *horizontal* if it lives in HY. A horizontal vector field takes the form

$$u = u^\lambda(y)(\partial_\lambda + \Gamma^i_\lambda \partial_i). \tag{2.1.5}$$

In particular, let τ be a vector field on the base X. By means of the tangent-valued form Γ (2.1.4), we obtain the projectable horizontal vector field

$$\boxed{\Gamma\tau = \tau \rfloor \Gamma = \tau^\lambda(\partial_\lambda + \Gamma^i_\lambda \partial_i)} \tag{2.1.6}$$

on Y, called the *horizontal lift* of τ by the connection Γ.

Given the splitting (2.1.1), the dual splitting of the exact sequence (1.1.17b) is

$$\Gamma : V^*Y \to T^*Y,$$
$$\boxed{\Gamma : \overline{d}y^i \mapsto dy^i - \Gamma^i_\lambda dx^\lambda.} \tag{2.1.7}$$

Hence, a connection Γ on $Y \to X$ is represented by the vertical-valued form

$$\boxed{\Gamma = (dy^i - \Gamma^i_\lambda dx^\lambda) \otimes \partial_i} \tag{2.1.8}$$

such that the morphism (2.1.7) reads

$$\Gamma : \overline{d}y^i \mapsto \Gamma \rfloor \overline{d}y^i = dy^i - \Gamma^i_\lambda dx^\lambda.$$

The corresponding horizontal splitting of the cotangent bundle T^*Y takes the form

$$T^*Y = T^*X \underset{Y}{\oplus} \Gamma(V^*Y), \tag{2.1.9}$$
$$\dot{x}_\lambda dx^\lambda + \dot{y}_i dy^i = (\dot{x}_\lambda + \dot{y}_i \Gamma^i_\lambda) dx^\lambda + \dot{y}_i (dy^i - \Gamma^i_\lambda dx^\lambda).$$

Then we have the projection

$$\Gamma = \mathrm{pr}_1 : T^*Y \to T^*X,$$
$$\boxed{\dot{x}_\lambda \circ \Gamma = \dot{x}_\lambda + \dot{y}_i \Gamma^i_\lambda,} \tag{2.1.10}$$

which also defines a connection on the fibre bundle $Y \to X$.

Remark 2.1.1. Treating a connection as the vertical-valued form (2.1.8), we come to the following important construction. Given a fibre bundle $Y \to X$, let $f : X' \to X$ be a map and $f^*Y \to X'$ the pull-back of Y by f. Any connection Γ (2.1.8) on $Y \to X$ induces the *pull-back connection*

$$f^*\Gamma = (dy^i - (\Gamma \circ f_Y)^i_\lambda \frac{\partial f^\lambda}{\partial x'^\mu} dx'^\mu) \otimes \partial_i \tag{2.1.11}$$

on $f^*Y \to X'$ (see Remark 1.2.4). Accordingly the curvature (see (2.3.3) later) of the pull-back connection (2.1.11) is the pull-back f^*R of the curvature of the connection Γ. •

2.2 Connections as jet bundle sections

The definition of connections as sections of the affine jet bundle enables one to say something more. This definition is based on the following canonical splittings.

Let $Y \to X$ be a fibre bundle, and J^1Y its first order jet manifold. Given the canonical morphisms (1.3.5) and (1.3.6), we have the corresponding morphisms

$$\widehat{\lambda}_1 : J^1Y \underset{X}{\times} TX \ni \partial_\lambda \mapsto d_\lambda = \partial_\lambda \rfloor \lambda_1 \in J^1Y \underset{Y}{\times} TY, \tag{2.2.1}$$

$$\widehat{\theta}_1 : J^1Y \underset{Y}{\times} V^*Y \ni \overline{d}y^i \mapsto \theta^i = \theta_1 \rfloor dy^i \in J^1Y \underset{Y}{\times} T^*Y \tag{2.2.2}$$

(see Remark 1.1.1). These morphisms yield the *canonical horizontal splittings* of the pull-backs

$$J^1Y \underset{Y}{\times} TY = \widehat{\lambda}_1(TX) \underset{J^1Y}{\oplus} VY, \tag{2.2.3}$$

$$\boxed{\dot{x}^\lambda \partial_\lambda + \dot{y}^i \partial_i = \dot{x}^\lambda (\partial_\lambda + y^i_\lambda \partial_i) + (\dot{y}^i - \dot{x}^\lambda y^i_\lambda)\partial_i,}$$

$$J^1Y \underset{Y}{\times} T^*Y = T^*X \underset{J^1Y}{\oplus} \widehat{\theta}_1(V^*Y), \tag{2.2.4}$$

$$\boxed{\dot{x}_\lambda dx^\lambda + \dot{y}_i dy^i = (\dot{x}_\lambda + \dot{y}_i y^i_\lambda) dx^\lambda + \dot{y}_i (dy^i - y^i_\lambda dx^\lambda).}$$

Remark 2.2.1. Let $u = u^\lambda \partial_\lambda + u^i \partial_i$ be a vector field on a fibre bundle $Y \to X$. In accordance with the formula (2.2.3), we have its *canonical horizontal splitting*

$$\boxed{u^\lambda \partial_\lambda + u^i \partial_i = u^\lambda (\partial_\lambda + y^i_\lambda \partial_i) + (u^i - u^\lambda y^i_\lambda)\partial_i.} \tag{2.2.5}$$

•

Let Γ be a global section of $J^1Y \to Y$. Substituting the tangent-valued form

$$\lambda_1 \circ \Gamma = dx^\lambda \otimes (\partial_\lambda + \Gamma^i_\lambda \partial_i)$$

in the canonical splitting (2.2.3), we obtain the familiar horizontal splitting (2.1.2) of TY by means of a connection Γ on $Y \to X$. Accordingly, substitution of the tangent-valued form

$$\theta_1 \circ \Gamma = (dy^i - \Gamma^i_\lambda dx^\lambda) \otimes \partial_i$$

in the canonical splitting (2.2.4) leads to the dual splitting (2.1.9) of T^*Y by means of a connection Γ.

PROPOSITION 2.2.1. [123, 274]. There is one-to-one correspondence between the connections Γ on a fibre bundle $Y \to X$ and the global sections

$$\Gamma : Y \to J^1Y,$$
$$\boxed{(x^\lambda, y^i, y^i_\lambda) \circ \Gamma = (x^\lambda, y^i, \Gamma^i_\lambda),} \tag{2.2.6}$$

of the affine jet bundle $J^1Y \to Y$. They are represented by the tangent-valued forms (2.1.4) and (2.1.8) in accordance with Remark 1.3.1. □

It follows at once from this correspondence that connections on a fibre bundle $Y \to X$ make up an affine space modelled over the vector space of soldering forms on $Y \to X$, i.e., sections of the vector bundle (1.3.4). One deduces immediately from (1.3.1) the coordinate transformation law

$$\Gamma'^i_\lambda = \frac{\partial x^\mu}{\partial x'^\lambda}(\partial_\mu + \Gamma^j_\mu \partial_j) y'^i$$

of connection parameters.

Every connection Γ (2.2.6) on a fibre bundle $Y \to X$ yields the first order differential operator

$$D_\Gamma : J^1Y \underset{Y}{\to} T^*X \underset{Y}{\otimes} VY, \tag{2.2.7}$$
$$\boxed{D_\Gamma = \lambda_1 - \Gamma \circ \pi^1_0 = (y^i_\lambda - \Gamma^i_\lambda) dx^\lambda \otimes \partial_i,}$$

called the *covariant differential* relative to the connection Γ. If $s : X \to Y$ is a (local) section, we obtain from (2.2.7) its covariant differential

$$\nabla^\Gamma s = D_\Gamma \circ J^1 s : X \to T^*X \otimes VY, \tag{2.2.8}$$
$$\boxed{\nabla^\Gamma s = (\partial_\lambda s^i - \Gamma^i_\lambda \circ s) dx^\lambda \otimes \partial_i,}$$

and the *covariant derivative* $\nabla^\Gamma_\tau s = \tau \rfloor \nabla^\Gamma s$ along a vector field τ on X.

A (local) section s is said to be an *integral section* of the connection Γ if s obeys the equivalent conditions

$$\nabla^\Gamma s = 0 \quad \text{or} \quad J^1 s = \Gamma \circ s. \tag{2.2.9}$$

Let $s : X \to Y$ be a global section. There exists a connection Γ such that s is an integral section of Γ. This connection Γ is an extension of the local section $s(x) \mapsto J^1 s(x)$ of the affine jet bundle $J^1Y \to Y$ over the closed imbedded submanifold $s(X) \subset Y$ in accordance with Theorem 1.1.2.

Treating connections as sections as sections of the affine jet bundle, we come naturally to the following two constructions.

Let Y and Y' be fibre bundles over the same base X. Given a connection Γ on $Y \to X$ and a connection Γ' on $Y' \to X$, the fibre bundle $Y \underset{X}{\times} Y' \to X$ is provided with the *product connection*

$$\begin{aligned}
&\Gamma \times \Gamma' : Y \underset{X}{\times} Y' \to J^1(Y \underset{X}{\times} Y') = J^1Y \underset{X}{\times} J^1Y', \\
&\Gamma \times \Gamma' = dx^\lambda \otimes (\partial_\lambda + \Gamma^i_\lambda \frac{\partial}{\partial y^i} + \Gamma'^j_\lambda \frac{\partial}{\partial y'^j}).
\end{aligned} \tag{2.2.10}$$

Let $i_Y : Y \to Y'$ be a subbundle of a fibre bundle $Y' \to X$ and Γ' a connection on $Y' \to X$. If there exists a connection Γ on $Y \to X$ such that the diagram

$$\begin{array}{ccc}
Y' & \xrightarrow{\Gamma'} & J^1Y \\
{\scriptstyle i_Y}\uparrow & & \uparrow{\scriptstyle J^1 i_Y} \\
Y & \xrightarrow{\Gamma} & J^1Y'
\end{array}$$

commutes, we say that Γ' is *reducible* to the connection Γ. The following conditions are equivalent:

- Γ' is reducible to Γ;
- $Ti_Y(HY) = HY'|_{i_Y(Y)}$, where $HY \subset TY$ and $HY' \subset TY'$ are the horizontal subbundles determined by Γ and Γ', respectively;
- for every vector field $\tau \in \mathcal{T}(X)$, the vector fields $\Gamma\tau$ and $\Gamma'\tau$ are i_Y-related, i.e.,

$$Ti_Y \circ \Gamma\tau = \Gamma'\tau \circ i_Y. \tag{2.2.11}$$

2.3 Curvature and torsion

Let Γ be a connection on a fibre bundle $Y \to X$. Given vector fields τ, τ' on X and their horizontal lifts $\Gamma\tau$ and $\Gamma\tau'$ (2.1.6) on Y, let us compute the vector field

$$R(\tau, \tau') = -\Gamma[\tau, \tau'] + [\Gamma\tau, \Gamma\tau'] \tag{2.3.1}$$

on Y. It is readily observed that this is the vertical vector field

$$R(\tau,\tau') = \tau^\lambda \tau'^\mu R^i_{\lambda\mu}\partial_i,$$
$$\boxed{R^i_{\lambda\mu} = \partial_\lambda\Gamma^i_\mu - \partial_\mu\Gamma^i_\lambda + \Gamma^j_\lambda\partial_j\Gamma^i_\mu - \Gamma^j_\mu\partial_j\Gamma^i_\lambda.} \tag{2.3.2}$$

The VY-valued horizontal 2-form on Y

$$R = \frac{1}{2}R^i_{\lambda\mu}dx^\lambda\wedge dx^\mu\otimes\partial_i \tag{2.3.3}$$

is called the *curvature of the connection* Γ.

In an equivalent way, the curvature (2.3.3) is defined as the Nijenhuis differential

$$R = \frac{1}{2}d_\Gamma\Gamma = \frac{1}{2}[\Gamma,\Gamma]_{\rm FN} : Y \to \overset{2}{\wedge} T^*X\otimes VY. \tag{2.3.4}$$

Then we obtain at once the identities

$$[R,R]_{\rm FN} \equiv 0, \tag{2.3.5}$$
$$d_\Gamma R = [\Gamma,R]_{\rm FN} \equiv 0. \tag{2.3.6}$$

The identity (2.3.5) results from the identity (1.2.32), while (2.3.6) is an immediate consequence of the graded Jacobi identity (1.2.33). The identity (2.3.6) is called the *(generalized) second Bianchi identity*. It takes the coordinate form

$$\sum_{(\lambda\mu\nu)}(\partial_\lambda R^i_{\mu\nu} + \Gamma^j_\lambda\partial_j R^i_{\mu\nu} - \partial_j\Gamma^i_\lambda R^j_{\mu\nu}) \equiv 0, \tag{2.3.7}$$

where the sum is cyclic over the indices λ, μ and ν.

In the same manner, given a soldering form σ, one defines the *soldered curvature*

$$\rho = \frac{1}{2}d_\sigma\sigma = \frac{1}{2}[\sigma,\sigma]_{\rm FN} : Y\to\overset{2}{\wedge} T^*X\otimes VY, \tag{2.3.8}$$
$$\rho = \frac{1}{2}\rho^i_{\lambda\mu}dx^\lambda\wedge dx^\mu\otimes\partial_i,$$
$$\rho^i_{\lambda\mu} = \sigma^j_\lambda\partial_j\sigma^i_\mu - \sigma^j_\mu\partial_j\sigma^i_\lambda,$$

which fulfills the identities

$$[\rho,\rho]_{\rm FN}\equiv 0, \qquad d_\sigma\rho = [\sigma,\rho]_{\rm FN}\equiv 0.$$

Given a connection Γ and a soldering form σ, the *torsion* of Γ with respect to σ is introduced as

$$\boxed{T = d_\Gamma \sigma = d_\sigma \Gamma : Y \to \overset{2}{\wedge} T^*X \otimes VY.}$$

Its coordinate expression is

$$T = (\partial_\lambda \sigma^i_\mu + \Gamma^j_\lambda \partial_j \sigma^i_\mu - \partial_j \Gamma^i_\lambda \sigma^j_\mu) dx^\lambda \wedge dx^\mu \otimes \partial_i. \tag{2.3.9}$$

There is the *(generalized) first Bianchi identity*

$$d_\Gamma T = d^2_\Gamma \sigma = [R, \sigma]_{\mathrm{FN}} = -d_\sigma R. \tag{2.3.10}$$

If $\Gamma' = \Gamma + \sigma$, we have the important relations

$$T' = T + 2\rho, \tag{2.3.11}$$

$$R' = R + \rho + T. \tag{2.3.12}$$

2.4 Linear connections

A connection Γ on a vector bundle $Y \to X$ is said to be a *linear connection* if the section

$$\Gamma : Y \to J^1Y,$$

$$\boxed{\Gamma = dx^\lambda \otimes (\partial_\lambda + \Gamma_\lambda{}^i{}_j(x) y^j \partial_i),} \tag{2.4.1}$$

is a linear bundle morphism over X. Note that linear connections are principal connections, and they always exist (see Section 6.1).

The curvature R (2.3.3) of a linear connection Γ (2.4.1) reads

$$R = \frac{1}{2} R_{\lambda\mu}{}^i{}_j(x) y^j dx^\lambda \wedge dx^\mu \otimes \partial_i,$$

$$\boxed{R_{\lambda\mu}{}^i{}_j = \partial_\lambda \Gamma_\mu{}^i{}_j - \partial_\mu \Gamma_\lambda{}^i{}_j + \Gamma_\lambda{}^h{}_j \Gamma_\mu{}^i{}_h - \Gamma_\mu{}^h{}_j \Gamma_\lambda{}^i{}_h.} \tag{2.4.2}$$

Due to the vertical splitting (1.1.15), we have the linear morphism

$$R : Y \ni y^i e_i \mapsto \frac{1}{2} R_{\lambda\mu}{}^i{}_j y^j dx^\lambda \wedge dx^\mu \otimes e_i \in \mathfrak{O}^2(X) \otimes Y. \tag{2.4.3}$$

Then one can write

$$R(\tau, \tau') \circ s = ([\nabla^\Gamma_\tau, \nabla^\Gamma_{\tau'}] - \nabla^\Gamma_{[\tau,\tau']}) s \tag{2.4.4}$$

for any section s of the vector bundle $Y \to X$ and any two vector fields τ and τ' on X. We will refer to the expressions (2.4.3) and (2.4.4) in order to introduce the curvature of connections on modules and sheaves (see Chapter 8).

Some standard operations with linear connections should be recalled.

(i) Let $Y \to X$ be a vector bundle and Γ a linear connection (2.4.1) on Y. Then there is a unique linear connection Γ^* on the dual vector bundle $Y^* \to X$ such that the diagram

$$\begin{array}{ccccc} & J^1Y \underset{X}{\times} J^1Y^* & \xrightarrow{J^1\langle\rangle} & T^*X \times \mathbb{R} & \\ {\scriptstyle \Gamma\times\Gamma^*} & \uparrow & & \uparrow & {\scriptstyle (\widehat{0},\mathrm{Id})} \\ & Y \underset{X}{\times} Y^* & \underset{\langle\rangle}{\longrightarrow} & X \times \mathbb{R} & \end{array}$$

commutes. The connection Γ^* is called the *dual connection* of Γ. It has the coordinate expression

$$\boxed{\Gamma^*_{\lambda i} = -\Gamma_\lambda{}^j{}_i y_j,} \tag{2.4.5}$$

where (x^λ, y_j) are the fibred coordinates on Y^* dual of those on Y.

(ii) Let $Y \to X$ and $Y' \to X$ be vector bundles with linear connections Γ and Γ', respectively. Then the product connection (2.2.10) is the *direct sum connection* $\Gamma \oplus \Gamma'$ on $Y \oplus Y'$.

(iii) Let $Y \to X$ and $Y' \to X$ be vector bundles with linear connections Γ and Γ', respectively. There is a unique linear connection $\Gamma \otimes \Gamma'$ on the tensor product $Y \underset{X}{\otimes} Y' \to X$ such that the diagram

$$\begin{array}{ccccc} & J^1Y \underset{X}{\times} J^1Y' & \xrightarrow{J^1\otimes} & J^1(Y \underset{X}{\otimes} Y') & \\ {\scriptstyle \Gamma\times\Gamma'} & \uparrow & & \uparrow & {\scriptstyle \Gamma\otimes\Gamma'} \\ & Y \underset{X}{\times} Y' & \underset{\otimes}{\longrightarrow} & Y \underset{X}{\otimes} Y' & \end{array}$$

commutes. It is called the *tensor product connection*, and has the coordinate expression

$$\boxed{(\Gamma \otimes \Gamma')^{ik}_\lambda = \Gamma_\lambda{}^i{}_j y^{jk} + \Gamma'_\lambda{}^k{}_r y^{ir},} \tag{2.4.6}$$

where (x^λ, y^{ik}) are linear bundle coordinates on $Y \underset{X}{\otimes} Y' \to X$.

An important example of linear connections is a linear connection

$$K = dx^\lambda \otimes (\partial_\lambda + K_\lambda{}^\mu{}_\nu \dot{x}^\nu \dot{\partial}_\mu) \tag{2.4.7}$$

on the tangent bundle TX of a manifold X. We will call it a *world connection* on a manifold X. The dual connection (2.4.5) on the cotangent bundle T^*X is

$$K^* = dx^\lambda \otimes (\partial_\lambda - K_\lambda{}^\mu{}_\nu \dot{x}_\mu \dot{\partial}^\nu). \tag{2.4.8}$$

Then, using the construction of the tensor product connection (2.4.6), one can introduce the corresponding linear connection on an arbitrary tensor bundle T (1.1.12).

Remark 2.4.1. It should be emphasized that the expressions (2.4.7) and (2.4.8) for a world connection differ in a minus sign from those usually used in the physical literature. •

The *curvature of a world connection* is defined as the curvature R (2.4.2) of the connection Γ (2.4.7) on the tangent bundle TX. This reads

$$R = \frac{1}{2} R_{\lambda\mu}{}^\alpha{}_\beta \dot{x}^\beta dx^\lambda \wedge dx^\mu \otimes \dot{\partial}_\alpha,$$
$$\boxed{R_{\lambda\mu}{}^\alpha{}_\beta = \partial_\lambda K_\mu{}^\alpha{}_\beta - \partial_\mu K_\lambda{}^\alpha{}_\beta + K_\lambda{}^\gamma{}_\beta K_\mu{}^\alpha{}_\gamma - K_\mu{}^\gamma{}_\beta K_\lambda{}^\alpha{}_\gamma.} \tag{2.4.9}$$

By the *torsion of a world connection* is meant the torsion (2.3.9) of the connection Γ (2.4.7) on the tangent bundle TX with respect to the canonical soldering form θ_J (1.2.30):

$$T = \frac{1}{2} T_\mu{}^\nu{}_\lambda dx^\lambda \wedge dx^\mu \otimes \dot{\partial}_\nu, \tag{2.4.10}$$
$$\boxed{T_\mu{}^\nu{}_\lambda = K_\mu{}^\nu{}_\lambda - K_\lambda{}^\nu{}_\mu.}$$

A world connection is said to be symmetric if its torsion (2.4.10) vanishes, i.e., $K_\mu{}^\nu{}_\lambda = K_\lambda{}^\nu{}_\mu$. Note that, due to the canonical vertical splitting (1.2.34), the torsion (2.4.10) can be seen as the tangent-valued 2-form

$$T = \frac{1}{2} T_\mu{}^\nu{}_\lambda dx^\lambda \wedge dx^\mu \otimes \partial_\nu \tag{2.4.11}$$

on X if there is no danger of confusion.

Remark 2.4.2. For any vector field τ on a manifold X, there exists a connection Γ on the tangent bundle $TX \to X$ such that τ is an integral section of K, but this connection is not necessarily linear. If a vector field τ is non-vanishing at a point $x \in X$, then there exists a local symmetric world connection K (2.4.7) around x for which τ is an integral section

$$\partial_\nu \tau^\alpha = K_\nu{}^\alpha{}_\beta \tau^\beta. \tag{2.4.12}$$

Then the canonical lift $\widetilde{\tau}$ (1.2.3) of τ onto TX can be seen locally as the horizontal lift $K\tau$ (2.1.6) of τ by means of this connection. •

Remark 2.4.3. Every manifold X can be provided with a non-degenerate fibre metric

$$g \in \overset{2}{\vee} \mathfrak{O}^1(X), \qquad g = g_{\lambda\mu} dx^\lambda \otimes dx^\mu,$$

in the tangent bundle TX, and with the corresponding metric

$$g \in \overset{2}{\vee} \mathcal{T}^1(X), \qquad g = g^{\lambda\mu} \partial_\lambda \otimes \partial_\mu,$$

in the cotangent bundle T^*X. We call it a *world metric* on X. For any world metric g, there exists a unique symmetric world connection Γ (2.4.7) with the components

$$K_\lambda{}^\nu{}_\mu = \{_\lambda{}^\nu{}_\mu\} = -\frac{1}{2} g^{\nu\rho}(\partial_\lambda g_{\rho\mu} + \partial_\mu g_{\rho\lambda} - \partial_\rho g_{\lambda\mu}), \tag{2.4.13}$$

called the *Christoffel symbols*, such that g is an integral section of Γ, i.e.,

$$\partial_\lambda g^{\alpha\beta} = g^{\alpha\gamma}\{_\lambda{}^\beta{}_\gamma\} + g^{\beta\gamma}\{_\lambda{}^\alpha{}_\gamma\}.$$

This is called the *Levi–Civita connection* associated with g. •

2.5 Affine connections

Let $Y \to X$ be an affine bundle modelled over a vector bundle $\overline{Y} \to X$. A connection Γ on $Y \to X$ is said to be an *affine connection* if the section $\Gamma : Y \to J^1Y$ (2.2.6) is an affine bundle morphism over X. Affine connections are associated with principal connections, and they always exist (see Section 6.1).

For any affine connection $\Gamma : Y \to J^1Y$, the corresponding linear derivative $\overline{\Gamma} : \overline{Y} \to J^1\overline{Y}$ (1.1.9) defines uniquely the associated linear connection on the vector bundle $\overline{Y} \to X$. Since every vector bundle has a natural structure of an affine bundle, any linear connection on a vector bundle is also an affine connection.

With affine bundle coordinates (x^λ, y^i) on Y, an affine connection Γ reads

$$\boxed{\Gamma^i_\lambda = \Gamma_\lambda{}^i{}_j(x)y^j + \sigma^i_\lambda(x).} \tag{2.5.1}$$

The coordinate expression of the associated linear connection is

$$\overline{\Gamma}^i_\lambda = \Gamma_\lambda{}^i{}_j(x)\overline{y}^j, \tag{2.5.2}$$

where $(x^\lambda, \overline{y}^i)$ are the linear bundle coordinates on $\overline{Y}$.

Affine connections on an affine bundle $Y \to X$ constitute an affine space modelled over the vector space of soldering forms on $Y \to X$. In view of the vertical splitting (1.1.14), these soldering forms can be seen as global sections of the vector bundle $T^*X \otimes \overline{Y} \to X$. If $Y \to X$ is a vector bundle, both the affine connection Γ (2.5.1) and the associated linear connection $\overline{\Gamma}$ are connections on the same vector bundle $Y \to X$, and their difference is a basic soldering form on Y. Thus, every affine connection on a vector bundle $Y \to X$ is the sum of a linear connection and a basic soldering form on $Y \to X$.

Given an affine connection Γ on a vector bundle $Y \to X$, let R and $\overline{R}$ be the curvatures of the connection Γ and the associated linear connection $\overline{\Gamma}$, respectively. It is readily observed that $R = \overline{R} + T$, where the VY-valued 2-form

$$\begin{aligned} &T = d_\Gamma\sigma = d_\sigma\Gamma : X \to \overset{2}{\wedge} T^*X \underset{X}{\otimes} VY, \\ &T = \frac{1}{2}T^i_{\lambda\mu}dx^\lambda \wedge dx^\mu \otimes \partial_i, \\ &T^i_{\lambda\mu} = \partial_\lambda\sigma^i_\mu - \partial_\mu\sigma^i_\lambda + \sigma^h_\lambda\Gamma_\mu{}^i{}_h - \sigma^h_\mu\Gamma_\lambda{}^i{}_h, \end{aligned} \tag{2.5.3}$$

is the torsion (2.3.9) of the connection Γ with respect to the basic soldering form σ.

In particular, let us consider the tangent bundle TX of a manifold X and the canonical soldering form $\sigma = \theta_J = \theta_X$ (1.2.30) on TX. Given an arbitrary world connection K (2.4.7) on TX, the corresponding affine connection

$$A = K + \theta_X, \qquad A^\mu_\lambda = K_\lambda{}^\mu{}_\nu\dot{x}^\nu + \delta^\mu_\lambda, \tag{2.5.4}$$

on TX is called the *Cartan connection*. Since the soldered curvature ρ (2.3.8) of θ_J equals to zero, the torsion (2.3.11) of the Cartan connection coincides with the torsion T (2.4.10) of the world connection K, while its curvature (2.3.12) is the sum $R + T$ of the curvature and the torsion of K.

2.6 Flat connections

By a *flat* or *curvature-free* connection is meant a connection which satisfies the following conditions.

PROPOSITION 2.6.1. Let Γ be a connection on a fibre bundle $Y \to X$. The following conditions are equivalent.

- The horizontal lift $\mathcal{T}(X) \ni \tau \mapsto \Gamma\tau \in \mathcal{T}(Y)$ is a Lie algebra morphism.
- The horizontal distribution is involutive.
- The curvature R of the connection Γ vanishes everywhere on Y.
- There exists a (local) integral section for the connection Γ through any point $y \in Y$.

□

By virtue of Theorem 1.2.1, a flat connection Γ on a fibre bundle $Y \to X$ yields the integrable horizontal distribution and, consequently, the *horizontal foliation* on Y, transversal to the fibration $Y \to X$. The leaf of this foliation through a point $y \in Y$ is defined locally by an integral section s_y for the connection Γ through y. Conversely, let a fibre bundle $Y \to X$ admit a horizontal foliation such that, for each point $y \in Y$, the leaf of this foliation through y is locally defined by a section s_y of $Y \to X$ through y. Then the map

$$\Gamma : Y \to J^1Y,$$
$$\Gamma(y) = j^1_x s_y, \qquad \pi(y) = x,$$

introduces a flat connection on $Y \to X$. Hence, there is one-to-one correspondence between the flat connections and the horizontal foliations on a fibre bundle $Y \to X$.

Given a horizontal foliation on a fibre bundle $Y \to X$, there exists the associated atlas of fibred coordinates (x^λ, y^i) of Y such that every leaf of this foliation is locally generated by the equations y^i =const., and the transition functions $y^i \to y'^i(y^j)$ are independent of the base coordinates x^λ [48, 123]. This is called the *atlas of constant local trivializations*. Two such atlases are said to be equivalent if their union is also an atlas of constant local trivializations. They are associated with the same horizontal foliation. Thus, we come to the following assertion.

PROPOSITION 2.6.2. There is one-to-one correspondence between the flat connections Γ on a fibre bundle $Y \to X$ and the equivalence classes of atlases of constant local trivializations of Y such that $\Gamma = dx^\lambda \otimes \partial_\lambda$ relative to these atlases. □

2.7 Composite connections

Let us consider the composition of fibre bundles

$$Y \to \Sigma \to X, \tag{2.7.1}$$

where

$$\pi_{Y\Sigma} : Y \to \Sigma \tag{2.7.2}$$

and

$$\pi_{\Sigma X} : \Sigma \to X \tag{2.7.3}$$

are fibre bundles. This is called a *composite fibre bundle*. It is provided with an atlas of fibred coordinates $(x^\lambda, \sigma^m, y^i)$, where (x^μ, σ^m) are fibred coordinates on the fibre bundle (2.7.3) and the transition functions $\sigma^m \to \sigma'^m(x^\lambda, \sigma^k)$ are independent of the coordinates y^i.

The following two assertions make composite fibre bundles useful for physical applications [123, 213, 268].

PROPOSITION 2.7.1. Given a composite fibre bundle (2.7.1), let h be a global section of the fibre bundle $\Sigma \to X$. Then the restriction

$$Y_h = h^* Y \tag{2.7.4}$$

of the fibre bundle $Y \to \Sigma$ to $h(X) \subset \Sigma$ is a subbundle $i_h : Y_h \hookrightarrow Y$ of the fibre bundle $Y \to X$. □

PROPOSITION 2.7.2. Given a section h of the fibre bundle $\Sigma \to X$ and a section s_Σ of the fibre bundle $Y \to \Sigma$, their composition

$$s = s_\Sigma \circ h \tag{2.7.5}$$

is a section of the composite fibre bundle $Y \to X$ (2.7.1). Conversely, every section s of the fibre bundle $Y \to X$ is the composition (2.7.5) of the section $h = \pi_{Y\Sigma} \circ s$ of the fibre bundle $\Sigma \to X$ and some section s_Σ of the fibre bundle $Y \to \Sigma$ over the closed submanifold $h(X) \subset \Sigma$. □

Let us consider the jet manifolds $J^1\Sigma$, $J^1_\Sigma Y$, and J^1Y of the fibre bundles $\Sigma \to X$, $Y \to \Sigma$ and $Y \to X$, respectively. They are parameterized respectively by the coordinates

$$(x^\lambda, \sigma^m, \sigma^m_\lambda), \quad (x^\lambda, \sigma^m, y^i, \widetilde{y}^i_\lambda, y^i_m), \quad (x^\lambda, \sigma^m, y^i, \sigma^m_\lambda, y^i_\lambda).$$

LEMMA 2.7.3. [274]. There is the canonical map

$$\varrho : J^1\Sigma \underset{\Sigma}{\times} J^1_\Sigma Y \underset{Y}{\longrightarrow} J^1Y, \tag{2.7.6}$$

$$\boxed{y^i_\lambda \circ \varrho = y^i_m \sigma^m_\lambda + \widetilde{y}^i_\lambda.}$$

□

Using this map, we can consider the relations between connections on the fibre bundles $Y \to X$, $Y \to \Sigma$ and $\Sigma \to X$ as follows.

Remark 2.7.1. Let

$$\gamma = dx^\lambda \otimes (\partial_\lambda + \gamma^m_\lambda \partial_m + \gamma^i_\lambda \partial_i)$$

be a connection on the composite fibre bundle $Y \to X$ and

$$\Gamma = dx^\lambda \otimes (\partial_\lambda + \Gamma^m_\lambda \partial_m) \tag{2.7.7}$$

a connection on the fibre bundle $\Sigma \to X$. We say that the connection γ is *projectable* over the connection Γ if the diagram

$$\begin{array}{ccc} Y & \xrightarrow{\gamma} & J^1Y \\ {\scriptstyle \pi_{Y\Sigma}} \downarrow & & \downarrow {\scriptstyle J^1\pi_{Y\Sigma}} \\ \Sigma & \xrightarrow{\Gamma} & J^1\Sigma \end{array}$$

is commutative. It is readily observed that the commutativity of this diagram is equivalent to the condition $\gamma^m_\lambda = \Gamma^m_\lambda$. •

Let

$$A_\Sigma = dx^\lambda \otimes (\partial_\lambda + A^i_\lambda \partial_i) + d\sigma^m \otimes (\partial_m + A^i_m \partial_i) \tag{2.7.8}$$

be a connection on the fibre bundle $Y \to \Sigma$. Given a connection Γ (2.7.7) on $\Sigma \to X$, the canonical morphism ϱ (2.7.6) enables one to obtain a connection γ on $Y \to X$ in accordance with the diagram

$$\begin{array}{ccc} J^1\Sigma \underset{\Sigma}{\times} J^1 Y_\Sigma & \xrightarrow{\varrho} & J^1 Y \\ {\scriptstyle (\Gamma,A)} \uparrow & & \uparrow {\scriptstyle \gamma} \\ \Sigma \underset{X}{\times} Y & \longleftarrow & Y \end{array}$$

This connection, called the *composite connection*, reads

$$\boxed{\gamma = dx^\lambda \otimes (\partial_\lambda + \Gamma^m_\lambda \partial_m + (A^i_\lambda + A^i_m \Gamma^m_\lambda)\partial_i).} \tag{2.7.9}$$

It is projectable over Γ.

An equivalent definition of a composite connection is the following. Let A_Σ and Γ be the connections as before. Then their composition

$$Y \underset{X}{\times} TX \xrightarrow{(\mathrm{Id}\,,\Gamma)} Y \underset{\Sigma}{\times} T\Sigma \xrightarrow{A_\Sigma} TY$$

is the composite connection γ (2.7.9) on the composite fibre bundle $Y \to X$. In brief, we will write

$$\boxed{\gamma = A_\Sigma \circ \Gamma.} \tag{2.7.10}$$

In particular, let us consider a vector field τ on the base X, its horizontal lift $\Gamma\tau$ onto Σ by means of the connection Γ and, in turn, the horizontal lift $A_\Sigma(\Gamma\tau)$ of $\Gamma\tau$ onto Y by means of the connection A_Σ. Then $A_\Sigma(\Gamma\tau)$ coincides with the horizontal lift $\gamma\tau$ of τ onto Y by means of the composite connection γ (2.7.10).

Let h be a section of the fibre bundle $\Sigma \to X$ and Y_h the subbundle (2.7.4) of the composite fibre bundle $Y \to X$, which is the restriction of the fibre bundle $Y \to \Sigma$ to $h(X)$. Every connection A_Σ (2.7.8) induces the pull-back connection

$$\boxed{A_h = i^*_h A_\Sigma = dx^\lambda \otimes [\partial_\lambda + ((A^i_m \circ h)\partial_\lambda h^m + (A \circ h)^i_\lambda)\partial_i]} \tag{2.7.11}$$

on $Y_h \to X$ (see (2.1.11)). Now, let Γ be a connection on $\Sigma \to X$ and let $\gamma = A_\Sigma \circ \Gamma$ be the composition (2.7.10). Then it follows from (2.2.11) that the connection γ is reducible to the connection A_h if and only if the section h is an integral section of Γ, i.e.,

$$\Gamma^m_\lambda \circ h = \partial_\lambda h^m.$$

Such a connection Γ always exists.

Given a composite fibre bundle Y (2.7.1), there are the following exact sequences of vector bundles over Y:

$$0 \to V_\Sigma Y \hookrightarrow VY \to Y \underset{\Sigma}{\times} V\Sigma \to 0, \tag{2.7.12a}$$

$$0 \to Y \underset{\Sigma}{\times} V^*\Sigma \hookrightarrow V^*Y \to V^*_\Sigma Y \to 0, \tag{2.7.12b}$$

where $V_\Sigma Y$ and $V^*_\Sigma Y$ are vertical tangent and cotangent bundles of the fibre bundle $Y \to \Sigma$, respectively. Every connection A (2.7.8) on the fibre bundle $Y \to \Sigma$ provides the splittings

$$VY = V_\Sigma Y \underset{Y}{\oplus} A_\Sigma(Y \underset{\Sigma}{\times} V\Sigma), \tag{2.7.13}$$

$$\dot y^i \partial_i + \dot\sigma^m \partial_m = (\dot y^i - A^i_m \dot\sigma^m)\partial_i + \dot\sigma^m(\partial_m + A^i_m \partial_i),$$

$$V^*Y = (Y \underset{\Sigma}{\times} V^*\Sigma) \underset{Y}{\oplus} A_\Sigma(V^*_\Sigma Y), \tag{2.7.14}$$

$$\dot y_i \overline{d}y^i + \dot\sigma_m \overline{d}\sigma^m = \dot y_i(\overline{d}y^i - A^i_m \overline{d}\sigma^m) + (\dot\sigma_m + A^i_m \dot y_i)\overline{d}\sigma^m,$$

of the exact sequences (2.7.12a) and (2.7.12b), respectively. Using the splitting (2.7.13), one can construct the first order differential operator

$$\widetilde{D} : J^1Y \to T^*X \underset{Y}{\otimes} V_\Sigma Y,$$

$$\boxed{\widetilde{D} = dx^\lambda \otimes (y^i_\lambda - A^i_\lambda - A^i_m \sigma^m_\lambda)\partial_i,} \tag{2.7.15}$$

called the *vertical covariant differential*, on the composite fibre bundle $Y \to X$. This operator can also be seen as the composition

$$\widetilde{D} = \mathrm{pr}_1 \circ D_\gamma : J^1Y \to T^*X \underset{Y}{\otimes} VY \to T^*X \underset{Y}{\otimes} VY_\Sigma,$$

where D_γ is the covariant differential (2.2.7) relative to some composite connection (2.7.9), but $\widetilde{D}$ does not depend on Γ.

The vertical covariant differential (2.7.15) possesses the following important property. Let h be a section of the fibre bundle $\Sigma \to X$ and Y_h the subbundle (2.7.4) of the composite fibre bundle $Y \to X$, which is the restriction of the fibre bundle $Y \to \Sigma$ to $h(X)$. Then the restriction of the vertical covariant differential $\widetilde{D}$ (2.7.15) to $J^1 i_h(J^1 Y_h) \subset J^1 Y$ coincides with the familiar covariant differential on Y_h relative to the pull-back connection A_h (2.7.11).

Let now $Y \to \Sigma \to X$ be a composite fibre bundle where $Y \to \Sigma$ is a vector bundle. Let a connection

$$\gamma = dx^\lambda \otimes (\partial_\lambda + \Gamma^m_\lambda \partial_m + A_\lambda{}^i{}_j y^j \partial_i) \tag{2.7.16}$$

on $Y \to X$ be a linear morphism over the connection Γ on $\Sigma \to X$. The following constructions generalize the notions of a dual connection and a tensor product connection on vector bundles.

(i) Let $Y^* \to \Sigma \to X$ be a composite fibre bundle where $Y^* \to \Sigma$ is the vector bundle dual of $Y \to \Sigma$. Given the projectable connection (2.7.16) on $Y \to X$ over Γ, there exists a unique connection

$$\boxed{\gamma^* = dx^\lambda \otimes (\partial_\lambda + \Gamma^m_\lambda \partial_m - A_\lambda{}^j{}_i y_j \partial^i)}$$

on $Y^* \to X$, projectable over Γ, such that the diagram

$$\begin{array}{ccccc} & J^1Y \underset{J^1\Sigma}{\times} J^1Y^* & \overset{J^1\langle\rangle}{\longrightarrow} & J^1\Sigma \underset{\Sigma}{\times} (T^*X \times \mathbb{R}) & \\ {\scriptstyle (\gamma,\gamma^*)} & \uparrow & & \uparrow & {\scriptstyle (\Gamma,\widehat{0},\mathrm{Id})} \\ & Y \underset{\Sigma}{\times} Y^* & \underset{\langle\rangle}{\longrightarrow} & \Sigma \times \mathbb{R} & \end{array}$$

is commutative. We call γ^* the *dual connection* of γ over Γ.

(ii) Let $Y \to \Sigma \to X$ and $Y' \to \Sigma \to X$ be composite fibre bundles where $Y \to \Sigma$ and $Y' \to \Sigma$ are vector bundles. Let γ and γ' be connections (2.7.16) on $Y \to X$ and $Y' \to X$, respectively, which are projectable over the same connection Γ on $\Sigma \to X$. There is a unique connection

$$\boxed{\gamma \otimes \gamma' = dx^\lambda \otimes [\partial_\lambda + \Gamma^m_\lambda \partial_m + (A_\lambda{}^i{}_j y^{jk} + A'_\lambda{}^k{}_j y^{ij})\partial_{ik}]} \tag{2.7.17}$$

on the tensor product $Y \underset{\Sigma}{\otimes} Y' \to X$, which is projectable over Γ, such that the diagram

$$\begin{array}{ccccc} & J^1Y \underset{J^1\Sigma}{\times} J^1Y' & \overset{J^1\otimes}{\longrightarrow} & J^1(Y \underset{J^1\Sigma}{\otimes} Y') & \\ {\scriptstyle (\gamma,\gamma')} & \uparrow & & \uparrow & {\scriptstyle \gamma\otimes\gamma'} \\ & Y \underset{\Sigma}{\times} Y' & \underset{\otimes}{\longrightarrow} & Y \underset{\Sigma}{\otimes} Y' & \end{array}$$

is commutative. This is called the *tensor product connection over* Γ.

Example 2.7.2. Let $\Gamma : Y \to J^1Y$ be a connection on a fibre bundle $Y \to X$. In accordance with the canonical isomorphism $VJ^1Y \cong J^1VY$ (1.3.11), the vertical tangent map $V\Gamma : VY \to VJ^1Y$ to Γ defines the connection

$$V\Gamma : VY \to J^1VY,$$

$$\boxed{V\Gamma = dx^\lambda \otimes (\partial_\lambda + \Gamma^i_\lambda \partial_i + \partial_j \Gamma^i_\lambda \dot{y}^j \dot{\partial}_i),} \tag{2.7.18}$$

on the composite vertical tangent bundle $VY \to Y \to X$. This is called the *vertical connection* to Γ. Of course, the connection $V\Gamma$ projects over Γ. Moreover, $V\Gamma$ is linear over Γ. Then the dual connection of $V\Gamma$ on the composite vertical cotangent bundle $V^*Y \to Y \to X$ reads

$$V^*\Gamma : V^*Y \to J^1V^*Y,$$

$$\boxed{V^*\Gamma = dx^\lambda \otimes (\partial_\lambda + \Gamma^i_\lambda \partial_i - \partial_j \Gamma^i_\lambda \dot{y}_i \dot{\partial}^j).} \tag{2.7.19}$$

It is called the *covertical connection* to Γ.

If $Y \to X$ is an affine bundle, the connection $V\Gamma$ (2.7.18) can be seen as the composite connection (2.7.8) generated by the connection Γ on $Y \to X$ and the linear connection

$$\widetilde{\Gamma} = dx^\lambda \otimes (\partial_\lambda + \partial_j \Gamma^i_\lambda \dot{y}^j \dot{\partial}_i) + dy^i \otimes \partial_i \tag{2.7.20}$$

on the vertical tangent bundle $VY \to Y$. •

Chapter 3

Connections in Lagrangian field theory

We will limit our study to first order Lagrangian formalism since the most contemporary field models are described by first order Lagrangians. This is not the case of General Relativity whose Hilbert–Einstein Lagrangian belongs to the particular class of second order Lagrangians leading to second order Euler–Lagrange equations (see [123, 187] for details).

As was mentioned above, we follow the geometric formulation of classical field theory, where fields are represented by sections of a fibre bundle $Y \to X$, coordinated by (x^λ, y^i). For example, matter fields, gauge fields, gravitational fields, Higgs fields are of this type. In this Chapter, we do not specify the type of fields and stand the collective notation y^i for all of them. The finite-dimensional *configuration space* of fields is the first order jet manifold J^1Y of $Y \to X$, coordinated by $(x^\lambda, y^i, y^i_\lambda)$. A first order *Lagrangian* is defined as a horizontal density

$$\begin{aligned} &L : J^1Y \to \overset{n}{\wedge} T^*X, \qquad n = \dim X, \\ &L = \mathcal{L}(x^\lambda, y^i, y^i_\lambda)\omega, \end{aligned} \tag{3.0.1}$$

on J^1Y (see the notation (1.2.9)).

In Lagrangian field theory, one deals with connections in the relation to the following three constructions.

- Every second order dynamic equation on a fibre bundle $Y \to X$ is a second order holonomic connection on $Y \to X$ (see Definition 3.1.2 below). The

Euler–Lagrange equations are not second order dynamic equations in general, but one introduces the notion of a Lagrangian connection which is a second order dynamic equation and whose integral sections are solutions of the Euler–Lagrange equations.

- In the physically relevant case of quadratic Lagrangians, a Lagrangian L factorizes always through the covariant differential D_Γ (2.2.7) for some connection Γ on $Y \to X$, i.e.,

$$L : J^1Y \overset{D_\Gamma}{\to} T^*X \underset{Y}{\otimes} VY \to \overset{n}{\wedge} T^*X \tag{3.0.2}$$

(see expressions (3.3.13) and (3.3.14) below).

- Different connections on a fibre bundle $Y \to X$ are responsible for different energy-momentum currents of fields, which differ from each other in Noether currents.

3.1 Connections and dynamic equations

In accordance with Definitions 1.3.3 and 1.3.4, a k-order dynamic equation on a fibre bundle $Y \to X$ is a section of the affine jet bundle $J^kY \to J^{k-1}Y$. Further on, we will restrict our consideration only to first and second order dynamic equations.

DEFINITION 3.1.1. A *first order dynamic equation* on a fibre bundle $Y \to X$ is a section of the affine jet bundle $J^1Y \to Y$, i.e., a connection

$$\Gamma = dx^\lambda \otimes (\partial_\lambda + \Gamma^i_\lambda \partial_i)$$

on a fibre bundle $Y \to X$. It characterizes the kernel of the covariant differential D_Γ (2.2.7) which is a closed subbundle

$$\boxed{y^i_\lambda = \Gamma^i_\lambda(x^\mu, y^j)} \tag{3.1.1}$$

of the jet bundle $J^1Y \to X$. This is a first order differential equation on the fibre bundle $Y \to X$ in accordance with Definition 1.3.4. □

Classical solutions of the first order dynamic equation (3.1.1) are integral sections s of the connection Γ. By virtue of Proposition 2.6.1, the dynamic equation (3.1.1) admits a solution through each point of an open subset $U \subset Y$ if and only if the curvature of the connection Γ vanishes on U.

To characterize second order dynamic equations on a fibre bundle $Y \to X$, the notion of a second order connection on $Y \to X$ is introduced.

A *second order connection* $\widehat{\Gamma}$ on a fibre bundle $Y \to X$ is defined as a connection

$$\widehat{\Gamma} = dx^\lambda \otimes (\partial_\lambda + \widehat{\Gamma}^i_\lambda \partial_i + \widehat{\Gamma}^i_{\lambda\mu} \partial^\mu_i) \tag{3.1.2}$$

on the jet bundle $J^1Y \to X$, i.e., this is a section of the affine bundle $\pi_{11} : J^1J^1Y \to J^1Y$.

Remark 3.1.1. Every connection on a fibre bundle $Y \to X$ gives rise to the second order one by means of a world connection on X. The first order jet prolongation $J^1\Gamma$ of a connection Γ on $Y \to X$ is a section of the repeated jet bundle $J^1\pi^1_0$ (1.3.13), but not of π_{11}. Given a world connection K (2.4.8) on X, one can construct the affine morphism

$$\begin{aligned} & s_K : J^1J^1Y \to J^1J^1Y, \\ & (x^\lambda, y^i, y^i_\lambda, \widehat{y}^i_\lambda, y^i_{\lambda\mu}) \circ s_K = (x^\lambda, y^i, \widehat{y}^i_\lambda, y^i_\lambda, y^i_{\mu\lambda} - K_\lambda{}^\nu{}_\mu (\widehat{y}^i_\nu - y^i_\nu)), \end{aligned}$$

such that $\pi_{11} = J^1\pi^1_0 \circ s_K$ [123]. Then Γ gives rise to the second order connection

$$\begin{aligned} & \widehat{\Gamma} = s_{\Gamma'} \circ J^1\Gamma : J^1Y \to J^1J^1Y, \\ & \widehat{\Gamma} = dx^\lambda \otimes (\partial_\lambda + \Gamma^i_\lambda \partial_i + [\partial_\lambda \Gamma^i_\mu + y^j_\lambda \partial_j \Gamma^i_\mu + K_\lambda{}^\nu{}_\mu (y^i_\nu - \Gamma^i_\nu)] \partial^\mu_i), \end{aligned} \tag{3.1.3}$$

which is an affine morphism

$$\begin{array}{rcc} J^1Y & \xrightarrow{\widehat{\Gamma}} & J^1J^1Y \\ {\scriptstyle \pi^1_0} \downarrow & & \downarrow {\scriptstyle \pi_{11}} \\ Y & \xrightarrow{\Gamma} & J^1Y \end{array}$$

over the connection Γ. Note that the curvature R (2.3.3) of a connection Γ on a fibre bundle $Y \to X$ can be seen as a soldering form

$$R = R^i_{\lambda\mu} dx^\lambda \otimes \partial^\mu_i$$

on the jet bundle $J^1Y \to X$. Therefore, $\widehat{\Gamma} - R$ is also a connection on $J^1Y \to X$. •

A second order connection Γ (3.1.2) is said to be *holonomic* if it takes its values into the subbundle J^2Y of J^1J^1Y. Such a connection is characterized by the coordinate conditions

$$\widehat{\Gamma}^i_\lambda = y^i_\lambda, \qquad \widehat{\Gamma}^i_{\lambda\mu} = \widehat{\Gamma}^i_{\mu\lambda},$$

and reads

$$\widehat{\Gamma} = dx^\lambda \otimes (\partial_\lambda + y^i_\lambda \partial_i + \widehat{\Gamma}^i_{\lambda\mu}\partial^\mu_i). \tag{3.1.4}$$

There is one-to-one correspondence between the global sections of the jet bundle $J^2Y \to J^1Y$ and the holonomic second order connections on $Y \to X$. Since the jet bundle $J^2Y \to J^1Y$ is affine, a holonomic second order connection on a fibre bundle $Y \to X$ always exists.

DEFINITION 3.1.2. A *second order dynamic equation* on a fibre bundle $Y \to X$ is a section of the affine jet bundle $J^2Y \to J^1Y$, i.e., a holonomic second order connection (3.1.4) on a fibre bundle $Y \to X$. It characterizes the closed subbundle

$$\boxed{y^i_{\lambda\mu} = \widehat{\Gamma}^i_{\lambda\mu}(x^\nu, y^j, y^j_\nu)} \tag{3.1.5}$$

of the second order jet bundle $J^2Y \to X$. □

By virtue of Proposition 1.3.1, every integral section $\overline{s}: X \to J^1Y$ of the holonomic second order connection (3.1.4) is holonomic, i.e., $\overline{s} = J^1s$ where $s: X \to Y$ provides a classical solution s of the second order dynamic equation (3.1.5).

3.2 The first variational formula

We will follow the standard formulation of the variational problem where deformations of sections of a fibre bundle $Y \to X$ are induced by local 1-parameter groups of local automorphisms of $Y \to X$ over $\mathrm{Id}\,X$ [25, 290]. We will not study the calculus of variations in depth, but apply in a straightforward manner the first variational formula [123]. In Section 11.2, Lagrangians, the first variational formula and Euler–Lagrange operators will be introduced in an algebraic way as elements of the variational cochain complex.

Remark 3.2.1. In the physical literature, automorphisms of a fibre bundle $Y \to X$ are called *gauge transformations* [123, 214]. Every projectable vector field u on a fibre bundle $Y \to X$ is a generator of a local 1-parameter group G_u of gauge transformations of $Y \to X$. Accordingly, the jet prolongation J^1u (1.3.10) of u onto J^1Y is the generator of the local 1-parameter group of the jet prolongations J^1g of local automorphisms $g \in G_u$. Recall that an exterior form ϕ on a fibre bundle Y is invariant under a local 1-parameter group G_u of gauge transformations of Y if and only if its Lie derivative $\mathbf{L}_u\phi$ along u vanishes. •

Let u be a projectable vector field on a fibre bundle $Y \to X$ and

$$\mathbf{L}_{J^1u}L = [\partial_\lambda u^\lambda \mathcal{L} + (u^\lambda\partial_\lambda + u^i\partial_i + (d_\lambda u^i - y^i_\mu\partial_\lambda u^\mu)\partial_i^\lambda)\mathcal{L}]\omega \tag{3.2.1}$$

the Lie derivative of a Lagrangian L along J^1u (or briefly along u). The *first variational formula* provides the canonical decomposition of the Lie derivative (3.2.1) in accordance with the variational problem. Bearing in mind the notation (1.3.15) and (1.3.16), this decomposition reads

$$\begin{aligned}\mathbf{L}_{J^1u}L &= u_V\rfloor\mathcal{E}_L + d_H h_0(u\rfloor H_L) \\ &= (u^i - y^i_\mu u^\mu)(\partial_i - d_\lambda\partial_i^\lambda)\mathcal{L}\omega - d_\lambda[\pi_i^\lambda(u^\mu y^i_\mu - u^i) - u^\lambda\mathcal{L}]\omega,\end{aligned} \tag{3.2.2}$$

where $u_V = (u\rfloor\theta^i)\partial_i$ is the vertical part of the canonical horizontal splitting (2.2.5) of the vector field u,

$$\mathcal{E}_L : J^2Y \to T^*Y \wedge (\overset{n}{\wedge} T^*X),$$

$$\boxed{\mathcal{E}_L = (\partial_i\mathcal{L} - d_\lambda\pi_i^\lambda)\theta^i \wedge \omega,} \qquad \pi_i^\lambda = \partial_i^\lambda\mathcal{L}, \tag{3.2.3}$$

is the *Euler–Lagrange operator* associated with the Lagrangian L, and

$$H_L : J^1Y \to Z_Y = T^*Y \wedge (\overset{n-1}{\wedge} T^*X), \tag{3.2.4}$$

$$\boxed{H_L = L + \pi_i^\lambda\theta^i \wedge \omega_\lambda = \pi_i^\lambda dy^i \wedge \omega_\lambda + (\mathcal{L} - \pi_i^\lambda y^i_\lambda)\omega,} \tag{3.2.5}$$

is the *Poincaré–Cartan form* (see the notation (1.2.9), (1.3.15) and (1.3.16)).

Remark 3.2.2. The Poincaré–Cartan form H_L (3.2.5) is a Lepagean equivalent of the Lagrangian L (i.e., $h_0(H_L) = L$) which is a horizontal form on the affine jet bundle $J^1Y \to Y$ (see, e.g., [123, 133]). The fibre bundle Z_Y (3.2.4), called the *homogeneous Legendre bundle*, is endowed with holonomic coordinates $(x^\lambda, y^i, p_i^\lambda, p)$ possessing the transition functions

$$p'^\lambda_i = \det(\frac{\partial x^\varepsilon}{\partial x'^\nu})\frac{\partial y^j}{\partial y'^i}\frac{\partial x'^\lambda}{\partial x^\mu}p_j^\mu, \qquad p' = \det(\frac{\partial x^\varepsilon}{\partial x'^\nu})(p - \frac{\partial y^j}{\partial y'^i}\frac{\partial y'^i}{\partial x^\mu}p_j^\mu). \tag{3.2.6}$$

Relative to these coordinates, the morphism (3.2.4) reads

$$(p_i^\mu, p) \circ H_L = (\pi_i^\mu, \mathcal{L} - \pi_i^\mu y_\mu^i).$$

A glance at the transition functions (3.2.6) shows that Z_Y is a 1-dimensional affine bundle

$$\pi_{Z\Pi} : Z_Y \to \Pi \tag{3.2.7}$$

over the *Legendre bundle*

$$\Pi = V^*Y \wedge (\overset{n-1}{\wedge} T^*X). \tag{3.2.8}$$

The latter is endowed with holonomic coordinates $(x^\lambda, y^i, p_i^\lambda)$. Then the composition

$$\widehat{L} = \pi_{Z\Pi} \circ H_L : J^1Y \underset{Y}{\to} \Pi, \tag{3.2.9}$$
$$(x^\lambda, y^i, p_i^\lambda) \circ \widehat{L} = (x^\lambda, y^i, \pi_i^\lambda),$$

is the well-known *Legendre map*. One can think of p_i^λ as being the covariant momenta of field functions, and the Legendre bundle Π (3.2.8) plays the role of a finite-dimensional momentum phase space of fields in the covariant Hamiltonian field theory (see Chapter 4). Recall that a Lagrangian L is said to be:

- *hyperregular* if the Legendre map $\widehat{L}$ is a diffeomorphism,
- *regular* if $\widehat{L}$ is of maximal rank, i.e., $\det(\partial_i^\mu \partial_j^\nu \mathcal{L}) \neq 0$,
- *almost regular* if the *Lagrangian constraint space* $N_L = \widehat{L}(J^1Y)$ is a closed imbedded subbundle of the Legendre bundle $\Pi \to Y$ and the Legendre map $\widehat{L} : J^1Y \to N_L$ is a fibred manifold with connected fibres.

•

The kernel of the Euler-Lagrange operator $\mathcal{E}_L$ (3.2.3), given by the coordinate relations

$$\boxed{(\partial_i - d_\lambda \partial_i^\lambda)\mathcal{L} = 0,} \tag{3.2.10}$$

defines the system of second order *Euler–Lagrange equations*. Classical solutions of these equations are section s of the fibre bundle $X \to Y$, whose second order jet prolongations J^2s live in (3.2.10). They satisfy the equations

$$\partial_i \mathcal{L} \circ s - (\partial_\lambda + \partial_\lambda s^j \partial_j + \partial_\lambda \partial_\mu s^j \partial_j^\mu)\partial_i^\lambda \mathcal{L} \circ s = 0. \tag{3.2.11}$$

Remark 3.2.3. The kernel (3.2.10) of the Euler–Lagrange operator $\mathcal{E}_L$ fails to be a closed subbundle of the second order jet bundle $J^2Y \to X$ in general. Therefore, it may happen that the Euler–Lagrange equations (3.2.10) are not differential equations in a strict sense (see Proposition 1.3.6). •

Remark 3.2.4. Different Lagrangians L and L' can lead to the same Euler–Lagrange operator if their difference $L_0 = L - L'$ is a *variationally trivial Lagrangian* whose Euler–Lagrange operator vanishes identically. A Lagrangian L_0 is variationally trivial if and only if

$$L_0 = h_0(\epsilon) \tag{3.2.12}$$

where ϵ is a closed n-form on Y [123, 186, 189]. We have locally $\epsilon = d\phi$ and

$$L_0 = h_0(d\phi) = d_H(h_0(\phi)) = d_\lambda h_0(\phi)^\lambda \omega, \qquad h_0(\phi) = h_0(\phi)^\lambda \omega_\lambda.$$

•

Given a Lagrangian L, a holonomic second order connection $\widehat{\Gamma}$ (3.1.4) on the fibre bundle $Y \to X$ is said to be a *Lagrangian connection* if it takes its values into the kernel of the Euler–Lagrange operator $\mathcal{E}_L$. A Lagrangian connection satisfies the pointwise algebraic equations

$$\partial_i \mathcal{L} - \partial_\lambda \pi_i^\lambda - y_\lambda^j \partial_j \pi_i^\lambda - \widehat{\Gamma}_{\lambda\mu}^j \partial_j^\mu \pi_i^\lambda = 0. \tag{3.2.13}$$

If a Lagrangian connection $\widehat{\Gamma}$ exists, it defines the second order dynamic equation (3.1.5) on $Y \to X$, whose solutions are also solutions of the Euler–Lagrange equations (3.2.10) for L. Conversely, since the jet bundle $J^2Y \to J^1Y$ is affine, every solution s on X of the Euler–Lagrange equations is also an integral section of a holonomic second order connection $\widehat{\Gamma}$ which is the global extension of the local section $J^1s(X) \to J^2s(X)$ of this jet bundle over the closed imbedded submanifold $J^1s(X) \subset J^1Y$. Hence, every solution on X of Euler–Lagrange equations is also a solution of some second order dynamic equation, but it is not necessarily a Lagrangian connection.

A glance at the equations (3.2.13) shows that a regular Lagrangian L admits a unique Lagrangian connection. In this case, Euler–Lagrange equations for L are equivalent to the second order dynamic equation associated with this Lagrangian connection.

3.3 Quadratic degenerate Lagrangians

This Section is devoted to the physically important case of almost regular quadratic Lagrangians. These Lagrangians describe almost all types of field interactions considered in contemporary field theory. We aim to show that such a Lagrangian factorizes as in (3.0.2) through the covariant differential relative to some connection on $Y \to X$.

Given a fibre bundle $Y \to X$, let us consider a quadratic Lagrangian L which has the coordinate expression

$$\mathcal{L} = \frac{1}{2} a^{\lambda\mu}_{ij}(x^\nu, y^k) y^i_\lambda y^j_\mu + b^\lambda_i(x^\nu, y^k) y^i_\lambda + c(x^\nu, y^k), \tag{3.3.1}$$

where a, b and c are local functions on Y. This property is coordinate-independent due to the affine transformation law of the coordinates y^i_λ. The associated Legendre map $\widehat{L}$ (3.2.9) is an affine morphism over Y given by the coordinate expression

$$p^\lambda_i \circ \widehat{L} = a^{\lambda\mu}_{ij} y^j_\mu + b^\lambda_i. \tag{3.3.2}$$

It defines the corresponding linear morphism

$$\overline{L} : T^*X \underset{Y}{\otimes} VY \underset{Y}{\to} \Pi,$$

$$p^\lambda_i \circ \overline{L} = a^{\lambda\mu}_{ij} \overline{y}^j_\mu, \tag{3.3.3}$$

where $\overline{y}^j_\mu$ are bundle coordinates on the vector bundle $T^*X \underset{Y}{\otimes} VY$.

Let the Lagrangian L (3.3.1) be almost regular, i.e., the matrix function $a^{\lambda\mu}_{ij}$ is of constant rank. Then the Lagrangian constraint space N_L (3.3.2) is an affine subbundle of the Legendre bundle $\Pi \to Y$, modelled over the vector subbundle $\overline{N}_L$ (3.3.3) of $\Pi \to Y$. Hence, $N_L \to Y$ has a global section s. For the sake of simplicity, let us assume that $s = \widehat{0}$ is the canonical zero section of $\Pi \to Y$. Then $\overline{N}_L = N_L$. Accordingly, the kernel of the Legendre map (3.3.2) is an affine subbundle of the affine jet bundle $J^1Y \to Y$, modelled over the kernel of the linear morphism $\overline{L}$ (3.3.3). Then there exists a connection

$$\Gamma : Y \to \operatorname{Ker} \widehat{L} \subset J^1Y, \tag{3.3.4}$$

$$\boxed{a^{\lambda\mu}_{ij} \Gamma^j_\mu + b^\lambda_i = 0,} \tag{3.3.5}$$

on $Y \to X$. Connections (3.3.4) constitute an affine space modelled over the linear space of soldering forms $\phi = \phi^i_\lambda dx^\lambda \otimes \partial_i$ on $Y \to X$, satisfying the conditions

$$a^{\lambda\mu}_{ij}\phi^j_\mu = 0 \tag{3.3.6}$$

and, as a consequence, the conditions $\phi^i_\lambda b^\lambda_i = 0$. If the Lagrangian (3.3.1) is regular, the connection (3.3.4) is unique.

Remark 3.3.1. If $s \neq \widehat{0}$, we can consider connections Γ with values into $\mathrm{Ker}\,{}_s\widehat{L}$. •

The matrix function a in the Lagrangian L (3.3.1) can be seen as a global section of constant rank of the tensor bundle

$$\overset{n}{\wedge} T^*X \underset{Y}{\otimes} [\overset{2}{\vee}(TX \underset{Y}{\otimes} V^*Y)] \to Y.$$

Then it satisfies the following corollary of Theorem 1.1.4.

COROLLARY 3.3.1. Given a k-dimensional vector bundle $E \to Z$, let a be a fibre metric of rank r in E. There is a splitting

$$E = \mathrm{Ker}\,a \underset{Z}{\oplus} E' \tag{3.3.7}$$

where $E' = E/\mathrm{Ker}\,a$ is the quotient bundle, and a is a non-degenerate fibre metric in E'. □

THEOREM 3.3.2. There exists a linear bundle map

$$\sigma : \Pi \underset{Y}{\to} T^*X \underset{Y}{\otimes} VY, \qquad \overline{y}^i_\lambda \circ \sigma = \sigma^{ij}_{\lambda\mu} p^\mu_j, \tag{3.3.8}$$

such that $\overline{L} \circ \sigma \circ i_N = i_N$. □

Proof. The map (3.3.8) is a solution of the algebraic equations

$$\boxed{a^{\lambda\mu}_{ij}\sigma^{jk}_{\mu\alpha}a^{\alpha\nu}_{kb} = a^{\lambda\nu}_{ib}.} \tag{3.3.9}$$

By virtue of Corollary 3.3.1, there exists the bundle splitting

$$TX^* \underset{Y}{\otimes} VY = \mathrm{Ker}\,a \underset{Y}{\oplus} E' \tag{3.3.10}$$

and a (non-holonomic) atlas of this bundle such that transition functions of Ker a and E' are independent. Since a is a non-degenerate section of $\overset{n}{\wedge} T^*X \underset{Y}{\otimes} (\overset{2}{\vee} E'^*) \to Y$, there is an atlas of E' such that a is brought into a diagonal matrix with non-vanishing components a^{AA}. Due to the splitting (3.3.10), we have the corresponding bundle splitting

$$TX \underset{Y}{\otimes} V^*Y = (\mathrm{Ker}\, a)^* \underset{Y}{\oplus} E'^*.$$

Then the desired map σ is represented by a direct sum $\sigma_1 \oplus \sigma_0$ of an arbitrary section σ_1 of the fibre bundle

$$\overset{n}{\wedge} TX \underset{Y}{\otimes} (\overset{2}{\vee} \mathrm{Ker}\, a) \to Y$$

and the section σ_0 of the fibre bundle

$$\overset{n}{\wedge} TX \underset{Y}{\otimes} (\overset{2}{\vee} E') \to Y$$

which has non-vanishing components $\sigma_{AA} = (a^{AA})^{-1}$ with respect to the above mentioned atlas of E'. Moreover, σ satisfies the additional relations

$$\boxed{\sigma_0 = \sigma_0 \circ \overline{L} \circ \sigma_0, \qquad a \circ \sigma_1 = 0, \qquad \sigma_1 \circ a = 0.} \tag{3.3.11}$$

QED

Remark 3.3.2. Using the relations (3.3.11), one can write the above assumption that the Lagrangian constraint space $N_L \to Y$ admits a global zero section in the form

$$b_i^\mu = a_{ij}^{\mu\lambda} \sigma_{\lambda\nu}^{jk} b_k^\nu.$$

•

With the relations (3.3.5), (3.3.9) and (3.3.11), we obtain the splitting

$$\boxed{J^1Y = \mathcal{S}(J^1Y) \underset{Y}{\oplus} \mathcal{F}(J^1Y) = \mathrm{Ker}\, \widehat{L} \underset{Y}{\oplus} \mathrm{Im}(\sigma \circ \widehat{L}),} \tag{3.3.12a}$$

$$y_\lambda^i = \mathcal{S}_\lambda^i + \mathcal{F}_\lambda^i = [y_\lambda^i - \sigma_{\lambda\alpha}^{ik} (a_{kj}^{\alpha\mu} y_\mu^j + b_k^\alpha)] + [\sigma_{\lambda\alpha}^{ik} (a_{kj}^{\alpha\mu} y_\mu^j + b_k^\alpha)]. \tag{3.3.12b}$$

Then with respect to the coordinates $\mathcal{S}^i_\lambda$ and $\mathcal{F}^i_\lambda$ (3.3.12b), the Lagrangian (3.3.1) reads

$$\boxed{\mathcal{L} = \frac{1}{2}a^{\lambda\mu}_{ij}\mathcal{F}^i_\lambda\mathcal{F}^j_\mu + c',} \tag{3.3.13}$$

where

$$\mathcal{F}^i_\lambda = \sigma^{ik}_{\lambda\alpha}a^{\alpha\mu}_{kj}(y^j_\mu - \Gamma^j_\mu) \tag{3.3.14}$$

for some (Ker $\widehat{L}$)-valued connection Γ (3.3.4) on $Y \to X$. Thus, the Lagrangian (3.3.1) written in the form (3.3.13) factorizes through the covariant differential relative to any such connection.

Note that, in gauge theory, we have the canonical splitting (3.3.12a) where $2\mathcal{F}$ is the strength tensor (see (6.2.21) below). The Yang–Mills Lagrangian (6.3.18) of gauge theory is exactly of the form (3.3.13) where $c' = 0$. The Lagrangian (6.4.16) of Proca fields is also of the form (3.3.13) where c' is the mass term. This is an example of a degenerate Lagrangian system without gauge symmetries.

3.4 Connections and Lagrangian conservation laws

Let L be again an arbitrary Lagrangian on the configuration space J^1Y. The first variational formula (3.2.2) provides the standard procedure for the study of differential conservation laws in Lagrangian field theory.

Let u be a projectable vector field on a fibre bundle $Y \to X$, treated as the generator of a local 1-parameter group of gauge transformations. On-shell, i.e., on the kernel (3.2.10) of the Euler–Lagrange operator $\mathcal{E}_L$, the first variational formula (3.2.2) leads to the weak identity

$$\begin{aligned}&\partial_\lambda u^\lambda\mathcal{L} + [u^\lambda\partial_\lambda + u^i\partial_i + (d_\lambda u^i - y^i_\mu\partial_\lambda u^\mu)\partial^\lambda_i]\mathcal{L} \approx \\ &\qquad -d_\lambda[\pi^\lambda_i(u^\mu y^i_\mu - u^i) - u^\lambda\mathcal{L}].\end{aligned} \tag{3.4.1}$$

Let the Lie derivative $\mathbf{L}_{J^1u}L$ (3.2.1) vanishes, i.e., a Lagrangian L is invariant under gauge transformations whose generator is the vector field u. Then we obtain the weak conservation law

$$\boxed{0 \approx -d_\lambda[\pi^\lambda_i(u^\mu y^i_\mu - u^i) - u^\lambda\mathcal{L}],} \tag{3.4.2}$$

of the *symmetry current*

$$\mathfrak{T} = \mathfrak{T}^\lambda\omega_\lambda, \qquad \boxed{\mathfrak{T}^\lambda = \pi^\lambda_i(u^\mu y^i_\mu - u^i) - u^\lambda\mathcal{L},} \tag{3.4.3}$$

along the vector field u.

Remark 3.4.1. It should be emphasized that, from the first variational formula, the symmetry current (3.4.3) is defined modulo the terms $d_\mu(c_i^{\mu\lambda}(y^i_\nu u^\nu - u^i))$, where $c_i^{\mu\lambda}$ are arbitrary skew-symmetric functions on Y [123]. Here we leave aside these boundary terms which are independent of a Lagrangian. •

The weak conservation law (3.4.2) leads to the differential conservation law

$$\partial_\lambda(\mathfrak{T}^\lambda \circ s) = 0 \tag{3.4.4}$$

on solutions of the Euler–Lagrange equations (3.2.11). This differential conservation law implies the *integral conservation law*

$$\int\limits_{\partial N} s^*\mathfrak{T} = 0, \tag{3.4.5}$$

where N is a compact n-dimensional submanifold of X with the boundary ∂N.

Remark 3.4.2. It may happen that a symmetry current $\mathfrak{T}$ (3.4.3) can be put into the form

$$\boxed{\mathfrak{T} = W + d_H U = (W^\lambda + d_\mu U^{\mu\lambda})\omega_\lambda,} \tag{3.4.6}$$

where the term W contains only the *variational derivatives* $\delta_i\mathcal{L} = (\partial_i - d_\lambda\partial_i^\lambda)\mathcal{L}$, i.e., $W \approx 0$ and

$$U = U^{\mu\lambda}\omega_{\mu\lambda} : J^1Y \to \overset{n-2}{\wedge} T^*X$$

is a horizontal $(n-2)$-form on $J^1Y \to X$. Then one says that $\mathfrak{T}$ reduces to the *superpotential* U [99, 123, 269]. On-shell, such a symmetry current is the d_H-exact form (3.4.6), while the equality

$$\mathfrak{T} - d_H U = W(\delta_i\mathcal{L}) = 0 \tag{3.4.7}$$

is a combination of the Euler–Lagrange equations $\delta_i\mathcal{L} = 0$. If a symmetry current $\mathfrak{T}$ reduces to a superpotential, the differential conservation law (3.4.40 and the integral conservation law (3.4.5) become tautological. At the same time, the superpotential form (3.4.6) of $\mathfrak{T}$ implies the following integral relation

$$\int\limits_{N^{n-1}} s^*\mathfrak{T} = \int\limits_{\partial N^{n-1}} s^*U, \tag{3.4.8}$$

where N^{n-1} is a compact oriented $(n-1)$-dimensional submanifold of X with the boundary ∂N^{n-1}. One can think of this relation as being a part of the Euler–Lagrange equations written in an integral form. Superpotentials are met in gauge theory (see Section 6.3) and in gravitation theory (see Section 7.4), where generators of gauge transformations depend on derivatives of gauge parameters. •

It is easy to see that the weak identity (3.4.1) is linear in a vector field u. Therefore, one can consider superposition of weak identities (3.4.1) associated with different vector fields. For instance, if u and u' are projectable vector fields on Y over the same vector field on X, the difference of the corresponding weak identities (3.4.1) results in the weak identity (3.4.1) associated with the vertical vector field $u - u'$. Every projectable vector field u on a fibre bundle $Y \to X$, which projects over a vector field τ on X, can be written as the sum $u = \widetilde{\tau} + \vartheta$ of a lift $\widetilde{\tau}$ of τ onto Y and a vertical vector field ϑ on Y. It follows that the weak identity (3.4.1) associated with a projectable vector field u can be represented as the superposition of those associated with $\widetilde{\tau}$ and ϑ.

In the case of a vertical vector field $\vartheta = \vartheta^i\partial_i$ on $Y \to X$, the weak identity (3.4.1) takes the form

$$[\vartheta^i\partial_i + d_\lambda\vartheta^i\partial_i^\lambda]\mathcal{L} \approx d_\lambda(\pi_i^\lambda\vartheta^i). \tag{3.4.9}$$

In field theory, vertical gauge transformations describe *internal symmetries*. If a Lagrangian is invariant under internal symmetries, we have the *Noether conservation law*

$$\boxed{0 \approx d_\lambda(\pi_i^\lambda\vartheta^i)}$$

of the *Noether current*

$$\boxed{\mathfrak{I}^\lambda = -\pi_i^\lambda\vartheta^i.} \tag{3.4.10}$$

The well-known example of a Noether conservation law is that in gauge theory of principal connections (see Section 6.3).

A vector field τ on X can be lifted onto Y by means of a connection Γ on a fibre bundle $Y \to X$. This lift is the horizontal vector field $\Gamma\tau$ (2.1.6). The weak identity (3.4.1) associated with such a vector field takes the form

$$\begin{aligned}\partial_\mu\tau^\mu\mathcal{L} + [\tau^\mu\partial_\mu + \tau^\mu\Gamma^i_\mu\partial_i + (d_\lambda(\tau^\mu\Gamma^i_\mu) - y^i_\mu\partial_\lambda\tau^\mu)\partial_i^\lambda]\mathcal{L} \approx \\ -d_\lambda[\pi_i^\lambda\tau^\mu(y^i_\mu - \Gamma^i_\mu) - \delta^\lambda_\mu\tau^\mu\mathcal{L}].\end{aligned} \tag{3.4.11}$$

The corresponding current (3.4.3) along $\Gamma\tau$ reads

$$\mathfrak{T}^\lambda_\Gamma = \tau^\mu(\pi^\lambda_i(y^i_\mu - \Gamma^i_\mu) - \delta^\lambda_\mu\mathcal{L}). \tag{3.4.12}$$

It is called the *energy-momentum current* relative to the connection Γ [104, 123, 269].

To discover energy-momentum conservation laws, one may choose different connections on $Y \to X$ (e.g., different connections for different vector fields τ on X and different connections for different solutions of the Euler–Lagrange equations). It is readily seen that the energy-momentum currents with respect to different connections Γ and Γ' differ from each other in the Noether current (3.4.10) along the vertical vector field

$$\vartheta = \tau^\mu(\Gamma^i_\mu - \Gamma'^i_\mu)\partial_i.$$

Example 3.4.3. Let all vector fields τ on X be lifted onto Y by means of the same connection Γ on $Y \to X$. The weak identity (3.4.11) can be rewritten as follows

$$\tau^\mu\{[\partial_\mu + \Gamma^i_\mu\partial_i + (\partial_\lambda\Gamma^i_\mu + y^j_\lambda\partial_j\Gamma^i_\mu)\partial^\lambda_i]\mathcal{L} - d_\lambda[\pi^\lambda_i(\Gamma^i_\mu - y^i_\mu) + \delta^\lambda_\mu\mathcal{L}]\} \approx 0.$$

Since this relation takes place for an arbitrary vector field τ on X, it is equivalent to the system of weak identities

$$(\partial_\mu + \Gamma^i_\mu\partial_i + d_\lambda\Gamma^i_\mu\partial^\lambda_i)\mathcal{L} + d_\lambda\mathfrak{T}_\Gamma{}^\lambda{}_\mu \approx 0,$$

where $\mathfrak{T}_\Gamma{}^\lambda{}_\mu$ are components of the tensor field

$$\overline{\mathfrak{T}}_\Gamma = \mathfrak{T}_\Gamma{}^\lambda{}_\mu dx^\mu \otimes \omega_\lambda,$$

$$\boxed{\mathfrak{T}_\Gamma{}^\lambda{}_\mu = \pi^\lambda_i(y^i_\mu - \Gamma^i_\mu) - \delta^\lambda_\mu\mathcal{L}.}$$

This is called the *energy-momentum tensor* relative to the connection Γ. •

Example 3.4.4. If we choose the local trivial connection $(\Gamma_0)^i_\mu = 0$, the weak identity (3.4.11) takes the form

$$\partial_\mu\mathcal{L} \approx -d_\lambda\mathfrak{T}_0{}^\lambda{}_\mu, \tag{3.4.13}$$
$$\mathfrak{T}_0{}^\lambda{}_\mu = \pi^\lambda_i y^i_\mu - \delta^\lambda_\mu\mathcal{L},$$

where $\mathfrak{T}_0{}^\lambda{}_\mu$ is called the *canonical energy-momentum tensor*. Though it is not a true tensor field, both the sides of the weak identity (3.4.14) on solutions s of the Euler–Lagrange equations are well defined:

$$\frac{\partial\mathcal{L}}{\partial x^\mu}\circ s \approx -\frac{\partial}{\partial x^\lambda}(\mathfrak{T}_0{}^\lambda{}_\mu \circ s). \tag{3.4.14}$$

This results from the weak identity (3.4.11) when, for every solution s, we choose the connection Γ which has s as its integral section. •

Note that, in gravitation theory on natural bundles $T \to X$, we have the canonical horizontal lift of vector fields on X onto T, e.g., the canonical lift (1.2.2) on a tensor bundle. This lift, under certain conditions, can be represented locally as a horizontal lift by means of a connection (see Remark 2.4.2). The corresponding current is also an energy-momentum current (see Section 7.4).

Remark 3.4.5. Let us consider conservation laws in the case of gauge transformations which preserve the Euler–Lagrange operator $\mathcal{E}_L$, but not necessarily a Lagrangian L. This is the case of BRST theory in Section 11.4 and the Chern–Simons topological field theory [123]. Let u be a projectable vector field on $Y \to X$, which is the generator of a local 1-parameter group of such transformations, i.e.,

$$\mathbf{L}_{J^2u}\mathcal{E}_L = 0,$$

where J^2u is the second order jet prolongation of the vector field u. There is the useful relation

$$\boxed{\mathbf{L}_{J^2u}\mathcal{E}_L = \mathcal{E}_{\mathbf{L}_{J^1u}L}} \tag{3.4.15}$$

[123]. Then, in accordance with (3.2.12), we have locally

$$\mathbf{L}_{J^1u}L = d_\lambda \epsilon^\lambda \omega. \tag{3.4.16}$$

In this case, the weak identity (3.4.1) reads

$$0 \approx d_\lambda(\epsilon^\lambda - \mathfrak{T}^\lambda), \tag{3.4.17}$$

where $\mathfrak{T}$ is the symmetry current (3.4.3) along the vector field u. •

Remark 3.4.6. Background fields, which do not live in the dynamic shell (3.2.10), violate conservation laws as follows. Let us consider the product

$$Y_{\rm tot} = Y \underset{X}{\times} Y' \tag{3.4.18}$$

of a fibre bundle Y with coordinates (x^λ, y^i), whose sections are dynamic fields, and a fibre bundle Y' with coordinates (x^λ, y^A), whose sections are background fields which take the background values

$$y^B = \phi^B(x), \qquad y^B_\lambda = \partial_\lambda \phi^B(x).$$

A Lagrangian L is defined on the total configuration space $J^1Y_{\rm tot}$. Let u be a projectable vector field on $Y_{\rm tot}$ which also projects onto Y' because gauge transformations of background fields do not depend on the dynamic ones. This vector field takes the coordinate form

$$u = u^\lambda(x^\mu)\partial_\lambda + u^A(x^\mu, y^B)\partial_A + u^i(x^\mu, y^B, y^j)\partial_i. \tag{3.4.19}$$

Substitution of (3.4.19) in (3.2.2) leads to the *first variational formula in the presence of background fields*

$$\begin{aligned}&\partial_\lambda u^\lambda \mathcal{L} + [u^\lambda\partial_\lambda + u^A\partial_A + u^i\partial_i + (d_\lambda u^A - y^A_\mu \partial_\lambda u^\mu)\partial^\lambda_A + \\ &\quad (d_\lambda u^i - y^i_\mu\partial_\lambda u^\mu)\partial^\lambda_i]\mathcal{L} = (u^A - y^A_\lambda u^\lambda)\partial_A\mathcal{L} + \pi^\lambda_A d_\lambda(u^A - y^A_\mu u^\mu) + \\ &\quad (u^i - y^i_\lambda u^\lambda)\delta_i\mathcal{L} - d_\lambda[\pi^\lambda_i(u^\mu y^i_\mu - u^i) - u^\lambda\mathcal{L}].\end{aligned} \tag{3.4.20}$$

Then the weak identity

$$\begin{aligned}&\partial_\lambda u^\lambda \mathcal{L} + [u^\lambda\partial_\lambda + u^A\partial_A + u^i\partial_i + (d_\lambda u^A - y^A_\mu \partial_\lambda u^\mu)\partial^\lambda_A + \\ &\quad (d_\lambda u^i - y^i_\mu\partial_\lambda u^\mu)\partial^\lambda_i]\mathcal{L} \approx (u^A - y^A_\lambda u^\lambda)\partial_A\mathcal{L} + \pi^\lambda_A d_\lambda(u^A - y^A_\mu u^\mu) - \\ &\quad d_\lambda[\pi^\lambda_i(u^\mu y^i_\mu - u^i) - u^\lambda\mathcal{L}]\end{aligned}$$

holds on the shell (3.2.10). If a total Lagrangian L is invariant under gauge transformations of the product (3.4.18), we obtain the *weak identity in the presence of background fields*

$$\boxed{(u^A - y^A_\mu u^\mu)\partial_A\mathcal{L} + \pi^\lambda_A d_\lambda(u^A - y^A_\mu u^\mu) \approx d_\lambda[\pi^\lambda_i(u^\mu y^i_\mu - u^i) - u^\lambda\mathcal{L}].} \tag{3.4.21}$$

The weak identity (3.4.21) can also be applied to the case of Euler–Lagrange equations plus external forces

$$(\partial_i - d_\lambda\partial^\lambda_i)\mathcal{L} + F_i = 0,$$

where F_i are local functions on J^1Y. Then it reads

$$(u^i - y^i_\mu u^\mu)F_i \approx -d_\lambda[\pi^\lambda_i(u^\mu y^i_\mu - u^i) - u^\lambda\mathcal{L}]. \tag{3.4.22}$$

•

Chapter 4

Connections in Hamiltonian field theory

Applied to field theory, the familiar symplectic Hamiltonian technique takes the form of instantaneous Hamiltonian formalism on an infinite-dimensional phase space, where canonical coordinates are field functions at a given instant of time (see, e.g., [134]). The Hamiltonian counterpart of Lagrangian formalism on fibre bundles is covariant Hamiltonian field theory where canonical momenta correspond to derivatives of fields with respect to all world coordinates (see [52, 123, 133, 268] for a survey). Covariant Hamiltonian formalism utilizes the Legendre bundle Π (3.2.8) as a finite-dimensional momentum phase space of fields, and is equivalent to Lagrangian formalism in the case of hyperregular Lagrangians. A degenerate Lagrangian requires a set of associated Hamiltonian forms in order to exhaust all solutions of the Euler–Lagrange equations. It is important for quantization that these Hamiltonians are not necessarily non-degenerate. In the case of fibre bundles over $\mathbb{R}$, the covariant Hamiltonian approach provides the adequate Hamiltonian formulation of non-relativistic time-dependent mechanics (see Section 5.10). Covariant Hamiltonian field theory is formulated in terms of Hamiltonian connections and Hamiltonian forms which contain the connection dependent terms [123, 268].

4.1 Hamiltonian connections and Hamiltonian forms

Given a fibre bundle $Y \to X$, let Π be the Legendre bundle (3.2.8) provided with the composite fibration

$$\pi_{\Pi X} = \pi \circ \pi_{\Pi Y} : \Pi \to Y \to X.$$

For the sake of convenience, the fibration $\Pi \to Y$ will be called a *vector Legendre bundle*, in contrast with the Legendre bundle $\Pi \to X$.

There are the canonical isomorphism

$$\Pi = \overset{n}{\wedge} T^*X \underset{Y}{\otimes} V^*Y \underset{Y}{\otimes} TX \tag{4.1.1}$$

and the canonical bundle monomorphism

$$\boldsymbol{\Theta}_Y : \Pi \underset{Y}{\hookrightarrow} \overset{n+1}{\wedge} T^*Y \underset{Y}{\otimes} TX,$$

$$\boxed{\boldsymbol{\Theta}_Y = p^\lambda_i dy^i \wedge \omega \otimes \partial_\lambda = p^\lambda_i \overline{d} y^i \wedge \omega_\lambda,} \tag{4.1.2}$$

called the *tangent-valued Liouville form* on Π. Since the exterior differential d can not be applied to the tangent-valued form (4.1.2), the polysymplectic form is defined as a unique TX-valued $(n+2)$-form $\boldsymbol{\Omega}_Y$ on Π such that the relation

$$\boldsymbol{\Omega}_Y \rfloor \phi = d(\boldsymbol{\Theta}_Y \rfloor \phi)$$

holds for arbitrary exterior 1-forms ϕ on X [123]. It is given by the coordinate expression

$$\boxed{\boldsymbol{\Omega}_Y = dp^\lambda_i \wedge dy^i \wedge \omega \otimes \partial_\lambda,} \tag{4.1.3}$$

which is maintained under holonomic coordinate transformations of Π.

Let $J^1\Pi$ be the first order jet manifold of the Legendre bundle $\Pi \to X$. It is equipped with the adapted coordinates $(x^\lambda, y^i, p^\lambda_i, y^i_\mu, p^\lambda_{\mu i})$ such that $y^i_\mu \circ J^1\pi_{\Pi Y} = y^i_\mu$. A connection

$$\gamma = dx^\lambda \otimes (\partial_\lambda + \gamma^i_\lambda \partial_i + \gamma^\mu_{\lambda i} \partial^i_\mu) \tag{4.1.4}$$

on $\Pi \to X$ is called a *Hamiltonian connection* if the exterior form $\gamma \rfloor \boldsymbol{\Omega}_Y$ is closed. Components of a Hamiltonian connection satisfy the conditions

$$\partial^i_\lambda \gamma^j_\mu - \partial^j_\mu \gamma^i_\lambda = 0, \quad \partial_i \gamma^\mu_{\mu j} - \partial_j \gamma^\mu_{\mu i} = 0, \quad \partial_j \gamma^i_\lambda + \partial^i_\lambda \gamma^\mu_{\mu j} = 0. \tag{4.1.5}$$

If the form $\gamma\rfloor\mathbf{\Omega}_Y$ is closed, there is a contractible neighbourhood U of each point of Π which belongs to a holonomic coordinate chart $(x^\lambda, y^i, p^\lambda_i)$ and where the local form $\gamma\rfloor\mathbf{\Omega}_Y$ is exact. We have

$$\gamma\rfloor\mathbf{\Omega}_Y = dH = dp^\lambda_i \wedge dy^i \wedge \omega_\lambda - (\gamma^i_\lambda dp^\lambda_i - \gamma^\lambda_{\lambda i} dy^i) \wedge \omega \tag{4.1.6}$$

on U. It is readily observed that the second term in the right-hand side of this equality is also an exact form on U. By virtue of the relative Poincaré lemma (see Remark 4.1.2 below), it can be brought into the form $d\mathcal{H}\wedge\omega$ where $\mathcal{H}$ is a local function on U. Then the form H in the expression (4.1.6) reads

$$H = p^\lambda_i dy^i \wedge \omega_\lambda - \mathcal{H}\omega. \tag{4.1.7}$$

Further on, we will restrict our consideration to Hamiltonian connections γ such that $\gamma\rfloor\mathbf{\Omega}_Y$ is the exterior differential of some global form (4.1.7).

Remark 4.1.1. In conservative Hamiltonian mechanics, any function on a momentum phase space can play the role of a Hamiltonian. It is not so in Hamiltonian field theory and time-dependent mechanics where Hamiltonians $\mathcal{H}$ obey a certain coordinate transformation law (see Remark 5.10.1 below). •

Remark 4.1.2. Relative Poincaré lemma. Let us consider the vector space $\mathbb{R}^m \times \mathbb{R}^n$ with the Cartesian coordinates (z^λ). Let

$$\phi = \varphi \wedge \omega, \qquad \omega = dz^1 \wedge \cdots \wedge dz^n,$$

be an exact $(r+n)$-form on $\mathbb{R}^m \times \mathbb{R}^n$. Then ϕ is brought into the form $\phi = d\sigma \wedge \omega$, where σ is an $(r-1)$-form on $\mathbb{R}^m \times \mathbb{R}^n$, defined by the relation

$$\sigma(z) \wedge \omega = \int_0^1 t^{r-1} z\rfloor\phi(tz)]dt. \tag{4.1.8}$$

Indeed, it is easy to check that

$$d(\sigma(z)\wedge\omega) = \int_0^1 \frac{d}{dt}(t^r\phi(tz))dt = \phi(z).$$

If $n = 0$, the expression (4.1.8) reduces to the well-known *homotopy operator*

$$\mathbf{h} : \mathfrak{O}^r(\mathbb{R}^m) \to \mathfrak{O}^{k-1}(\mathbb{R}^m),$$

$$\mathbf{h}(\phi) = \int_0^1 t^{r-1} z\rfloor\phi(tz)]dt, \tag{4.1.9}$$

(see, e.g., [73, 280]). It is readily observed that the homotopy operator (4.1.9) obeys the relation

$$d \circ \mathbf{h} + \mathbf{h} \circ d = \operatorname{Id} \mathfrak{O}^*(\mathbb{R}^m), \qquad \mathbf{h}^2 = 0.$$

•

Let us consider the homogeneous Legendre bundle Z_Y (3.2.4) and the affine bundle $Z_Y \to \Pi$ (3.2.7). This affine bundle is modelled over the pull-back vector bundle $\Pi \underset{X}{\times} \overset{n}{\wedge} T^*X \to \Pi$ in accordance with the exact sequence

$$0 \longrightarrow \Pi \underset{X}{\times} \overset{n}{\wedge} T^*X \hookrightarrow Z_Y \longrightarrow \Pi \longrightarrow 0. \tag{4.1.10}$$

The homogeneous Legendre bundle Z_Y is provided with the canonical exterior n-form

$$\Xi_Y = p\omega + p_i^\lambda dy^i \wedge \omega_\lambda, \tag{4.1.11}$$

whose exterior differential $d\Xi_Y$ is a *multisymplectic form*. Given any section h of the affine bundle $Z_Y \to \Pi$ (3.2.7), the pull-back

$$\boxed{H = h^*\Xi_Y = p_i^\lambda dy^i \wedge \omega_\lambda - \mathcal{H}\omega} \tag{4.1.12}$$

is a called a *Hamiltonian form* on the Legendre bundle Π.

As an immediate consequence of this definition, we obtain that:

(i) Hamiltonian forms constitute a non-empty affine space modelled over the linear space of horizontal densities $\widetilde{H} = \widetilde{\mathcal{H}}\omega$ on $\Pi \to X$;

(ii) every connection Γ on the fibre bundle $Y \to X$ yields the splitting (2.1.7) of the exact sequence (4.1.10) and defines the Hamiltonian form

$$\boxed{H_\Gamma = \Gamma^*\Xi_Y = p_i^\lambda dy^i \wedge \omega_\lambda - p_i^\lambda \Gamma_\lambda^i \omega;} \tag{4.1.13}$$

(iii) given a connection Γ on $Y \to X$, every Hamiltonian form H admits the decomposition

$$\boxed{H = H_\Gamma - \widetilde{H}_\Gamma = p_i^\lambda dy^i \wedge \omega_\lambda - p_i^\lambda \Gamma_\lambda^i \omega - \widetilde{\mathcal{H}}_\Gamma \omega.} \tag{4.1.14}$$

Remark 4.1.3. The physical meaning of the splitting (4.1.14) is illustrated by the fact that, in the case of $X = \mathbb{R}$ of time-dependent mechanics, $\widetilde{\mathcal{H}}_\Gamma$ is exactly the energy of a mechanical system with respect to the reference frame Γ (see Section 5.11). If a reference frame is fixed, the Hamiltonian form (4.1.12) coincides with the well-known integral invariant of Poincaré–Cartan in mechanics, where $\mathcal{H}$ is a Hamiltonian. In covariant Hamiltonian field theory, $\mathcal{H}$ is also called a Hamiltonian. The splitting (4.1.14) shows that it is not a density under holonomic coordinate transformations of Π. •

The assertion (ii) can be generalized as follows. By a *Hamiltonian map* is meant any bundle morphism

$$\Phi = dx^\lambda \otimes (\partial_\lambda + \Phi^i_\lambda \partial_i) : \Pi \underset{Y}{\to} J^1Y. \tag{4.1.15}$$

In particular, let Γ be a connection on $Y \to X$. Then the composition

$$\widehat{\Gamma} = \Gamma \circ \pi_{\Pi Y} = dx^\lambda \otimes (\partial_\lambda + \Gamma^i_\lambda \partial_i) : \Pi \to Y \to J^1Y \tag{4.1.16}$$

is a Hamiltonian map.

PROPOSITION 4.1.1. Every Hamiltonian map (4.1.15) defines the Hamiltonian form

$$H_\Phi = -\Phi\rfloor\Theta = p^\lambda_i dy^i \wedge \omega_\lambda - p^\lambda_i \Phi^i_\lambda \omega. \tag{4.1.17}$$

□

Proof. Given an arbitrary connection Γ on the fibre bundle $Y \to X$, the corresponding Hamiltonian map (4.1.16) defines the form $-\widehat{\Gamma}\rfloor\Theta$ which is exactly the Hamiltonian form H_Γ (4.1.13). Since $\Phi - \widehat{\Gamma}$ is a VY-valued basic 1-form on $\Pi \to X$, $H_\Phi - H_\Gamma$ is a horizontal density on Π. Then the result follows from the assertion (i). **QED**

Now we will show that every Hamiltonian form H (4.1.12) admits a Hamiltonian connection γ_H which obeys the condition

$$\boxed{\gamma_H\rfloor\mathbf{\Omega}_Y = dH,} \tag{4.1.18}$$

$$\gamma^i_\lambda = \partial^i_\lambda \mathcal{H}, \qquad \gamma^\lambda_{\lambda i} = -\partial_i \mathcal{H}. \tag{4.1.19}$$

It is readily observed that a Hamiltonian form H (4.1.12) is the Poincaré–Cartan form (3.2.5) of the Lagrangian

$$\boxed{L_H = h_0(H) = (p^\lambda_i y^i_\lambda - \mathcal{H})\omega} \tag{4.1.20}$$

on the jet manifold $J^1\Pi$. The Euler–Lagrange operator (3.2.3) associated with this Lagrangian reads

$$\mathcal{E}_H : J^1\Pi \to T^*\Pi \wedge (\overset{n}{\wedge} T^*X),$$
$$\boxed{\mathcal{E}_H = [(y^i_\lambda - \partial^i_\lambda \mathcal{H})dp^\lambda_i - (p^\lambda_{\lambda i} + \partial_i \mathcal{H})dy^i] \wedge \omega.} \tag{4.1.21}$$

It is called the *Hamilton operator* for H. A glance at the expression (4.1.21) shows that this operator is an affine morphism over Π of constant rank. It follows that its kernel (i.e., the Euler–Lagrange equations for L_H)

$$y^i_\lambda = \partial^i_\lambda \mathcal{H}, \tag{4.1.22a}$$
$$p^\lambda_{\lambda i} = -\partial_i \mathcal{H}, \tag{4.1.22b}$$

is an affine closed imbedded subbundle of the jet bundle $J^1\Pi \to \Pi$. By virtue of Definition 1.3.4, it is a system of first order differential equations on the Legendre bundle $\Pi \to X$, called the *covariant Hamilton equations*. Being an affine subbundle, $\operatorname{Ker}\mathcal{E}_H \to \Pi$ has a global section γ_H which is a desired Hamiltonian connection obeying the relation (4.1.18).

If $n > 1$, there is a set of Hamiltonian connections associated with the same Hamiltonian form H. They differ from each other in soldering forms σ on $\Pi \to X$ which fulfill the equation $\sigma\rfloor\mathbf{\Omega}_Y = 0$. In particular, a Hamiltonian connection for the Hamiltonian form H_Γ (4.1.13) reads

$$\widetilde{\Gamma} = dx^\lambda \otimes [\partial_\lambda + \Gamma^i_\lambda \partial_i + (K_\lambda{}^\mu{}_\nu p^\nu_j - K_\lambda{}^\alpha{}_\alpha p^\mu_j - \partial_j \Gamma^i_\lambda p^\mu_i)\partial^j_\mu], \tag{4.1.23}$$

where K is a symmetric linear connection on X.

Every Hamiltonian form H defines the Hamiltonian map

$$\widehat{H} = J^1\pi_{\Pi Y} \circ \gamma_H : \Pi \to J^1\Pi \to J^1Y, \tag{4.1.24}$$
$$\boxed{y^i_\lambda \circ \widehat{H} = \partial^i_\lambda \mathcal{H},}$$

which is the same for all Hamiltonian connection γ_H associated with H.

Every integral section $J^1r = \gamma \circ r$ of a Hamiltonian connection γ_H associated with a Hamiltonian form H is obviously a solution of the Hamilton equations (4.1.22a)

– (4.1.22b). If $r : X \to \Pi$ is a global solution, there exists an extension of the local section $J^1r : r(X) \to \operatorname{Ker}\mathcal{E}_H$ to a Hamiltonian connection γ_H which has r as an integral section. Substituting J^1r in (4.1.24), we obtain the equality

$$J^1(\pi_{\Pi Y} \circ r) = \widehat{H} \circ r, \tag{4.1.25}$$

which is the coordinate-free form of the Hamilton equations (4.1.22a).

4.2 Lagrangian and Hamiltonian degenerate systems

Let us study the relations between the Euler–Lagrange equations (3.2.10) and the covariant Hamilton equations (4.1.22a) – (4.1.22a). The key point is that different solutions of the Euler–Lagrange equations for a degenerate Lagrangian may appear to be solutions of different Hamilton equations for different Hamiltonian forms.

We will start from the case of a hyperregular Lagrangian L. Then $\widehat{L}^{-1}$ is a Hamiltonian map. Let us consider the Hamiltonian form

$$\boxed{H = H_{\widehat{L}^{-1}} + \widehat{L}^{-1*}L,} \tag{4.2.1}$$

$$\mathcal{H} = p^\mu_i \widehat{L}^{-1i}{}_\mu - \mathcal{L}(x^\lambda, y^j, \widehat{L}^{-1j}{}_\lambda),$$

where $H_{\widehat{L}^{-1}}$ is the Hamiltonian form (4.1.17) associated with the Hamiltonian map $\widehat{L}^{-1}$. Let s be a solution of the Euler–Lagrange equations (3.2.10) for the Lagrangian L. A direct computation shows that $\widehat{L} \circ J^1s$ is a solution of the Hamilton equations (4.1.22a) – (4.1.22a) for the Hamiltonian form H (4.2.1). Conversely, if r is a solution of the Hamilton equations (4.1.22a) – (4.1.22a) for the Hamiltonian form H (4.2.1), then $s = \pi_{\Pi Y} \circ r$ is a solution of the Euler–Lagrange equations (3.2.10) for L (see the equality (4.1.25)). It follows that, in the case of hyperregular Lagrangians, covariant Hamiltonian formalism is equivalent to the Lagrangian one.

Let now L be an arbitrary Lagrangian on the configuration space J^1Y. A Hamiltonian form H is said to be *associated* with a Lagrangian L if H satisfies the relations

$$\boxed{\widehat{L} \circ \widehat{H} \circ \widehat{L} = \widehat{L},} \tag{4.2.2a}$$

$$\boxed{H = H_{\widehat{H}} + \widehat{H}^* L.} \tag{4.2.2b}$$

A glance at the relation (4.2.2a) shows that $\widehat{L} \circ \widehat{H}$ is the projector

$$\boxed{p^\mu_i(p) = \partial^\mu_i \mathcal{L}(x^\lambda, y^i, \partial^j_\lambda \mathcal{H}(p)),} \qquad p \in N_L, \tag{4.2.3}$$

from Π onto the Lagrangian constraint space $N_L = \widehat{L}(J^1Y)$. Accordingly, $\widehat{H} \circ \widehat{L}$ is the projector from J^1Y onto $\widehat{H}(N_L)$. A Hamiltonian form is called *weakly associated* with a Lagrangian L if the condition (4.2.2b) holds on the Lagrangian constraint space N_L.

PROPOSITION 4.2.1. If a Hamiltonian map Φ (4.1.15) obeys the relation (4.2.2a), then the Hamiltonian form $H = H_\Phi + \Phi^* L$ is weakly associated with the Lagrangian L. If $\Phi = \widehat{H}$, then H is associated with L. □

PROPOSITION 4.2.2. Any Hamiltonian form H weakly associated with a Lagrangian L fulfills the relation

$$H\mid_{N_L} = \widehat{H}^* H_L \mid_{N_L}, \tag{4.2.4}$$

where H_L is the Poincaré–Cartan form (3.2.5). □

Proof. The relation (4.2.2b) takes the coordinate form

$$\boxed{\mathcal{H}(p) = p^\mu_i \partial^i_\mu \mathcal{H} - \mathcal{L}(x^\lambda, y^i, \partial^j_\lambda \mathcal{H}(p)),} \qquad p \in N_L. \tag{4.2.5}$$

Substituting (4.2.3) and (4.2.5) in (4.1.12), we obtain the relation (4.2.4). **QED**

The difference between associated and weakly associated Hamiltonian forms lies in the following. Let H be an associated Hamiltonian form, i.e., the equality (4.2.5) holds everywhere on Π. Acting on this equality by the exterior differential, we obtain the relations

$$\partial_\mu \mathcal{H}(p) = -(\partial_\mu \mathcal{L}) \circ \widehat{H}(p), \qquad p \in N_L,$$

$$\partial_i \mathcal{H}(p) = -(\partial_i \mathcal{L}) \circ \widehat{H}(p), \qquad p \in N_L, \tag{4.2.6}$$

$$(p^\mu_i - (\partial^\mu_i \mathcal{L})(x^\lambda, y^i, \partial^j_\lambda \mathcal{H}))\partial^i_\mu \partial^a_\alpha \mathcal{H} = 0. \tag{4.2.7}$$

The relation (4.2.7) shows that the associated Hamiltonian form is not regular outside the Lagrangian constraint space N_L. In particular, any Hamiltonian form is weakly associated with the Lagrangian $L = 0$, while the associated Hamiltonian forms are only of the form H_Γ (4.1.13).

A hyperregular Lagrangian has a unique weakly associated Hamiltonian form (4.2.1). In the case of a regular Lagrangian L, the Lagrangian constraint space N_L is an open subbundle of the vector Legendre bundle $\Pi \to Y$. If $N_L \neq \Pi$, a weakly associated Hamiltonian form fails to be defined everywhere on Π in general. At the same time, the open subbundle N_L can be provided with the pull-back polysymplectic structure with respect to the imbedding $N_L \hookrightarrow \Pi$, so that one may consider Hamiltonian forms on N_L.

Contemporary field models are almost never regular. Hereafter, we will restrict our consideration to almost regular Lagrangians. From the physical viewpoint, Lagrangians of the most of field models are of this type. From the mathematical one, this notion of degeneracy is particularly appropriate in order to study the relations between solutions of Euler–Lagrange and Hamilton equations.

The following assertion is a corollary of Proposition 4.2.1.

PROPOSITION 4.2.3. A Hamiltonian form H weakly associated with an almost regular Lagrangian L exists if and only if the fibred manifold $J^1Q \to N_L$ admits a global section. □

There is another useful fact.

LEMMA 4.2.4. The Poincaré–Cartan form H_L for an almost regular Lagrangian L is constant on the connected pre-image $\widehat{L}^{-1}(p)$ of any point $p \in N_L$. □

Proof. Let u be a vertical vector field on the affine jet bundle $J^1Y \to Y$ which takes its values into the kernel of the tangent map $T\widehat{L}$ to $\widehat{L}$. Then $\mathbf{L}_u H_L = 0$. **QED**

Then we come to the following assertion.

PROPOSITION 4.2.5. All Hamiltonian forms weakly associated with an almost regular Lagrangian L coincide with each other on the Lagrangian constraint space N_L, and the Poincaré–Cartan form H_L (3.2.5) for L is the pull-back

$$H_L = \widehat{L}^* H, \tag{4.2.8}$$

$$\boxed{(\pi_i^\lambda y_\lambda^i - \mathcal{L})\omega = \mathcal{H}(x^\mu, y^j, \pi_j^\mu)\omega,}$$

of any such a Hamiltonian form H. □

Proof. Given a vector $v \in T_p\Pi$, the value $T\widehat{H}(v)\rfloor H_L(\widehat{H}(p))$ is the same for all Hamiltonian maps $\widehat{H}$ satisfying the relation (4.2.2a). Then the results follow from the relation (4.2.4). **QED**

Proposition 4.2.5 enables one to connect Euler–Lagrange equations for an almost regular Lagrangian L with the Hamilton equations for Hamiltonian forms weakly associated with L [123, 126, 266, 267, 268].

THEOREM 4.2.6. Let a section r of $\Pi \to X$ be a solution of the Hamilton equations (4.1.22a) – (4.1.22b) for a Hamiltonian form H weakly associated with an almost regular Lagrangian L. If r lives in the Lagrangian constraint space N_L, the section $s = \pi_{\Pi Y} \circ r$ of $Y \to X$ satisfies the Euler–Lagrange equations (3.2.10). □

The proof is based on the relation $\mathcal{E}_L = (J^1\widehat{L})^* \mathcal{E}_H \mid_{J^2Y}$. The converse assertion is more intricate.

THEOREM 4.2.7. Given an almost regular Lagrangian L, let a section s of the fibre bundle $Y \to X$ be a solution of the Euler–Lagrange equations (3.2.10). Let H be a Hamiltonian form weakly associated with L, and let H satisfy the relation

$$\widehat{H} \circ \widehat{L} \circ J^1 s = J^1 s. \tag{4.2.9}$$

Then the section $r = \widehat{L} \circ J^1 s$ of the Legendre bundle $\Pi \to X$ is a solution of the Hamilton equations (4.1.22a) – (4.1.22b) for H. □

We will say that a set of Hamiltonian forms H weakly associated with an almost regular Lagrangian L is *complete* if, for each solution s of the Euler-Lagrange equations, there exists a solution r of the Hamilton equations for a Hamiltonian form H from this set such that $s = \pi_{\Pi Y} \circ r$. By virtue of Theorem 4.2.7, a set of weakly associated Hamiltonian forms is complete if, for every solution s of the Euler-Lagrange equations for L, there is a Hamiltonian form H from this set which fulfills the relation (4.2.9).

In accordance with Proposition 4.2.3, on an open neighbourhood in Π of each point $p \in N_L$, there exists a complete set of local Hamiltonian forms weakly associated with an almost regular Lagrangian L. Moreover, one can always construct a complete set of associated Hamiltonian forms [268, 311]

4.3 Quadratic and affine degenerate systems

Lagrangians of field models are almost always quadratic or affine in the derivatives of field functions. Gauge theory exemplifies a model with a degenerate quadratic

Lagrangian (see Section 6.4), whereas fermion fields are described by the affine one (see Section 7.5). In this Section, we obtain the complete sets of Hamiltonian forms weakly associated with almost regular quadratic and affine Lagrangians. Hamiltonian forms in such a complete set are parametrised by different connections on the fibre bundle $Y \to X$.

Let L (3.3.1) be an almost regular quadratic Lagrangian, σ a linear map (3.3.8) and Γ a connection (3.3.4). Similarly to the splitting (3.3.12a) of the configuration space J^1Y, we have the splitting of the momentum phase space

$$\boxed{\Pi = \mathcal{R}(\Pi) \underset{Y}{\oplus} \mathcal{P}(\Pi) = \mathrm{Ker}\,\sigma_0 \underset{Y}{\oplus} N_L,} \tag{4.3.1a}$$

$$p_i^\lambda = \mathcal{R}_i^\lambda + \mathcal{P}_i^\lambda = [p_i^\lambda - a_{ij}^{\lambda\mu}\sigma_{\mu\alpha}^{jk}p_k^\alpha] + [a_{ij}^{\lambda\mu}\sigma_{\mu\alpha}^{jk}p_k^\alpha]. \tag{4.3.1b}$$

In the coordinates (4.3.1b), the Lagrangian constraint space (3.3.2) is given by the reducible constraints

$$\boxed{\mathcal{R}_i^\lambda = [p_i^\lambda - a_{ij}^{\lambda\mu}\sigma_{\mu\alpha}^{jk}p_k^\alpha] = 0.} \tag{4.3.2}$$

Let us consider the affine Hamiltonian map

$$\Phi = \widehat{\Gamma} + \sigma : \Pi \to J^1Y, \qquad \Phi_\lambda^i = \Gamma_\lambda^i + \sigma_{\lambda\mu}^{ij}p_j^\mu, \tag{4.3.3}$$

and the Hamiltonian form

$$\begin{aligned} H = H_\Phi + \Phi^*L &= p_i^\lambda dy^i \wedge \omega_\lambda - [\Gamma_\lambda^i p_i^\lambda + \frac{1}{2}\sigma_0{}_{\lambda\mu}^{ij}p_i^\lambda p_j^\mu + \sigma_1{}_{\lambda\mu}^{ij}p_i^\lambda p_j^\mu - c']\omega \\ &= (\mathcal{R}_i^\lambda + \mathcal{P}_i^\lambda)dy^i \wedge \omega_\lambda - [(\mathcal{R}_i^\lambda + \mathcal{P}_i^\lambda)\Gamma_\lambda^i + \frac{1}{2}\sigma_0{}_{\lambda\mu}^{ij}\mathcal{P}_i^\lambda \mathcal{P}_j^\mu + \sigma_1{}_{\lambda\mu}^{ij}p_i^\lambda p_j^\mu - c']\omega. \end{aligned} \tag{4.3.4}$$

PROPOSITION 4.3.1. The Hamiltonian forms (4.3.4) parametrised by connections Γ (3.3.4) are weakly associated with the Lagrangian (3.3.1) and constitute a complete set. □

Proof. By the very definitions of Γ and σ, the Hamiltonian map (4.3.3) satisfies the condition (4.2.2a). Then H is weakly associated with L (3.3.1) in accordance with Proposition 4.2.1. Let us write the corresponding Hamilton equations (4.1.22a) for a section r of the Legendre bundle $\Pi \to X$. They are

$$J^1s = (\widehat{\Gamma} + \sigma) \circ r, \qquad s = \pi_{\Pi Y} \circ r. \tag{4.3.5}$$

Due to the surjections $\mathcal{S}$ and $\mathcal{F}$ (3.3.12a), the Hamilton equations (4.3.5) break in two parts

$$\mathcal{S}\circ J^1s = \Gamma\circ s, \tag{4.3.6}$$
$$\partial_\lambda r^i - \sigma^{ik}_{\lambda\alpha}(a^{\alpha\mu}_{kj}\partial_\mu r^j + b^\alpha_k) = \Gamma^i_\lambda\circ s,$$
$$\mathcal{F}\circ J^1s = \sigma\circ r, \tag{4.3.7}$$
$$\sigma^{ik}_{\lambda\alpha}(a^{\alpha\mu}_{kj}\partial_\mu r^j + b^\alpha_k) = \sigma^{ik}_{\lambda\alpha}r^\alpha_k.$$

Let s be an arbitrary section of $Y\to X$, e.g., a solution of the Euler–Lagrange equations. There exists a connection Γ (3.3.4) such that the relation (4.3.6) holds, namely, $\Gamma = \mathcal{S}\circ\Gamma'$ where Γ' is a connection on $Y\to X$ which has s as an integral section. It is easily seen that, in this case, the Hamiltonian map (4.3.3) satisfies the relation (4.2.9) for s. Hence, the Hamiltonian forms (4.3.4) constitute a complete set. **QED**

It is readily observed that, if $\sigma_1 = 0$, then $\Phi = \widehat{H}$ and the Hamiltonian forms (4.3.4) are associated with the Lagrangian (3.3.1). If σ_1 is non-degenerate, so is the Hamiltonian form (4.3.4). Thus, for different σ_1, we have different complete sets of Hamiltonian forms (4.3.4). Hamiltonian forms H (4.3.4) from such a complete set differ from each other in the term $\phi^i_\lambda\mathcal{R}^\lambda_i$, where ϕ are the soldering forms (3.3.6). It follows from the splitting (4.3.1a) that this term vanishes on the Lagrangian constraint space. Accordingly, the Hamilton equations for different Hamiltonian forms (4.3.4) weakly associated with the same quadratic Lagrangian (3.3.1) differ from each other in the equations (4.3.6). These equations are independent of momenta and play the role of gauge-type conditions.

Let us turn now to an *affine Lagrangian*

$$\mathcal{L} = b^\lambda_i y^i_\lambda + c, \tag{4.3.8}$$

where b and c are local functions on Y. The associated Legendre map takes the form

$$p^\lambda_i\circ\widehat{L} = b^\lambda_i. \tag{4.3.9}$$

Clearly, the Lagrangian (4.3.8) is almost regular.

Let Γ be an arbitrary connection on the fibre bundle $Y\to X$ and $\widehat{\Gamma}$ the associated Hamiltonian map (4.1.16). Let us consider the Hamiltonian form

$$H = H_\Gamma + L\circ\Gamma = p^\lambda_i dy^i\wedge\omega_\lambda - (p^\lambda_i - b^\lambda_i)\Gamma^i_\lambda\omega + c\omega. \tag{4.3.10}$$

It is associated with the affine Lagrangian (4.3.8). The corresponding Hamiltonian map is

$$y^i_\lambda \circ \widehat{H} = \Gamma^i_\lambda. \tag{4.3.11}$$

It follows that the Hamilton equations (4.1.22a) for the Hamiltonian form H reduce to the gauge-type condition

$$\partial_\lambda r^i = \Gamma^i_\lambda,$$

whose solutions are integral sections of the connection Γ.

Conversely, for each section s of the fibre bundle $Y \to X$, there exists a connection Γ on Y whose integral section is s. Then the corresponding Hamiltonian map (4.3.11) obeys the condition (4.2.9). It follows that the Hamiltonian forms (4.3.10) parameterized by connections Γ on $Y \to X$ constitute a complete set.

4.4 Connections and Hamiltonian conservation laws

As in Lagrangian field theory, different connections are responsible for different energy-momentum currents in Hamiltonian field theory.

To obtain the conservation laws within the framework of covariant Hamiltonian formalism, let use the fact that a Hamiltonian form H (4.1.12) is the Poincaré–Cartan form for the Lagrangian L_H (4.1.20) and that the Hamilton equations for H, by definition, are the Euler–Lagrange equations for L_H.

Due to the canonical lift (1.2.2), every projectable vector field u on the fibre bundle $Y \to X$ gives rise to the vector field

$$\boxed{\widetilde{u} = u^\mu\partial_\mu + u^i\partial_i + (-\partial_i u^j p^\lambda_j - \partial_\mu u^\mu p^\lambda_i + \partial_\mu u^\lambda p^\mu_i)\partial^i_\lambda} \tag{4.4.1}$$

on the Legendre bundle $\Pi \to Y$. We have

$$\boxed{\mathbf{L}_{\widetilde{u}} H = \mathbf{L}_{J^1\widetilde{u}} L_H = (-u^i\partial_i\mathcal{H} - \partial_\mu(u^\mu\mathcal{H}) - u^\lambda_i\partial^i_\lambda\mathcal{H} + p^\lambda_i\partial_\lambda u^i)\omega.} \tag{4.4.2}$$

It follows that the Hamiltonian form H and the Lagrangian L_H have the same gauge symmetries. Then one can follow the standard procedure of describing differential conservation laws in Lagrangian formalism (see Chapter 3), and apply the first variational formula (3.2.2) to the Lagrangian (4.1.20) [123]. We have

$$\begin{aligned}
-u^i\partial_i\mathcal{H} - \partial_\mu(u^\mu\mathcal{H}) - u^\lambda_i\partial^i_\lambda\mathcal{H} + p^\lambda_i\partial_\lambda u^i \equiv -(u^i - y^i_\mu u^\mu)(p^\lambda_{\lambda i} + \partial_i\mathcal{H}) + \\
(-\partial_i u^j p^\lambda_j - \partial_\mu u^\mu p^\lambda_i + \partial_\mu u^\lambda p^\mu_i - p^\lambda_{\mu i} u^\mu)(y^i_\lambda - \partial^i_\lambda\mathcal{H}) - \\
d_\lambda[p^\lambda_i(\partial^i_\mu\mathcal{H}u^\mu - u^i) - u^\lambda(p^\mu_i\partial^i_\mu\mathcal{H} - \mathcal{H})].
\end{aligned}$$

On the shell (4.1.22a) – (4.1.22b), this identity takes the form

$$-u^i\partial_i\mathcal{H} - \partial_\mu(u^\mu\mathcal{H}) - u_i^\lambda\partial_\lambda^i\mathcal{H} + p_i^\lambda\partial_\lambda u^i \approx - \tag{4.4.3}$$
$$d_\lambda[p_i^\lambda(\partial_\mu^i\mathcal{H}u^\mu - u^i) - u^\lambda(p_i^\mu\partial_\mu^i\mathcal{H} - \mathcal{H})].$$

If $\mathbf{L}_{J^1\widetilde{u}}L_H = 0$, we obtain the weak conservation law

$$0 \approx -d_\lambda[p_i^\lambda(u^\mu\partial_\mu^i\mathcal{H} - u^i) - u^\lambda(p_i^\mu\partial_\mu^i\mathcal{H} - \mathcal{H})]\omega \tag{4.4.4}$$

of the current

$$\boxed{\widetilde{\mathfrak{T}}^\lambda = p_i^\lambda(u^\mu\partial_\mu^i\mathcal{H} - u^i) - u^\lambda(p_i^\mu\partial_\mu^i\mathcal{H} - \mathcal{H}).} \tag{4.4.5}$$

On solutions r of the covariant Hamilton equations (4.1.22a) – (4.1.22b), the weak equality (4.4.4) leads to the weak differential conservation law

$$0 \approx -\frac{\partial}{\partial x^\lambda}\widetilde{\mathfrak{T}}^\lambda(r).$$

There is the following relation between differential conservation laws in Lagrangian and Hamiltonian formalisms.

PROPOSITION 4.4.1. Let a Hamiltonian form H be associated with an almost regular Lagrangian L. Let r be a solution of the Hamilton equations (4.1.22a) – (4.1.22b) for H which lives in the Lagrangian constraint space N_L, and $s = \pi_{\Pi Y} \circ r$ the corresponding solution of the Euler–Lagrange equations for L so that the relation (4.2.9) holds. Then, for any projectable vector field u on the fibre bundle $Y \to X$, we have

$$\boxed{\widetilde{\mathfrak{T}}(r) = \mathfrak{T}(\widehat{H} \circ r), \qquad \widetilde{\mathfrak{T}}(\widehat{L} \circ J^1 s) = \mathfrak{T}(s),} \tag{4.4.6}$$

where $\mathfrak{T}$ is the current (3.4.3) on J^1Y and $\widetilde{\mathfrak{T}}$ is the current (4.4.5) on Π. □

In particular, let $u = u^i\partial_i$ be a vertical vector field on $Y \to X$. Given the splitting (4.1.14) of a Hamiltonian form H, the Lie derivative (4.4.2) takes $\mathbf{L}_{\widetilde{u}}H$ takes the form

$$\mathbf{L}_{\widetilde{u}}H = (p_j^\lambda[\partial_\lambda + \Gamma_\lambda^i\partial_i, u]^j - u\rfloor d\widetilde{\mathcal{H}}_\Gamma)\omega,$$

where $[.,.]$ is the Lie bracket of vector fields. The corresponding Noether current (4.4.5) reads

$$\widetilde{\mathfrak{T}}_u = \widetilde{\mathfrak{T}}^\lambda\omega_\lambda = -u\rfloor\mathbf{\Theta}_Y = -u\rfloor H, \qquad \boxed{\widetilde{\mathfrak{T}}^\lambda = -u^ip_i^\lambda.} \tag{4.4.7}$$

Remark 4.4.1. The Noether currents (4.4.7), taken with the sign minus, constitute a Lie algebra with respect to the bracket

$$[-\widetilde{\mathfrak{T}}_u, -\widetilde{\mathfrak{T}}_{u'}] \stackrel{\text{def}}{=} -\widetilde{\mathfrak{T}}_{[u,u']}. \tag{4.4.8}$$

If $Y \to X$ is a vector bundle and X is provided with a non-degenerate metric g, the bracket (4.4.8) can be extended to any horizontal $(n-1)$-forms $\phi = \phi^\lambda \omega_\lambda$ on Π by the law

$$[\phi, \phi'] = g_{\alpha\beta} g^{\mu\nu} (\partial^i_\mu \phi^\alpha \partial_i \phi'^\beta - \partial^i_\mu \phi'^\beta \partial_i \phi) \omega_\nu.$$

•

Let $\tau = \tau^\lambda \partial_\lambda$ be a vector field on X and

$$\tau_\Gamma = \tau^\lambda (\partial_\lambda + \Gamma^i_\lambda \partial_i)$$

its horizontal lift onto Y by means of a connection Γ on $Y \to X$. In this case, the weak identity (4.4.3) takes the form

$$-(\partial_\mu + \Gamma^j_\mu \partial_j - p^\lambda_i \partial_j \Gamma^i_\mu \partial^j_\lambda) \widetilde{\mathcal{H}}_\Gamma + p^\lambda_i R^i_{\lambda\mu} \approx -d_\lambda \widetilde{\mathfrak{T}}_\Gamma{}^\lambda{}_\mu,$$

where the current (4.4.5) reads

$$\boxed{\widetilde{\mathfrak{T}}_\Gamma{}^\lambda = \tau^\mu \widetilde{\mathfrak{T}}_\Gamma{}^\lambda{}_\mu = \tau^\mu (p^\lambda_i \partial^i_\mu \widetilde{\mathcal{H}}_\Gamma - \delta^\lambda_\mu (p^\nu_i \partial^i_\nu \widetilde{\mathcal{H}}_\Gamma - \widetilde{\mathcal{H}}_\Gamma)).} \tag{4.4.9}$$

The relations (4.4.6) show that, on the Lagrangian constraint space N_L, the current (4.4.9) can be treated as the Hamiltonian energy-momentum current relative to the connection Γ.

In particular, let us consider the weak identity (4.4.3) when the vector field $\widetilde{u}$ on Π is the horizontal lift of a vector field τ on X by means of a Hamiltonian connection on $\Pi \to X$ which is associated with the Hamiltonian form H. We have

$$\widetilde{u} = \tau^\mu (\partial_\mu + \partial^i_\mu \mathcal{H} \partial_i + \gamma^\lambda_{\mu i} \partial^i_\lambda).$$

In this case, the corresponding energy-momentum current is

$$\widetilde{\mathfrak{T}}^\lambda = -\tau^\lambda (p^\mu_i \partial^i_\mu \mathcal{H} - \mathcal{H}), \tag{4.4.10}$$

and the weak identity (4.4.3) takes the form

$$-\partial_\mu \mathcal{H} + d_\lambda (p^\lambda_i \partial^i_\mu \mathcal{H}) \approx \partial_\mu (p^\lambda_i \partial^i_\lambda \mathcal{H} - \mathcal{H}). \tag{4.4.11}$$

A glance at the expression (4.4.11) shows that the energy-momentum current (4.4.10) is not conserved, but we can write the weak identity

$$-\partial_\mu \mathcal{H} + d_\lambda[p_i^\lambda \partial_\mu^i \mathcal{H} - \delta_\mu^\lambda (p_i^\nu \partial_\nu^i \mathcal{H} - \mathcal{H})] \approx 0.$$

This is exactly the Hamiltonian form of the canonical energy-momentum conservation law (3.4.13) in Lagrangian formalism.

4.5 The vertical extension of Hamiltonian formalism

The vertical extension of of field theory on a fibre bundle $Y \to X$ to the vertical tangent bundle VY of $Y \to X$ describe linear deviations of fields. It is also a preliminary step toward its SUSY extension in Section 9.6.

Let us start from the vertical extension of Lagrangian formalism. The configuration space of a field theory on $VY \to X$ is the jet manifold J^1VY. Due to the canonical isomorphism $J^1VY = VJ^1Y$ (1.3.11), it is provided with the coordinates $(x^\lambda, y^i, y_\lambda^i, \dot{y}^i, \dot{y}_\lambda^i)$. It follows that Lagrangian formalism on J^1VY can be developed as the vertical extension of Lagrangian formalism on J^1Y. Let L be a Lagrangian on J^1Y. Its prolongation onto the *vertical configuration space* J^1VY can be defined as the vertical tangent morphism

$$L_V = \mathrm{pr}_2 \circ VL : VJ^1Y \to \overset{n}{\wedge} T^*X, \tag{4.5.1}$$
$$\boxed{\mathcal{L}_V = \partial_V \mathcal{L} = (\dot{y}^i \partial_i + \dot{y}_\lambda^i \partial_i^\lambda)\mathcal{L},}$$

to the morphism L (3.0.1). The corresponding Euler–Lagrange equations (3.2.10) read

$$\dot{\delta}_i \mathcal{L}_V = \delta_i \mathcal{L} = 0, \tag{4.5.2a}$$
$$\delta_i \mathcal{L}_V = \partial_V \delta_i \mathcal{L} = 0, \qquad \partial_V = \dot{y}^i \partial_i + \dot{y}_\lambda^i \partial_i^\lambda + \dot{y}_{\mu\lambda}^i \partial_i^{\mu\lambda}. \tag{4.5.2b}$$

The equations (4.5.2a) are exactly the Euler–Lagrange equations for the Lagrangian L. In order to clarify the physical meaning of the equations (4.5.2b), let us suppose that $Y \to X$ is a vector bundle. Given a solution s of the Euler–Lagrange equations (3.2.10), let δs be a *Jacobi field*, i.e., $s + \varepsilon \delta s$ is also a solution of the same Euler–Lagrange equations modulo the terms of order > 1 in the parameter ε. Then it is readily observed that the Jacobi field δs satisfies the Euler–Lagrange equations (4.5.2b).

The momentum phase space of a field theory on VY is the *vertical Legendre bundle*

$$\Pi_{VY} = V^*VY \underset{VY}{\wedge} (\overset{n-1}{\wedge} T^*X).$$

LEMMA 4.5.1. There exists the bundle isomorphism

$$\Pi_{VY} \underset{VY}{\cong} V\Pi, \qquad p_i^\lambda \longleftrightarrow \dot{p}_i^\lambda, \qquad q_i^\lambda \longleftrightarrow p_i^\lambda, \tag{4.5.3}$$

written with respect to the holonomic coordinates $(x^\lambda, y^i, \dot{y}^i, p_i^\lambda, q_i^\lambda)$ on Π_{VY} and $(x^\lambda, y^i, p_i^\lambda, \dot{y}^i, \dot{p}_i^\lambda)$ on $V\Pi$. □

Proof. Similarly to the well-known isomorphism between the fibre bundles TT^*X and T^*TX [174], the isomorphism

$$VV^*Y \underset{VY}{\cong} V^*VY, \quad p_i \longleftrightarrow \dot{v}_i, \quad \dot{p}_i \longleftrightarrow \dot{y}_i,$$

can be established by inspection of the transformation laws of the holonomic coordinates (x^λ, y^i, p_i) on V^*Y and (x^λ, y^i, v^i) on VY. **QED**

It follows that Hamiltonian formalism on the vertical Legendre bundle Π_{VY} can be developed as the vertical extension onto $V\Pi$ of covariant Hamiltonian formalism on Π, where the canonical conjugate pairs are $(y^i, \dot{p}_i^\lambda)$ and $(\dot{y}^i, p_i^\lambda)$. Any Lagrangian (4.5.1) yields the vertical Legendre map

$$\widehat{L}_V = V\widehat{L} : VJ^1Y \underset{VY}{\to} V\Pi, \tag{4.5.4}$$

$$p_i^\lambda = \dot{\partial}_i^\lambda \mathcal{L}_V = \pi_i^\lambda, \qquad \dot{p}_i^\lambda = \partial_V \pi_i^\lambda, \tag{4.5.5}$$

$$\partial_V = \dot{y}^i \partial_i + \dot{p}_i^\lambda \partial_\lambda^i.$$

Due to the isomorphism (4.5.3), $V\Pi$ is endowed with the canonical polysymplectic form (4.1.3) which reads

$$\boxed{\Omega_{VY} = [d\dot{p}_i^\lambda \wedge dy^i + dp_i^\lambda \wedge d\dot{y}^i] \wedge \omega \otimes \partial_\lambda.} \tag{4.5.6}$$

Let Z_{VY} be the homogeneous Legendre bundle (3.2.4) over VY with the corresponding coordinates $(x^\lambda, y^i, \dot{y}^i, p_i^\lambda, q_i^\lambda, p)$. It can be endowed with the canonical form Ξ_{VY} (4.1.11). Sections of the affine bundle

$$Z_{VY} \to V\Pi, \tag{4.5.7}$$

by definition, provide Hamiltonian forms on $V\Pi$. Let us consider the Hamiltonian forms on Π which are related to those on the Legendre bundle Π. Due to the fibre bundle

$$\begin{aligned}&\zeta : VZ_Y \to Z_{VY}, \\ &(x^\lambda, y^i, \dot{y}^i, p_i^\lambda, q_i^\lambda, p) \circ \zeta = (x^\lambda, y^i, \dot{y}^i, \dot{p}_i^\lambda, p_i^\lambda, \dot{p}),\end{aligned} \tag{4.5.8}$$

the vertical tangent bundle VZ_Y of $Z_Y \to X$ is provided with the exterior form

$$\Xi_V = \zeta^*\Xi_{VY} = \dot{p}\omega + (\dot{p}_i^\lambda dy^i + p_i^\lambda d\dot{y}^i) \wedge \omega_\lambda.$$

Given the affine bundle $Z_Y \to \Pi$ (3.2.7), we have the fibre bundle

$$V\pi_{Z\Pi} : VZ_Y \to V\Pi, \tag{4.5.9}$$

where $V\pi_{Z\Pi}$ is the vertical tangent map to $\pi_{Z\Pi}$. The fibre bundles (4.5.7), (4.5.8) and (4.5.9) make up the commutative diagram.

Let h be a section of the affine bundle $Z_Y \to \Pi$ and $H = h^*\Xi$ the corresponding Hamiltonian form (4.1.12) on Π. Then the section Vh of the fibre bundle (4.5.9) and the corresponding section $\zeta \circ Vh$ of the affine bundle (4.5.7) defines the Hamiltonian form

$$\begin{aligned}&\boxed{H_V = (Vh)^*\Xi_V = (\dot{p}_i^\lambda dy^i + p_i^\lambda d\dot{y}^i) \wedge \omega_\lambda - \mathcal{H}_V\omega,} \\ &\mathcal{H}_V = \partial_V\mathcal{H} = (\dot{y}^i\partial_i + \dot{p}_i^\lambda\partial_\lambda^i)\mathcal{H},\end{aligned} \tag{4.5.10}$$

on $V\Pi$. It is called the *vertical extension* of H. In particular, given the splitting (4.1.14) of H with respect to a connection Γ on $Y \to X$, we have the corresponding splitting

$$\mathcal{H}_V = \dot{p}_i^\lambda\Gamma_\lambda^i + \dot{y}^j p_i^\lambda \partial_j \Gamma_\lambda^i + \partial_V\widetilde{\mathcal{H}}_\Gamma$$

of H_V with respect to the vertical connection $V\Gamma$ (2.7.18) on $VY \to X$.

PROPOSITION 4.5.2. Let γ (4.1.4) be a Hamiltonian connection on Π associated with a Hamiltonian form H. Then its vertical prolongation $V\gamma$ (2.7.18) on $V\Pi \to X$ is a Hamiltonian connection associated with the vertical Hamiltonian form H_V (4.5.10). □

Proof. The proof follows from a direct computation. We have

$$V\gamma = \gamma + dx^\mu \otimes [\partial_V\gamma_\mu^i\dot{\partial}_i + \partial_V\gamma_{\mu i}^\lambda\dot{\partial}_\lambda^i]. \tag{4.5.11}$$

Components of this connection obey the Hamilton equations (4.1.19) and the equations

$$\dot\gamma^i_\mu = \partial^i_\mu \mathcal{H}_V = \partial_V \partial^i_\mu \mathcal{H}, \qquad \dot\gamma^\lambda_{\lambda i} = -\partial_i \mathcal{H}_V = -\partial_V \partial_i \mathcal{H}. \tag{4.5.12}$$

QED

The Hamiltonian form H_V (4.5.10) defines the Lagrangian L_{H_V} (4.1.20) on $J^1V\Pi$ which takes the form

$$\boxed{\mathcal{L}_{H_V} = h_0(H_V) = \dot p^\lambda_i (y^i_\lambda - \partial^i_\lambda \mathcal{H}) - \dot y^i (p^\lambda_{\lambda i} + \partial_i \mathcal{H}) + d_\lambda (p^\lambda_i \dot y^i).} \tag{4.5.13}$$

The corresponding Hamilton equations contain the Hamilton equations (4.1.22a) – (4.1.22b) and the equations

$$\dot y^i_\lambda = \partial^i_\lambda \mathcal{H}_V = \partial_V \partial^i_\lambda \mathcal{H}, \qquad \dot p^\lambda_{\lambda i} = -\partial_i \mathcal{H}_V = -\partial_V \partial_i \mathcal{H}$$

for Jacobi fields $\delta y^i = \varepsilon \dot y^i$, $\delta p^\lambda_i = \varepsilon \dot p^\lambda_i$.

In conclusion, let us study the relationship between the vertical extensions of Lagrangian and Hamiltonian formalisms. The Hamiltonian form H_V (4.5.10) on $V\Pi$ yields the vertical Hamiltonian map

$$\widehat{H}_V = V\widehat{H} : V\Pi \underset{VY}{\to} VJ^1Y, \tag{4.5.14}$$

$$y^i_\lambda = \dot\partial^i_\lambda \mathcal{H}_V = \partial^i_\lambda \mathcal{H}, \qquad \dot y^i_\lambda = \partial_V \partial^i_\lambda \mathcal{H}. \tag{4.5.15}$$

PROPOSITION 4.5.3. Let H on Π be a Hamiltonian form associated with a Lagrangian L on J^1Y. Then the vertical Hamiltonian form H_V (4.5.10) is weakly associated with the Lagrangian L_V (4.5.1). □

Proof. If the morphisms $\widehat{H}$ and $\widehat{L}$ obey the relation (4.2.2a), then the corresponding vertical tangent morphisms satisfy the relation

$$V\widehat{L} \circ V\widehat{H} \circ V\widehat{L} = V\widehat{L}.$$

The condition (4.2.2b) for H_V reduces to the equality (4.2.6) which is fulfilled if H is associated with L.

QED

Chapter 5

Connections in classical mechanics

The technique of Poisson and symplectic manifolds is well known to provide the adequate Hamiltonian formulation of conservative mechanics. This formulation, however, cannot be extended to time-dependent mechanics because the standard symplectic form

$$\Omega = dp_i \wedge dq^i \tag{5.0.1}$$

is not invariant under time-dependent transformations of canonical coordinates and momenta

$$q^i \mapsto q'^i(t, q^j, p_j), \qquad p_i \mapsto p'_i(t, q^j, p_j),$$

including inertial frame transformations. Non-relativistic time-dependent mechanics can be formulated as a particular field theory on fibre bundles $Q \to \mathbb{R}$ over a time axis $\mathbb{R}$ [121, 190, 199, 213, 216, 271]. At the same time, there is the essential difference between field theory and time-dependent mechanics. In contrast with gauge potentials in field theory, connections on a configuration bundle $Q \to \mathbb{R}$ of time-dependent mechanics fail to be dynamic variables since their curvature vanishes identically. They characterize non-relativistic reference frames (see Section 5.5). Connections play a prominent role in the formulation of time-dependent mechanics. In particular, dynamic equations of time-dependent mechanics are given by connections.

Throughout this Chapter,

$$\pi : Q \to \mathbb{R} \tag{5.0.2}$$

is a fibre bundle whose typical fibre M is an m-dimensional manifold. This is not the case of relativistic mechanics whose configuration space does not imply any preferable fibration over a time (see Section 7.3). A fibre bundle $Q \to \mathbb{R}$ is endowed with bundle coordinates (t, q^i) where t is a Cartesian coordinate on $\mathbb{R}$ with the transition functions $t' = t+$ const. The base $\mathbb{R}$ is provided with the standard vector field ∂_t and the standard 1-form dt which are invariant under the coordinate transformations $t' = t+$ const. The same symbol dt also stands for any pull-back of the standard 1-form dt onto fibre bundles over $\mathbb{R}$. For the sake of convenience, we also use the compact notation (q^λ) where $q^0 = t$.

5.1 Fibre bundles over $\mathbb{R}$

In this Section, we point out some peculiarities of fibre bundles over $\mathbb{R}$. The most important one is that connections on these fibre bundles are represented by vector fields.

Since $\mathbb{R}$ is contractible, any fibre bundle over $\mathbb{R}$ is trivial. Different trivializations

$$\psi : Q \cong \mathbb{R} \times M \tag{5.1.1}$$

differ from each other in the projections $Q \to M$, while the fibration $Q \to \mathbb{R}$ is one for all.

Let J^1Q be the first order jet manifold of a fibre bundle $Q \to \mathbb{R}$ (5.0.2). It is provided with the adapted coordinates (t, q^i, q^i_t). Every trivialization (5.1.1) yields the corresponding trivialization of the jet manifold

$$J^1Q \cong \mathbb{R} \times TM. \tag{5.1.2}$$

The canonical imbedding (1.3.5) of J^1Q takes the form

$$\begin{aligned} &\lambda_1 : J^1Q \hookrightarrow TQ, \\ &\boxed{\lambda_1 : (t, q^i, q^i_t) \mapsto (t, q^i, \dot{t} = 1, \dot{q}^i = q^i_t),} \\ &\lambda_1 = d_t = \partial_t + q^i_t \partial_i. \end{aligned} \tag{5.1.3}$$

Following Remark 1.3.1, we will identify the jet manifold J^1Q with its image in TQ. In particular, the jet prolongation Jc of a section $c : \mathbb{R} \to Q$ can be identified with the tangent prolongation $\dot{c}$ (1.2.1) of the curve c.

A glance at the morphism λ_1 (5.1.3) shows that the affine jet bundle $J^1Q \to Q$ is modelled over the vertical tangent bundle VQ of the fibre bundle of $Q \to \mathbb{R}$.

As a consequence, we have the following canonical splitting (1.1.14) of the vertical tangent bundle V_QJ^1Q of the affine jet bundle $J^1Q \to Q$:

$$\alpha : V_QJ^1Q \cong J^1Q \underset{Q}{\times} VQ, \qquad \alpha(\partial_i^t) = \partial_i, \tag{5.1.4}$$

together with the corresponding splitting of the vertical cotangent bundle $V_Q^*J^1Q$ of $J^1Q \to Q$:

$$\alpha^* : V_Q^*J^1Q \cong J^1Q \underset{Q}{\times} V^*Q, \qquad \alpha^*(\overline{d}q_t^i) = \overline{d}q^i, \tag{5.1.5}$$

where $\overline{d}q_t^i$ and $\overline{d}q^i$ are the holonomic bases for $V_Q^*J^1Q$ and V^*Q, respectively.

There is the following endomorphism, called the *vertical endomorphism*, of the tangent bundle TJ^1Q:

$$\begin{gathered} \widehat{v} : TJ^1Q \to TJ^1Q, \\ \widehat{v}(\partial_t) = -q_t^i\partial_i^t, \qquad \widehat{v}(\partial_i) = \partial_i^t, \qquad \widehat{v}(\partial_i^t) = 0. \end{gathered} \tag{5.1.6}$$

This endomorphism obeys the nilpotency rule $\widehat{v} \circ \widehat{v} = 0$. The transpose of the vertical endomorphism $\widehat{v}$ (5.1.6) is

$$\begin{gathered} \widehat{v}^* : T^*J^1Q \to T^*J^1Q, \\ \boxed{\widehat{v}^*(dt) = 0, \qquad \widehat{v}^*(dq^i) = 0, \qquad \widehat{v}^*(dq_t^i) = \theta^i,} \end{gathered} \tag{5.1.7}$$

where $\theta^i = dq^i - q_t^i dt$ are the contact forms (1.3.7).

In view of the morphism λ_1 (5.1.3), any connection

$$\Gamma = dt \otimes (\partial_t + \Gamma^i\partial_i) \tag{5.1.8}$$

on a fibre bundle $Q \to \mathbb{R}$ can be identified with a nowhere vanishing *horizontal vector field*

$$\boxed{\Gamma = \partial_t + \Gamma^i\partial_i} \tag{5.1.9}$$

on Q which is the horizontal lift $\Gamma\partial_t$ (2.1.6) of the standard vector field ∂_t on $\mathbb{R}$ by means of the connection (5.1.8). Conversely, any vector field Γ on Q such that $dt\rfloor\Gamma = 1$ defines a connection on $Q \to \mathbb{R}$. As a consequence, connections on a fibre bundle $Q \to \mathbb{R}$ constitute an affine space modelled over the vector space of vertical vector fields on $Q \to \mathbb{R}$. Accordingly, the covariant differential (2.2.7) associated with a connection Γ on $Q \to \mathbb{R}$ takes its values into the vertical tangent bundle VQ of $Q \to \mathbb{R}$:

$$D_\Gamma : J^1Q \underset{Q}{\to} VQ, \qquad \dot{q}^i \circ D_\Gamma = q^i_t - \Gamma^i. \tag{5.1.10}$$

A connection Γ (5.1.8) is obviously flat. By virtue of Proposition 2.6.2, it defines an atlas of local constant trivializations of $Q \to \mathbb{R}$ such that the associated bundle coordinates (t, q^i) on Q possess the transition function $q^i \to q'^i(q^j)$ independent of t, and $\Gamma = \partial_t$ with respect to these coordinates. Conversely, every atlas of local constant trivializations of the fibre bundle $Q \to \mathbb{R}$ determines a connection on $Q \to \mathbb{R}$ which is equal to ∂_t relative to this atlas.

A connection Γ on a fibre bundle $Q \to \mathbb{R}$ is said to be *complete* if the horizontal vector field (5.1.9) is complete.

PROPOSITION 5.1.1. [213]. Every trivialization of a fibre bundle $Q \to \mathbb{R}$ yields a complete connection on this fibre bundle. Conversely, every complete connection Γ on $Q \to \mathbb{R}$ defines its trivialization (5.1.1) such that the vector field (5.1.9) equals ∂_t relative to the bundle coordinates associated with this trivialization. □

Let J^1J^1Q be the repeated jet manifold of a fibre bundle $Q \to \mathbb{R}$, provided with the adapted coordinates $(t, q^i, q^i_t, q^i_{(t)}, q^i_{tt})$. For a fibre bundle $Q \to \mathbb{R}$, we have the canonical isomorphism k between the affine fibrations π_{11} (1.3.12) and $J^1\pi^1_0$ (1.3.13) of J^1J^1Q over J^1Q (see Remark 3.1.1), i.e.,

$$\pi_{11} \circ k = J^1_0\pi_{01}, \qquad k \circ k = \mathrm{Id}\, J^1J^1Q,$$

where

$$q^i_t \circ k = q^i_{(t)}, \qquad q^i_{(t)} \circ k = q^i_t, \qquad q^i_{tt} \circ k = q^i_{tt}. \tag{5.1.11}$$

By $J^1_QJ^1Q$ throughout is meant the first order jet manifold of the affine jet bundle $J^1Q \to Q$, equipped with the adapted coordinates $(q^\lambda, q^i_t, q^i_{\lambda t})$.

For a fibre bundle $Q \to \mathbb{R}$, the sesquiholonomic jet manifold $\widehat{J}^2Q$ coincides with the second order jet manifold J^2Q, coordinated by $(t, q^i, q^i_t, q^i_{tt})$. The affine bundle $J^2Q \to J^1Q$ is modelled over the vertical tangent bundle

$$V_QJ^1Q \cong J^1Q \underset{Q}{\times} VQ \to J^1Q \tag{5.1.12}$$

of the affine jet bundle $J^1Q \to Q$. There are the imbeddings

$$J^2Q \overset{\lambda_2}{\hookrightarrow} TJ^1Q \overset{T\lambda_1}{\hookrightarrow} V_QTQ \cong T^2Q \subset TTQ,$$

$$\lambda_2 : (t, q^i, q^i_t, q^i_{tt}) \mapsto (t, q^i, q^i_t, \dot{t} = 1, \dot{q}^i = q^i_t, \dot{q}^i_t = q^i_{tt}), \tag{5.1.13}$$

$$T\lambda_1 \circ \lambda_2 : (t, q^i, q^i_t, q^i_{tt}) \mapsto (t, q^i, \dot{t} = \mathrm{\dot{t}} = 1, \dot{q}^i = \mathrm{\dot{q}}^i = q^i_t, \ddot{t} = 0, \ddot{q}^i = q^i_{tt}), \tag{5.1.14}$$

where $(t, q^i, \dot{t}, \dot{q}^i, \mathrm{\dot{t}}, \mathrm{\dot{q}}^i, \ddot{t}, \ddot{q}^i)$ are the holonomic coordinates on the double tangent bundle TTQ, by V_QTQ is meant the vertical tangent bundle of $TQ \to Q$, and $T^2Q \subset TTQ$ is a second order tangent space, given by the coordinate relation $\dot{t} = \mathrm{\dot{t}}$.

Due to the imbedding (5.1.13), any connection ξ on the jet bundle $J^1Q \to \mathbb{R}$ (i.e., a second order connection on $Q \to \mathbb{R}$) is represented by a horizontal vector field on J^1Q such that $\xi\rfloor dt = 1$. Every connection Γ on a fibre bundle $Q \to \mathbb{R}$ has the jet prolongation to the section $J^1\Gamma$ of the affine bundle $J^1\pi^1_0$ and, by virtue of the isomorphism k (5.1.11), gives rise to the connection

$$J\Gamma \overset{\mathrm{def}}{=} k \circ J^1\Gamma : J^1Q \to J^1J^1Q,$$

$$\boxed{J\Gamma = \partial_t + \Gamma^i\partial_i + d_t\Gamma^i\partial^t_i,} \tag{5.1.15}$$

on the jet bundle $J^1Q \to \mathbb{R}$. It is holonomic if

$$\boxed{\xi = dt \otimes (\partial_t + q^i_t\partial_i + \xi^i\partial^t_i).}$$

In view of the imbedding (5.1.13), a holonomic second order connection on $Q \to \mathbb{R}$ is represented by a horizontal vector field

$$\boxed{\xi = \partial_t + q^i_t\partial_i + \xi^i\partial^t_i} \tag{5.1.16}$$

on J^1Q. Conversely, every vector field ξ on J^1Q which fulfills the conditions

$$dt\rfloor\xi = 1, \qquad \widehat{v}(\xi) = 0,$$

where $\widehat{v}$ is the vertical endomorphism (5.1.6), is a holonomic connection on the jet bundle $J^1Q \to \mathbb{R}$. As a consequence, holonomic connections (5.1.16) make up an affine space modelled over the linear space of vertical vector fields on the affine jet bundle $J^1Q \to Q$. Therefore, the covariant differential (5.1.10) relative to a holonomic connection

$$D_\xi : J^1J^1Q \underset{J^1Q}{\longrightarrow} V_QJ^1Q \subset VJ^1Q,$$

$$\dot{q}^i \circ D_\xi = 0, \qquad \dot{q}^i_t \circ D_\xi = q^i_{tt} - \xi^i,$$

takes its values into the vertical tangent bundle V_QJ^1Q of the affine jet bundle $J^1Q \to Q$.

5.2 Connections in conservative mechanics

Conservative mechanics is characterized by first order dynamic equations on a momentum phase space in Hamiltonian mechanics and by second order dynamic equations on a configuration space in Lagrangian and Newtonian mechanics.

A generic example of conservative Hamiltonian mechanics is a regular Poisson manifold (Z, w) and a Hamiltonian $\mathcal{H}$ seen as a real function on Z (by w is meant a Poisson bivector). Given the corresponding Hamiltonian vector field $\vartheta_{\mathcal{H}} = w^\sharp(df)$, the closed subbundle $\vartheta_{\mathcal{H}}(Z)$ of the tangent bundle TZ is a first order dynamic equation on a manifold Z (see Definition 1.3.3), called the Hamilton equations. This is also the case of presymplectic Hamiltonian systems. Since every presymplectic form can be represented as a pull-back of a symplectic form by a coisotropic imbedding [132, 213], a presymplectic Hamiltonian system can be seen as a Dirac constraint system [50, 213]. An autonomous Lagrangian system also exemplifies a presymplectic Hamiltonian system where a presymplectic form is the exterior differential of the Poincaré–Cartan form, while a Hamiltonian is the energy function [51, 198, 213, 229].

Lagrangian conservative mechanics implies the existence of a configuration manifold M of a mechanical system. In this case, the momentum phase space is the cotangent bundle T^*M, while the velocity phase space is the tangent bundle TM. Let us consider second order dynamic equations on a configuration manifold M and their relations with connections on the tangent bundle $TM \to M$ (see [142, 196, 213, 226]).

DEFINITION 5.2.1. An *autonomous second order dynamic equation* on a manifold M is defined as a first order dynamic equation (1.3.25) on the tangent bundle TM which is associated with a *holonomic vector field*

$$\boxed{\Xi = \dot{q}^i\partial_i + \Xi^i(q^j, \dot{q}^j)\dot{\partial}_i} \tag{5.2.1}$$

on TM. This vector field, by definition, obeys the condition $J(\Xi) = u_{TM}$, where J is the endomorphism (1.2.29) and u_{TM} is the Liouville vector field (1.2.7) on TM. □

With respect to the holonomic coordinates $(q^i, \dot{q}^i, \dot{\mathrm{q}}^i, \ddot{q}^i)$ on the double tangent bundle TTM, the second order dynamic equation given by the holonomic vector field Ξ (5.2.1) reads

$$\boxed{\dot{\mathrm{q}}^i = \dot{q}^i, \qquad \ddot{q}^i = \Xi^i(q^j, \dot{q}^j).} \tag{5.2.2}$$

By its classical *solutions* are meant the curves $c : () \to M$ in a manifold M whose tangent prolongations $\dot{c} : () \to TM$ (1.2.1) are integral curves of the holonomic vector field Ξ or, equivalently, whose second order tangent prolongations $\ddot{c}$ live in the subbundle (5.2.2). They satisfy the equations

$$\ddot{c}^i(t) = \Xi(c^j(t), \dot{c}^j(t)).$$

A particular second order dynamic equation on a manifold M is a geodesic equation on the tangent bundle TM. Given a connection

$$K = dq^j \otimes (\partial_j + K^i_j \dot{\partial}_i)$$

on the tangent bundle $TM \to M$, let

$$\widehat{K} : TM \underset{M}{\times} TM \to TTM \tag{5.2.3}$$

be the corresponding linear bundle morphism (2.1.1) over TM which splits the exact sequence

$$0 \longrightarrow V_M TM \hookrightarrow TTM \longrightarrow TM \underset{M}{\times} TM \longrightarrow 0.$$

DEFINITION 5.2.2. A *geodesic equation* on TM with respect to the connection K is defined as the image

$$\dot{\mathrm{q}}^i = \dot{q}^i, \qquad \ddot{q}^i = K^i_j \dot{q}^j \tag{5.2.4}$$

of the morphism (5.2.3) restricted to the diagonal $TM \subset TM \times TM$. □

By a solution of a geodesic equation on TM is meant a *geodesic line* c in M, whose tangent prolongation $\dot{c}$ is an integral section (a *geodesic vector field*) over $c \subset M$ for the connection K.

It is readily observed that the morphism $\widehat{K}\,|_{TM}$ is a holonomic vector field on TM. It follows that any geodesic equation (5.2.3) on TM is a second order equation on M. The converse is not true in general. There is the following theorem.

THEOREM 5.2.3. [226]. Every second order dynamic equation (5.2.2) on a manifold M defines a connection K_Ξ on the tangent bundle $TM \to M$ whose components are

$$K^i_j = \frac{1}{2} \dot{\partial}_j \Xi^i. \tag{5.2.5}$$

□

However, the second order dynamic equation (5.2.2) fails to be a geodesic equation with respect to the connection (5.2.5) in general. In particular, the geodesic equation (5.2.4) with respect to a connection K determines the connection (5.2.5) on $TM \to M$ which does not necessarily coincide with K. A second order equation Ξ on M is a geodesic equation for the connection (5.2.5) if and only if

$$\Xi^i = a^i_{kj}(q)\dot{q}^k\dot{q}^j$$

is a *spray*, i.e., $[u_{TM}, \Xi] = \Xi$, where u_{TM} is the Liouville vector field (1.2.7) on TM. In this case, the connection K (5.2.5) is a symmetric linear connection (2.4.7) on $TM \to M$.

In the next Section, we will improve Theorem 5.2.3 (see Proposition 5.3.4 below).

5.3 Dynamic connections in time-dependent mechanics

Turn now to second order dynamic equations in time-dependent mechanics on the configuration bundle $Q \to \mathbb{R}$, coordinated by (t, q^i).

DEFINITION 5.3.1. In accordance with Definition 3.1.2, a second order dynamic equation on a fibre bundle $Q \to \mathbb{R}$ is the kernel of the covariant differential D_ξ corresponding to a holonomic second order connection ξ (5.1.16) on $Q \to \mathbb{R}$. This is a closed subbundle of the second order jet bundle $J^2Q \to \mathbb{R}$, given by the coordinate relations

$$\boxed{q^i_{tt} = \xi^i(t, q^j, q^j_t).} \tag{5.3.1}$$

Throughout this Chapter, we will call (5.3.1) simply a *dynamic equation* if there is no danger of confusion. The corresponding horizontal vector field ξ (5.1.16) is also termed a dynamic equation. □

A solution of the dynamic equation (5.3.1), called a *motion*, is a curve c in Q whose second order jet prolongation $\ddot{c}$ lives in (5.3.1). It is clear that any integral section $\overline{c}$ for the holonomic connection ξ is the jet prolongation $\dot{c}$ of a solution c of the dynamic equation (5.3.1), i.e.,

$$\ddot{c}^i = \xi^i \circ \dot{c}, \tag{5.3.2}$$

and *vice versa.*

One can easily find the transformation law

$$q'^i_{tt} = \xi'^i, \qquad \boxed{\xi'^i = (\xi^j \partial_j + q^j_t q^k_t \partial_j \partial_k + 2q^j_t \partial_j \partial_t + \partial^2_t) q'^i(t, q^j),} \tag{5.3.3}$$

of a second order dynamic equation under fibred coordinate transformations $q^i \to q'^i(t, q^j)$.

Remark 5.3.1. There are the following relations between second order dynamic equations on fibre bundles and autonomous second order dynamic equations on manifolds. A dynamic equation ξ on a fibre bundle $Q \to \mathbb{R}$ is said to be *conservative* if there exists a trivialization (5.1.1) of Q and the corresponding trivialization (5.1.2) of J^1Q such that the vector field ξ (5.1.16) on J^1Q is projectable over M. Then this projection

$$\Xi_\xi = \dot{q}^i \partial_i + \xi^i(q^j, \dot{q}^j) \dot{\partial}_i$$

is an autonomous second order dynamic equation on the typical fibre M of $Q \to \mathbb{R}$ in accordance with Definition 5.2.1. Conversely, every autonomous second order dynamic equation Ξ on a manifold M can be seen as a conservative dynamic equation

$$\xi_\Xi = \partial_t + \dot{q}^i \partial_i + u^i \dot{\partial}_i \tag{5.3.4}$$

on the fibre bundle $\mathbb{R} \times M \to \mathbb{R}$ in accordance with the isomorphism (5.1.2).

One can also show [213] that any dynamic equation (5.3.1) given by a holonomic second order connection ξ (5.1.16) on a fibre bundle $Q \to \mathbb{R}$ is equivalent to the autonomous second order dynamic equation

$$\ddot{t} = 0, \qquad \dot{t} = 1, \qquad \ddot{q}^i = \Xi^i, \tag{5.3.5}$$

on a manifold Q defined by the holonomic vector field Ξ on TQ which satisfies the relations

$$\Xi^0 = 0, \qquad \xi^i = \Xi^i(t, q^j, \dot{t} = 1, \dot{q}^j = q^j_t).$$

This vector field Ξ makes the diagram

$$\begin{array}{ccc} J^2Q & \longrightarrow & T^2Q \\ {\scriptstyle \xi} \uparrow & & \uparrow {\scriptstyle \Xi} \\ J^1Q & \xrightarrow{\lambda_1} & TQ \end{array}$$

commutative (see the morphism (5.1.14)). Theorem 5.4.2 below will improve this result. •

The fact that ξ is a curvature-free connection places a limit on the geometric analysis of dynamic equations by holonomic second order connections. Therefore, we will consider the relationship between the holonomic connections on the jet bundle $J^1Q \to \mathbb{R}$ and the connections on the affine jet bundle $J^1Q \to Q$. We aim to show that any connection on $J^1Q \to Q$ defines a dynamic equation, and *vice versa* [68, 196, 213].

Let $\gamma : J^1Q \to J^1_QJ^1Q$ be a connection on the affine jet bundle $J^1Q \to Q$. It takes the coordinate form

$$\gamma = dq^\lambda \otimes (\partial_\lambda + \gamma^i_\lambda \partial^t_i) \tag{5.3.6}$$

with the transformation law

$$\gamma'^i_\lambda = (\partial_j q'^i \gamma^j_\mu + \partial_\mu q'^i_t)\frac{\partial q^\mu}{\partial q'^\lambda}. \tag{5.3.7}$$

Let us consider the composite fibre bundle

$$J^1Q \to Q \to \mathbb{R} \tag{5.3.8}$$

and the morphism ϱ (2.7.6) which reads

$$\varrho : J^1_QJ^1Q \ni (q^\lambda, q^i_t, q^i_{\lambda t}) \mapsto (q^\lambda, q^i_t, q^i_{(t)} = q^i_t, q^i_{tt} = q^i_{0t} + q^j_t q^i_{jt}) \in J^2Q. \tag{5.3.9}$$

PROPOSITION 5.3.2. Any connection γ (5.3.6) on the affine jet bundle $J^1Q \to Q$ defines the holonomic connection

$$\xi_\gamma = \varrho\circ\gamma : J^1Q \to J^1_QJ^1Q \to J^2Q, \tag{5.3.10}$$
$$\xi_\gamma = \partial_t + q^i_t\partial_i + (\gamma^i_0 + q^j_t\gamma^i_j)\partial^t_i,$$

on the jet bundle $J^1Q \to \mathbb{R}$. □

It follows that every connection γ (5.3.6) on the affine jet bundle $J^1Q \to Q$ yields the dynamic equation

$$\boxed{q^i_{tt} = \gamma^i_0 + q^j_t\gamma^i_j} \tag{5.3.11}$$

on the configuration bundle $Q \to \mathbb{R}$. This is precisely the restriction to J^2Q of the kernel Ker $\widetilde{D}_\gamma$ of the vertical covariant differential $\widetilde{D}_\gamma$ (2.7.15) relative the connection γ:

$$\widetilde{D}_\gamma : J^1J^1Q \to V_QJ^1Q,$$
$$\dot{q}^i_t \circ \widetilde{D}_\gamma = q^i_{tt} - \gamma^i_0 - q^j_t\gamma^i_j. \tag{5.3.12}$$

Therefore, connections on the jet bundle $J^1Q \to Q$ are called the *dynamic connections*. The corresponding equation (5.3.2) can be written in the form

$$\ddot{c}^i = \varrho \circ \gamma \circ \dot{c},$$

where ϱ is the morphism (5.3.9). Of course, different dynamic connections can lead to the same dynamic equation (5.3.11).

A converse of Proposition 5.3.2 is the following.

PROPOSITION 5.3.3. Any holonomic connection ξ (5.1.16) on the jet bundle $J^1Q \to \mathbb{R}$ defines the dynamic connection

$$\boxed{\gamma_\xi = dt \otimes [\partial_t + (\xi^i - \tfrac{1}{2}q^j_t\partial^t_j\xi^i)\partial^t_i] + dq^j \otimes [\partial_j + \tfrac{1}{2}\partial^t_j\xi^i\partial^t_i]} \tag{5.3.13}$$

on the affine jet bundle $J^1Q \to Q$. □

It is readily observed that the dynamic connection γ_ξ (5.3.13) possesses the property

$$\boxed{\gamma^k_i = \partial^t_i\gamma^k_0 + q^j_t\partial^t_i\gamma^k_j} \tag{5.3.14}$$

which implies the relation $\partial^t_j\gamma^k_i = \partial^t_i\gamma^k_j$. Therefore, a dynamic connection γ obeying the condition (5.3.14) is said to be *symmetric*. The *torsion of a dynamic connection* γ is defined as the tensor field

$$T = T^k_i\overline{d}q^i \otimes \partial_k : J^1Q \to V^*Q \underset{Q}{\otimes} VQ,$$
$$T^k_i = \gamma^k_i - \partial^t_i\gamma^k_0 - q^j_t\partial^t_i\gamma^k_j. \tag{5.3.15}$$

It follows at once that a dynamic connection is symmetric if and only if its torsion vanishes.

Let γ be a dynamic connection (5.3.6) and ξ_γ the corresponding dynamic equation (5.3.10). Then the dynamic connection (5.3.13) associated with the dynamic equation ξ_γ takes the form

$$\gamma_{\xi_\gamma}{}_i^k = \frac{1}{2}(\gamma_i^k + \partial_i^t\gamma_0^k + q_t^j\partial_i^t\gamma_j^k), \qquad \gamma_{\xi_\gamma}{}_0^k = \xi^k - q_t^i\gamma_{\xi_\gamma}{}_i^k.$$

It is readily observed that $\gamma = \gamma_{\xi_\gamma}$ if and only if the torsion T (5.3.15) of the dynamic connection γ vanishes.

Example 5.3.2. The affine jet bundle $J^1Q \to Q$ admits an affine connection

$$\gamma = dq^\lambda \otimes [\partial_\lambda + (\gamma_{\lambda 0}^i(q^\mu) + \gamma_{\lambda j}^i(q^\mu)q_t^j)\partial_i^t]. \tag{5.3.16}$$

This connection is symmetric if and only if $\gamma_{\lambda\mu}^i = \gamma_{\mu\lambda}^i$. One can easily justify that an affine dynamic connection generates a quadratic dynamic equation, and *vice versa*. A non-affine dynamic connection whose symmetric part is affine also defines a quadratic dynamic equation. •

Using the notion of a dynamic connection, we can modify Theorem 5.2.3 as follows. Let Ξ be an autonomous second order dynamic equation on a manifold M, and ξ_Ξ (5.3.4) the corresponding conservative dynamic equation on the fibre bundle $\mathbb{R}\times M \to \mathbb{R}$. The latter yields the dynamic connection γ (5.3.13) on the fibre bundle $\mathbb{R}\times TM \to \mathbb{R}\times M$. Its components γ_j^i are exactly those of the connection (5.2.5) on the tangent bundle $TM \to M$ in Theorem 5.2.3, while γ_0^i make up a vertical vector field

$$e = \gamma_0^i\dot{\partial}_i = (\Xi^i - \frac{1}{2}\dot{q}^j\dot{\partial}_j\Xi^i)\dot{\partial}_i \tag{5.3.17}$$

on $TM \to M$. Thus, we have shown the following.

PROPOSITION 5.3.4. Every autonomous second order dynamic equation Ξ (5.2.2) on a manifold M admits the decomposition

$$\Xi^i = K_j^i\dot{q}^j + e^i$$

where K is the connection (5.2.5) on the tangent bundle $TM \to M$, and e is the vertical vector field (5.3.17) on $TM \to M$. □

Remark 5.3.3. Every dynamic equation ξ on Q and the corresponding dynamic connection γ_ξ (5.3.13) also define a linear connection on the tangent bundle $TJ^1Q \to J^1Q$ [68, 213, 216]. •

5.4 Non-relativistic geodesic equations

In this Section, we aim to show that every dynamic equation in non-relativistic mechanics on a configuration bundle $Q \to \mathbb{R}$ is equivalent to a geodesic equation on the tangent bundle $TQ \to Q$. Treated in such a way, non-relativistic dynamic equations can be examined by means of the standard geometric methods (see Section 5.9).

We start from the relation between the dynamic connections γ on the affine jet bundle $J^1Q \to Q$ and the connections

$$K = dq^\lambda \otimes (\partial_\lambda + K^\mu_\lambda \dot{\partial}_\mu) \tag{5.4.1}$$

on the tangent bundle $TQ \to Q$. Let us consider the diagram

$$\begin{array}{ccc} J^1_Q J^1 Q & \xrightarrow{J^1\lambda_1} & J^1_Q TQ \\ {\scriptstyle\gamma}\uparrow & & \uparrow{\scriptstyle K} \\ J^1Q & \xrightarrow{\lambda_1} & TQ \end{array} \tag{5.4.2}$$

where $J^1_Q TQ$ is the first order jet manifold of the tangent bundle $TQ \to Q$, coordinated by $(t, q^i, \dot{t}, \dot{q}^i, (\dot{t})_\mu, (\dot{q}^i)_\mu)$. The jet prolongation over Q of the canonical imbedding λ_1 (5.1.3) reads

$$J^1\lambda_1 : (t, q^i, q^i_t, q^i_{\mu t}) \mapsto (t, q^i, \dot{t} = 1, \dot{q}^i = q^i_t, (\dot{t})_\mu = 0, (\dot{q}^i)_\mu = q^i_{\mu t}).$$

Then we have

$$J^1\lambda_1 \circ \gamma : (t, q^i, q^i_t) \mapsto (t, q^i, \dot{t} = 1, \dot{q}^i = q^i_t, (\dot{t})_\mu = 0, (\dot{q}^i)_\mu = \gamma^i_\mu),$$
$$K \circ \lambda_1 : (t, q^i, q^i_t) \mapsto (t, q^i, \dot{t} = 1, \dot{q}^i = q^i_t, (\dot{t})_\mu = K^0_\mu, (\dot{q}^i)_\mu = K^i_\mu).$$

It follows that the diagram (5.4.2) can be commutative only if the components K^0_μ of the connection K (5.4.1) on the tangent bundle $TQ \to Q$ vanish.

Since the transition functions $t \to t'$ are independent of q^i, a connection

$$\widetilde{K} = dq^\lambda \otimes (\partial_\lambda + K^i_\lambda \dot{\partial}_i) \tag{5.4.3}$$

with $K^0_\mu = 0$ can exist on the tangent bundle $TQ \to Q$ in accordance with the transformation law

$$K'^i_\lambda = (\partial_j q'^i K^j_\mu + \partial_\mu \dot{q}'^i)\frac{\partial q^\mu}{\partial q'^\lambda}. \tag{5.4.4}$$

Now the diagram (5.4.2) becomes commutative if the connections γ and $\widetilde{K}$ fulfill the relation

$$\boxed{\gamma^i_\mu = K^i_\mu \circ \lambda_1 = K^i_\mu(t, q^i, \dot{t} = 1, \dot{q}^i = q^i_t).} \tag{5.4.5}$$

It is easily seen that this relation holds globally because the substitution of $\dot{q}^i = q^i_t$ in (5.4.4) restates the transformation law (5.3.7) of a connection on the affine jet bundle $J^1Q \to Q$. In accordance with the relation (5.4.5), the desired connection $\widetilde{K}$ is an extension of the section $J^1\lambda_1 \circ \gamma$ of the affine jet bundle $J^1_Q TQ \to TQ$ over the closed submanifold $J^1Q \subset TQ$ to a global section. Such an extension always exists, but is not unique. Thus, we have proved the following.

PROPOSITION 5.4.1. In accordance with the relation (5.4.5), every dynamic equation on a configuration bundle $Q \to \mathbb{R}$ can be written in the form

$$\boxed{q^i_{tt} = K^i_0 \circ \lambda_1 + q^j_t K^i_j \circ \lambda_1,} \tag{5.4.6}$$

where $\widetilde{K}$ is a connection (5.4.3) on the tangent bundle $TQ \to Q$. Conversely, each connection $\widetilde{K}$ (5.4.3) on $TQ \to Q$ defines the dynamic connection γ (5.4.5) on the affine jet bundle $J^1Q \to Q$ and the dynamic equation (5.4.6) on a configuration bundle $Q \to \mathbb{R}$. □

Then we come to the following theorem.

THEOREM 5.4.2. Every dynamic equation (5.3.1) on a configuration bundle $Q \to \mathbb{R}$ is equivalent to the geodesic equation

$$\begin{aligned} &\ddot{q}^0 = 0, \qquad \dot{q}^0 = 1, \\ &\ddot{q}^i = K^i_\lambda(q^\mu, \dot{q}^\mu)\dot{q}^\lambda, \end{aligned} \tag{5.4.7}$$

on the tangent bundle TQ relative to a connection $\widetilde{K}$ with the components $K^0_\lambda = 0$ and K^i_λ (5.4.5). Its solution is a geodesic curve in Q which also obeys the dynamic equation (5.4.6), and *vice versa.* □

In accordance with this theorem, the second order equation (5.3.5) can be chosen as a geodesic equation. It should be emphasized that, written in the bundle coordinates (t, q^i), the geodesic equation (5.4.7) and the connection $\widetilde{K}$ (5.4.5) are well defined with respect to any coordinates on Q.

From the physical viewpoint, the most interesting dynamic equations are the quadratic ones

$$\xi^i = a^i_{jk}(q^\mu)q^j_t q^k_t + b^i_j(q^\mu)q^j_t + f^i(q^\mu). \tag{5.4.8}$$

This property is coordinate-independent due to the transformation law (5.3.3). Then one can use the following two facts.

PROPOSITION 5.4.3. There is one-to-one correspondence between the affine connections γ on the affine jet bundle $J^1Q \to Q$ and the linear connections K (5.4.3) on the tangent bundle $TQ \to Q$. □

This correspondence is given by the relation (5.4.5), written in the form

$$\gamma^i_\mu = \gamma^i_{\mu 0} + \gamma^i_{\mu j}q^j_t = K_\mu{}^i{}_0(q^\nu)\dot t + K_\mu{}^i{}_j(q^\nu)\dot q^j|_{\dot t=1,\dot q^i=q^i_t} = K_\mu{}^i{}_0(q^\nu) + K_\mu{}^i{}_j(q^\nu)q^j_t,$$

i.e.,

$$\boxed{\gamma^i_{\mu\lambda} = K_\mu{}^i{}_\lambda.}$$

In particular, if an affine dynamic connection γ is symmetric, so is the corresponding linear connection K.

COROLLARY 5.4.4. Every quadratic dynamic equation (5.4.8) on a configuration bundle $Q \to \mathbb{R}$ of non-relativistic mechanics gives rise to the geodesic equation

$$\begin{aligned} &\ddot q^0 = 0, \qquad \dot q^0 = 1, \\ &\ddot q^i = a^i_{jk}(q^\mu)\dot q^j\dot q^k + b^i_j(q^\mu)\dot q^j\dot q^0 + f^i(q^\mu)\dot q^0\dot q^0 \end{aligned} \tag{5.4.9}$$

on the tangent bundle TQ with respect to the symmetric linear connection

$$K_\lambda{}^0{}_\nu = 0, \quad K_0{}^i{}_0 = f^i, \quad K_0{}^i{}_j = K_j{}^i{}_0 = \frac{1}{2}b^i_j, \quad K_k{}^i{}_j = a^i_{kj} \tag{5.4.10}$$

on the tangent bundle $TQ \to Q$. □

The geodesic equation (5.4.9), however, is not unique for the dynamic equation (5.4.8).

PROPOSITION 5.4.5. Any quadratic dynamic equation (5.4.8), being equivalent to the geodesic equation with respect to the symmetric linear connection $\widetilde K$ (5.4.10), is also equivalent to the geodesic equation with respect to an affine connection K' on $TQ \to Q$ which differs from $\widetilde{K}$ (5.4.10) in a soldering form σ on $TQ \to Q$ with the components

$$\sigma^0_\lambda = 0, \qquad \sigma^i_k = h^i_k + (s-1)h^i_k\dot{q}^0, \qquad \sigma^i_0 = -sh^i_k\dot{q}^k - h^i_0\dot{q}^0 + h^i_0,$$

where s and h^i_λ are local functions on Q. □

The proof follows from direct computation.

Now let us extend our inspection of dynamic equations to connections on the tangent bundle $TM \to M$ of the typical fibre M of a configuration bundle $Q \to \mathbb{R}$. In this case, the relationship fails to be canonical, but depends on a trivialization (5.1.1) of $Q \to \mathbb{R}$.

Given such a trivialization, let $(t, \overline{q}^i)$ be the associated coordinates on Q, where $\overline{q}^i$ are coordinates on M with transition functions independent of t. The corresponding trivialization (5.1.2) of $J^1Q \to \mathbb{R}$ takes place in the coordinates $(t, \overline{q}^i, \dot{\overline{q}}^i)$, where $\dot{\overline{q}}^i$ are coordinates on TM. With respect to these coordinates, the transformation law (5.3.7) of a dynamic connection γ on the affine jet bundle $J^1Q \to Q$ reads

$$\gamma'^i_0 = \frac{\partial \overline{q}'^i}{\partial \overline{q}^j}\gamma^j_0 \qquad \gamma'^i_k = \left(\frac{\partial \overline{q}'^i}{\partial \overline{q}^j}\gamma^j_n + \frac{\partial \dot{\overline{q}}'^i}{\partial \overline{q}^n}\right)\frac{\partial \overline{q}^n}{\partial \overline{q}'^k}.$$

It follows that, given a trivialization of $Q \to \mathbb{R}$, a connection γ on $J^1Q \to Q$ defines the time-dependent vertical vector field

$$\gamma^i_0(t, \overline{q}^j, \dot{\overline{q}}^j)\frac{\partial}{\partial \dot{\overline{q}}^i} : \mathbb{R} \times TM \to VTM$$

and the time-dependent connection

$$d\overline{q}^k \otimes \left(\frac{\partial}{\partial \overline{q}^k} + \gamma^i_k(t, \overline{q}^j, \dot{\overline{q}}^j)\frac{\partial}{\partial \dot{\overline{q}}^i}\right) : \mathbb{R} \times TM \to J^1TM \subset TTM \tag{5.4.11}$$

on the tangent bundle $TM \to M$.

Conversely, let us consider a connection

$$\overline{K} = d\overline{q}^k \otimes \left(\frac{\partial}{\partial \overline{q}^k} + \overline{K}^i_k(\overline{q}^j, \dot{\overline{q}}^j)\frac{\partial}{\partial \dot{\overline{q}}^i}\right)$$

on the tangent bundle $TM \to M$. Given the above-mentioned trivialization of the configuration bundle $Q \to \mathbb{R}$, the connection $\overline{K}$ defines the connection $\widetilde{K}$ (5.4.3), with the components

$$K_0^i = 0, \qquad K_k^i = \overline{K}_k^i$$

on the tangent bundle $TQ \to Q$. The corresponding dynamic connection γ on the affine jet bundle $J^1Q \to Q$ reads

$$\gamma_0^i = 0, \qquad \gamma_k^i = \overline{K}_k^i. \tag{5.4.12}$$

Using the transformation law (5.3.7), one can extend the expression (5.4.12) to arbitrary bundle coordinates (t, q^i) on the configuration space Q as follows:

$$\gamma_k^i = \left[\frac{\partial q^i}{\partial \overline{q}^j}\overline{K}_n^j(\overline{q}^j(q^r), \dot{\overline{q}}^j(q^r, q_t^r)) + \frac{\partial^2 q^i}{\partial \overline{q}^n \partial \overline{q}^j}\dot{\overline{q}}^j + \frac{\partial \Gamma^i}{\partial \overline{q}^n}\right]\partial_k \overline{q}^n, \tag{5.4.13}$$
$$\gamma_0^i = \partial_t \Gamma^i + \partial_j \Gamma^i q_t^j - \gamma_k^i \Gamma^k,$$

where $\Gamma^i = \partial_t q^i(t, \overline{q}^j)$ is the connection on $Q \to \mathbb{R}$, corresponding to a given trivialization of Q, i.e., $\Gamma^i = 0$ relative to $(t, \overline{q}^i)$. The dynamic equation on Q defined by the dynamic connection (5.4.13) takes the form

$$q_{tt}^i = \partial_t \Gamma^i + q_t^j \partial_j \Gamma^i + \gamma_k^i (q_t^k - \Gamma^k). \tag{5.4.14}$$

By construction, it is a conservative dynamic equation. Thus, we have proved the following.

PROPOSITION 5.4.6. A connection $\overline{K}$ on the tangent bundle $TM \to M$ of the typical fibre M of a configuration bundle $Q \to \mathbb{R}$ and a connection Γ on $Q \to \mathbb{R}$ yield a conservative dynamic equation (5.4.14) on Q. □

5.5 Connections and reference frames

From the physical viewpoint, a reference frame in non-relativistic mechanics determines a tangent vector at each point of a configuration space Q, which characterizes the velocity of an "observer" at this point. These speculations lead to the following notion of a reference frame in non-relativistic mechanics [213, 216, 271].

DEFINITION 5.5.1. In non-relativistic mechanics, a *reference frame* is a connection Γ on the configuration bundle $Q \to \mathbb{R}$. $\square$

In accordance with this definition, one can think of the horizontal vector field (5.1.9) associated with a connection Γ on $Q \to \mathbb{R}$ as being a family of "observers", while the corresponding covariant differential

$$\boxed{\dot{q}^i_\Gamma \stackrel{\text{def}}{=} D_\Gamma(q^i_t) = q^i_t - \Gamma^i}$$

determines the *relative velocities* with respect to the reference frame Γ.

In particular, given a motion $c : \mathbb{R} \to Q$, its covariant derivative $\nabla^\Gamma c$ with respect to a connection Γ is the velocity of this motion relative to the reference frame Γ. For instance, if c is an integral section for the connection Γ, the velocity of the motion c relative to the reference frame Γ is equal to 0. Conversely, every motion $c : \mathbb{R} \to Q$, defines a reference frame Γ_c such that the velocity of c relative to Γ_c vanishes. This reference frame Γ_c is a global extension of the section $c(\mathbb{R}) \to J^1Q$ of the affine jet bundle $J^1Q \to Q$ over the closed submanifold $c(\mathbb{R}) \subset Q$.

By virtue of Proposition 2.6.2, any reference frame Γ on a configuration bundle $Q \to \mathbb{R}$ is associated with an atlas of local constant trivializations, and *vice versa.* The connection Γ reduces to $\Gamma = \partial_t$ with respect to the corresponding coordinates $(t, \overline{q}^i)$, whose transition functions $\overline{q}^i \to \overline{q}'^i$ are independent of time. One can think of these coordinates as being also the reference frame, corresponding to the connection $\Gamma = \partial_t$. They are called *adapted* to the reference frame Γ. Thus, we come to the following notion of a reference frame, equivalent to Definition 5.5.1.

DEFINITION 5.5.2. In non-relativistic mechanics, a reference frame is an atlas of local constant trivializations of a configuration bundle $Q \to \mathbb{R}$. $\square$

In particular, with respect to the coordinates $\overline{q}^i$ adapted to a reference frame Γ, the velocities relative to this reference frame are equal to the absolute ones

$$D_\Gamma(\overline{q}^i_t) = \dot{\overline{q}}^i_\Gamma = \overline{q}^i_t.$$

A reference frame is said to be *complete* if the associated connection Γ is complete. By virtue of Proposition 5.1.1, every complete reference frame defines a trivialization of a bundle $Q \to \mathbb{R}$, and *vice versa.*

Remark 5.5.1. Given a reference frame Γ, one should solve the equations

$$\Gamma^i(t,q^j(t,\overline{q}^a)) = \frac{\partial q^i(t,\overline{q}^a)}{\partial t}, \tag{5.5.1a}$$

$$\frac{\partial \overline{q}^a(t,q^j)}{\partial q^i}\Gamma^i(t,q^j) + \frac{\partial \overline{q}^a(t,q^j)}{\partial t} = 0 \tag{5.5.1b}$$

in order to find the coordinates $(t,\overline{q}^a)$ adapted to Γ.

Let (t,q_1^a) and (t,q_2^i) be the adapted coordinates for reference frames Γ_1 and Γ_2, respectively. In accordance with the equality (5.5.1b), the components Γ_1^i of the connection Γ_1 with respect to the coordinates (t,q_2^i) and the components Γ_2^a of the connection Γ_2 with respect to the coordinates (t,q_1^a) fulfill the relation

$$\frac{\partial q_1^a}{\partial q_2^i}\Gamma_1^i + \Gamma_2^a = 0.$$

•

Using the relations (5.5.1a) – (5.5.1b), one can rewrite the coordinate transformation law (5.3.3) of dynamic equations as follows. Let

$$\overline{q}_{tt}^a = \overline{\xi}^a \tag{5.5.2}$$

be a dynamic equation on a configuration space Q, written with respect to a reference frame $(t,\overline{q}^n)$. Then, relative to arbitrary bundle coordinates (t,q^i) on $Q \to \mathbb{R}$, the dynamic equation (5.5.2) takes the form

$$q_{tt}^i = d_t\Gamma^i + \partial_j\Gamma^i(q_t^j - \Gamma^j) - \frac{\partial q^i}{\partial \overline{q}^a}\frac{\partial \overline{q}^a}{\partial q^j \partial q^k}(q_t^j - \Gamma^j)(q_t^k - \Gamma^k) + \frac{\partial q^i}{\partial \overline{q}^a}\overline{\xi}^a, \tag{5.5.3}$$

where Γ is the connection corresponding to the reference frame $(t,\overline{q}^n)$. The dynamic equation (5.5.3) can be expressed in the relative velocities $\dot{q}_\Gamma^i = q_t^i - \Gamma^i$ with respect to the initial reference frame $(t,\overline{q}^a)$. We have

$$d_t\dot{q}_\Gamma^i = \partial_j\Gamma^i\dot{q}_\Gamma^j - \frac{\partial q^i}{\partial \overline{q}^a}\frac{\partial \overline{q}^a}{\partial q^j \partial q^k}\dot{q}_\Gamma^j\dot{q}_\Gamma^k + \frac{\partial q^i}{\partial \overline{q}^a}\overline{\xi}^a(t,q^j,\dot{q}_\Gamma^j). \tag{5.5.4}$$

Accordingly, any dynamic equation (5.3.1) can be expressed in the relative velocities $\dot{q}_\Gamma^i = q_t^i - \Gamma^i$ with respect to an arbitrary reference frame Γ as follows:

$$d_t\dot{q}_\Gamma^i = (\xi - J\Gamma)_t^i = \xi^i - d_t\Gamma^i, \tag{5.5.5}$$

where $J\Gamma$ is the prolongation (5.1.15) of the connection Γ onto $J^1Q \to \mathbb{R}$.

Remark 5.5.2. Let us consider the following particular reference frame Γ for a dynamic equation ξ. The covariant differential of a reference frame Γ with respect to the corresponding dynamic connection γ_ξ (5.3.13) reads

$$\begin{aligned}&\nabla^\gamma\Gamma = \nabla^\gamma_\lambda\Gamma^k dq^\lambda\otimes\partial_k : Q\to T^*Q\times V_QJ^1Q,\\ &\nabla^\gamma_\lambda\Gamma^k = \partial_\lambda\Gamma^k - \gamma^k_\lambda\circ\Gamma.\end{aligned} \tag{5.5.6}$$

A connection Γ is called a *geodesic reference frame* for the dynamic equation ξ if

$$\Gamma\rfloor\nabla^\gamma\Gamma = \Gamma^\lambda(\partial_\lambda\Gamma^k - \gamma^k_\lambda\circ\Gamma) = (d_t\Gamma^i - \xi^i\circ\Gamma)\partial_i = 0. \tag{5.5.7}$$

It is readily observed that integral sections c of a reference frame Γ are solutions of a dynamic equation ξ if and only if Γ is a geodesic reference frame for ξ. •

With a reference frame, we obtain a converse of Theorem 5.4.2.

THEOREM 5.5.3. Given a reference frame Γ, any connection K (5.4.1) on the tangent bundle $TQ\to Q$ defines a dynamic equation

$$\xi^i = (K^i_\lambda - \Gamma^i K^0_\lambda)\dot q^\lambda\mid_{\dot q^0=1,\dot q^j=q^j_t}.$$

□

This theorem is a corollary of Proposition 5.4.1 and the following lemma proved by the inspection of transition functions.

LEMMA 5.5.4. Given a connection Γ on the fibre bundle $Q\to\mathbb{R}$ and a connection K on the tangent bundle $TQ\to Q$, there is the connection $\widetilde K$ on $TQ\to Q$ with the components

$$\widetilde K^0_\lambda = 0, \qquad \widetilde K^i_\lambda = K^i_\lambda - \Gamma^i K^0_\lambda.$$

□

5.6 The free motion equation

We say that the dynamic equation (5.3.1) is a *free motion equation* if there exists a reference frame $(t, \overline{q}^i)$ on the configuration space Q such that this equation reads

$$\boxed{\overline{q}^i_{tt} = 0.} \tag{5.6.1}$$

With respect to arbitrary bundle coordinates (t, q^i), a free motion equation takes the form

$$q^i_{tt} = d_t\Gamma^i + \partial_j\Gamma^i(q^j_t - \Gamma^j) - \frac{\partial q^i}{\partial \overline{q}^m}\frac{\partial \overline{q}^m}{\partial q^j \partial q^k}(q^j_t - \Gamma^j)(q^k_t - \Gamma^k), \tag{5.6.2}$$

where $\Gamma^i = \partial_t q^i(t, \overline{q}^j)$ is the connection associated with the initial reference frame $(t, \overline{q}^i)$ (cf. (5.5.3)). One can think of the right-hand side of the equation (5.6.2) as being the general coordinate expression for an *inertial force* in non-relativistic mechanics. The corresponding dynamic connection γ_ξ on the affine jet bundle $J^1Q \to Q$ reads

$$\begin{aligned} \gamma^i_k &= \partial_k\Gamma^i - \frac{\partial q^i}{\partial \overline{q}^m}\frac{\partial \overline{q}^m}{\partial q^j \partial q^k}(q^j_t - \Gamma^j), \\ \gamma^i_0 &= \partial_t\Gamma^i + \partial_j\Gamma^i q^j_t - \gamma^i_k\Gamma^k. \end{aligned} \tag{5.6.3}$$

It is affine. By virtue of Proposition 5.4.3, this dynamic connection defines a linear connection K on the tangent bundle $TQ \to Q$, whose curvature necessarily vanishes. Thus, we come to the following criterion of a dynamic equation to be a free motion equation.

- If ξ is a free motion equation on a configuration space Q, it is quadratic, and the corresponding symmetric linear connection (5.4.10) on the tangent bundle $TQ \to Q$ is a curvature-free connection.

This criterion is not a sufficient condition because it may happen that the components of a curvature-free symmetric linear connection on $TQ \to Q$ vanish with respect to the coordinates on Q which are not compatible with the fibration $Q \to \mathbb{R}$. The similar criterion involves the curvature of a dynamic connection (5.6.3) of a free motion equation.

- If ξ is a free motion equation, then the curvature R of the corresponding dynamic connection γ_ξ is equal to 0.

This criterion also fails to be a sufficient condition. If the curvature R of a dynamic connection γ_ξ vanishes, it may happen that components of γ_ξ are equal to 0 with respect to non-holonomic bundle coordinates on the affine jet bundle $J^1Q \to Q$.

Nevertheless, one can formulate the following necessary and sufficient condition of the existence of a free motion equation on a configuration space Q [68, 213].

- A free motion equation on a fibre bundle $Q \to \mathbb{R}$ exists if and only if the typical fibre M of Q admits a curvature-free symmetric linear connection.

Indeed, let a free motion equation take the form (5.6.1) with respect to some atlas of local constant trivializations of a fibre bundle $Q \to \mathbb{R}$. By virtue of Proposition 5.3.3, there exists an affine dynamic connection γ on the affine jet bundle $J^1Q \to Q$ whose components relative to this atlas are equal to 0. Given a trivialization chart of this atlas, the connection γ defines the curvature-free symmetric linear connection (5.4.11) on M. The converse statement follows at once from Proposition 5.4.6.

The free motion equation (5.6.2) is simplified if the coordinate transition functions $\overline{q}^i \to q^i$ are affine in the coordinates $\overline{q}^i$. Then we have

$$q^i_{tt} = \partial_t\Gamma^i - \Gamma^j\partial_j\Gamma^i + 2q^j_t\partial_j\Gamma^i. \tag{5.6.4}$$

The following lemma shows that the free motion equation (5.6.4) is affine in the coordinates q^i and q^i_t.

LEMMA 5.6.1. [213]. Let $(t,\overline{q}^a)$ be a reference frame on a configuration bundle $Q \to \mathbb{R}$ and Γ the corresponding connection. Components Γ^i of this connection with respect to another coordinate system (t,q^i) are affine functions in the coordinates q^i if and only if the transition functions between the coordinates $\overline{q}^a$ and q^i are affine. □

As an example, let us consider a free motion on a plane $\mathbb{R}^2$. The corresponding configuration bundle is $\mathbb{R}^3 \to \mathbb{R}$, coordinated by $(t,\overline{\mathbf{r}})$. The dynamic equation of this motion is

$$\ddot{\overline{\mathbf{r}}} = 0. \tag{5.6.5}$$

Let us choose the rotatory reference frame with the adapted coordinates

$$\mathbf{r} = A\overline{\mathbf{r}}, \qquad A = \begin{pmatrix} \cos\omega t & -\sin\omega t \\ \sin\omega t & \cos\omega t \end{pmatrix}. \tag{5.6.6}$$

Relative to these coordinates, the connection Γ corresponding to the initial reference frame reads

$$\Gamma = \partial_t \mathbf{r} = \partial_t A \cdot A^{-1}\mathbf{r}.$$

Then the free motion equation (5.6.5) with respect to the rotatory reference frame (5.6.6) takes the familiar form

$$\mathbf{r}_{tt} = \omega^2 \mathbf{r} + 2\begin{pmatrix} 0 & -1 \\ 1 & 0 \end{pmatrix}\mathbf{r}_t. \qquad (5.6.7)$$

The first term in the right-hand side of the equation (5.6.7) is the *centrifugal force* $(-\Gamma^j\partial_j\Gamma^i)$, while the second one is the *Coriolis force* $(2q^j_t\partial_j\Gamma^i)$.

One can easily find the geodesic reference frames for a free motion equation. They are $\Gamma^i = v^i =$ const. By virtue of Lemma 5.6.1, these reference frames define the adapted coordinates

$$\overline{q}^i = k^i_j q^j - v^i t - a^i, \qquad k^i_j = \text{const.}, \qquad v^i = \text{const.}, \qquad a^i = \text{const.} \qquad (5.6.8)$$

The equation (5.6.1) keeps obviously its free motion form under the transformations (5.6.8) between the geodesic reference frames. It is readily observed that these transformations are precisely the elements of the *Galilei group*.

5.7 The relative acceleration

It should be emphasized that, taken separately, the left- and right-hand sides of the dynamic equation (5.5.5) are not well-behaved objects. This equation can be brought into the covariant form if we introduce the notion of a relative acceleration.

To consider a relative acceleration with respect to a reference frame Γ, one should prolong the connection Γ on the configuration bundle $Q \to \mathbb{R}$ to a holonomic connection ξ_Γ on the jet bundle $J^1Q \to \mathbb{R}$. Note that the jet prolongation $J\Gamma$ (5.1.15) of Γ onto $J^1Q \to \mathbb{R}$ is not holonomic. We can construct the desired prolongation by means of a dynamic connection γ on the affine jet bundle $J^1Q \to Q$.

LEMMA 5.7.1. [213]. Let us consider the composite bundle (5.3.8). Given a frame Γ on $Q \to \mathbb{R}$ and a dynamic connections γ on $J^1Q \to Q$, there exists a dynamic connection $\widetilde{\gamma}$ on $J^1Q \to Q$ with the components

$$\widetilde{\gamma}^i_k = \gamma^i_k, \qquad \widetilde{\gamma}^i_0 = d_t\Gamma^i - \gamma^i_k\Gamma^k. \qquad (5.7.1)$$

□

Now, we construct a certain soldering form on the affine jet bundle $J^1Q \to Q$ and add it to this connection. Let us apply the canonical projection $T^*Q \to V^*Q$ and then the imbedding $\Gamma : V^*Q \to T^*Q$ to the covariant differential (5.5.6) of the reference frame Γ with respect to the dynamic connection γ. We obtain the $V_Q J^1Q$-valued 1-form

$$\sigma = [-\Gamma^i(\partial_i\Gamma^k - \gamma_i^k \circ \Gamma)dt + (\partial_i\Gamma^k - \gamma_i^k \circ \Gamma)dq^i] \otimes \partial_k^t$$

on Q whose pull-back onto J^1Q is the desired soldering form. The sum

$$\gamma_\Gamma \stackrel{\text{def}}{=} \widetilde{\gamma} + \sigma,$$

called the *frame connection*, reads

$$\begin{aligned} \gamma_\Gamma{}_0^i &= d_t\Gamma^i - \gamma_k^i\Gamma^k - \Gamma^k(\partial_k\Gamma^i - \gamma_k^i \circ \Gamma), \\ \gamma_\Gamma{}_k^i &= \gamma_k^i + \partial_k\Gamma^i - \gamma_k^i \circ \Gamma. \end{aligned} \tag{5.7.2}$$

This connection yields the desired holonomic connection

$$\boxed{\xi_\Gamma^i = d_t\Gamma^i + (\partial_k\Gamma^i + \gamma_k^i - \gamma_k^i \circ \Gamma)(q_t^k - \Gamma^k)}$$

on the jet bundle $J^1Q \to \mathbb{R}$.

Let ξ be a dynamic equation and $\gamma = \gamma_\xi$ the connection (5.3.13) associated with ξ. Then one can think of the vertical vector field

$$\boxed{a_\Gamma \stackrel{\text{def}}{=} \xi - \xi_\Gamma = (\xi^i - \xi_\Gamma^i)\partial_i^t} \tag{5.7.3}$$

on the affine jet bundle $J^1Q \to Q$ as being a *relative acceleration* with respect to the reference frame Γ in comparison with the *absolute acceleration* ξ.

Example 5.7.1. Let us consider a reference frame which is geodesic for the dynamic equation ξ, i.e., the relation (5.5.7) holds. Then the relative acceleration of a motion c with respect to the reference frame Γ is

$$(\xi - \xi_\Gamma) \circ \Gamma = 0.$$

•

Let ξ now be an arbitrary dynamic equation, written with respect to coordinates (t, q^i) adapted to the reference frame Γ, i.e., $\Gamma^i = 0$. In these coordinates, the relative acceleration with respect to the reference frame Γ is

$$a^i_\Gamma = \xi^i(t, q^j, q^j_t) - \frac{1}{2} q^k_t (\partial_k \xi^i - \partial_k \xi^i \mid_{q^j_t = 0}). \tag{5.7.4}$$

Given another bundle coordinates (t, q'^i) on $Q \to \mathbb{R}$, this dynamic equation takes the form (5.5.4), while the relative acceleration (5.7.4) with respect to the reference frame Γ reads $a'^i_\Gamma = \partial_j q'^i a^j_\Gamma$. Then we can write a dynamic equation (5.3.1) in the form which is covariant under coordinate transformations:

$$\boxed{\widetilde{D}_{\gamma_\Gamma} q^i_t = d_t q^i_t - \xi^i_\Gamma = a_\Gamma,} \tag{5.7.5}$$

where $\widetilde{D}_{\gamma_\Gamma}$ is the vertical covariant differential (5.3.12) with respect to the frame connection γ_Γ (5.7.2) on the affine jet bundle $J^1Q \to Q$.

In particular, if ξ is a free motion equation which takes the form (5.6.1) with respect to a reference frame Γ, then

$$\boxed{\widetilde{D}_{\gamma_\Gamma} q^i_t = 0}$$

relative to arbitrary bundle coordinates on the configuration bundle $Q \to \mathbb{R}$.

The left-hand side of the dynamic equation (5.7.5) can also be expressed into the relative velocities such that this dynamic equation takes the form

$$d_t \dot q^i_\Gamma - {\gamma_\Gamma}^i_k \dot q^k_\Gamma = a_\Gamma \tag{5.7.6}$$

which is the covariant form of the equation (5.5.5).

The concept of a relative acceleration is understood better when we deal with the quadratic dynamic equation ξ, and the corresponding dynamic connection γ is affine. If a dynamic connection γ is affine, i.e.,

$$\gamma^i_\lambda = \gamma^i_{\lambda 0} + \gamma^i_{\lambda k} q^k_t,$$

so is the frame connection γ_Γ for any frame Γ:

$$\begin{aligned}
&{\gamma_\Gamma}^i_{jk} = \gamma^i_{jk}, \\
&{\gamma_\Gamma}^i_{0k} = \partial_k \Gamma^i - \gamma^i_{jk}\Gamma^j, \qquad {\gamma_\Gamma}^i_{k0} = \partial_k \Gamma^i - \gamma^i_{kj}\Gamma^j, \\
&{\gamma_\Gamma}^i_{00} = \partial_t \Gamma^i - \Gamma^j \partial_j \Gamma^i + \gamma^i_{jk}\Gamma^j \Gamma^k.
\end{aligned} \tag{5.7.7}$$

In particular, we obtain

$$\gamma_\Gamma{}^i_{jk} = \gamma^i_{jk}, \qquad \gamma_\Gamma{}^i_{0k} = \gamma_\Gamma{}^i_{k0} = \gamma_\Gamma{}^i_{00} = 0$$

relative to the coordinates adapted to a reference frame Γ. A glance at the expression (5.7.7) shows that, if a dynamic connection γ is symmetric, so is a frame connection γ_Γ.

COROLLARY 5.7.2. If a dynamic equation ξ is quadratic, the relative acceleration a_Γ (5.7.3) is always affine, and it admits the decomposition

$$a^i_\Gamma = -(\Gamma^\lambda \nabla^\gamma_\lambda \Gamma^i + 2\dot{q}^\lambda_\Gamma \nabla^\gamma_\lambda \Gamma^i), \tag{5.7.8}$$

where $\gamma = \gamma_\xi$ is the dynamic connection (5.3.13), and

$$\dot{q}^\lambda_\Gamma = q^\lambda_t - \Gamma^\lambda, \qquad q^0_t = 1, \qquad \Gamma^0 = 1$$

is the relative velocity with respect to the reference frame Γ. □

Note that the splitting (5.7.8) gives a *generalized Coriolis theorem*. In particular, the well-known analogy between inertial and electromagnetic forces is restated. Corollary 5.7.2 shows that this analogy can be extended to an arbitrary quadratic dynamic equation.

5.8 Lagrangian and Newtonian systems

Lagrangian formalism of time-dependent mechanics is the repetition of Lagrangian field theory (see Section 3.2) in the particular case of fibre bundles $Q \to \mathbb{R}$. By a *Lagrangian system* is meant a mechanical system whose motions are solutions of Euler–Lagrange equations for some Lagrangian

$$L = \mathcal{L}dt, \qquad \mathcal{L} : J^1Q \to \mathbb{R}, \tag{5.8.1}$$

on the *velocity phase space* J^1Q. We also consider more general notion of a Newtonian system. A Newtonian system is characterised both by a dynamic equation (i.e., a dynamic connection) and a mass tensor which satisfy a certain relation. At first, we will obtain this relation in the framework of Lagrangian formalism.

As in field theory, by gauge transformations in time-dependent mechanics are meant automorphism of the configuration bundle $Q \to \mathbb{R}$, but only over translations of the base $\mathbb{R}$. Therefore, we will restrict our consideration to projectable vector fields

$$u = u^t\partial_t + u^i\partial_i, \qquad u\rfloor dt = u^t = \text{const.}, \tag{5.8.2}$$

on $Q \to \mathbb{R}$ which are generators of local 1-parameter groups of such gauge transformations. The jet prolongation (1.3.10) of u (5.8.2) is

$$J^1u = u^t\partial_t + u^i\partial_i + d_tu^i\partial_i^t.$$

Given a Lagrangian L (5.8.1), its Lie derivative (3.2.1) along u reads

$$\mathbf{L}_{J^1u}L = (\overline{u}\rfloor d\mathcal{L})dt = (u^t\partial_t + u^i\partial_i + d_tu^i\partial_i^t)\mathcal{L}dt. \tag{5.8.3}$$

Then the first variational formula (3.2.2) takes the form

$$\boxed{J^1u\rfloor d\mathcal{L} = (u^i - u^tq_t^i)\mathcal{E}_i + d_t(u\rfloor H_L),} \tag{5.8.4}$$

where

$$H_L = \widehat{v}^*(dL) + L = \pi_i dq^i - (\pi_i q_t^i - \mathcal{L})dt \tag{5.8.5}$$

is the Poincaré–Cartan form (3.2.5) and

$$\begin{aligned} &\mathcal{E}_L : J^2Q \to V^*Q, \\ &\mathcal{E}_L = \mathcal{E}_i\overline{d}q^i = (\partial_i - d_t\partial_i^t)\mathcal{L}\overline{d}q^i, \end{aligned} \tag{5.8.6}$$

is the Euler–Lagrange operator (3.2.3) for a Lagrangian L. We will use the notation

$$\pi_i = \partial_i^t\mathcal{L}, \qquad \pi_{ji} = \partial_j^t\partial_i^t\mathcal{L}.$$

The kernel $\operatorname{Ker}\mathcal{E}_L \subset J^2Q$ of the Euler–Lagrange operator (5.8.6) defines the Euler–Lagrange equations

$$\boxed{(\partial_i - d_t\partial_i^t)\mathcal{L} = 0} \tag{5.8.7}$$

in time-dependent mechanics.

As in field theory, a holonomic connection ξ (5.1.16) on the jet bundle $J^1Q \to \mathbb{R}$ is said to be a Lagrangian connection for the Lagrangian L if it takes its values in the kernel of the Euler–Lagrange operator, i.e., obeys the equation

$$\partial_i\mathcal{L} - \partial_t\pi_i - q_t^j\partial_j\pi_i - \xi^j\pi_{ji} = 0 \tag{5.8.8}$$

(cf. (3.2.13)). If a Lagrangian connection ξ_L exists, it defines a dynamic equation whose solutions are solutions of the Euler–Lagrange equations (5.8.7). Different Lagrangian connections ξ_L lead to different dynamic equations associated with the same system of Euler–Lagrange equations. In particular, if L is a regular Lagrangian, there exists a unique Lagrangian connection

$$\xi_L^j = (\pi^{-1})^{ij}[-\partial_i\mathcal{L} + \partial_t\pi_i + q_t^k\partial_k\pi_i] \tag{5.8.9}$$

for L. In this case, Euler–Lagrange equations are equivalent to a dynamic equation.

Turn now to the notion of a mass tensor.

Every Lagrangian L on the jet manifold J^1Q yields the Legendre map (3.2.9):

$$\widehat{L} : J^1Q \to V^*Q, \qquad p_i \circ \widehat{L} = \pi_i, \tag{5.8.10}$$

where (t, q^i, p_i) are coordinates on the vertical cotangent bundle V^*Q which plays the role of a momentum phase space of time-dependent mechanics (see Section 5.10. Due to the vertical splitting (1.1.14) of VV^*Q, the vertical tangent map $V\widehat{L}$ to $\widehat{L}$ reads

$$V\widehat{L} : V_QJ^1Q \to V^*Q \underset{Q}{\times} V^*Q.$$

It yields the linear bundle morphism

$$\flat \overset{\text{def}}{=} (\operatorname{Id} J^1Q, \operatorname{pr}_2 \circ V\widehat{L}) : V_QJ^1Q \underset{J^1Q}{\to} V_Q^*J^1Q, \tag{5.8.11}$$

$$\flat : \partial_i^t \mapsto \pi_{ij}\overline{d}q_t^j,$$

where $\{\overline{d}q_t^j\}$ are bases for the fibres of the vertical tangent bundle $V_Q^*J^1Q \to J^1Q$. The morphism (5.8.11) defines both the mapping

$$J^1Q \to V_Q^*J^1Q \underset{J^1Q}{\otimes} V_Q^*J^1Q$$

and, due to the splitting (5.1.5), the mapping

$$\widehat{m} : J^1Q \underset{Q}{\longrightarrow} V^*Q \underset{Q}{\otimes} V^*Q,$$

$$\boxed{m_{ij} = p_{ij} \circ \widehat{m} = \pi_{ij},}$$

where (t, q^i, p_{ij}) are holonomic coordinates on $V^*Q \underset{Q}{\otimes} V^*Q$. Thus, $\pi_{ij} = m_{ij}$ are components of the $\overset{2}{\vee} V^*Q$-valued field $\widehat{m}$ on the velocity phase space J^1Q. It is called the *mass tensor*.

Let a Lagrangian L be regular. Then the mass tensor is non-degenerate, and defines a fibre metric, called *mass metric*, in the vertical tangent bundle $V_QJ^1Q \to J^1Q$. Since a Lagrangian L is regular, there exists a unique Lagrangian connection ξ_L for L which obeys the equation

$$m_{ik}\xi_L^k = -\partial_t\pi_i - \partial_j\pi_i q_t^j + \partial_i\mathcal{L}. \tag{5.8.12}$$

This holonomic connection defines the dynamic equation (5.8.9). At the same time, the equation (5.8.12) leads to the commutative diagram

$$\begin{array}{ccc} V_QJ^1Q & \xrightarrow{\flat} & V_Q^*J^1Q \\ & {}_{D_{\xi_L}}\nwarrow \nearrow_{\mathcal{E}_L} & \\ & J^2Q & \end{array}$$

where

$$\begin{aligned} &\mathcal{E}_L = \flat\circ D_{\xi_L}, \\ &\mathcal{E}_i = m_{ik}(q_{tt}^k - \xi_L^k), \end{aligned} \tag{5.8.13}$$

and D_{ξ_L} is the covariant differential (5.1.10) relative to the connection ξ_L. Furthermore, the derivation of (5.8.12) with respect to q_t^j results in the relation

$$\boxed{\xi_L\rfloor dm_{ij} + m_{ik}\gamma_j^k + m_{jk}\gamma_i^k = 0,} \tag{5.8.14}$$

where

$$\gamma_i^k = \frac{1}{2}\partial_i^t\xi_L^k$$

are coefficients of the symmetric dynamic connection γ_{ξ_L} (5.3.13) corresponding to the dynamic equation ξ_L.

Thus, each regular Lagrangian L defines both the dynamic equation ξ_L, related to the Euler–Lagrange operator $\mathcal{E}_L$ by means of the equality (5.8.13), and the non-degenerate mass tensor m_{ij}, related to the dynamic equation ξ_L by means of the relation (5.8.14). This is a Newtonian system in accordance with the following definition [213].

DEFINITION 5.8.1. Let $Q \to \mathbb{R}$ be a fibre bundle together with

- a non-degenerate fibre metric $\widehat{m}$ in the fibre bundle $V_QJ^1Q \to J^1Q$:

$$\widehat{m} = \frac{1}{2}m_{ij}\overline{d}q^i \vee \overline{d}q^j : J^1Q \to V^*Q \underset{Q}{\otimes} V^*Q,$$

satisfying the symmetry condition

$$\boxed{\partial_k^t m_{ij} = \partial_j^t m_{ik},} \tag{5.8.15}$$

- and a holonomic connection ξ (5.1.16) on the jet bundle $J^1Q \to \mathbb{R}$, related to the fibre metric $\widehat{m}$ by the compatibility condition (5.8.14).

The triple $(Q, \widehat{m}, \xi)$ is called a *Newtonian system.* □

This Definition generalizes the second Newton law of particle mechanics. Indeed, the dynamic equation for a Newtonian system is equivalent to the equation

$$m_{ik}(q_{tt}^k - \xi^k) = 0. \tag{5.8.16}$$

There are two main reasons for considering Newtonian systems.

From the physical viewpoint, with a mass tensor, we can introduce the notion of an *external force* which can be defined as a section of the vertical cotangent bundle $V_Q^*J^1Q \to J^1Q$. Let us also bear in mind the isomorphism (5.1.5). Note that there are no canonical isomorphisms between the vertical cotangent bundle $V_Q^*J^1Q$ and the vertical tangent bundle V_QJ^1Q of J^1Q. Therefore, one should distinguish forces and accelerations which are related with each other by means of a mass metric.

Remark 5.8.1. Let $(Q, \widehat{m}, \xi)$ be a Newtonian system and f an external force. Then

$$\xi_f^i \overset{\text{def}}{=} \xi^i + (m^{-1})^{ik} f_k \tag{5.8.17}$$

is a dynamic equation, but the triple $(Q, \widehat{m}, \xi_f)$ is not a Newtonian system in general. As follows from direct computations, only if an external force possesses the property

$$\partial_i^t f_j + \partial_j^t f_i = 0, \tag{5.8.18}$$

then ξ_f (5.8.17) fulfills the relation (5.8.14), and $(Q, \widehat{m}, \xi_f)$ is also a Newtonian system. For instance, the *Lorentz force*

$$\boxed{f_i = eF_{\lambda i}q_t^\lambda,} \qquad q_t^0 = 1, \tag{5.8.19}$$

where $F_{\lambda\mu} = \partial_\lambda A_\mu - \partial_\mu A_\lambda$ is the electromagnetic strength, obeys the condition (5.8.18). •

From the mathematical viewpoint, one may hope to bring the equation (5.8.16) into a system of Euler–Lagrange equations by a choice of an appropriate mass tensor. This is the well-known inverse problem formulated for time-dependent mechanics (see [213] for details).

Example 5.8.2. Let us consider the 1-dimensional motion of a point mass m_0 subject to friction. It is described by the equation

$$m_0 q_{tt} = -k q_t, \qquad k > 0, \tag{5.8.20}$$

on the configuration space $\mathbb{R}^2 \to \mathbb{R}$, coordinated by (t, q). This mechanical system is characterized by the mass function $m = m_0$ and the holonomic connection

$$\xi = \partial_t + q_t \partial_q - \frac{k}{m} q_t \partial_q^t, \tag{5.8.21}$$

but it is neither a Newtonian nor a Lagrangian system. Nevertheless, there is the mass function

$$m = m_0 \exp\left[\frac{k}{m_0} t\right] \tag{5.8.22}$$

such that m and the holonomic connection (5.8.21) is both a Newtonian and a Lagrangian system with the Havas Lagrangian

$$\mathcal{L} = \frac{1}{2} m_0 \exp\left[\frac{k}{m_0} t\right] q_t^2 \tag{5.8.23}$$

[253]. The corresponding Euler–Lagrange equations are equivalent to the equation of motion (5.8.20). •

Example 5.8.3. Let us consider a non-degenerate quadratic Lagrangian

$$\mathcal{L} = \frac{1}{2} m_{ij}(q^\mu) q_t^i q_t^j + k_i(q^\mu) q_t^i + f(q^\mu), \tag{5.8.24}$$

where the mass tensor m_{ij} is a Riemannian metric in the vertical tangent bundle $VQ \to Q$ (see the isomorphism (5.1.4)). Then the Lagrangian L (5.8.24) can be written as

$$\mathcal{L} = \frac{1}{2} g_{\alpha\mu} q_t^\alpha q_t^\mu, \qquad q_t^0 = 1, \tag{5.8.25}$$

where g is the fibre metric

$$g_{00} = 2\phi, \qquad g_{0i} = k_i, \qquad g_{ij} = m_{ij} \tag{5.8.26}$$

in the tangent bundle TQ. The corresponding Euler–Lagrange equations take the form

$$q^i_{tt} = (m^{-1})^{ik}\{_{\lambda k\nu}\}q^\lambda_t q^\nu_t, \qquad q^0_t = 1, \tag{5.8.27}$$

where $\{_{\lambda\mu\nu}\}$ are the Christoffel symbols (2.4.13) of the metric (5.8.26). Let us assume that this metric is non-degenerate. By virtue of Corollary 5.4.4, the dynamic equation (5.8.27) gives rise to the geodesic equation (5.4.9)

$$\ddot q^0 = 0, \qquad \dot q^0 = 1,$$
$$\ddot q^i = (m^{-1})^{ik}\{_{\lambda k\nu}\}\dot q^\lambda \dot q^\nu$$

on the tangent bundle TQ with respect to the linear connection $\widetilde K$ (5.4.3) with the components

$$K_\lambda{}^0{}_\nu = 0, \qquad K_\lambda{}^i{}_\nu = (m^{-1})^{ik}\{_{\lambda k\nu}\}. \tag{5.8.28}$$

We have the relation

$$\nabla_\lambda m_{ij} = \partial_\lambda m_{ij} + m_{ik}K_\lambda{}^k{}_j + m_{jk}K_\lambda{}^k{}_i = 0. \tag{5.8.29}$$

•

A Newtonian system $(Q, \widehat m, \xi)$ is said to be *standard*, if $\widehat m$ is the pull-back on $V_Q J^1Q$ of a fibre metric in the vertical tangent bundle $VQ \to Q$ in accordance with the isomorphisms (5.1.4) and (5.1.5), i.e., the mass tensor $\widehat m$ is independent of the velocity coordinates q^i_t. It is readily observed that any fibre metric $\widehat m$ in $VQ \to Q$ can be seen as a mass metric of a standard Newtonian system, given by the Lagrangian

$$\mathcal{L} = \frac{1}{2}m_{ij}(q^\mu)(q^i_t - \Gamma^i)(q^j_t - \Gamma^j), \tag{5.8.30}$$

where Γ is a reference frame. If $\widehat m$ is a Riemannian metric, one can think of the Lagrangian (5.8.30) as being a *kinetic energy* with respect to the reference frame Γ.

Example 5.8.4. Let us consider a system of n distinguishable particles with masses $(m_1, \ldots, m_n)$ in a 3-dimensional Euclidean space $\mathbb{R}^3$. Their positions $(\mathbf{r}_1, \ldots, \mathbf{r}_n)$ span the configuration space $\mathbb{R}^{3n}$. The total kinetic energy is

$$T_{\text{tot}} = \frac{1}{2}\sum_{A=1}^{n} m_A \mid \dot{\mathbf{r}}_A \mid^2$$

that corresponds to the mass tensor

$$m_{ABij} = \delta_{AB}\delta_{ij}m_A, \qquad A, B = 1, \ldots n, \qquad i, j = 1, 2, 3,$$

on the configuration space $\mathbb{R}^{3n}$. To separate the translation degrees of freedom, one performs a linear coordinate transformation

$$(\mathbf{r}_1, \ldots, \mathbf{r}_n) \mapsto (\vec{\rho}_1, \ldots, \vec{\rho}_{n-1}, \mathbf{R}),$$

where $\mathbf{R}$ is the centre of mass, while the $n-1$ vectors $(\vec{\rho}_1, \ldots, \vec{\rho}_{n-1})$ are *mass-weighted Jacobi vectors* (see their definition below) [200, 201]. The Jacobi vectors ρ_A are chosen so that the kinetic energy about the centre of mass has the form

$$T = \frac{1}{2}\sum_{A=1}^{n-1} \mid \dot{\vec{\rho}}_A \mid^2 \tag{5.8.31}$$

that corresponds to the Euclidean mass tensor

$$m_{ABij} = \delta_{AB}\delta_{ij}, \qquad A, B = 1, \ldots n-1, \qquad i, j = 1, 2, 3,$$

on the *translation-reduced* configuration space $\mathbb{R}^{3n-3}$. The usual procedure for defining *Jacobi vectors* involves organizing the particles into a hierarchy of clusters, in which every cluster consists of one or more particles, and where each Jacobi vector joins the centres of mass of two clusters, thereby creating a larger cluster. A Jacobi vector, weighted by the square root of the reduced mass of the two clusters it joins, is the above-mentioned mass-weighted Jacobi vector. For example, in the four-body problem, one can use the following clustering of the particles:

$$\vec{\rho}_1 = \sqrt{\mu_1}(\mathbf{r}_2 - \mathbf{r}_1),$$
$$\vec{\rho}_2 = \sqrt{\mu_2}(\mathbf{r}_4 - \mathbf{r}_3),$$
$$\vec{\rho}_3 = \sqrt{\mu_3}\left(\frac{m_3\mathbf{r}_3 + m_4\mathbf{r}_4}{m_3 + m_4} - \frac{m_1\mathbf{r}_1 + m_2\mathbf{r}_2}{m_1 + m_2}\right),$$
$$\frac{1}{\mu_1} = \frac{1}{m_1} + \frac{1}{m_2}, \qquad \frac{1}{\mu_2} = \frac{1}{m_3} + \frac{1}{m_4}, \qquad \frac{1}{\mu_3} = \frac{1}{m_1 + m_2} + \frac{1}{m_3 + m_4}.$$

Different clusterings lead to different collections of Jacobi vectors, which are related by linear transformations. Since these transformations maintain the Euclidean form (5.8.31) of the kinetic energy, they are elements of the group $O(n-1)$, called the *"democracy group"*. •

5.9 Non-relativistic Jacobi fields

Return now to Section 5.4. In accordance with Proposition 5.4.1, any non-relativistic dynamic equation (5.3.1) on a configuration bundle $Q \to \mathbb{R}$ is equivalent to a geodesic equation with respect to a connection $\widetilde{K}$ (5.4.3) on the tangent bundle $TQ \to Q$. Let us consider the generic quadratic dynamic equation (5.4.8). Then the connection $\widetilde{K}$ is linear and symmetric (see the expression (5.4.10), and the equation for Jacobi vector fields along the geodesics of this connection can be written as follows.

The curvature R (2.4.2) of the connection $\widetilde{K}$ (5.4.10) has the temporal component

$$R_{\lambda\mu}{}^0{}_\beta = 0. \tag{5.9.1}$$

Then the equation for a sl Jacobi vector field u along a geodesic c reads

$$\dot{q}^\beta \dot{q}^\mu (\nabla_\beta (\nabla_\mu u^\alpha) - R_{\lambda\mu}{}^\alpha{}_\beta u^\lambda) = 0, \qquad \nabla_\beta \dot{q}^\alpha = 0, \tag{5.9.2}$$

where ∇_μ denote the covariant derivatives with respect to the connection $\widetilde{K}$ [177]. Due to the relation (5.9.1), the equation (5.9.2) for the temporal component u^0 of a Jakobi field takes the form

$$\dot{q}^\beta \dot{q}^\mu (\partial_\mu \partial_\beta u^0 + K_\mu{}^\gamma{}_\beta \partial_\gamma u^0) = 0.$$

We choose its solution

$$u^0 = 0 \tag{5.9.3}$$

because all non-relativistic geodesics obey the constraint $\dot{q}^0 = 0$.

Let us consider a quadratic Newtonian system with a Riemannian mass metric m_{ij}. Given a reference frame (t, q^i), this mass metric is extended to the Riemannian metric

$$\overline{g}_{00} = 1, \qquad \overline{g}_{0i} = 0, \qquad \overline{g}_{ij} = m_{ij} \tag{5.9.4}$$

on Q. However, the covariant derivative of this metric with respect to the connection $\widetilde{K}$ (5.8.28) does not vanish in general, namely, $\nabla_\lambda g_{0i} \neq 0$. Nevertheless, due to the relations (5.8.29) and (5.9.3), the well-known formula

$$\int_a^b (m_{ij}(\dot{q}^\alpha \nabla_\alpha u^i)(\dot{q}^\beta \nabla_\beta u^j) + R_{i\mu j\nu} u^i u^j \dot{q}^\mu \dot{q}^\nu) dt + \tag{5.9.5}$$
$$m_{ij}\dot{q}^\alpha \nabla_\alpha u^i u^j \mid_{t=a} - m_{ij}\dot{q}^\alpha \nabla_\alpha u^i u^j \mid_{t=b} = 0$$

for a Jacobi vector field u, which is perpendicular to a geodesic c, takes place. Note that this expression is independent of the components $\overline{g}_{0\lambda}$ of the metric (5.9.4), i.e., is frame-independent. It differs from that obtained by the variational methods where a metric is independent of a dynamic equation [215].

Remark 5.9.1. Note that, in the case of a quadratic Lagrangian L, the equation (5.9.2) coincides with the *Jacobi equation*

$$u^j d_t(\partial_j \dot{\partial}_i L) + d_t(\dot{u}^j \dot{\partial}_i \dot{\partial}_j L) - u^j \partial_i \partial_j L = 0 \tag{5.9.6}$$

for a Jacobi vector field on solutions of the Euler–Lagrange equations for L. This equation is the Euler–Lagrange equation

$$(\partial_i - d_t \partial_i^t)\mathcal{L}_V = 0$$

for the vertical extension L_V (4.5.1) of the Lagrangian L [81, 213]. •

With the formula (5.9.5), the conjugate points of solutions of the dynamic equation ξ can be examined in accordance with the well-known geometric criteria [177].

PROPOSITION 5.9.1. If the sectional curvature $R_{i\mu j\nu} u^i u^j \dot{q}^\mu \dot{q}^\nu$ is non-negative on a solution c, this geodesic has no conjugate points. □

PROPOSITION 5.9.2. If the sectional curvature $R_{i\mu j\nu} u^i u^j v^\mu v^\nu$, where v is an arbitrary unit vector field on a Riemannian manifold Q does not exceed $-k_0 < 0$, then, for any solution c, the distance between two consecutive conjugate points is at most $\pi/\sqrt{k_0}$. □

For instance, let us consider a one-dimensional time-dependent oscillator described by the Lagrangian

$$L = \frac{1}{2}[m(t)\dot{q}^1)^2 - k(t)(q^1)^2].$$

The corresponding Lagrange equation is the well-known Sturm equation. In this case, the metric (5.8.26) reads

$$g_{00} = -kx^2, \qquad g_{01} = 0, \qquad g_{11} = m.$$

Its Christoffel symbols are

$$\{_{111}\} = 0, \qquad \{_{011}\} = \{_{110}\} = -\frac{1}{2}\dot{m},$$
$$\{_{010}\} = \{_{100}\} = kx, \qquad \{_{101}\} = \frac{1}{2}\dot{m}, \qquad \{_{000}\} = -\frac{1}{2}\dot{k}.$$

The connection $\widetilde{K}$ (5.8.28) takes the form

$$K_\lambda{}^0{}_\mu = 0, \qquad K_0{}^1{}_1 = K_1{}^1{}_0 = -\frac{\dot{m}}{2m}, \qquad K_0{}^1{}_0 = \frac{kx}{m}.$$

Its curvature has the nonzero component

$$R_{1010} = -\frac{k}{m} - \frac{\ddot{m}}{2m} + \frac{3}{4}\left(\frac{\dot{m}}{m}\right)^2.$$

Let us apply the Propositions 5.9.1 and 5.9.2 to this case. If $R_{1010} \geq 0$ on a solution c, this solution has no conjugate points. This condition reads

$$\frac{(3\dot{m})^2}{4m} \geq k + \frac{1}{2}\ddot{m}.$$

If $R_{1010} \leq -k_0 < 0$, the distance between two consecutive conjugate points is at most $\pi/\sqrt{k}$. This condition takes the form

$$\frac{k}{m} \geq k_0 - \frac{\ddot{m}}{2m} + \frac{3}{4}\left(\frac{\dot{m}}{m}\right)^2.$$

For instance, let us consider an oscillator where m and k are independent of t. In this case, $R_{0101} = -k$, while the half-period of this oscillator is exactly $\pi/\sqrt{k_0}$ in accordance with Proposition 5.9.2. Similarly, solutions for the oscillator with a constant mass m and a function $k(t) \geq k_0 > 0$ also have conjugate points, and the distance between two consecutive conjugate points is at most $\pi/\sqrt{k_0}$.

5.10 Hamiltonian time-dependent mechanics

This Section is devoted to the Hamiltonian mechanics subject to time-dependent transformations [213, 271]. Hamiltonian time-dependent mechanics is formulated as the particular Hamiltonian field theory on a fibre bundle $Q \to \mathbb{R}$. The corresponding *momentum phase space* Π (3.2.8) is the vertical cotangent bundle

$$\pi_\Pi : V^*Q \to Q, \tag{5.10.1}$$

equipped with the holonomic coordinates $(t, q^i, p_i = \dot{q}_i)$. The main peculiarity of Hamiltonian time-dependent mechanics in comparison with the general polysymplectic case lies in the fact that V^*Q is equipped with the canonical Poisson structure. In contrast with conservative mechanics, this Poisson structure however does not define dynamic equations.

Remark 5.10.1. Note on the widely spread formulation of time-dependent mechanics which implies a preliminary splitting of a momentum phase space $\Pi = \mathbb{R} \times Z$, where Z is a Poisson manifold [53, 57, 96, 144, 197, 226]. From the physical viewpoint, this means that a certain reference frame is chosen. In this case, the momentum phase space Π is endowed with the product of the zero Poisson structure on $\mathbb{R}$ and the Poisson structure on Z. A Hamiltonian is defined as a real function $\mathcal{H}$ on Π. The corresponding Hamiltonian vector field $\vartheta_\mathcal{H}$ on Π is vertical with respect to the fibration $\Pi \to \mathbb{R}$. Due to the canonical imbedding

$$\Pi \times T\mathbb{R} \to T\Pi, \tag{5.10.2}$$

one introduces the vector field

$$\gamma_\mathcal{H} = \partial_t + \vartheta_\mathcal{H}, \tag{5.10.3}$$

where ∂_t is the standard vector field on $\mathbb{R}$ [144]. The first order dynamic equation $\gamma_\mathcal{H}(\Pi) \subset T\Pi$ on the manifold Π plays the role of Hamilton equations, while the evolution equation on the Poisson algebra $C^\infty(\Pi)$ of smooth functions on Π is given by the Lie derivative

$$\mathbf{L}_{\gamma_\mathcal{H}} f = \partial_t f + \{\mathcal{H}, f\}. \tag{5.10.4}$$

This is not the case of mechanical systems subject to time-dependent transformations. These transformations violate the splitting $\mathbb{R} \times Z$. As a consequence, there is no canonical imbedding (5.10.2). The vector field (5.10.3) and the splitting in the right-hand side of the evolution equation (5.10.4) are ill-defined. A Hamiltonian $\mathcal{H}$ is not scalar under time-dependent transformations. Its Poisson bracket with functions $f \in C^\infty(\Pi)$ is not maintained under these transformations. •

A generic momentum phase space of time-dependent mechanics is a fibre bundle $\Pi \to \mathbb{R}$ endowed with a regular Poisson structure whose characteristic distribution belongs to the vertical tangent bundle $V\Pi$ of $\Pi \to \mathbb{R}$ [144]. It can be seen locally as the Poisson product over $\mathbb{R}$ of the Legendre bundle $V^*Q \to \mathbb{R}$ and a fibre bundle over $\mathbb{R}$, equipped with the zero Poisson structure. Such a Poisson structure however cannot provide dynamic equations. A first order dynamic equation on $\Pi \to \mathbb{R}$, by definition, is a section of the affine jet bundle $J^1\Pi \to \Pi$, i.e., a connection on $\Pi \to \mathbb{R}$. Being a horizontal vector field, such a connection cannot be a Hamiltonian vector field with respect to the above mentioned Poisson structure on Π.

To endow the momentum phase space V^*Q of time-dependent mechanics with a Poisson structure, let us consider the cotangent bundle T^*Q of the configuration space Q, equipped with the coordinates (t, q^i, p, p_i). This is a particular case of the homogeneous Legendre bundle Z_Y (3.2.4). This cotangent bundle admits the canonical Liouville form

$$\Xi = pdt + p_i dq^i \tag{5.10.5}$$

and the canonical symplectic form

$$d\Xi = dp \wedge dt + dp_i \wedge dq^i. \tag{5.10.6}$$

The corresponding Poisson bracket on the ring $C^\infty(T^*Q)$ of functions on T^*Q reads

$$\{f, g\}_T = \partial^p f \partial_t g - \partial^p g \partial_t f + \partial^i f \partial_i g - \partial^i g \partial_i f. \tag{5.10.7}$$

Let us consider the subring of $C^\infty(T^*Q)$ which comprises the pull-backs $\zeta^* f$ onto T^*Q of functions f on the vertical cotangent bundle V^*Q by the canonical projection ζ (1.1.16). This subring is closed under the Poisson bracket (5.10.7). Then, by virtue of the well-known theorem [213, 299], there exists the degenerate Poisson structure

$$\boxed{\{f, g\}_V = \partial^i f \partial_i g - \partial^i g \partial_i f} \tag{5.10.8}$$

on the momentum phase space V^*Q such that

$$\zeta^*\{f, g\}_V = \{\zeta^* f, \zeta^* g\}_T. \tag{5.10.9}$$

A glance at the expression (5.10.8) shows that the holonomic coordinates of V^*Q are canonical for the Poisson structure (5.10.8).

Given the Poisson bracket (5.10.8), the *Hamiltonian vector field* ϑ_f for a function f on the momentum phase space V^*Q is defined by the relation

$$\{f,g\}_V = \vartheta_f \rfloor dg, \qquad \forall g \in \mathfrak{O}^0(V^*Q).$$

It is the vertical vector field

$$\boxed{\vartheta_f = \partial^i f \partial_i - \partial_i f \partial^i} \tag{5.10.10}$$

on the fibre bundle $V^*Q \to \mathbb{R}$. Hamiltonian vector fields generate the characteristic distribution of the Poisson structure (5.10.8) which is precisely the vertical tangent bundle $VV^*Q \subset TV^*Q$ of the fibre bundle $V^*Q \to \mathbb{R}$. This distribution is involutive and defines the symplectic foliation on the momentum phase space V^*Q, which coincides with the fibration $V^*Q \to \mathbb{R}$. The symplectic forms on the fibres of $V^*Q \to \mathbb{R}$ are the pull-backs

$$\Omega_t = dp_i \wedge dq^i$$

of the canonical symplectic form on the typical fibre T^*M of the fibre bundle $V^*Q \to \mathbb{R}$ with respect to trivialization morphisms.

The Poisson structure (5.10.8) can be introduced in a different way [213, 271]. Given a section h of the fibre bundle (1.1.16), let us consider the pull-back forms

$$\mathbf{\Theta} = h^*(\Xi \wedge dt), \qquad \mathbf{\Omega} = h^*(d\Xi \wedge dt)$$

on V^*Q. They are independent of a section h and are canonical exterior forms on V^*Q:

$$\boxed{\mathbf{\Theta} = p_i dq^i \wedge dt,} \tag{5.10.11}$$

$$\boxed{\mathbf{\Omega} = dp_i \wedge dq^i \wedge dt.} \tag{5.10.12}$$

These are the particular tangent-valued Liouville form (4.1.2) and the polysymplectic form (4.1.3), respectively. With $\mathbf{\Omega}$, the Hamiltonian vector field ϑ_f (5.10.10) for a function f on V^*Q is given by the relation

$$\vartheta_f \rfloor \mathbf{\Omega} = -df \wedge dt, \tag{5.10.13}$$

while the Poisson bracket (5.10.9) is written as

$$\{f,g\}_V dt = \vartheta_g\rfloor\vartheta_f\rfloor\boldsymbol{\Omega}.$$

Remark 5.10.2. It is easily seen that an automorphism ρ of the Legendre bundle $V^*Q \to \mathbb{R}$ is a canonical transformation of the Poisson structure (5.10.8) if and only if it preserves the canonical 3-form $\boldsymbol{\Omega}$ (5.10.12). A vector field u on V^*Q is canonical for the Poisson structure $\{,\}_V$ if and only if the form $u\rfloor\boldsymbol{\Omega}$ is exact. Let us emphasize that canonical transformations are compatible with the fibration $V^*Q \to \mathbb{R}$, but not necessarily with the fibration $\pi_Q : V^*Q \to Q$. Unless otherwise stated, we will restrict ourselves to the holonomic coordinates on V^*Y and holonomic transformations which are obviously canonical. •

The 3-form $\boldsymbol{\Omega}$ (5.10.12) gives something more than a Poisson structure on V^*Q. Let us follow the general scheme of Section 4.1.

A Hamiltonian vector field, by definition, is canonical. A converse is the following.

PROPOSITION 5.10.1. Every vertical canonical vector field on the Legendre bundle $V^*Q \to \mathbb{R}$ is locally a Hamiltonian vector field. □

The proof is based on the following fact.

LEMMA 5.10.2. Let σ be a 1-form on V^*Q. If $\sigma \wedge dt$ is closed form, it is exact. Then $\sigma \wedge dt = dg \wedge dt$ locally □

Proof. Since V^*Q is diffeomorphic to $\mathbb{R} \times T^*M$, we have the De Rham cohomology group

$$H^2(V^*Q) = H^0(\mathbb{R}) \otimes H^2(T^*M) \oplus H^1(\mathbb{R}) \otimes H^1(T^*M)$$

(see (6.8.4) below). The form $\sigma \wedge dt$ belongs to its second item which is zero. Then the relative Poincaré lemma (see, e.g., [123]) can be applied. **QED**

A connection γ on the Legendre bundle $V^*Q \to \mathbb{R}$ is called *canonical* if the corresponding horizontal vector field is canonical for the Poisson structure on V^*Q. We will prove that such a form is necessarily exact (see Proposition 5.10.3 below). A canonical connection γ is a said to be a *Hamiltonian connection* if

$$\boxed{\gamma\rfloor\boldsymbol{\Omega} = dH} \tag{5.10.14}$$

where $H = h^*\Xi$ is a Hamiltonian form. Given a local reference frame (t, q^i) on $Q \to \mathbb{R}$, a Hamiltonian form reads

$$\boxed{H = p_i dq^i - \mathcal{H} dt.} \tag{5.10.15}$$

This is the well-known integral invariant of Poincaré–Cartan, where $\mathcal{H}$ is a *Hamiltonian*. A glance at the expression (5.10.15) shows that $\mathcal{H}$ fails to be a scalar under time-dependent transformations. We will show that every Hamiltonian form H admits a unique Hamiltonian connection γ_H (see Proposition 5.10.5 below). The kernel of the covariant differential D_{γ_H}, relative to this Hamiltonian connection γ_H provides the Hamilton equations in time-dependent mechanics.

Let $\gamma = \partial_t + \gamma^i \partial_i + \gamma_i \partial^i$ be a canonical connection on the Legendre bundle $V^*Q \to \mathbb{R}$. Its components obey the relations

$$\partial^i \gamma^j - \partial^j \gamma^i = 0, \qquad \partial_i \gamma_j - \partial_j \gamma_i = 0, \qquad \partial_j \gamma^i + \partial^i \gamma_j = 0. \tag{5.10.16}$$

Canonical connections constitute an affine space modelled over the vector space of vertical canonical vector fields on $V^*Q \to \mathbb{R}$.

PROPOSITION 5.10.3. If γ is a canonical connection, then the form $\gamma \rfloor \mathbf{\Omega}$ is exact. □

Proof. Every connection Γ on $Q \to \mathbb{R}$ gives rise to the covertical connection

$$V^*\Gamma = \partial_t + \Gamma^i \partial_i - p_i \partial_j \Gamma^i \partial^j \tag{5.10.17}$$

(2.7.20) on $V^*Q \to \mathbb{R}$ which is a Hamiltonian connection for the Hamiltonian form H_Γ. Let us consider the decomposition $\gamma = V^*\Gamma + \vartheta$, where Γ is a connection on $Q \to \mathbb{R}$. The assertion follows from Proposition 5.10.1. **QED**

Thus, every canonical connection γ on V^*Q defines an exterior 1-form H modulo closed forms so that $dH = \gamma \rfloor \mathbf{\Omega}$. Such a form is called a *locally Hamiltonian form*.

PROPOSITION 5.10.4. Every locally Hamiltonian form on the momentum phase space V^*Q is locally a Hamiltonian form modulo closed forms. □

Proof. Given locally Hamiltonian forms H_γ and $H_{\gamma'}$, their difference $\sigma = H_\gamma - H_{\gamma'}$ is a 1-form on V^*Q such that the 2-form $\sigma \wedge dt$ is closed. By virtue of Lemma 5.10.2, the form $\sigma \wedge dt$ is exact and $\sigma = f dt + dg$ locally. Put $H_{\gamma'} = H_\Gamma$ where Γ

is a connection on $V^*Q \to \mathbb{R}$. Then H_γ modulo closed forms takes the local form $H_\gamma = H_\Gamma + f dt$, and coincides with the pull-back of the Liouville form Ξ on T^*Q by the local section $p = -p_i\Gamma^i + f$ of the fibre bundle (1.1.16). **QED**

PROPOSITION 5.10.5. Conversely, each Hamiltonian form H on the momentum phase space V^*Q admits a unique Hamiltonian connection γ_H on $V^*Q \to \mathbb{R}$ such that the relation (5.10.14) holds. □

Proof. Given a Hamiltonian form H, its exterior differential

$$dH = h^* d\Xi = (dp_i + \partial_i \mathcal{H} dt) \wedge (dq^i - \partial^i \mathcal{H} dt) \tag{5.10.18}$$

is a presymplectic form of constant rank $2m$ since the form

$$(dH)^m = (dp_i \wedge dq^i)^m - m(dp_i \wedge dq^i)^{m-1} \wedge d\mathcal{H} \wedge dt \tag{5.10.19}$$

is nowhere vanishing. It is also seen that $(dH)^m \wedge dt \neq 0$. It follows that the kernel of dH is a 1-dimensional distribution. Then the desired Hamiltonian connection

$$\boxed{\gamma_H = \partial_t + \partial^i \mathcal{H} \partial_i - \partial_i \mathcal{H} \partial^i} \tag{5.10.20}$$

is a unique vector field γ_H on V^*Q such that $\gamma_H \rfloor dH = 0$, $\gamma_H \rfloor dt = 1$. **QED**

Remark 5.10.3. Hamiltonian forms constitute an affine space modelled over the vector space of horizontal densities $f dt$ on $V^*Q \to \mathbb{R}$, i.e., over $C^\infty(V^*Q)$. Accordingly, Hamiltonian connections γ_H form an affine space modelled over the vector space of Hamiltonian vector fields. •

Given a Hamiltonian connection γ_H (5.10.20), the corresponding Hamilton equations $D_{\gamma_H} = 0$ take the coordinate form

$$q_t^i = \partial^i \mathcal{H}, \tag{5.10.21a}$$

$$p_{ti} = -\partial_i \mathcal{H}. \tag{5.10.21b}$$

Their classical solutions $r(t)$ are integral sections of the Hamiltonian connection γ_H, i.e., $\dot r = \gamma_H \circ r$. Thus, evolution in Hamiltonian time-dependent mechanics is described as a parallel transport along time. In Section 10.4, this description is extended to evolution in quantum time-dependent mechanics.

Remark 5.10.4. The Hamilton equations (5.10.21a) – (5.10.21b) are the particular case of the first order dynamic equations

$$q_t^i = \gamma^i, \qquad p_{ti} = \gamma_i \tag{5.10.22}$$

on the momentum phase space $V^*Q \to \mathbb{R}$, where γ is a connection on $V^*Q \to \mathbb{R}$.

The first order reduction of the equation of a motion of a point mass m subject to friction in Example 5.8.2 exemplifies first order dynamic equations which are not Hamilton ones. These equations read

$$q_t = \frac{1}{m_0}p, \qquad p_t = -\frac{k}{m_0}p. \tag{5.10.23}$$

The connection

$$\gamma = \partial_t + \frac{1}{m}p\partial_q - \frac{k}{m}p\partial_p$$

does not obey the third condition (5.10.16) for a Hamiltonian connection. At the same time, the equations (5.10.23) are equivalent to the Hamilton equations for the Hamiltonian associated with the Lagrangian (5.8.23). •

As in the case of field theory, a Hamiltonian form H (5.10.15) is the Poincaré–Cartan form for the Lagrangian

$$\boxed{L_H = h_0(H) = (p_i q_t^i - \mathcal{H})\omega} \tag{5.10.24}$$

on the jet manifold J^1V^*Q such that the Hamilton equations (5.10.21a) – (5.10.21b) are equivalent to the Euler–Lagrange equations for L_H. They characterize the kernel of the Euler–Lagrange operator

$$\mathcal{E}_H = (q_t^i - \partial^i\mathcal{H})\overline{d}p_i - (p_{ti} + \partial_i\mathcal{H})\overline{d}q^i : J^1V^*Q \to V^*V^*Q \tag{5.10.25}$$

for the Lagrangian L_H which is the *Hamilton operator* for the Hamiltonian form H.

Furthermore, given a function $f \in C^\infty(V^*Q)$ and its pull-back onto J^1V^*Q, let us consider the bracket

$$(f, L_H) = \delta^i f \delta_i L_H - \delta_i f \delta^i L_H = \mathbf{L}_{\gamma_H} f - d_t f,$$

where δ^i, δ_i are variational derivatives. Then the equation $(f, L_H) = 0$ is the evolution equation

$$d_t f = \mathbf{L}_{\gamma_H} f = \partial_t f + \{\mathcal{H}, f\}_V \tag{5.10.26}$$

in time-dependent mechanics. Note that, taken separately, the terms in its right-hand side are ill-behaved objects under time-dependent transformations. In particular, the equality $\{\mathcal{H}, f\}_V = 0$ is not preserved under time-dependent transformations.

To obtain the covariant splitting of the evolution equation, let consider a reference frame Γ and the corresponding splittings

$$\boxed{H = H_\Gamma - \widetilde{\mathcal{H}}_\Gamma dt,} \tag{5.10.27}$$
$$\boxed{\gamma_H = V^*\Gamma + \vartheta_{\widetilde{\mathcal{H}}_\Gamma},}$$

where $\vartheta_{\widetilde{\mathcal{H}}_\Gamma}$ is the vertical Hamiltonian field for the function $\widetilde{\mathcal{H}}_\Gamma$. With these splitting, the evolution equation (5.10.26) takes the form

$$\boxed{\mathbf{L}_{\gamma_H} f = V^*\Gamma\rfloor H + \{\widetilde{\mathcal{H}}_\Gamma, f\}_V.} \tag{5.10.28}$$

Remark 5.10.5. The following construction enables us to represent the right-hand side of the evolution equation (5.10.28) as a pure Poisson bracket. Given a Hamiltonian form $H = h^*\Xi$, let us consider its pull-back ζ^*H onto the cotangent bundle T^*Q. It is readily observed that the difference $\Xi - \zeta^*H$ is a horizontal 1-form on $T^*Q \to \mathbb{R}$, while

$$\mathcal{H}^* = \partial_t\rfloor(\Xi - \zeta^*H) = p + \mathcal{H} \tag{5.10.29}$$

is a function on T^*Q. Then the relation

$$\boxed{\zeta^*(\mathbf{L}_{\gamma_H} f) = \{\mathcal{H}^*, \zeta^*f\}_T} \tag{5.10.30}$$

holds for every function $f \in C^\infty(V^*Q)$. In particular, f is an integral of motion if and only if its bracket (5.10.30) vanishes. Moreover, let $\vartheta_{\mathcal{H}^*}$ be the Hamiltonian vector field for the function $\mathcal{H}^*$ (5.10.29) with respect to the canonical Poisson structure $\{,\}_T$ on T^*Q. Then

$$T\zeta(\vartheta_{\mathcal{H}^*}) = \gamma_H. \tag{5.10.31}$$

•

We complete this Section by consideration of canonical transformations in time-dependent mechanics (see Remark 5.10.2).

PROPOSITION 5.10.6. Let γ be a Hamiltonian connection on $V^*Q \to \mathbb{R}$. There exist canonical coordinates on V^*Q such that $\gamma = \partial_t$. □

The proof is based on the fact that, in accordance with the relation (5.10.14), every locally Hamiltonian connection γ is the generator of a local 1-parameter group of canonical automorphisms of the Legendre bundle $V^*Q \to \mathbb{R}$. Let V_0^*Q be the fibre of $V^*Q \to \mathbb{R}$ at $0 \in \mathbb{R}$. Then canonical coordinates of the symplectic manifold $V_0^*Q \cong T^*M$ dragged along integral curves of the complete vector field γ determine the desired canonical coordinates on V^*Q. In other words, a complete Hamiltonian connection γ on the momentum phase space V^*Q in accordance with Proposition 5.1.1 defines a trivialization

$$\psi : V^*Q \to \mathbb{R} \times V_0^*Q \tag{5.10.32}$$

of the fibre bundle $V^*Q \to \mathbb{R}$ such that the corresponding coordinates (q^A, p_A) of V^*Q (where q^A are not coordinates on Q in general) are canonical, i.e.,

$$\Omega = dp_A \wedge dq^A \wedge dt, \qquad \gamma_H = \partial_t, \qquad dH = dp_A \wedge dq^A,$$

and H reduces to the Hamiltonian form

$$H = p_A dq^A$$

with the Hamiltonian $\mathcal{H} = 0$. Then the corresponding Hamilton equations take the form of the *equilibrium equations*

$$q_t^A = 0, \qquad p_{tA} = 0 \tag{5.10.33}$$

such that $q^A(t, q^i, p_i)$ and $p_A(t, q^i, p_i)$ are constants of motion.

Example 5.10.6. Let us consider the 1-dimensional motion with constant acceleration a with respect to the coordinates (t, q). Its Hamiltonian on the momentum phase space $\mathbb{R}^3 \to \mathbb{R}$ reads

$$\mathcal{H} = \frac{p^2}{2} - aq.$$

The associated Hamiltonian connection is

$$\gamma_H = \partial_t + p\partial_q + a\partial_p.$$

This Hamiltonian connection is complete. The canonical coordinate transformations

$$q' = q - pt + \frac{at^2}{2}, \qquad p' = p - at \tag{5.10.34}$$

bring it into $\gamma_H = \partial_t$. Then the functions $q'(t, q, p)$ and $p'(t, p)$ (5.10.34) are constants of motion. •

Example 5.10.7. Let us consider the 1-dimensional oscillator with respect to the same coordinates as in the previous Example. Its Hamiltonian on the momentum phase space $\mathbb{R}^3 \to \mathbb{R}$ reads

$$\mathcal{H} = \frac{1}{2}(p^2 + q^2).$$

The associated Hamiltonian connection is

$$\gamma_H = \partial_t + p\partial_q - q\partial_p,$$

which is complete. The canonical coordinate transformations

$$q' = q\cos t - p\sin t, \qquad p' = p\cos t + q\sin t \tag{5.10.35}$$

bring it into $\gamma_H = \partial_t$. Then the functions $q'(t, q, p)$ and $p'(t, q, p)$ (5.10.35) are constants of motion. •

Note that the kinematic term in the evolution equation (5.10.28) can be eliminated at least locally by means of canonical transformations. Let a connection Γ in the splitting of a Hamiltonian form (5.10.27) be complete. With respect to the coordinate system (t, q^i) adapted to the reference frame Γ, the configuration bundle Q is trivialized, and the corresponding holonomic coordinates (t, q^i, p_i) on the momentum phase space V^*Q are canonical. With respect to these coordinates, the evolution equation (5.10.28) takes the form (5.10.4).

5.11 Connections and energy conservation laws

Following the methods of field theory in order to obtain conservation laws in time-dependent mechanics, we aim to show that, in time-dependent mechanics on a fibre bundle $Q \to \mathbb{R}$, the energy-momentum current $\mathfrak{T}_\Gamma$ with respect to a connection Γ

on $Q \to \mathbb{R}$ is the energy function of a mechanical system relative to the reference frame Γ.

Let L be a Lagrangian (5.8.1) on the velocity phase space J^1Q and u a projectable vector field (5.8.2) on the configuration bundle $Q \to \mathbb{R}$. On the shell (5.8.7), the first variational formula (5.8.4) is brought into the weak identity

$$\begin{aligned} &J^1u\rfloor d\mathcal{L} \approx -d_t\mathfrak{T}, \\ &(u^t\partial_t + u^i\partial_i + d_tu^i\partial_i^t)\mathcal{L} \approx -d_t(\pi_i(u^tq_t^i - u^i) - u^t\mathcal{L}), \end{aligned} \tag{5.11.1}$$

where, by analogy with field theory,

$$\mathfrak{T} = -u\rfloor H_L = \pi_i(u^tq_t^i - u^i) - u^t\mathcal{L} \tag{5.11.2}$$

is said to be the symmetry current along the vector field u. If the Lie derivative $\mathbf{L}_{J^1u}L$ (5.8.3) vanishes, we obtain the weak conservation law

$$\boxed{0 \approx -d_t[\pi_i(u^tq_t^i - u^i) - u^t\mathcal{L}].} \tag{5.11.3}$$

It is brought into the differential conservation law

$$0 \approx -\frac{d}{dt}[(\pi_i \circ \dot c)(u^t\partial_t c^i - u^i \circ c) - u^t\mathcal{L} \circ \dot c]$$

on solutions c of the Euler–Lagrange equations. A glance at this expression shows that, in time-dependent mechanics, the conserved current (5.11.2) plays the role of an *integral of motion*.

As in field theory, since the weak identity (5.11.1) is linear in the vector field u, one can consider superposition of weak identities (5.11.1). Every vector field u (5.8.2), projected onto ∂_t, can be written as the sum $u = \Gamma + \vartheta$ of some reference frame $\Gamma = \partial_t + \Gamma^i\partial_i$ and a vertical vector field ϑ on $Q \to \mathbb{R}$. It follows that the weak identity (5.11.1) associated with an arbitrary vector field u (5.8.2) can be represented as the superposition of those associated with a reference frame Γ and a vertical vector field ϑ.

If $u = \vartheta = \vartheta)\partial_i$, the weak identity (5.11.1) reads

$$(\vartheta^i\partial_i + d_t\vartheta^i\partial_i^t)\mathcal{L} \approx d_t(\pi_i\vartheta^i).$$

If $\mathbf{L}_\vartheta L = 0$, we obtain from (5.11.3) the weak conservation law

$$0 \approx d_t(\pi_i\vartheta^i) \tag{5.11.4}$$

and the integral of motion

$$\boxed{\mathfrak{T} = -\pi_i\vartheta^i.} \tag{5.11.5}$$

By analogy with field theory, (5.11.4) is called the Noether conservation law for the Noether current (5.11.5).

Example 5.11.1. Let assume that, given a trivialization $Q \cong \mathbb{R} \times M$ in coordinates (t, q^i), a Lagrangian L is independent of a coordinate q^1. Then the Lie derivative of L along the vertical vector field $\vartheta = \partial_1$ equals zero, and we have the conserved Noether current (5.11.5) which reduces to the momentum $\mathfrak{T} = -\pi_1$. With respect to arbitrary coordinates (t, q'^i), this Noether current takes the form

$$\mathfrak{T} = -\frac{\partial q'^i}{\partial q^1}\pi'_i.$$

•

In the case of a reference frame Γ, where $u^t = 1$, the weak identity (5.11.1) reads

$$(\partial_t + \Gamma^i\partial_i + d_t\Gamma^i\partial_i^t)\mathcal{L} \approx -d_t(\pi_i(q_t^i - \Gamma^i) - \mathcal{L}), \tag{5.11.6}$$

where

$$\boxed{\mathfrak{T}_\Gamma = \pi_i(q_t^i - \Gamma^i) - \mathcal{L}} \tag{5.11.7}$$

is said to be the *energy function* relative to the reference frame Γ [97, 213, 271].

Example 5.11.2. With respect to the coordinates adapted to the reference frame Γ, the weak identity (5.11.6) takes the form of the familiar energy conservation law

$$\partial_t\mathcal{L} \approx -d_t(\pi_i q_t^i - \mathcal{L}), \tag{5.11.8}$$

and $\mathfrak{T}_\Gamma$ coincides with the *canonical energy function*

$$E_L = \pi_i q_t^i - \mathcal{L}.$$

•

Example 5.11.3. Let us consider a free motion on a configuration space $Q \to \mathbb{R}$. It is described by the Lagrangian

$$\mathcal{L} = \frac{1}{2}\overline{m}_{ij}\overline{q}_t^i\overline{q}_t^j, \qquad \overline{m} = \text{const.}, \tag{5.11.9}$$

written with respect to a reference frame $(t,\overline{q}^i)$ such that the free motion equation takes the form (5.6.1). Let Γ be the associated connection. Then the conserved energy function $\mathfrak{T}_\Gamma$ (5.11.7) relative to this reference frame Γ is precisely the kinetic energy of this free motion. Relative to arbitrary coordinates (t,q^i) on Q, it takes the form

$$\mathfrak{T}_\Gamma = \pi_i(q^i_t - \Gamma^i) - \mathcal{L} = \frac{1}{2}m_{ij}(q^\mu)(q^i_t - \Gamma^i)(q^j_t - \Gamma^j).$$

Now we generalize this example for a motion described by the equation

$$(\partial_i - d_t\partial^t_i)\mathcal{L} + F_i(t,q^j,q^j_t) = 0$$

where $\mathcal{L}$ is the free motion Lagrangian (5.11.9) and F is an external force. The Lie derivative of the Lagrangian (5.11.9) along the reference frame Γ vanishes, and we have the weak equality (3.4.22) which reads

$$\dot{q}^i_\Gamma F_i \approx d_t\mathfrak{T}_\Gamma, \tag{5.11.10}$$

where $\dot{q}^i_\Gamma$ is the relative velocity. This is the well-known physical law whose left-hand side is the power of an external force. •

Example 5.11.4. Let us consider a 1-dimensional motion of a point mass m_0 subject to friction on the configuration space $\mathbb{R}^2 \to \mathbb{R}$, coordinated by (t,q) (see Example 5.8.2). It is described by the dynamic equation (5.8.20) which coincides with the Euler–Lagrange equations for the Lagrangian L (5.8.23). It is readily observed that the Lie derivative (5.8.3) of this Lagrangian along the vector field

$$\Gamma = \partial_t - \frac{1}{2}\frac{k}{m_0}q\partial_q \tag{5.11.11}$$

vanishes. Hence, we have the conserved energy function (5.11.7) with respect to the reference frame Γ (5.11.11). This energy function reads

$$\mathfrak{T}_\Gamma = \frac{1}{2}m_0\exp\left[\frac{k}{m_0}t\right]q_t(q_t + \frac{k}{m_0}q) = \frac{1}{2}m\dot{q}^2_\Gamma - \frac{mk^2}{8m_0^2}q^2,$$

where m is the mass function (5.8.22). •

These examples show that the energy function $\mathfrak{T}_\Gamma$ (5.11.7) characterizes a physical energy of a mechanical system with respect to a reference frame Γ. Energy functions relative to different reference frames Γ and Γ' differ from each other in the Noether current (5.11.5) along the vertical vector field $\Gamma - \Gamma'$.

Let us consider conservation laws in the case of gauge transformations which preserve the Euler–Lagrange operator $\mathcal{E}_L$, but not necessarily a Lagrangian L. We use the formula (3.4.15) where

$$J^2u = u^t\partial_t + u^i\partial_i + d_tu^i\partial_i^t + d_td_tu^i\partial_i^{tt}$$

is the second order jet prolongation of a vector field (5.8.2). Since

$$\mathbf{L}_{J^2u}\mathcal{E}_L = 0, \tag{5.11.12}$$

we have the equality $\mathbf{L}_{\overline{u}}L = d_tf$ (3.4.16), where f is a function on Q. In this case, we obtain the weak equality (3.4.17) which reads

$$0 \approx d_t(\pi_i(u^i - q_t^i) + u^t\mathcal{L} - f). \tag{5.11.13}$$

Example 5.11.5. Let L be the free motion Lagrangian (5.11.9). The corresponding Euler–Lagrange operator

$$\mathcal{E}_L = -m_{ij}q_{tt}^j\overline{d}q^i$$

is invariant under the Galilei transformations with the generator

$$u^i = v^it + a^i, \qquad v^i = \text{const.}, \qquad a^i = \text{const.}, \tag{5.11.14}$$

(see (5.6.8)). At the same time, the Lie derivative of the free motion Lagrangian (5.11.9) along the vector field (5.11.14) does not vanish, and we have

$$\mathbf{L}_{\overline{u}}L = m_{ij}v^iq_t^j = d_t(m_{ij}v^iq^j + c), \qquad c = \text{const.}$$

Then the weak equality (3.4.17) shows that $(q_t^it - q^i)$ is a constant of motion. •

Turn now to conservation laws in Hamiltonian time-dependent mechanics.

Let H be a Hamiltonian form H (5.10.15) on the momentum phase space V^*Q and the corresponding Lagrangian L_H (5.10.24) on the jet manifold J^1V^*Q. Given a vector field u (5.8.2) on the configuration space Q and its lift

$$\widetilde{u} = u^t\partial_t + u^i\partial_i - \partial_iu^jp_j\partial^i \tag{5.11.15}$$

(4.4.1) onto the Legendre bundle $V^*Q \to \mathbb{R}$, we obtain the Lie derivative

$$\mathbf{L}_{\widetilde{u}}H = \mathbf{L}_{J^1\widetilde{u}}L_H = (-u^t\partial_t\mathcal{H} + p_i\partial_t u^i - u^i\partial_i\mathcal{H} + \partial_j u^i p_i \partial^j\mathcal{H})dt. \tag{5.11.16}$$

The first variational formula (5.8.4) applied to the Lagrangian L_H (4.1.20) leads to the weak identity

$$\mathbf{L}_{\widetilde{u}}H \approx d_t(u\rfloor H)dt.$$

If the Lie derivative (5.11.16) vanishes, we have the conserved current

$$\boxed{\widetilde{\mathfrak{T}}_u = -u\rfloor dH = -p_i u^i + u^t\mathcal{H}} \tag{5.11.17}$$

along u.

If u is a vertical vector field, $\widetilde{\mathfrak{T}}$ (5.11.17) is the Noether current

$$\widetilde{\mathfrak{T}}_u = -u\rfloor q = -p_i u^i, \qquad q = p_i\overline{d}q^i \in V^*Q. \tag{5.11.18}$$

The current $\widetilde{\mathfrak{T}}$ (5.11.17) along a reference frame Γ reads

$$\widetilde{\mathfrak{T}}_\Gamma = \mathcal{H} - p_i\Gamma^i = \widetilde{\mathcal{H}}_\Gamma \tag{5.11.19}$$

(see the splitting (5.10.27)). In the case of almost regular Lagrangians, we have the relationship between Lagrangian and Hamiltonian conserved currents in accordance with Proposition 4.4.1. In particular, the Hamiltonian counterpart of the Lagrangian energy function $\mathfrak{T}_\Gamma$ (5.11.7) relative to a reference frame Γ is just the function $\widetilde{\mathcal{H}}_\Gamma$ (5.11.19), called the *Hamiltonian energy function* relative to the reference frame Γ. For instance, if $\Gamma^i = 0$, we obtain the well-known energy conservation law

$$\partial_t\mathcal{H} \approx d_t\mathcal{H}$$

relative to the coordinates adapted to the reference frame Γ. This is the Hamiltonian variant of the Lagrangian energy conservation law (5.11.8).

It is readily observed that, given a Hamiltonian form H, the energy functions $\widetilde{\mathcal{H}}_\Gamma$ constitute an affine space modelled over the vector space of Noether currents.

PROPOSITION 5.11.1. The conserved currents (5.11.17), taken with the sign minus, form a Lie algebra with respect to the Poisson bracket

$$\{-\widetilde{\mathfrak{T}}_u, -\widetilde{\mathfrak{T}}_{u'}\}_V = -\widetilde{\mathfrak{T}}_{[u,u']}. \tag{5.11.20}$$

□

In accordance with Remark 4.4.1, all Noether currents (5.11.18), taken with the sign minus, constitute a Lie algebra with respect to the bracket (5.11.20).

5.12 Systems with time-dependent parameters

Let us consider a configuration space which is a composite fibre bundle

$$Q \to \Sigma \to \mathbb{R}, \tag{5.12.1}$$

coordinated by (t, σ^m, q^i) where (t, σ^m) are coordinates of the fibre bundle $\Sigma \to \mathbb{R}$. We will treat sections h of the fibre bundle $\Sigma \to \mathbb{R}$ as time-dependent parameters, and call $\Sigma \to \mathbb{R}$ the *parameter bundle*. Then the configuration space (5.12.1) describes a mechanical system with time-dependent parameters. Note that the fibre bundle $Q \to \Sigma$ is not necessarily trivial.

Let us recall that, by virtue of Proposition 2.7.1, every section h of the parameter bundle $\Sigma \to \mathbb{R}$ defines the restriction

$$Q_h = h^*Q \tag{5.12.2}$$

of the fibre bundle $Q \to \Sigma$ to $h(\mathbb{R}) \subset \Sigma$, which is a subbundle $i_h : Q_h \hookrightarrow Q$ of the fibre bundle $Q \to \mathbb{R}$. One can think of the fibre bundle $Q_h \to \mathbb{R}$ as being a configuration space of a mechanical system with the background parameter function $h(t)$.

The velocity momentum space of a mechanical system with parameters is the jet manifold J^1Q of the composite fibre bundle (5.12.1) which is equipped with the adapted coordinates $(t, \sigma^m, q^i, \sigma^m_t, q^i_t)$.

Let the fibre bundle $Q \to \Sigma$ be provided with a connection

$$A_\Sigma = dt \otimes (\partial_t + A^i_t \partial_i) + d\sigma^m \otimes (\partial_m + A^i_m \partial_i). \tag{5.12.3}$$

Then the corresponding vertical covariant differential

$$\widetilde{D} : J^1Q \to V_\Sigma Q,$$

$$\boxed{\widetilde{D} = (q^i_t - A^i_t - A^i_m \sigma^m_t)\partial_i,} \tag{5.12.4}$$

(2.7.15) is defined on the configuration space Q. Given a section h of the parameter bundle $\Sigma \to \mathbb{R}$, its restriction to $J^1 i_h(J^1 Q_h) \subset J^1 Q$ is precisely the familiar covariant differential on Q_h corresponding to the restriction

$$A_h = \partial_t + ((A^i_m \circ h)\partial_t h^m + (A \circ h)^i_t)\partial_i \tag{5.12.5}$$

of the connection A_Σ to $h(\mathbb{R}) \subset \Sigma$. Therefore, one may use the vertical covariant differential $\widetilde{D}$ in order to construct a Lagrangian for a mechanical system with parameters on the configuration space Q (5.12.1).

We will suppose that such a Lagrangian L depends on derivatives of parameters σ_t^m only through the vertical covariant differential $\widetilde{D}$ (5.12.4), i.e.,

$$L = \mathcal{L}(t, \sigma_m, q^i, q_t^i - A^i - A_m^i \sigma_t^m)dt. \tag{5.12.6}$$

This Lagrangian is obviously degenerate because of the constraint condition

$$\boxed{\pi_m + A_m^i \pi_i = 0.}$$

As a consequence, the total system of the Euler–Lagrange equations

$$(\partial_i - d_t \partial_i^t)\mathcal{L} = 0, \tag{5.12.7}$$

$$(\partial_m - d_t \partial_m^t)\mathcal{L} = 0 \tag{5.12.8}$$

admits a solution only if the very particular relation

$$(\partial_m + A_m^i \partial_i)\mathcal{L} + \pi_i d_t A_m^i = 0 \tag{5.12.9}$$

holds. However, we believe that parameter functions are background, i.e., independent of a motion. In this case, only the Euler–Lagrange equations (5.12.7) should be considered. One can think of these equations as being the Euler–Lagrange equations for the Lagrangian $L_h = J^1 h^* L$ on the velocity phase space $J^1 Q_h$.

In particular, let us apply the first variational formula (3.4.20) in order to obtain conservation laws for a mechanical system with time-dependent parameters. Let

$$u = u^t \partial_t + u^m(t, \sigma^k)\partial_m + u^i(t, \sigma^k, q^j)\partial_i \tag{5.12.10}$$

be a vector field which is projectable with respect to both the fibration $Q \to \mathbb{R}$ and the fibration $Q \to \Sigma$. If the Lie derivative $\mathbf{L}_{J^1\overline{u}} L$ vanishes, then, on the shell (5.12.7), we obtain the conservation law

$$0 \approx (u^m - \sigma_t^m u^t)\partial_m \mathcal{L} + \pi_m d_t(u^m - \sigma_t^m u^t) - d_t[\pi_i(u^t q_t^i - u^i) - u^t \mathcal{L}] \tag{5.12.11}$$

for a system with time-dependent parameters.

Now, let us describe such a system in the framework of Hamiltonian formalism. Its momentum phase space is the vertical cotangent bundle V^*Q. Given a connection (5.12.3), we have the splitting (2.7.14) which reads

$$V^*Q = A_\Sigma(V_\Sigma^* Q) \underset{Q}{\oplus} (Q \underset{Q}{\times} V^*\Sigma),$$

$$p_i \overline{d}q^i + p_m \overline{d}\sigma^m = p_i(\overline{d}q^i - A_m^i \overline{d}\sigma^m) + (p_m + A_m^i p_i)\overline{d}\sigma^m.$$

Then V^*Q can be provided with the coordinates

$$\overline{p}_i = p_i, \qquad \overline{p}_m = p_m + A^i_m p_i$$

compatible with this splitting. However, these coordinates fail to be canonical in general. Given a section h of the parameter bundle $\Sigma \to \mathbb{R}$, the submanifold

$$\boxed{\{\sigma = h(t),\ \overline{p}_m = 0\}}$$

of the momentum phase space V^*Q is isomorphic to the Legendre bundle V^*Q_h of the subbundle (5.12.2) of the fibre bundle $Q \to \mathbb{R}$, which is the configuration space of a mechanical system with the parameter function $h(t)$.

Let us consider Hamiltonian forms on the momentum phase space V^*Q which are associated with the Lagrangian L (5.12.6). Given a connection

$$\Gamma = \partial_t + \Gamma^m \partial_m \tag{5.12.12}$$

on the parameter bundle $\Sigma \to \mathbb{R}$, the desired Hamiltonian form

$$\boxed{H = (p_i dq^i + p_m d\sigma^m) - [p_i(A^i + A^i_m \Gamma^m) + p_m \Gamma^m + \widetilde{\mathcal{H}}]dt} \tag{5.12.13}$$

can be found, where $\partial_t + \Gamma^m \partial_m + (A^i + A^i_m \Gamma^m)\partial_i$ is the composite connection (2.7.9) on the fibre bundle $Q \to \mathbb{R}$, which is defined by the connection A_Σ (5.12.3) on $Q \to \Sigma$ and the connection Γ (5.12.12) on $\Sigma \to \mathbb{R}$. The Hamiltonian function $\widetilde{H}$ is independent of the momenta p_m and satisfies the conditions

$$\begin{aligned}
&\pi_i(t, q^i, \sigma^m, \partial^i \widetilde{\mathcal{H}}(t, \sigma^m, q^i, \pi_i)) \equiv \pi_i, \\
&p_i \partial^i \widetilde{\mathcal{H}} - \widetilde{\mathcal{H}} \equiv \mathcal{L}(t, q^i, \sigma^m, \partial^i \widetilde{\mathcal{H}})
\end{aligned}$$

which are obtained by substitution of the expression (5.12.13) in the conditions (4.2.2a) – (4.2.2b).

The Hamilton equations for the Hamiltonian form (5.12.13) read

$$q^i_t = A^i + A^i_m \Gamma^m + \partial^i \widetilde{\mathcal{H}}, \tag{5.12.14a}$$

$$p_{ti} = -p_j(\partial_i A^j + \partial_i A^j_m \Gamma^m) - \partial_i \widetilde{\mathcal{H}}, \tag{5.12.14b}$$

$$\sigma^m_t = \Gamma^m, \tag{5.12.14c}$$

$$p_{tm} = -p_i(\partial_m A^i + \Gamma^n \partial_m A^i_n) - \partial_m \widetilde{\mathcal{H}}, \tag{5.12.14d}$$

whereas the Lagrangian constraint space is given by the equations

$$p_i = \pi_i(t, q^i, \sigma^m, \partial^i \widetilde{\mathcal{H}}(t, \sigma^m, q^i, p_i)), \tag{5.12.15}$$

$$p_m + A^i_m p_i = 0. \tag{5.12.16}$$

The system of equations (5.12.14a) – (5.12.16) are related in the sense of Proposition 4.2.6 to the Euler–Lagrange equations (5.12.7) – (5.12.8). Because of the equations (5.12.14d) and (5.12.16), this system is overdetermined. Therefore, the Hamilton equations (5.12.14a) – (5.12.14d) admit a solution living in the Lagrangian constraint space (5.12.15) – (5.12.16) if the very particular condition, similar to the condition (5.12.9), holds. Since the Euler–Lagrange equations (5.12.8) are not considered, we can also ignore the equation (5.12.14d). Note that the equations (5.12.14d) and (5.12.16) determine only the momenta p_m and do not influence other equations.

Therefore, let us consider the system of equations (5.12.14a) – (5.12.14c) and (5.12.15) – (5.12.16). Let the connection Γ in the equation (5.12.14c) be complete and admit the integral section $h(t)$. This equation together with the equation (5.12.16) defines a submanifold V^*Q_h of the momentum phase space V^*Q, which is the momentum phase space of a mechanical system with the parameter function $h(t)$. The remaining equations (5.12.14a) – (5.12.14b) and (5.12.15) are the equations of this system on the momentum phase space V^*Q_h, which correspond to the Euler–Lagrange equations (5.12.7) in the presence of the background parameter function $h(t)$.

Conversely, whenever $h(t)$ is a parameter function, there exists a connection Γ on the parameter bundle $\Sigma \to \mathbb{R}$ such that $h(t)$ is its integral section. Then the system of equations (5.12.14a) – (5.12.14b) and (5.12.15) – (5.12.16) describes a mechanical system with the background parameter function $h(t)$. Moreover, we can locally restrict our consideration to the equations (5.12.14a) – (5.12.14b) and (5.12.15). These are the Hamilton equation for the Hamiltonian form

$$\boxed{H_h = p_i dq^i - [p_i(A^i + A^i_m \partial_t h^m) + \widetilde{\mathcal{H}}]dt} \tag{5.12.17}$$

on V^*Q_h associated with the Lagrangian L_h on J^1Q_h.

The following examples illustrate the above construction.

Example 5.12.1. Let us consider the 1-dimensional motion of a probe particle in the presence of a force field whose centre moves. The configuration space of this system is the composite fibre bundle

$$\mathbb{R}^3 \to \mathbb{R}^2 \to \mathbb{R},$$

coordinated by (t, σ, q) where σ is a coordinate of the field centre with respect to some inertial frame and q is a coordinate of the probe particle with respect to the field centre. There is the natural inclusion

$$Q \underset{\Sigma}{\times} T\Sigma \ni (t, \sigma, q, \dot{t}, \dot{\sigma}) \mapsto (t, \sigma, \dot{t}, \dot{\sigma}, \dot{y} = -\dot{\sigma}) \in TQ$$

which defines the connection

$$A_\Sigma = dt \otimes \partial_t + d\sigma \otimes (\partial_\sigma - \partial_q)$$

on the fibre bundle $Q \to \Sigma$. The corresponding vertical covariant differential (5.12.4) reads

$$\widetilde{D} = (q_t + \sigma_t)\partial_q.$$

This is precisely the velocity of the probe particle with respect to the inertial frame. Then the Lagrangian of this particle takes the form

$$L = [\frac{1}{2}(q_t + \sigma_t)^2 - V(q)]dt. \tag{5.12.18}$$

In particular, we can obtain the energy conservation law for this system. Let us consider the vector field $u = \partial_t$. The Lie derivative of the Lagrangian (5.12.18) along this vector field vanishes. Using the formula (5.12.11), we obtain

$$0 \approx -\pi_q \sigma_{tt} - d_t[\pi_q q_t - \mathcal{L}]$$

or

$$0 \approx \partial_q \mathcal{L} \sigma_t - d_t[\pi_q(q_t + \sigma_t) - \mathcal{L}],$$

where

$$\mathfrak{T} = \pi_q(q_t + \sigma_t) - \mathcal{L}$$

is an energy function of the probe particle with respect to the inertial reference frame. •

Example 5.12.2. Let us consider an *n-body* as in Example 5.8.4. Let $\mathbb{R}^{3n-3}$ be the translation-reduced configuration space of the mass-weight Jacobi vectors $\{\vec{\rho}_A\}$. Two configurations $\{\vec{\rho}_A\}$ and $\{\vec{\rho}'_A\}$ are said to define the same shape of the n-body if $\vec{\rho}'_A = R\vec{\rho}_A$ for some rotation $R \in SO(3)$. This introduces the equivalence relation between configurations, and the *shape space* $\mathcal{S}$ of the n-body is defined as the quotient

$$\mathcal{S} = \mathbb{R}^{3n-3}/SO(3)$$

[200, 201]. Then we have the composite fibre bundle

$$\mathbb{R} \times \mathbb{R}^{3n-3} \to \mathbb{R} \times \mathcal{S} \to \mathbb{R}, \tag{5.12.19}$$

where the fibre bundle

$$\mathbb{R}^{3n-3} \to \mathcal{S} \tag{5.12.20}$$

has the structure group $SO(3)$. The composite fibre bundle (5.12.19) is provided with the bundle coordinates(t, σ^m, q^i), where q^i, $i = 1, 2, 3$, are some angle coordinates, e.g., the Eulerian angles, while σ^m, $m = 1, \ldots, 3n-6$, are said to be the *shape coordinates* on $\mathcal{S}$. A section $\vec{\varrho}_A(\sigma^m)$ of the fibre bundle (5.12.20), called a *gauge convention*, determines an orientation of the n-body in a space. Given such a section, any point $\{\rho_\alpha\}$ of the translation-reduced configuration space $\mathbb{R}^{3n-3}$ is written as

$$\vec{\rho}_A = R(q^i)\vec{\varrho}_A(\sigma^m).$$

This relation yields the splitting

$$\dot{\vec{\rho}}_A = \partial_i R \dot{q}^i \varrho_A + \partial_m \vec{\varrho}_A \dot{\sigma}^m$$

of the tangent bundle $T\mathbb{R}^{3n-3}$, which determines a connection A_Σ on the fibre bundle (5.12.20). This is also a connection on the fibre bundle

$$\mathbb{R} \times \mathbb{R}^{3n-3} \to \mathbb{R} \times \mathcal{S}$$

with the components $A_t = 0$. Then the corresponding vertical covariant differential (5.12.4) reads

$$\widetilde{D} = (q^i_t - A^i_m \sigma^m_t)\partial_i.$$

With this vertical covariant differential, the total angular velocity of the n-body takes the form

$$\vec{\Omega} = \mathbf{a}_i \widetilde{D}^i = \vec{\omega} + \mathbf{A}_m \sigma_t^m, \tag{5.12.21}$$

where $\mathbf{a}_i$ are certain kinematic coefficients and $\vec{\omega}$ is the angular velocity of the n-body as a rigid one. In particular, we obtain the phenomenon of a falling cat if $\vec{\Omega} = 0$ so that

$$\vec{\omega} = -\mathbf{A}_m \sigma_t^m.$$

•

In Section 10.5, we will extend the above description of mechanical systems with time-dependent parameters to quantum systems in order to reproduce the phenomenon of Berry's geometric phase.

Chapter 6

Gauge theory of principal connections

The literature on the geometric gauge theory is extensive. We refer the reader, e.g., to [177] for the standard exposition of geometry of principal bundles and to [214] for a survey on geometric foundations of gauge theory. In this Chapter, we presents gauge theory of principal connections as a particular case of geometric field theory formulated in terms of jet manifolds. The main ingredient in our consideration is the bundle of principal connections $C = J^1P/G$ whose sections are principal connections on a principal bundle P with a structure group G. The first order jet manifold J^1C plays the role of a finite-dimensional configuration space of gauge theory.

6.1 Principal connections

By $\pi_P : P \to X$ throughout is meant a *principal bundle* whose structure group is a real Lie group G, $\dim G > 0$. For short, P is called a G-principal bundle. By definition, $P \to X$ is a fibre bundle provided with the free transitive action of G on P on the right:

$$\begin{aligned} &R_G : P \underset{X}{\times} G \to P, \\ &R_g : p \mapsto pg, \quad p \in P, \quad g \in G. \end{aligned} \tag{6.1.1}$$

A G-principal bundle P is equipped with a bundle atlas $\Psi_P = \{(U_\alpha, \psi^P_\alpha)\}$ whose trivialization morphisms

$$\psi^P_\alpha : \pi_P^{-1}(U_\alpha) \to U_\alpha \times G$$

obey the condition

$$(\mathrm{pr}_2 \circ \psi_\alpha^P)(pg) = (\mathrm{pr}_2 \circ \psi_\alpha^P)(p)g, \qquad \forall g \in G, \qquad \forall p \in \pi_P^{-1}(U_\alpha).$$

Due to this property, every trivialization morphism ψ_α^P determines a unique local section $z_\alpha : U_\alpha \to P$ such that

$$\mathrm{pr}_2 \circ \psi_\alpha^P \circ z_\alpha = \mathbf{1},$$

where $\mathbf{1}$ is the unit element of G. The transformation rules for z_α read

$$z_\beta(x) = z_\alpha(x)\rho_{\alpha\beta}(x), \qquad x \in U_\alpha \cap U_\beta, \tag{6.1.2}$$

where $\rho_{\alpha\beta}(x,g) = \rho_{\alpha\beta}(x)g$ are transition functions (1.1.3) of the atlas Ψ_P. Conversely, the family $\{(U_\alpha, z_\alpha)\}$ of local sections of P obeying (6.1.2) uniquely determines a bundle atlas Ψ_P of P.

Note that there is the pull-back operation of a principal bundle structure. The pull-back f^*P (1.1.6) of a principal bundle is also a principal bundle with the same structure group.

Remark 6.1.1. We recall some notions related to tangent and cotangent bundles of Lie groups. Let G be a real Lie group with $\dim G > 0$ and $\mathfrak{g}_l$ [$\mathfrak{g}_r$] its *left* [*right*] Lie algebra of *left-invariant vector fields* $\xi_l(g) = TL_g(\xi_l(\mathbf{1}))$ [*right-invariant vector fields* $\xi_r(g) = TR_g(\xi_r(\mathbf{1}))$] on the group G. Here L_g and R_g denote the action of G on itself on the left and on the right, respectively. Every left-invariant vector field $\xi_l(g)$ [right-invariant vector field $\xi_r(g)$] corresponds to the element $v = \xi_l(\mathbf{1})$ [$v = \xi_r(\mathbf{1})$] of the tangent space $T_\mathbf{1}G$ provided with both left and right Lie algebra structures. For instance, given $v \in T_\mathbf{1}G$, let $v_l(g)$ and $v_r(g)$ be the corresponding left-invariant and right-invariant vector fields. There is the relation

$$v_l(g) = TL_g \circ TR_g^{-1}(v_r(g)).$$

Let $\{\epsilon_m = \epsilon_m(\mathbf{1})\}$ [$\{\varepsilon_m = \varepsilon_m(\mathbf{1})\}$] denote the basis for the left [right] Lie algebra, and let c_{mn}^k be the right structure constants, i.e.,

$$[\varepsilon_m, \varepsilon_n] = c_{mn}^k \varepsilon_k.$$

The mapping $g \mapsto g^{-1}$ yields the isomorphism

$$\mathfrak{g}_l \ni \epsilon_m \mapsto \varepsilon_m = -\epsilon_m \in \mathfrak{g}_r$$

of left and right Lie algebras.

The tangent bundle $\pi_G : TG \to G$ of the Lie group G is trivial. There are the isomorphisms

$$\varrho_l : TG \ni q \mapsto (g = \pi_G(q), TL_g^{-1}(q)) \in G \times \mathfrak{g}_l,$$
$$\varrho_r : TG \ni q \mapsto (g = \pi_G(q), TR_g^{-1}(q)) \in G \times \mathfrak{g}_r.$$

The left action L_g of a Lie group G on itself defines its adjoint representation $g \mapsto \mathrm{Ad}g$ in the right Lie algebra $\mathfrak{g}_r$ and its identity representation in the left Lie algebra $\mathfrak{g}_l$. Correspondingly, there is the *adjoint representation*

$$\varepsilon' : \varepsilon \mapsto \mathrm{ad}\,\varepsilon'(\varepsilon) = [\varepsilon', \varepsilon],$$
$$\mathrm{ad}\,\varepsilon_m(\varepsilon_n) = c_{mn}^k \varepsilon_k,$$

of the right Lie algebra $\mathfrak{g}_r$ in itself.

An action

$$G \times Z \ni (g, z) \mapsto gz \in Z$$

of a Lie group G on a manifold Z on the left yields the homomorphism

$$\mathfrak{g}_r \ni \varepsilon \to \xi_\varepsilon \in \mathcal{T}(Z)$$

of the right Lie algebra $\mathfrak{g}_r$ of G into the Lie algebra of vector fields on Z such that

$$\xi_{\mathrm{Ad}\,g(\varepsilon)} = Tg \circ \xi_\varepsilon \circ g^{-1} \tag{6.1.3}$$

[177]. Vector fields ξ_{ε_m} are said to be the *generators* of a representation of the Lie group G in Z.

Let $\mathfrak{g}^* = T_e^* G$ be the vector space dual of the tangent space $T_1 G$. It is called the *dual Lie algebra* (or the *Lie coalgebra*), and is provided with the basis $\{\varepsilon^m\}$ dual of the basis $\{\varepsilon_m\}$ for $T_1 G$. The group G and the right Lie algebra $\mathfrak{g}_r$ act on $\mathfrak{g}^*$ by the *coadjoint representation*

$$\langle \mathrm{Ad}^* g(\varepsilon^*), \varepsilon \rangle \stackrel{\mathrm{def}}{=} \langle \varepsilon^*, \mathrm{Ad}\, g^{-1}(\varepsilon) \rangle, \qquad \varepsilon^* \in \mathfrak{g}^*, \qquad \varepsilon \in \mathfrak{g}_r, \tag{6.1.4}$$
$$\langle \mathrm{ad}^* \varepsilon'(\varepsilon^*), \varepsilon \rangle = -\langle \varepsilon^*, [\varepsilon', \varepsilon] \rangle, \qquad \varepsilon' \in \mathfrak{g}_r,$$
$$\mathrm{ad}^* \varepsilon_m(\varepsilon^n) = -c_{mk}^n \varepsilon^k.$$

Note that, in the literature, one can meet another definition of the coadjoint representation in accordance with the relation

$$\langle \mathrm{Ad}^* g(\varepsilon^*), \varepsilon \rangle = \langle \varepsilon^*, \mathrm{Ad}\, g(\varepsilon) \rangle.$$

An exterior form ϕ on the group G is said to be *left-invariant* [*right-invariant*] if $\phi(\mathbf{1}) = L^*_g(\phi(g))$ $[\phi(\mathbf{1}) = R^*_g(\phi(g))]$. The exterior differential of a left-invariant [right-invariant] form is left-invariant [right-invariant]. In particular, left-invariant 1-forms satisfy the *Maurer–Cartan equations*

$$d\phi(\epsilon,\epsilon') = -\frac{1}{2}\phi([\epsilon,\epsilon']), \qquad \epsilon,\epsilon'\in\mathfrak{g}_l.$$

There is the *canonical* $\mathfrak{g}_l$*-valued left-invariant 1-form*

$$\theta_l : T_\mathbf{1}G \ni \epsilon \mapsto \epsilon \in \mathfrak{g}_l \tag{6.1.5}$$

on a Lie group G. The components θ^m_l of its decomposition $\theta_l = \theta^m_l\epsilon_m$ with respect to the basis for the left Lie algebra $\mathfrak{g}_l$ make up the basis for the space of left-invariant exterior 1-forms on G:

$$\epsilon_m\rfloor\theta^n_l = \delta^n_m.$$

The Maurer–Cartan equations, written with respect to this basis, read

$$d\theta^m_l = \frac{1}{2}c^m_{nk}\theta^n_l\wedge\theta^k_l.$$

•

The canonical action (6.1.1) of G on P on the right defines the canonical trivial vertical splitting

$$\alpha : VP \underset{P}{\cong} P\times\mathfrak{g}_l$$

such that $\alpha^{-1}(\epsilon_m)$ are the familiar *fundamental vector fields* on P corresponding to the basis elements ϵ_m of the Lie algebra $\mathfrak{g}_l$.

Taking the quotient of the tangent bundle $TP \to P$ and the vertical tangent bundle VP of P by TR_G (or simply by G), we obtain the vector bundles

$$T_GP = TP/G \quad\text{and}\quad V_GP = VP/G \tag{6.1.6}$$

over X. Sections of $T_GP\to X$ are G-invariant vector fields on P, while sections of $V_GP\to X$ are G-invariant vertical vector fields on P. Hence, the typical fibre of $V_GP\to X$ is the *right Lie algebra* $\mathfrak{g}_r$ of the right-invariant vector fields on the group G. The group G acts on this typical fibre by the adjoint representation.

The Lie bracket of G-invariant vector fields on P goes to the quotient by G and defines the Lie bracket of sections of the vector bundles $T_GP \to X$ and $V_GP \to X$. It follows that $V_GP \to X$ is a Lie algebra bundle (the *gauge algebra bundle* in the terminology of gauge theories) whose fibres are Lie algebras isomorphic to the right Lie algebra $\mathfrak{g}_r$ of G.

Given a local bundle splitting of P, there are the corresponding local bundle splittings of T_GP and V_GP. Given the basis $\{\varepsilon_p\}$ for the Lie algebra $\mathfrak{g}_r$, we obtain the local fibre bases $\{\partial_\lambda, e_p\}$ for $T_GP \to X$ and $\{e_p\}$ for V_GP. If

$$\xi = \xi^\lambda \partial_\lambda + \xi^p e_p, \qquad \eta = \eta^\mu \partial_\mu + \eta^q e_q$$

are sections of T_GP, the coordinate expression of their bracket is

$$[\xi, \eta] = (\xi^\mu \partial_\mu \eta^\lambda - \eta^\mu \partial_\mu \xi^\lambda)\partial_\lambda + (\xi^\lambda \partial_\lambda \eta^r - \eta^\lambda \partial_\lambda \xi^r + c^r_{pq} \xi^p \eta^q) e_r. \tag{6.1.7}$$

Let J^1P be the first order jet manifold of a G-principal bundle $P \to X$. Bearing in mind that $J^1P \to P$ is an affine bundle modelled over the vector bundle

$$T^*X \underset{P}{\otimes} VP \to P,$$

let us consider the quotient of the jet bundle $J^1P \to P$ by the jet prolongation J^1R_G of the canonical action (6.1.1). We obtain the affine bundle

$$\boxed{C = J^1P/G \to X} \tag{6.1.8}$$

modelled over the vector bundle

$$\overline{C} = T^*X \otimes V_GP \to X. \tag{6.1.9}$$

Hence, there is the canonical vertical splitting

$$VC \cong C \underset{X}{\times} \overline{C}.$$

It is easily seen that the fibre bundle $J^1P \to C$ is a principal bundle with the structure group G. It is canonically isomorphic to the pull-back

$$\boxed{J^1P \cong P_C = C \underset{X}{\times} P \to C.} \tag{6.1.10}$$

Turn now to connections on a principal bundle $P \to X$. In this case, the exact sequence (1.1.17a) can be reduced to the exact sequence

$$\boxed{0 \to V_GP \underset{X}{\hookrightarrow} T_GP \to TX \to 0} \tag{6.1.11}$$

by taking the quotient with respect to the action of the group G.

DEFINITION 6.1.1. A *principal connection* A on a principal bundle $P \to X$ is defined as a section $A : P \to J^1P$ which is equivariant under the action (6.1.1) of the group G on P, that is,

$$J^1R_g \circ A = A \circ R_g, \qquad \forall g \in G. \tag{6.1.12}$$

□

Such a connection yields the splitting of the exact sequence (6.1.11), and can be represented by the T_G-valued form

$$\begin{array}{ccc} & & T^*X \otimes T_GP \\ & {\scriptstyle A} \nearrow & \downarrow \\ X & \xrightarrow{\theta_X} & T^*X \otimes TX \end{array}$$

$$\boxed{A = dx^\lambda \otimes (\partial_\lambda + A^q_\lambda e_q),} \tag{6.1.13}$$

where A^p_λ are local functions on X. On the other hand, due to the property (6.1.12), there is one-to-one correspondence between the principal connection on a principal bundle $P \to X$ and the global sections of the fibre bundle $C \to X$ (6.1.8), called the *bundle of principal connections*. Since this fibre bundle is affine, principal connections on a principal bundle always exist.

Given a bundle atlas of P, the bundle of principal connections C is equipped with the associated bundle coordinates (x^λ, a^q_λ) such that, for any section A of $C \to X$, the local functions

$$A^q_\lambda = a^q_\lambda \circ A$$

are coefficients of the connection 1-form (6.1.13). In gauge theory, these coefficients are treated as *gauge potentials*. We will use this term to refer to sections A of the fibre bundle $C \to X$.

Let a principal connection on the principal bundle $P \to X$ be represented by the vertical-valued form A (2.1.8). Then the form

$$\overline{A} : P \xrightarrow{A} T^*P \underset{P}{\otimes} VP \xrightarrow{\mathrm{Id}\otimes\alpha} T^*P \otimes \mathfrak{g}_l \tag{6.1.14}$$

is the familiar $\mathfrak{g}_l$-valued *connection form* on the principal bundle P. Given a local bundle splitting (U_ζ, z_ζ) of P, this form reads

$$\overline{A} = \psi_\zeta^*(\theta_l - \overline{A}_\lambda^q dx^\lambda \otimes \epsilon_q), \tag{6.1.15}$$

where θ_l is the canonical $\mathfrak{g}_l$-valued 1-form (6.1.5) on G and $\overline{A}_\lambda^p$ are local functions on P such that

$$\overline{A}_\lambda^q(pg)\epsilon_q = \overline{A}_\lambda^q(p)\mathrm{ad}g^{-1}(\epsilon_q).$$

The pull-back $z_\zeta^*\overline{A}$ of the connection form $\overline{A}$ over U_ζ is the well-known *local connection 1-form*

$$A_\zeta = -A_\lambda^q dx^\lambda \otimes \epsilon_q = A_\lambda^q dx^\lambda \otimes \varepsilon_q, \tag{6.1.16}$$

where $A_\lambda^q = \overline{A}_\lambda^q \circ z_\zeta$ are local functions on X. It is readily observed that the coefficients A_λ^q of this form are precisely the coefficients of the form (6.1.13). It should be emphasized that the local connection form (6.1.16) is $\mathfrak{g}_l$-valued, while the vertical part $A - \theta_X$ of A (6.1.13) is a V_G-valued form. There is the relation $A_\zeta = \psi_z eta(A - \theta_X)$. In local expressions, we will mean by a local connection form the V_G-valued form

$$A = A_\lambda^q dx^\lambda \otimes e_q. \tag{6.1.17}$$

In the case of principal bundles, there are both pull-back and push-forward operations of principal connections [177].

THEOREM 6.1.2. Let P be a principal fibre bundle and f^*P (1.1.6) the pull-back principal bundle with the same structure group. Let f_P be the canonical morphism (1.1.7) of f^*P to P. If A is a principal connection on P, then the pull-back connection f^*A (2.1.11) on f^*P is a principal connection. □

THEOREM 6.1.3. Let $P' \to X$ and $P \to X$ be principle bundles with structure groups G' and G, respectively. Let $\Phi : P' \to P$ be a principal bundle morphism over X with the corresponding homomorphism $G' \to G$. For every principal connection A' on P', there exists a unique principal connection A on P such that $T\Phi$ sends the horizontal subspaces of A' onto the horizontal subspaces of A. □

The *curvature of a principal connection* A (or the *strength* of A) is defined to be the V_GP-valued 2-form on X

$$F_A : X \to \overset{2}{\wedge} T^*X \otimes V_GP,$$
$$F_A = \frac{1}{2} F^r_{\lambda\mu} dx^\lambda \wedge dx^\mu \otimes e_r, \tag{6.1.18}$$
$$\boxed{F^r_{\lambda\mu} = [\partial_\lambda + A^p_\lambda e_p, \partial_\mu + A^q_\mu e_q]^r = \partial_\lambda A^r_\mu - \partial_\mu A^r_\lambda + c^r_{pq} A^p_\lambda A^q_\mu,} \tag{6.1.19}$$

whose coordinate expression follows from (6.1.7). We have locally

$$F_A = dA + \frac{1}{2}[A, A] = dA + A \wedge A, \tag{6.1.20}$$

where A is the local connection form (6.1.17). It should be emphasized that the form F_A (6.1.19) is not the standard curvature (2.3.3) of a connection (see (6.1.25) below). It also differs from the $\mathfrak{g}_l$-valued curvature form

$$\Omega = d\overline{A} + \frac{1}{2}[\overline{A}, \overline{A}].$$

Given a local bundle splitting $(U_\varsigma, psi_\varsigma)$ of P, we have the relation $z^*_\varsigma \Omega = -\psi_\varsigma(F_A)$. In local expressions, by F_A we will also mean the $\mathfrak{g}_r$-valued 2-form

$$\psi_\varsigma(F_A) = dA + \frac{1}{2}[A, A] = dA + A \wedge A, \tag{6.1.21}$$

where A is the local connection form (6.1.16). It is given by the expression (6.1.18) where the fibre basis $\{e_r\}$ are replaced with the right Lie algebra basis $\{\varepsilon_r\}$.

Let now

$$Y = (P \times V)/G \tag{6.1.22}$$

be a *fibre bundle associated with the principal bundle* $P \to X$ whose structure group G acts on the typical fibre V of Y on the left. For short, we will say that (6.1.22) is a *P-associated fibre bundle.*

Remark 6.1.2. Let us recall that the quotient in (6.1.22) is defined by identification of the elements (p, v) and $(pg, g^{-1}v)$ for all $g \in G$. By $[p]$ we will denote the restriction of the canonical morphism

$$P \times V \to (P \times V)/G \tag{6.1.23}$$

to $\{p\} \times V$, and write

$$[p](v) = (p,v) \cdot G, \qquad v \in V.$$

Then we have $[p](v) = [pg](g^{-1}v)$. •

Remark 6.1.3. In fact, Y (6.1.22) is the fibre bundle *canonically associated* with the principal bundle P. A fibre bundle $Y \to X$, given by the triple (X, V, Ψ) of a base X, a typical fibre V and a bundle atlas Ψ, is called a *fibre bundle with a structure group G* if G acts effectively on V on the left and the transition functions $\rho_{\lambda\beta}$ of the atlas Ψ take their values into the group G. The set $\{(U_\alpha \cap U_\beta, \rho_{\lambda\beta}\}$ of these transition functions, satisfying the cocycle condition (1.1.4), form a cocycle (see Remark 6.9.2 below). If atlases are equivalent the cocycles of their transition functions are equivalent. The set of equivalent cocycles are elements of the first cohomology set $H^1(X; G_\infty)$. Fibre bundles (X, V, G, Ψ) and (X, V', G, Ψ') with the same structure group G, which may have different typical fibres, are called *associated* if the transition functions $\{\rho_{\alpha\beta}\}$ and $\{\rho'_{\mu\nu}\}$ of the atlases Ψ and Ψ', respectively, belong to the same element of the the cohomology set $H^1(X; G_\infty)$. Any two associated fibre bundles with the same typical fibre are isomorphic to each other, but their isomorphism is not canonical in general. A fibre bundle $Y \to X$ with a structure group G is associated with some G-principal bundle $P \to X$. If Y is canonically associated with P as in (6.1.22), then

- every atlas $\Psi_P = \{(U_\alpha, z_\alpha)\}$ of P determines canonically the *associated atlas* $\Psi = \{(U_\alpha, \psi_\alpha(x) = [z_\alpha(x)]^{-1})$ of Y;
- every automorphism of a principal bundle P yields the corresponding automorphism of the P-associated fibre bundle (6.1.22) (see Section 6.3).

Unless otherwise stated (see Section 6.7), by a P-associated fibre bundle we mean the quotient (6.1.22).

It should be emphasized that the notions introduced in this Remark are extended in a straightforward manner to topological fibre bundles over topological spaces when morphisms are continuous, but not necessarily smooth. •

Let Y be a P-associated fibre bundle (6.1.22). Every principal connection on $P \to X$ induces canonically the corresponding connection on the P-associated fibre bundle (6.1.22) as follows. Given a principal connection A (6.1.13) on P and the corresponding horizontal splitting of the tangent bundle TP, the tangent map to the canonical morphism (6.1.23) defines the horizontal splitting of the tangent bundle TY and the corresponding connection on the P-associated fibre bundle $Y \to X$ [177]. The latter is called the *associated principal connection* or simply a principal connection on $Y \to X$. If Y is a vector bundle, this connection takes the form

$$\boxed{A = dx^\lambda \otimes (\partial_\lambda - A^p_\lambda I^i_p \partial_i),} \tag{6.1.24}$$

where I_p are generators of the representation of the Lie algebra $\mathfrak{g}_r$ in V. The curvature (2.3.3) of this connection reads

$$F = -\frac{1}{2} F^p_{\lambda\mu} I^i_p dx^\lambda \wedge dx^\mu \otimes \partial_i. \tag{6.1.25}$$

In particular, a principal connection A yields an associated linear connection on the gauge algebra bundle $V_G P \to X$. The corresponding covariant differential $\nabla^A \xi$ of a section $\xi = \xi^p e_p$ of $V_G P \to X$ reads

$$\begin{aligned} &\nabla^A \xi : X \to T^*X \otimes V_G P, \\ &\boxed{\nabla^A \xi = (\partial_\lambda \xi^r + c^r_{pq} A^p_\lambda \xi^q) dx^\lambda \otimes e_r.} \end{aligned} \tag{6.1.26}$$

If u is a vector field on X, the covariant derivative $\nabla^A_u \xi$ of ξ along u is given by

$$\nabla^A_u \xi = u \rfloor \nabla^A \xi = [u \rfloor A, \xi],$$

where A is the T_G-valued form (6.1.13). In particular, we have

$$\nabla^A_\lambda e_q = c^r_{pq} A^p_\lambda e_r. \tag{6.1.27}$$

The covariant derivative ∇^A_u is compatible with the Lie bracket of sections of $V_G P \to X$, i.e.,

$$\nabla^A_u [\xi, \eta] = [\nabla^A_u \xi, \eta] + [\xi, \nabla^A_u \eta]$$

for any vector field $u : X \to TX$ and sections $\xi, \eta : X \to V_G P$.

Remark 6.1.4. Let $P \to X$ be a G-principal fibre bundle. Then the F-N bracket on $\mathfrak{O}^*(P) \otimes \mathcal{T}(P)$ is compatible with the canonical action R_G, and we obtain the induced F-N bracket on $\mathfrak{O}^*(X) \otimes T_G P(X)$, where $T_G P(X)$ is the vector space of sections of the vector bundle $T_G P \to X$. Recall that $T_G P(X)$ projects onto $\mathcal{T}(X)$.

If $A \in \mathfrak{O}^1(X) \otimes T_G P(X)$ is a principal connection as in (6.1.13), the associated Nijenhuis differential is

$$d_A : \mathfrak{O}^r(X) \otimes T_G P(X) \to \mathfrak{O}^{r+1}(X) \otimes V_G P(X),$$
$$d_A \phi = [A, \phi]_{\rm FN}, \quad \phi \in \mathfrak{O}^r(X) \otimes T_G P(X). \tag{6.1.28}$$

On $V_G P(X)$, the differential d_A coincides with the covariant differential ∇^A (6.1.26), i.e.,

$$d_a \xi = \nabla^A \xi.$$

We also have the local expression

$$\nabla^A \xi = d\xi + [A, \xi], \tag{6.1.29}$$

where A is the local connection form (6.1.17). If $\phi = \alpha \otimes \xi \in \mathfrak{O}^r(X) \otimes V_G P(X)$ where $\alpha \in \mathfrak{O}^r(X)$ and $\xi \in V_G P(X)$, we have the formula

$$\boxed{d_A \phi = d\alpha \otimes \xi + (-1)^r \alpha \wedge \nabla^A \xi} \tag{6.1.30}$$

which follows from (1.2.31).

By means of the Nijenhuis differential (6.1.28), the strength F_A of the connection A is defined as

$$F_A = \frac{1}{2} d_A A = \frac{1}{2} [A, A]_{\rm FN} \in \mathfrak{O}^2(X) \otimes V_G P(X). \tag{6.1.31}$$

•

6.2 The canonical principal connection

This Section is devoted to vector fields and connections on the bundle of principal connections $C \to X$. To introduce them, we will use the canonical connection on the pull-back principal bundle $P_C \to C$ (6.1.10).

Given a G-principal bundle $P \to X$ and its jet manifold $J^1 C$ coordinated by $(x^\lambda, a^p_\mu, a^p_{\lambda\mu})$, let us consider the canonical morphism

$$\theta : J^1 P \underset{P}{\times} TP \to VP$$

(1.3.8). Taking its quotient with respect to G, we obtain the morphism

$$\begin{array}{ccc} C\underset{X}{\times} T_GP & \xrightarrow{\theta} & V_GP \\ & \searrow\swarrow & \\ & X & \end{array} \tag{6.2.1}$$

$$\theta(\partial_\lambda) = -a^p_\lambda e_p, \qquad \theta(e_p) = e_p.$$

It follows that the exact sequence (6.1.11) admits the canonical splitting over C [116].

Let us now consider the pull-back principal bundle P_C (6.1.10) whose structure group is G. Since

$$V_G(C\underset{X}{\times} P) = C\underset{X}{\times} V_GP, \qquad T_G(C\underset{X}{\times} P) = TC\underset{X}{\times} T_GP, \tag{6.2.2}$$

the exact sequence (6.1.11) for this principal bundle P_C reads

$$0 \to C\underset{X}{\times} V_GP \underset{C}{\hookrightarrow} TC\underset{X}{\times} T_GP \to TC \to 0. \tag{6.2.3}$$

It is readily observed that the morphism (6.2.1) yields the horizontal splitting (2.1.3)

$$TC\underset{X}{\times} T_GP \to C\underset{X}{\times} T_GP \to C\underset{X}{\times} V_GP$$

of the exact sequence (6.2.3) and, consequently, defines the non-flat principal connection

$$\begin{gathered} \mathcal{A} : TC \to TC\underset{X}{\times} T_GP, \\ \boxed{\mathcal{A} = dx^\lambda \otimes (\partial_\lambda + a^p_\lambda e_p) + da^r_\lambda \otimes \partial^\lambda_r,} \\ \mathcal{A} \in \mathfrak{O}^1(C) \otimes T_G(C\underset{X}{\times} P)(X), \end{gathered} \tag{6.2.4}$$

on the principal bundle

$$\boxed{P_C = C\underset{X}{\times} P \to C.} \tag{6.2.5}$$

It follows that the principal bundle P_C carries the *canonical principal connection* (6.2.4).

Accordingly, the vector bundle

$$C\underset{X}{\times} V_GP \to C$$

is provided with the canonical linear connection such that the corresponding covariant differential (6.1.27) reads

$$\partial_\lambda\rfloor\nabla^{\mathcal{A}} e_q = c^r_{pq} a^p_\lambda e_r, \qquad \partial^\lambda_r\rfloor\nabla^{\mathcal{A}} e_q = 0. \tag{6.2.6}$$

Following (6.1.31), we define the *canonical curvature*

$$F_{\mathcal{A}} = \frac{1}{2} d_{\mathcal{A}}\mathcal{A} = \frac{1}{2}[\mathcal{A},\mathcal{A}]_{\mathrm{FN}} \in \mathfrak{O}^2(C)\otimes V_G P(X),$$
$$\boxed{F_{\mathcal{A}} = (da^r_\mu \wedge dx^\mu + \tfrac{1}{2} c^r_{pq} a^p_\lambda a^q_\mu dx^\lambda \wedge dx^\mu)\otimes e_r.} \tag{6.2.7}$$

Its meaning is the following. Let $A : X \to C$ be a principal connection on the principal bundle $P \to X$. Then the pull-back

$$A^* F_{\mathcal{A}} = F_A \tag{6.2.8}$$

is the curvature of the principal connection A.

Example 6.2.1. In particular, let us consider the trivial principal bundle $P = X \times \mathbb{R} \to X$. Then $C = T^*X \to X$ is the affine cotangent bundle, and principal connections on P are precisely 1-forms on X. The canonical connection $\mathcal{A}$, the F-N covariant differential $d_{\mathcal{A}}$ and the curvature reduce to

$$\theta = \dot{x}_\lambda dx^\lambda,$$
$$d_{\mathcal{A}} = d,$$
$$F_{\mathcal{A}} = \Omega = d\theta \in \overset{2}{\wedge}(T^*X), \qquad \Omega = d\dot{x}_\lambda \wedge dx^\lambda. \tag{6.2.9}$$

•

This Example shows that the bundle $C \to X$ is a generalization of the cotangent bundle $T^*X \to X$ in a sense. Indeed, just as T^*X carries the canonical symplectic form (6.2.9), C does the canonical $V_G P$-valued 2-form (6.2.7). In particular, given a vector field τ on a manifold X, its canonical lift $\widetilde{\tau}$ (1.2.4) onto the cotangent bundle T^*X can be determined by the equation

$$\widetilde{\tau}\rfloor\Omega = d(\tau\rfloor\theta). \tag{6.2.10}$$

The generalization of this equation by means of the canonical curvature $F_{\mathcal{A}}$ is of basic importance in gauge theories.

Let $\xi = \tau^\lambda\partial_\lambda + \xi^p e_p$ be a section of the fibre bundle $T_GP \to X$, projected onto a vector field τ on X. One can think of ξ as being a generator of a 1-parameter group of general gauge transformations of the principal bundle $P \to X$ (see the next Section). Using (6.2.1), we obtain the morphism over X

$$\xi\rfloor\theta : C \to V_GP,$$

i.e., as (6.2.2) shows, a section of $V_G(C \underset{X}{\times} P) \to C$. Then the equation

$$\xi_C\rfloor F_A = d_A(\xi\rfloor\theta) \tag{6.2.11}$$

determines uniquely a vector field ξ_C on C projectable over τ. Simple computations lead to

$$\begin{aligned} &\xi_C = \tau^\lambda\partial_\lambda + u^r_\lambda\partial^\lambda_r, \\ &u^r_\lambda = \partial_\lambda\xi^r + c^r_{pq}a^p_\lambda\xi^q - a^r_\mu\partial_\lambda\tau^\mu. \end{aligned} \tag{6.2.12}$$

The vector field ξ_C (6.2.12) is the generator of the associated gauge transformations of the bundle of principal connections C. In particular, if $\xi \in V_GP(X)$, we obtain the vertical vector field

$$\boxed{\xi_C = u^r_\lambda\partial^\lambda_r, \quad u^r_\lambda = \partial_\lambda\xi^r + c^r_{pq}a^p_\lambda\xi^q.} \tag{6.2.13}$$

In this case, since $VC = C \underset{X}{\times} T^*X \otimes V_GP \subset TC$, we can write

$$\xi_C = \nabla\xi : C \to VC \tag{6.2.14}$$

where ∇ is the covariant differential (6.2.6).

Remark 6.2.2. The jet prolongation (1.3.10) of the vector field ξ_C reads

$$\begin{aligned} &J^1\xi_C = \tau^\lambda\partial_\lambda + u^r_\lambda\partial^\lambda_r + u^r_{\lambda\mu}\partial^{\lambda\mu}_r, \\ &u^r_{\lambda\mu} = \partial_{\lambda\mu}\xi^r - a^r_\nu\partial_{\lambda\mu}\tau^\nu + c^r_{pq}a^p_\mu\partial_\lambda\xi^q - \\ &\qquad a^r_{\lambda\nu}\partial_\mu\tau^\nu + c^r_{pq}a^p_{\lambda\mu}\xi^q - a^r_{\nu\mu}\partial_\lambda\tau^\nu, \end{aligned} \tag{6.2.15}$$

where u^r_λ is given as in (6.2.12). •

Example 6.2.3. Let A (6.1.13) be a principal connection. For any vector field τ on X, this connection yields the section

$$\begin{aligned} &\xi = \tau\rfloor A : X \to T_GP, \\ &\xi = \tau^\lambda\partial_\lambda + A^p_\lambda\tau^\lambda e_p, \end{aligned}$$

which, in turn, defines the vector field (6.2.12):

$$\widetilde{\tau}_A = \tau^\lambda \partial_\lambda + u^r_\lambda \partial^\lambda_r,$$

$$\boxed{u^r_\lambda = \partial_\lambda A^r_\mu \tau^\mu + c^r_{pq} a^p_\lambda A^q_\mu \tau^\mu - (a^r_\mu - A^r_\mu)\partial_\lambda \tau^\mu,} \qquad (6.2.16)$$

on the bundle of principal connection C. •

One can obtain the vector field (6.2.16) as the horizontal lift $u\rfloor\Gamma$ of τ with respect to some connection Γ on the fibre bundle $C \to X$. For this purpose, let us consider a symmetric world connection K on X and a principal connection A on $P \to X$. They define a connection on the fibre bundle $C \to X$ as follows. Given the linear connection (6.1.27) induced by A on the fibre bundle $V_G P \to X$, let us consider the tensor product connection $\overline{\Gamma}$ on $T^*X \otimes V_G P \to X$ induced by K and A. Given the coordinates $(x^\lambda, \overline{a}^r_\mu, \overline{a}^r_{\lambda\mu})$ of $J^1(T^*X \otimes V_G P)$, we have

$$\overline{\Gamma} : T^*X \otimes V_G P \to J^1(T^*X \otimes V_G P),$$

$$\overline{a}^r_{\lambda\mu} \circ \overline{\Gamma} = -K_\lambda{}^\nu{}_\mu \overline{a}^r_\nu + c^r_{pq} \overline{a}^p_\mu A^q_\lambda.$$

Using the fibred morphism

$$D_A : C \to T^*X \otimes V_G P$$

(1.1.10), we obtain the section

$$\Gamma : C \to J^1 C,$$

$$\boxed{a^r_{\lambda\mu} \circ \Gamma = \Gamma^r_{\lambda\mu} = \partial_\lambda A^r_\mu + c^r_{pq} a^p_\mu A^q_\lambda - K_\lambda{}^\nu{}_\mu (a^r_\nu - A^r_\nu) + c^r_{pq} A^p_\lambda A^q_\mu,} \qquad (6.2.17)$$

in accordance with the commutative diagram

$$\begin{array}{ccc} J^1 C & \overset{J^1 D_A}{\longrightarrow} & J^1(T^*X \otimes V_G P) \\ \Gamma \uparrow & & \uparrow \overline{\Gamma} \\ C & \overset{D_A}{\longrightarrow} & T^*X \otimes V_G P \end{array}$$

Of course, Γ is an affine morphism over X, i.e., an affine connection on the affine bundle $C \to X$, while the associated linear connection is $\overline{\Gamma}$. Moreover, it is easily seen that A is an integral section of Γ, i.e., $J^1 A = \Gamma \circ A$. The connection (6.2.17) is not a unique one defined by a symmetric world connection K and a principal connection A. Since the strength F_A of A can be seen as a soldering form

$$F_A = F^r_{\lambda\mu} dx^\lambda \otimes \partial^\mu_r, \qquad (6.2.18)$$

one can obtain another connection $\Gamma' = \Gamma - F_A$ on the fibre bundle $C \to X$. Now let us assume that a vector field τ on X is an integral section of the symmetric world connection K (see Remark 2.4.2). Then it is readily observed that the horizontal lift $\Gamma'\tau$ of τ by means of the connection Γ' coincides with the vector field $\widetilde{\tau}_A$ (6.2.16) on the fibre bundle C.

Let us return to the canonical curvature F_A (6.2.7). It can be seen in a slightly different way. Namely, there is a horizontal V_GP-valued 2-form

$$\mathcal{F} = \frac{1}{2}\mathcal{F}^r_{\lambda\mu} dx^\lambda \wedge dx^\mu \otimes e_r,$$
$$\mathcal{F}^r_{\lambda\mu} = a^r_{\lambda\mu} - a^r_{\mu\lambda} + c^r_{pq} a^p_\lambda a^q_\mu, \tag{6.2.19}$$

on J^1C which satisfies the condition

$$\mathcal{F} \circ J^1A = h_0(F_A)$$

for each principal connection $A : X \to C$. It is readily observed that

$$\mathcal{F}/2 \to \overset{2}{\wedge} T^*X \otimes V_GP \tag{6.2.20}$$

is an affine surjection over C and, hence, its kernel $C_+ = \operatorname{Ker} \mathcal{F}$ is an affine subbundle of $J^1C \to C$ (see Proposition 1.1.3). Thus, we have the canonical splitting over C:

$$\boxed{J^1C = C_+ \underset{C}{\oplus} C_- = C_+ \oplus (C \underset{X}{\times} \overset{2}{\wedge} T^*X \otimes V_GP),} \tag{6.2.21}$$
$$a^r_{\lambda\mu} = \frac{1}{2}(a^r_{\lambda\mu} + a^r_{\mu\lambda} - c^r_{pq} a^p_\lambda a^q_\mu) + \frac{1}{2}(a^r_{\lambda\mu} - a^r_{\mu\lambda} + c^r_{pq} a^p_\lambda a^q_\mu).$$

The corresponding canonical projections are $\mathrm{pr}_2 = \mathcal{F}/2$ (6.2.20) and

$$\mathrm{pr}_1 = \mathcal{S} : J^1C \to C_+, \tag{6.2.22}$$
$$\mathcal{S}^r_{\lambda\mu} = \frac{1}{2}(a^r_{\lambda\mu} + a^r_{\mu\lambda} - c^r_{pq} a^p_\lambda a^q_\mu).$$

In particular, let $\Gamma : C \to J^1C$ be a connection on the bundle of principal connections $C \to X$. Then $\mathcal{S} \circ \Gamma$ is a C_+-valued connection on $C \to X$ which satisfies the condition

$$(\mathcal{S} \circ \Gamma)^r_{\lambda\mu} - (\mathcal{S} \circ \Gamma)^r_{\mu\lambda} + c^r_{pq} a^p_\lambda a^q_\mu = 0.$$

For instance, let us consider the affine connection Γ (6.2.17). Then we obtain the connection

$$\begin{aligned}\Gamma_A = \mathcal{S}\circ\Gamma : C \to C_+ \subset J^1C,\\ \Gamma_A{}^r_{\lambda\mu} = \frac{1}{2}[\partial_\lambda A^r_\mu + \partial_\mu A^r_\lambda - c^r_{pq}a^p_\lambda a^q_\mu +\\ c^r_{pq}(a^p_\lambda A^q_\mu + a^p_\mu A^q_\lambda)] - K_\lambda{}^\nu{}_\mu(a^r_\nu - A^r_\nu),\end{aligned} \tag{6.2.23}$$

which has the property

$$\Gamma_A \circ A = \mathcal{S}\circ J^1A. \tag{6.2.24}$$

As in the general case (3.3.14) of quadratic degenerate systems, one can write

$$\mathcal{F}^r_{\lambda\mu} = (a^r_{\lambda\mu} - \Gamma_A{}^r_{\lambda\mu}) + (a^r_{\mu\lambda} - \Gamma_A{}^r_{\mu\lambda}). \tag{6.2.25}$$

6.3 Gauge conservation laws

The main peculiarities of conservation laws in gauge theory of principal connections consists in the following.

- Noether currents reduce to superpotentials (see Remark 3.4.2) because generators of gauge transformations depend on derivatives of gauge parameters.
- Noether conservation laws and Noether currents depend on gauge parameters, but this is not the case of an Abelian gauge model.
- An energy-momentum conservation law implies the gauge invariance of a Lagrangian.

Let $P\to X$ be a G-principal bundle. In a gauge model with a symmetry group G, gauge potentials are identified with principal connections on the principal bundle $P\to X$, i.e., with global sections of the bundle of principal connections $C\to X$ (6.1.8), while *matter fields* are represented by global sections of a P-associated vector bundle Y (6.1.22), called a *matter bundle*. The total configuration space of a gauge model with unbroken symmetries is the product

$$J^1Y_{\rm tot} = J^1Y \underset{X}{\times} J^1C. \tag{6.3.1}$$

In gauge theory, several particular classes of gauge transformations are considered [123, 214, 283]. By a gauge transformation of a principal bundle P is meant its automorphism Φ_P which is equivariant under the canonical action (6.1.1), i.e.,

$$R_g \circ \Phi_P = \Phi_P \circ R_g, \qquad \forall g \in G.$$

This is called a *general principal automorphism* of P or simply an automorphism of P if there is no danger of confusion.

Every general principal automorphism of P yields the corresponding automorphisms

$$\Phi_Y : (p, v) \cdot G \mapsto (\Phi_P(p), v) \cdot G, \qquad p \in P, \qquad v \in V, \tag{6.3.2}$$

of the P-associated bundle Y (6.1.22). For the sake of brevity, we will write

$$\boxed{\Phi_Y : (P \times V)/G \to (\Phi_P(P) \times V)/G.}$$

General principal automorphisms Φ of the principal bundle P also determine the corresponding automorphisms

$$\Phi_C : J^1P/G \to J^1\Phi_P(J^1P)/G \tag{6.3.3}$$

of the bundle of principal connections C [123, 172].

To obtain the Noether conservation laws, we will consider only vertical automorphisms of the principal bundle P, which are called *principal automorphisms* or simply gauge transformations if there is no danger of confusion.

Every principal automorphism of a principal bundle P is represented as

$$\Phi_P(p) = pf(p), \qquad p \in P, \tag{6.3.4}$$

where f is a G-valued equivariant function on P, i.e.,

$$f(pg) = g^{-1}f(p)g, \qquad \forall g \in G. \tag{6.3.5}$$

There is one-to-one correspondence between the functions f (6.3.5) and the global sections s of the *group bundle*

$$P^G = (P \times G)/G, \tag{6.3.6}$$

whose typical fibre is the group G which acts on itself by the adjoint representation. There is the canonical fibre-to-fibre action of the group bundle P^G on any P-associated bundle Y:

$$P^G \underset{X}{\times} Y \to Y,$$
$$((p,g)\cdot G, (p,v)\cdot G) \to (p, gv)\cdot G, \qquad \forall g \in G, \quad \forall v \in V.$$

Then the above-mentioned correspondence is defined by the relation

$$(s(\pi_{PX}(p)), p) \mapsto pf(p).$$

It follows that principal automorphisms of a G-principal bundle $P \to X$ form the group $\mathrm{Gau}(P)$, called the *gauge group*, which is isomorphic to the group of global sections of the group bundle (6.3.6).

Here we restrict our consideration to (local) 1-parameter subgroups $[\Phi_P]$ of the gauge group. Their generators are G-invariant vertical vector fields ξ on a principal bundle P. We will call ξ a *principal vector field*. Recall one-to-one correspondence between the principal vector fields on P and the sections

$$\xi = \xi^p e_p \tag{6.3.7}$$

of the gauge algebra bundle $V_G P \to X$ (6.1.6). Therefore, one can think of the components $\xi^p(x)$ of a principal vector field (6.3.7) as being *gauge parameters*. The principal vector fields (6.3.7) are transformed under the generators of gauge transformations by the adjoint representation given by the Lie bracket

$$\xi' : \xi \to [\xi', \xi] = c^p_{rq} \xi'^r \xi^q e_p, \qquad \xi, \xi' \in V_G P(X).$$

Accordingly, gauge parameters are changed by the coadjoint representation

$$\xi' : \xi^p \mapsto -c^p_{rq} \xi'^r \xi^q. \tag{6.3.8}$$

Given a principal vector field ξ (6.3.7) on P, the corresponding principal vector field on the P-associated vector bundle $Y \to X$, which corresponds to the (local) 1-parameter group $[\Phi_Y]$ of principal automorphisms (6.3.2) of Y, reads

$$\xi_Y = \xi^p I^i_p \partial_i,$$

where I_p are generators of the group G acting on the typical fibre V of Y. Accordingly, the principal vector field on the bundle of principal connections C, which corresponds to the local 1-parameter group $[\Phi_C]$ of principal automorphisms (6.3.3) of C, takes the form

$$\xi_C = (\partial_\mu \xi^r + c^r_{qp} a^q_\mu \xi^p)\partial^\mu_r \tag{6.3.9}$$

(see (6.2.13)). Then a principal vector field on the product $C \underset{X}{\times} Y$ reads

$$\boxed{\xi_{YC} = (\partial_\mu \xi^r + c^r_{qp} a^q_\mu \xi^p)\partial^\mu_r + \xi^p I^i_p \partial_i = (u^{A\mu}_p \partial_\mu \xi^p + u^A_p \xi^p)\partial_A,} \tag{6.3.10}$$

where the collective index A is used:

$$u^{A\mu}_p \partial_A = \delta^r_p \partial^\mu_r, \qquad u^A_p \partial_A = c^r_{qp} a^q_\mu \partial^\mu_r + I^i_p \partial_i.$$

Remark 6.3.1. Let us consider a local 1-parameter group of general principal automorphisms $[\Phi_P]$ of the principal bundle $P \to X$ whose generator is a projectable G-invariant vector field on P, given by a section

$$\xi = \tau^\lambda \partial_\lambda + \xi^p e_p$$

of the fibre bundle $T_G P \to X$. Let $[\Phi_C]$ be the corresponding 1-parameter group of automorphisms (6.3.3) of the fibre bundle $C \to X$. The generator of $[\Phi_C]$ is the vector field ξ_C (6.2.12) on C which takes the coordinate form

$$\boxed{\xi_C = \tau^\lambda \partial_\lambda + (\partial_\mu \xi^r + c^r_{qp} a^q_\mu \xi^p - a^r_\lambda \partial_\mu \tau^\lambda)\partial^\mu_r,} \tag{6.3.11}$$

where ξ^p are gauge parameters. •

A Lagrangian L on the configuration space (6.3.1) is said to be *gauge-invariant* if the strong equality

$$\mathbf{L}_{J^1\xi_{YC}} L = 0$$

holds for every principal vector field ξ (6.3.7).

In this case, the first variational formula (3.2.2) leads to the strong equality

$$0 = (u^A_p \xi^p + u^{A\mu}_p \partial_\mu \xi^p)\delta_A \mathcal{L} + d_\lambda[(u^A_p \xi^p + u^{A\mu}_p \partial_\mu \xi^p)\pi^\lambda_A], \tag{6.3.12}$$

where $\delta_A \mathcal{L}$ are the variational derivatives of L and

$$d_\lambda = \partial_\lambda + a^p_{\lambda\mu} \partial^\mu_p + y^i_\lambda \partial_i.$$

Due to the arbitrariness of gauge parameters $\xi^p(x)$, this equality is equivalent to the following system of strong equalities:

$$u^A_p\delta_A\mathcal{L} + d_\mu(u^A_p\pi^\mu_A) = 0, \tag{6.3.13a}$$
$$u^{A\mu}_p\delta_A\mathcal{L} + d_\lambda(u^{A\mu}_p\pi^\lambda_A) + u^A_p\pi^\mu_A = 0, \tag{6.3.13b}$$
$$u^{A\lambda}_p\pi^\mu_A + u^{A\mu}_p\pi^\lambda_A = 0. \tag{6.3.13c}$$

One can regard these strong equalities as the conditions of a Lagrangian L to be gauge-invariant. Let us study these equations in the case of a Lagrangian

$$L : J^1C \to \overset{n}{\wedge} T^*X \tag{6.3.14}$$

for free gauge fields. Then the equations (6.3.13a) – (6.3.13c) read

$$c^r_{pq}(a^p_\mu\partial^\mu_r\mathcal{L} + a^p_{\lambda\mu}\partial^{\lambda\mu}_r\mathcal{L}) = 0, \tag{6.3.15a}$$
$$\partial^\mu_q\mathcal{L} + c^r_{pq}a^p_\lambda\partial^{\mu\lambda}_r\mathcal{L} = 0, \tag{6.3.15b}$$
$$\partial^{\mu\lambda}_p\mathcal{L} + \partial^{\lambda\mu}_p\mathcal{L} = 0. \tag{6.3.15c}$$

Let us utilize the coordinates $(a^q_\mu, \mathcal{S}^r_{\mu\lambda}, \mathcal{F}^r_{\mu\lambda})$ (6.2.19), (6.2.22), which correspond to the canonical splitting (6.2.21) of the jet manifold J^1C.

With respect to these coordinates, the equation (6.3.15c) reads

$$\frac{\partial\mathcal{L}}{\partial\mathcal{S}^r_{\mu\lambda}} = 0. \tag{6.3.16}$$

Then the equation (6.3.15b) takes the form

$$\partial^\mu_q a^q_\mu = 0. \tag{6.3.17}$$

A glance at the equations (6.3.16) and (6.3.17) shows that the gauge-invariant Lagrangian (6.3.14) factorizes through the strength $\mathcal{F}$ of gauge potentials, i.e.,

$$\begin{array}{ccc} J^1C & \xrightarrow{\mathcal{F}} & C_- \\ {\scriptstyle L}\searrow & & \swarrow{\scriptstyle \overline{L}} \\ & \overset{n}{\wedge} T^*X & \end{array}$$

[43, 116]. Then the equation (6.3.15a) is written as

$$c^r_{pq}\mathcal{F}^p_{\lambda\mu}\frac{\partial\mathcal{L}}{\partial\mathcal{F}^r_{\lambda\mu}} = 0,$$

which is an equivalent of the gauge invariance of the Lagrangian L (6.3.14). As a result, we obtain the conventional *Yang-Mills Lagrangian* $L_{\rm YM}$ of gauge potentials on the configuration space J^1C in the presence of a background world metric g on the base X. It reads

$$L_{\rm YM} = \frac{1}{4\varepsilon^2} a^G_{pq} g^{\lambda\mu} g^{\beta\nu} \mathcal{F}^p_{\lambda\beta} \mathcal{F}^q_{\mu\nu} \sqrt{|\, g\,|}\omega, \qquad g = \det(g_{\mu\nu}), \tag{6.3.18}$$

where a^G is a non-degenerate G-invariant metric in the Lie algebra of $\mathfrak{g}_r$ and ε is a coupling constant. The expression (6.2.25) shows that, as in the general case (3.3.13) of quadratic Lagrangians, the Yang–Mills Lagrangian (6.3.18) factorizes through the covariant differential relative to the connection (6.2.23) on the bundle of principal connections $C \to X$.

Remark 6.3.2. Substituting (6.3.13b) and (6.3.13c) in (6.3.13a), we obtain the well-known constraint conditions of the variational derivatives of a gauge-invariant Lagrangian:

$$\boxed{u^A_p \delta_A \mathcal{L} - d_\mu(u^{A\mu}_p \delta_A \mathcal{L}) = 0.} \tag{6.3.19}$$

•

On-shell, the strong equality (6.3.12) becomes the weak conservation law

$$\boxed{0 \approx d_\lambda[(u^A_p \xi^p + u^{A\mu}_p \partial_\mu \xi^p)\pi^\lambda_A]} \tag{6.3.20}$$

of the Noether current

$$\boxed{\mathfrak{T}^\lambda = -(u^A_p \xi^p + u^{A\mu}_p \partial_\mu \xi^p)\pi^\lambda_A.} \tag{6.3.21}$$

Accordingly, the equalities (6.3.13a) – (6.3.13c) on-shell lead to the familiar Noether identities for a gauge-invariant Lagrangian L:

$$d_\mu(u^A_p \pi^\mu_A) \approx 0, \tag{6.3.22a}$$
$$d_\lambda(u^{A\mu}_p \pi^\lambda_A) + u^A_p \pi^\mu_A \approx 0, \tag{6.3.22b}$$
$$u^{A\lambda}_p \pi^\mu_A + u^{A\mu}_p \pi^\lambda_A = 0. \tag{6.3.22c}$$

They are equivalent to the weak equality (6.3.20) due to the arbitrariness of the gauge parameters $\xi^p(x)$.

A glance at the expressions (6.3.20) and (6.3.21) shows that the Noether conservation law and the Noether current depend on gauge parameters. The weak identities (6.3.22a) – (6.3.22c) play the role of the necessary and sufficient conditions in order that the weak conservation law (6.3.20) be *gauge-covariant*, i.e., form-invariant under changing gauge parameters. This means that, if the equality (6.3.20) takes place for gauge parameters ξ, it does so for arbitrary deviations $\xi + \delta\xi$ of ξ. Then the conservation law (6.3.20) is also covariant under gauge transformations, when gauge parameters are transformed by the coadjoint representation (6.3.8).

Thus, dependence of the Noether current on gauge parameters guarantees that the Noether conservation law is maintained under gauge transformations.

It is easily seen that the equalities (6.3.22a) – (6.3.22c) are not mutually independent, but (6.3.22a) is a corollary of (6.3.22b) and (6.3.22c). This property reflects the fact that, in accordance with the strong equalities (6.3.13b) and (6.3.13c), the Noether current (6.3.21) is brought into the superpotential form (3.4.6):

$$\mathfrak{T}^\lambda = \xi^p u_p^{A\lambda}\delta_A\mathcal{L} - d_\mu(\xi^p u_p^{A\mu}\pi_A^\lambda)$$

where the superpotential is $U^{\mu\lambda} = -\xi^p u_p^{A\mu}\pi_A^\lambda$. Since a matter field Lagrangian does not depend on the derivative coordinates $a_{\lambda\mu}^p$, the *Noether superpotential*

$$\boxed{U^{\mu\lambda} = \xi^p \pi_p^{\mu\lambda}} \qquad (6.3.23)$$

depends on the gauge potentials only.

We have the corresponding integral relation (3.4.8), which reads

$$\int\limits_{N^{n-1}} s^*\mathfrak{T}^\lambda\omega_\lambda = \int\limits_{\partial N^{n-1}} s^*(\xi^p\pi_p^{\mu\lambda})\omega_{\mu\lambda}, \qquad (6.3.24)$$

where N^{n-1} is a compact oriented $(n-1)$-dimensional submanifold of X with the boundary ∂N^{n-1}. One can think of (6.3.24) as being the integral relation between the symmetry current (6.3.21) and the gauge field generated by this current. In the electromagnetic theory, the similar relation between an electric current and the electromagnetic field generated by this current is well known. In comparison with (6.3.24), this relation is free from gauge parameters due to the peculiarity of Abelian gauge models.

Remark 6.3.3. It should be emphasized that the superpotential form of the Noether current (6.3.21) is caused by the fact that principal vector fields (6.3.10) depend on derivatives of gauge parameters. •

Example 6.3.4. The Abelian gauge model. In gauge theory with an Abelian symmetry group G, one can take the Noether current and the Noether conservation law independent of gauge parameters.

Let us consider the electromagnetic theory, where

$$G = U(1), \qquad I^j(y) = iy^j.$$

In this case, a gauge parameter ξ is not changed under gauge transformations as follows from the coadjoint representation law (6.3.8). Therefore, one can put, e.g., $\xi = 1$. Then the Noether current (6.3.21) takes the form

$$\mathfrak{T}^\lambda = -u^A \pi^\lambda_A.$$

Since the group G is Abelian, this current (6.3.25) does not depend on gauge potentials and it is invariant under gauge transformations. We have

$$\mathfrak{T}^\lambda = -iy^j \pi^\lambda_j. \tag{6.3.25}$$

It is easy to see that $\mathfrak{T}$, under the sign change, is the familiar electric current of matter fields, while the Noether conservation law (6.3.20) is precisely the *equation of continuity*. The corresponding integral equation of continuity (3.4.5) reads

$$\int\limits_{\partial N} s^*(y^j \pi^\lambda_j)\omega_\lambda = 0,$$

where N is a compact n-dimensional submanifold of X with the boundary ∂N.

Though the Noether current $\mathfrak{T}$ (6.3.25) is expressed in the superpotential form

$$\mathfrak{T}^\lambda = -\delta^\lambda \mathcal{L} + d_\mu U^{\mu\lambda},$$

the equation of continuity is not tautological. This equation is independent of an electromagnetic field generated by the electric current (6.3.25) and it is therefore treated as the strong conservation law.

When $\xi = 1$, the electromagnetic superpotential (6.3.23) takes the form

$$U^{\mu\lambda} = \pi^{\mu\lambda} = -\frac{1}{4\pi}\mathcal{F}^{\mu\lambda},$$

where $\mathcal{F}$ is the electromagnetic strength. The corresponding equality (3.4.7) is precisely the system of Maxwell equations

$$\frac{1}{4\pi} d_\mu \mathcal{F}^{\mu\lambda} = iy^j \pi^\lambda_j.$$

Accordingly, the integral relation (6.3.24) is the integral form of the Maxwell equations. In particular, the well-known relation between the flux of an electric field through a closed surface and the total electric charge inside this surface is recovered. •

Let us turn now to energy-momentum conservation laws in gauge theory. For the sake of simplicity, we will consider only gauge theory without matter fields. The corresponding Lagrangian is the Yang–Mills Lagrangian (6.3.18) on the jet manifold J^1C.

Given a vector field τ on X, let B be a principal connection on the principal bundle $P \to X$ and

$$\tau_B = \tau^\lambda(\partial_\lambda + B^p_\lambda \varepsilon_p)$$

the horizontal lift of τ onto P by means of the connection B. This vector field, in turn, gives rise to the vector field $\widetilde{\tau}_B$ (6.2.16) on the bundle of principal connections C, which reads

$$\boxed{\widetilde{\tau}_B = \tau^\lambda\partial_\lambda + [\tau^\lambda(\partial_\mu B^r_\lambda + c^r_{pq} a^p_\mu B^q_\lambda) - \partial_\mu\tau^\beta(a^r_\beta - B^r_\beta)]\partial^\mu_r.} \tag{6.3.26}$$

Let us discover the energy-momentum current along the vector field $\widetilde{\tau}_B$ (6.3.26) [120, 123, 269].

Since the Yang–Mills Lagrangian (6.3.18) depends also on a background world metric g, we will consider the total Lagrangian

$$L = \frac{1}{4\varepsilon^2} a^G_{pq} \sigma^{\lambda\mu}\sigma^{\beta\nu}\mathcal{F}^p_{\lambda\beta}\mathcal{F}^q_{\mu\nu}\sqrt{|\sigma|}\omega, \qquad \sigma = \det(\sigma_{\mu\nu}), \tag{6.3.27}$$

on the total configuration space

$$J^1(C \underset{X}{\times} \overset{2}{\vee} TX)$$

(see Remark 2.4.3), where the tensor bundle $\overset{2}{\vee} TX$ is provided with the holonomic coordinates $(x^\lambda, \sigma^{\mu\nu})$.

Given a vector field τ on X, there exists its canonical lift (1.2.2)

$$\widetilde{\tau} = \tau^\lambda\partial_\lambda + (\partial_\nu\tau^\alpha\sigma^{\nu\beta} + \partial_\nu\tau^\beta\sigma^{\nu\alpha})\partial_{\alpha\beta}$$

onto the tensor bundle $\overset{2}{\vee} T^*X$, which is the generator of a local 1-parameter group of general covariant transformations of $\overset{2}{\vee} T^*X$ (see Section 7.1). Thus, we have the lift

$$\widetilde{\tau}_B = \tau^\lambda \partial_\lambda + [\tau^\lambda(\partial_\mu B^r_\lambda + c^r_{pq} a^p_\mu B^q_\lambda) - \partial_\mu \tau^\beta (a^r_\beta - B^r_\beta)]\partial^\mu_r + (\partial_\nu \tau^\alpha \sigma^{\nu\beta} + \partial_\nu \tau^\beta \sigma^{\nu\alpha})\partial_{\alpha\beta} \tag{6.3.28}$$

of a vector field τ on X onto the product $C \underset{X}{\times} \overset{2}{\vee} T^*X$. For the sake of simplicity, we denote it by the same symbol $\widetilde{\tau}_B$.

The total Lagrangian (6.3.27), by construction, is invariant under gauge transformations and general covariant transformations. Hence, its Lie derivative along the vector field $\widetilde{\tau}_B$ (6.3.28) equals zero. Then one can use the formula (3.4.21). On the Yang–Mills shell and on the background field $\sigma^{\mu\nu} = g^{\mu\nu}(x)$, this reads

$$0 \approx (\partial_\nu \tau^\alpha g^{\nu\beta} + \partial_\nu \tau^\beta g^{\nu\alpha} - \partial_\lambda g^{\alpha\beta}\tau^\lambda)\partial_{\alpha\beta}\mathcal{L} - d_\lambda \mathfrak{T}^\lambda_B, \tag{6.3.29}$$

where

$$\mathfrak{T}^\lambda_B = \pi^{\lambda\nu}_r[-\tau^\mu(\partial_\nu B^r_\mu + c^r_{pq} a^p_\nu B^q_\mu - a^r_{\mu\nu}) + \partial_\nu \tau^\mu (a^r_\mu - B^r_\mu)] - \delta^\lambda_\mu \tau^\mu \mathcal{L}_{\rm YM} \tag{6.3.30}$$

is the energy-momentum current along the vector field (6.3.26). The weak identity (6.3.29) can be written in the form

$$0 \approx \partial_\lambda \tau^\mu t^\lambda_\mu \sqrt{\mid g \mid} - \tau^\mu \{_\mu{}^\beta{}_\lambda\} t^\lambda_\beta \sqrt{\mid g \mid} - d_\lambda \mathfrak{T}^\lambda_B, \tag{6.3.31}$$

where $\{_\mu{}^\beta{}_\lambda\}$ are the Christoffel symbols (2.4.13) of g and

$$t^\mu_\beta \sqrt{\mid g \mid} = 2 g^{\mu\alpha}\partial_{\alpha\beta}\mathcal{L}_{\rm YM}$$

is the metric energy-momentum tensor of gauge potentials. We have the relation

$$t^\lambda_\mu \sqrt{\mid g \mid} = \pi^{\lambda\nu}_q \mathcal{F}^q_{\mu\nu} - \delta^\lambda_\mu \mathcal{L}_{\rm YM}.$$

In particular, let A be a solution of the Yang-Mills equations. Let us consider the lift (6.3.26) of the vector field τ on X onto C by means of the principal connection $B = A$. In this case, the energy-momentum current (6.3.30) reads

$$\mathfrak{T}^\lambda_A \circ A = \tau^\mu (t^\lambda_\mu \circ A)\sqrt{\mid g \mid}.$$

Then the weak identity (6.3.31) on the solution A takes the form

$$0 \approx -\{_\mu{}^\beta{}_\lambda\}(t^\lambda_\beta \circ A)\sqrt{|\,g\,|} - d_\lambda[(t^\lambda_\mu \circ A)\sqrt{|\,g\,|}].$$

Thus, it leads to the familiar *covariant conservation law*

$$\nabla_\lambda((t^\lambda_\mu \circ A)\sqrt{|\,g\,|}) = 0, \tag{6.3.32}$$

where ∇_λ are the covariant derivatives with respect to the Levi–Civita connection $\{_\mu{}^\beta{}_\lambda\}$ of the background metric g.

Note that, in the case of an arbitrary principal connection B, the corresponding weak identity (6.3.31) differs from (6.3.32) in the Noether conservation law

$$0 \approx d_\lambda(\xi^r_\nu \pi^{\lambda\nu}_r), \tag{6.3.33}$$

where

$$\xi_C = \xi^r_\nu \partial^\nu_r = (\partial_\nu \xi^r + c^r_{qp} a^q_\nu \xi^p)\partial^\nu_r, \qquad \xi^r = \tau^\mu(B^r_\mu - A^r_\mu),$$

is the principal vector field (6.3.9) on C. Since Noether currents in gauge theory reduce to a superpotential and, consequently, the conservation law (6.3.33) is tautological, one can always bring the weak identity (6.3.31) into the covariant conservation law (6.3.32) by cutting down Noether currents (see [135]). Though the physical motivation of this operation is under a question. It follows that differential conservation laws in gauge invariant models are not sufficient.

6.4 Hamiltonian gauge theory

This Section is devoted to Hamiltonian formulation of gauge theory in the framework of the covariant Hamiltonian formalism in Chapter 4. As was mentioned above, the main ingredients in gauge theory are not connected directly with the gauge invariance property, but are common for field models with degenerate quadratic Lagrangians. In order to illustrate this fact clearly, we will compare the gauge-invariant model of electromagnetic fields with that of Proca fields. We will follow the general scheme for models with degenerate quadratic Lagrangians in Section 4.3. The peculiarity of gauge theory consists in the fact that the splittings (3.3.12a) and (4.3.1a) of configuration and phase spaces are canonical.

Given a bundle $C \to X$ of principal connections, the corresponding Legendre bundle Π (3.2.8) is

$$\pi_{\Pi C} : \Pi \to C,$$
$$\Pi = \overset{n}{\wedge} T^*X \underset{C}{\otimes} TX \underset{C}{\otimes} [C \times \overline{C}]^*. \tag{6.4.1}$$

It is coordinated by $(x^\lambda, a^p_\lambda, p^{\mu\lambda}_m)$, and admits the canonical splitting

$$\Pi = \Pi_+ \underset{C}{\oplus} \Pi_-, \tag{6.4.2}$$
$$p^{\mu\lambda}_m = p^{(\mu\lambda)}_m + p^{[\mu\lambda]}_m = \frac{1}{2}(p^{\mu\lambda}_m + p^{\lambda\mu}_m) + \frac{1}{2}(p^{\mu\lambda}_m - p^{\lambda\mu}_m).$$

The Legendre map defined by the Yang-Mills Lagrangian L_{YM} (6.3.18) takes the form

$$p^{(\mu\lambda)}_m \circ \widehat{L}_{YM} = 0, \tag{6.4.3a}$$
$$p^{[\mu\lambda]}_m \circ \widehat{L}_{YM} = \varepsilon^{-2} a^G_{mn} g^{\mu\alpha} g^{\lambda\beta} \mathcal{F}^n_{\alpha\beta} \sqrt{\mid g \mid}. \tag{6.4.3b}$$

A glance at these expressions shows that $\operatorname{Ker} \widehat{L}_{YM} = C_+$, and the Lagrangian constraint space is

$$N_L = \widehat{L}_{YM}(J^1C) = \Pi_-.$$

Obviously, N_L is an imbedded submanifold of Π, and the Lagrangian $L_{\rm YM}$ is almost regular. Accordingly, the canonical splittings (6.2.21) and (6.4.2) are similar to the splittings (3.3.12a) and (4.3.1a), respectively.

Therefore, we can follow the general procedure in Section 4.3 in order to construct a complete set of Hamiltonian forms associated with the Yang-Mills Lagrangian (6.3.18).

Let us consider connections Γ on the fibre bundle $C \to X$ which take their values into $\operatorname{Ker} \widehat{L}$, i.e.,

$$\Gamma : C \to C_+, \tag{6.4.4}$$
$$\Gamma^r_{\lambda\mu} - \Gamma^r_{\mu\lambda} + c^r_{pq} a^p_\lambda a^q_\mu = 0.$$

Given a symmetric world connection K on X, every principal connection B on the principal bundle $P \to X$ gives rise to the connection $\Gamma_B : C \to C_+$ (6.2.23) which has the property (6.2.24). With this connection, the Hamiltonian form (4.3.4) (where $\sigma_1 = 0$) is

$$H_B = p_r^{\lambda\mu} da_\mu^r \wedge \omega_\lambda - p_r^{\lambda\mu} \Gamma_B{}^r_{\lambda\mu} \omega - \widetilde{\mathcal{H}}_{YM} \omega, \tag{6.4.5}$$
$$\widetilde{\mathcal{H}}_{YM} = \frac{\varepsilon^2}{4} a_G^{mn} g_{\mu\nu} g_{\lambda\beta} p_m^{[\mu\lambda]} p_n^{[\nu\beta]} \sqrt{\mid g \mid}.$$

It is associated with the Lagrangian $L_{\rm YM}$. The corresponding covariant Hamilton equations for sections r of the Legendre bundle $\Pi \to X$ consist of the equations (6.4.3b) and the equations

$$\partial_\lambda r_\mu^m + \partial_\mu r_\lambda^m = 2\Gamma_B{}^m_{(\lambda\mu)}, \tag{6.4.6}$$
$$\partial_\lambda r_r^{\lambda\mu} = c_{pr}^q r_\lambda^p r_q^{[\lambda\mu]} - c_{rp}^q B_\lambda^p r_q^{(\lambda\mu)} + K_\lambda{}^\mu{}_\nu r_r^{(\lambda\nu)}. \tag{6.4.7}$$

The Hamilton equations (6.4.6) and (6.4.3b) are similar to the equations (4.3.6) and (4.3.7), respectively. The Hamilton equations (6.4.3b) and (6.4.7) restricted to the constraint space (6.4.3a) are equivalent to the Yang-Mills equations for a gauge potential $A = \pi_{\Pi C} \circ r$.

Different Hamiltonian forms H_B lead to different equations (6.4.6). The equations (6.4.6) are independent of canonical momenta, and take the form of a gauge-type condition (4.3.6):

$$\Gamma_B \circ A = \mathcal{S} \circ J^1 A.$$

A glance at this condition shows that, given a solution A of the Yang-Mills equations, there always exists a Hamiltonian form H_B (e.g., $H_{B=A}$) which obeys the condition (4.2.9), i.e.,

$$\widehat{H}_B \circ \widehat{L}_{YM} \circ J^1 A = J^1 A.$$

It follows that the Hamiltonian forms H_B (6.4.5) parameterized by principal connections B constitute a complete family.

Remark 6.4.1. It should be emphasized that the gauge-type condition (6.4.6) differs from the familiar gauge conditions in gauge theory which single out a representative of each gauge coset (with the accuracy to Gribov's ambiguity). Namely, if a gauge potential A is a solution of the Yang-Mills equations, there exists a gauge conjugate potential A' which is also a solution of the same Yang-Mills equations and satisfies a given gauge condition. At the same time, not every solution of the

Yang-Mills equations is a solution of the system of the Yang-Mills equations and a certain gauge condition. In other words, there are solutions of Yang-Mills equations which are not singled out by the gauge conditions known in gauge theory. In this sense, this set of gauge conditions is not complete. In gauge theory, this lack is not essential since one can think of all gauge conjugate potentials as being physically equivalent, but not in the case of other constraint field theories, e.g., that of Proca fields. Within the framework of the covariant Hamiltonian description of quadratic Lagrangian systems, there is a complete set of gauge-type conditions in the sense that, for any solution of the Euler–Lagrange equations, there exists a system of Hamilton equations equivalent to these Euler-Lagrange equations and a supplementary gauge-type condition which this solution satisfies. •

Example 6.4.2. Electromagnetic fields. Let us consider the particular Abelian case of an electromagnetic theory. In gauge theory, electromagnetic potentials are identified with principal connections on a principal bundle $P \to X$ with the structure group $U(1)$. In this case, the gauge algebra bundle (6.1.6) is equivalent to the trivial line bundle

$$V_G P \cong X \times \mathbb{R}.$$

The corresponding bundle of principal connections C (6.1.8) coordinated by (x^λ, a_μ) is the cotangent bundle T^*X provided with the natural affine structure. The configuration space J^1C of electromagnetic theory admits the canonical splitting (6.2.21) which takes the form

$$J^1C = C_+ \underset{C}{\oplus} (\overset{2}{\wedge} T^*X \underset{X}{\times} C), \tag{6.4.8}$$

where $C_+ \to C$ is an affine bundle modelled over the pull-back symmetric tensor bundle

$$\overline{C}_+ = \overset{2}{\vee} T^*X \underset{X}{\times} C.$$

Relative to the adapted coordinates $(x^\lambda, a_\mu, a_{\lambda\mu})$ on J^1C, the splitting (6.4.8) reads

$$a_{\lambda\mu} = \frac{1}{2}(\mathcal{S}_{\lambda\mu} + \mathcal{F}_{\lambda\mu}) = a_{(\lambda\mu)} + a_{[\lambda\mu]}.$$

For any section A of $C \to X$, we find that

$$F_{\mu\lambda} = \mathcal{F}_{\mu\lambda} \circ J^1A = \partial_\mu A_\lambda - \partial_\lambda A_\mu$$

is the familiar strength of an electromagnetic field.

For the sake of simplicity, let $X = \mathbb{R}^4$ be the Minkowski space with the *Minkowski metric* $\eta = \text{diag}(1, -1, -1, -1)$. Then the conventional Lagrangian of electromagnetic fields on the configuration space (6.4.8) is written as

$$L_E = -\frac{1}{16\pi}\eta^{\lambda\mu}\eta^{\beta\nu}\mathcal{F}_{\lambda\beta}\mathcal{F}_{\mu\nu}\omega. \tag{6.4.9}$$

The momentum phase space of electromagnetic theory is the Legendre bundle

$$\Pi = (\overset{4}{\wedge} T^*X \otimes TX \otimes TX) \underset{X}{\times} C, \tag{6.4.10}$$

coordinated by $(x^\lambda, a_\mu, p^{\lambda\mu})$. With respect to these coordinates, the Legendre map defined by the Lagrangian (6.4.9) reads

$$p^{(\lambda\mu)} \circ \widehat{L}_E = 0, \tag{6.4.11a}$$

$$p^{[\lambda\mu]} \circ \widehat{L}_E = -\frac{1}{4\pi}\eta^{\lambda\alpha}\eta^{\mu\beta}\mathcal{F}_{\alpha\beta}. \tag{6.4.11b}$$

On the Legendre bundle Π (6.4.10, we have a complete set of Hamiltonian forms

$$H_B = p^{\lambda\mu}da_\mu \wedge \omega_\lambda - p^{\lambda\mu}\Gamma_{B\lambda\mu}\omega - \widetilde{\mathcal{H}}_E\omega, \tag{6.4.12}$$

$$\Gamma_{B\lambda\mu} = \frac{1}{2}(\partial_\mu B_\lambda + \partial_\lambda B_\mu),$$

$$\widetilde{\mathcal{H}}_E = -\pi\eta_{\mu\nu}\eta_{\lambda\beta}p^{[\mu\lambda]}p^{[\nu\beta]},$$

which are parametrised by electromagnetic potentials B, and are associated with the Lagrangian (6.4.9). Given such a Hamiltonian form H_B (6.4.12), the corresponding Hamilton equations consist of the equations (6.4.11b) and the equations

$$\partial_\lambda r_\mu + \partial_\mu r_\lambda = \partial_\lambda B_\mu + \partial_\mu B_\lambda, \tag{6.4.13}$$

$$\partial_\lambda r^{\lambda\mu} = 0. \tag{6.4.14}$$

On the constraint space (6.4.11a), the equations (6.4.11b) and (6.4.14) reduce to the Maxwell equations in the absence of matter sources, while the equations (6.4.13), independent of canonical momenta, play the role of a gauge-type condition. •

Example 6.4.3. Proca fields. The model of massive vector Proca fields exemplifies a degenerate field theory which is similar to the electromagnetic one, but without the gauge invariance property.

Proca fields (see Example 7.4.1 below) are represented by sections of the cotangent bundle T^*X (in contrast with electromagnetic potentials given by sections of the affine cotangent bundle). The configuration space of Proca fields is the jet manifold $J^1(T^*X)$ with coordinates $(x^\lambda, k_\mu, k_{\lambda\mu})$, modelled over the pull-back tensor bundle

$$\overset{2}{\otimes} T^*X \times TX \to TX. \tag{6.4.15}$$

On the Minkowski space X, the Lagrangian of Proca fields looks like the electromagnetic one (6.4.9) minus the mass term, i.e.,

$$L_P = L_E - \frac{1}{8\pi} m^2 \eta^{\mu\lambda} k_\mu k_\lambda \omega. \tag{6.4.16}$$

It is almost regular.

The momentum phase space of Proca fields is the Legendre bundle

$$\Pi = \overset{4}{\wedge} T^*X \otimes TX \otimes TX \underset{X}{\times} T^*X,$$

equipped with the holonomic coordinates $(x^\lambda, k_\mu, p^{\lambda\mu})$. With respect to these coordinates, the Legendre map defined by the Lagrangian (6.4.16) takes the form

$$p^{(\lambda\mu)} \circ \widehat{L}_P = 0, \tag{6.4.17a}$$
$$p^{[\lambda\mu]} \circ \widehat{L}_P = -\frac{1}{4\pi} \eta^{\lambda\alpha} \eta^{\mu\beta} \mathcal{F}_{\alpha\beta}. \tag{6.4.17b}$$

We have

$$\operatorname{Ker} \widehat{L}_P = \overset{2}{\vee} T^*X \underset{X}{\times} T^*X$$

and

$$N_L = \overset{4}{\wedge} T^*X \underset{X}{\otimes} (\overset{2}{\wedge} TX) \underset{X}{\times} T^*X,$$
$$p^{(\lambda\mu)} = 0.$$

Following the general procedure of describing quadratic degenerate systems, let us consider the map σ (3.3.8):

$$\overline{k}_{\lambda\mu} \circ \sigma = -2\pi \eta_{\lambda\nu} \eta_{\mu\beta} p^{[\nu\beta]},$$

where $\overline{k}_{\lambda\mu}$ are the fibred coordinates on the fibre bundle (6.4.15). Since

$$\operatorname{Im}\sigma = \overset{2}{\wedge} T^*X \underset{X}{\times} T^*X,$$
$$\operatorname{Ker}\sigma = \overset{4}{\wedge} T^*X \otimes (\overset{2}{\vee} TX) \underset{X}{\times} T^*X,$$

one can perform the corresponding splitting (6.2.21) of the configuration space

$$J^1T^*X = S_+ \underset{T^*X}{\oplus} \overset{2}{\wedge} T^*X,$$
$$k_{\lambda\mu} = \frac{1}{2}(\mathcal{S}_{\lambda\mu} + \mathcal{F}_{\lambda\mu}) = k_{(\lambda\mu)} + k_{[\lambda\mu]},$$

(see (7.4.7) below) and the splitting (4.3.1a) of the phase space

$$\Pi = [\overset{4}{\wedge} T^*X \otimes (\overset{2}{\vee} TX)] \underset{T^*X}{\oplus} Q,$$
$$p^{\lambda\mu} = p^{(\lambda\mu)} + p^{[\lambda\mu]}.$$

Let us consider connections on the cotangent bundle T^*X taking their values into $\operatorname{Ker}\widehat{L}_P$. Bearing in mind that $K = 0$ on the Minkowski space X, we can write every such connection as

$$\Gamma = dx^\lambda \otimes (\partial_\lambda + \phi_{\lambda\mu}\partial^\mu),$$

where $\phi = \phi_{\lambda\mu} dx^\lambda \otimes \partial^\mu$ is a symmetric soldering form on T^*X. By analogy with the case of electromagnetic fields, it suffices to take the connections

$$\Gamma_B = dx^\lambda \otimes [\partial_\lambda + \frac{1}{2}(\partial_\mu B_\lambda + \partial_\lambda B_\mu)\partial^\mu],$$

where B are sections of $T^*X \to X$. Then it is readily observed that the Hamiltonian forms

$$H_B = p^{\lambda\mu} dk_\mu \wedge \omega_\lambda - p^{\lambda\mu}\Gamma_{B\lambda\mu}\omega - \widetilde{\mathcal{H}}_P\omega,$$
$$\widetilde{\mathcal{H}}_P = \widetilde{\mathcal{H}}_E + \frac{1}{8\pi} m^2 \eta^{\mu\nu} k_\mu k_\nu,$$

are associated with the Lagrangian L_P (6.4.9) and constitute a complete set.

Given the Hamiltonian form H_B, the corresponding Hamilton equations for sections r of the fibre bundle $\Pi \to X$ consist of the equations (6.4.17b) and the equations

$$\partial_\lambda r_\mu + \partial_\mu r_\lambda = \partial_\lambda B_\mu + \partial_\mu B_\lambda, \tag{6.4.18}$$
$$\partial_\lambda r^{\lambda\mu} = -\frac{1}{4\pi} m^2 \eta^{\mu\nu} r_\nu. \tag{6.4.19}$$

On the constraint space (6.4.17a), the equations (6.4.17b) and (6.4.19) are equivalent to the Euler–Lagrange equations, and they are supplemented by the gauge-type condition (6.4.18). •

6.5 Geometry of symmetry breaking

Spontaneous symmetry breaking is a quantum phenomenon. In classical field theory, spontaneous symmetry breaking is modelled by classical Higgs fields. In gauge theory on a principal bundle $P \to X$, a symmetry breaking is said to occur when the structure group G of this principal bundle is reducible to a closed subgroup H of exact symmetries [123, 161, 172, 236, 264, 295]. This reduction of a structure group takes place if and only if a global section h of the quotient bundle $P/H \to X$ exists (see Theorem 6.5.2 below). In gauge theory, such a global section h is treated as a *Higgs field*. From the mathematical viewpoint, one talks on the *Klein–Chern geometry* [312] or the reduced G-structure.

This Section provides a brief exposition of geometry of G-structures.

We will start from some basic notions. Let $\pi_{PX} : P \to X$ be a G-principal bundle and H a closed Lie subgroup of G. We assume that $\dim H > 0$. Recall that a closed subgroup of a Lie group is a Lie group. There is the composite fibre bundle

$$\boxed{P \to P/H \to X,} \tag{6.5.1}$$

where

$$P_\Sigma = P \xrightarrow{\pi_{P\Sigma}} P/H \tag{6.5.2}$$

is a principal bundle with the structure group H and

$$\boxed{\Sigma = P/H \xrightarrow{\pi_{\Sigma X}} X} \tag{6.5.3}$$

is a P-associated fibre bundle with the typical fibre G/H on which the structure group G acts naturally on the left. Note that the canonical surjection $G \to G/H$ is an H-principal bundle.

One says that the structure group G of a principal bundle P is *reducible* to a Lie subgroup H if there exists a H-principal subbundle P^h of P with the structure group H. This subbundle is called a *reduced* $G^\downarrow H$*-structure* [123, 131, 178, 312].

Two reduced $G^{\downarrow}H$-structures P^h and $P^{h'}$ on a G-principal bundles are said to be *isomorphic* if there is an automorphism Φ of P which provides an isomorphism of P^h and $P^{h'}$. If Φ is a vertical automorphism of P, reduced structures P^h and $P^{h'}$ are called *equivalent*.

Remark 6.5.1. Note that, in [131, 178] (see also [66]), reduced structures on the frame bundle LX are considered. Therefore, the class of isomorphisms of such reduced structures is restricted to holonomic automorphisms of LX, i.e., the canonical lifts onto LX of diffeomorphisms of the base X (see Section 7.1). •

Remark 6.5.2. Reduction of a structure group is a particular changing a structure group. Let $\phi : H \to G$ be a Lie group homomorphism. There are two variants of this problem [49].

(i) If $P_H \to X$ is an H-principal bundle, there is always exists a G-principal bundle $P_G \to X$ together with the principal bundle morphism $\Phi : P_H \to P_G$ over X. This is the P_H-associated fibre bundle $P_G = (P_H \times G)/H$ with the typical fibre G on which H acts on the left by the rule $h(g) = \phi(h)g$, while G acts on P_G as

$$G \ni g' : (p, g) \cdot H \mapsto (p, gg') \cdot H.$$

(ii) More intricate is an inverse problem to this. If $P_G \to X$ is a G-principal bundle, can we find an H-principal bundle $P_H \to X$ together with the principal bundle morphism $P_H \to P_G$? If $H \to G$ is a subgroup, we have the structure group reduction discussed in this Section. If $H \to G$ is a group epimorphism (extension), one says that H lifts to G. We will study such a lift when G is a central extension of H with kernels $\mathbb{Z}_2$ (see Section 7.5). •

Let us recall the following two theorems [177].

THEOREM 6.5.1. A structure group G of a principal bundle P is reducible to its closed subgroup H if and only if P has an atlas Ψ_P with H-valued transition functions. □

Given a reduced subbundle P^h of P, such an atlas Ψ_P is defined by a family of local sections $\{z_\alpha\}$ which take their values into P^h.

THEOREM 6.5.2. There is one-to-one correspondence

$$\boxed{P^h = \pi_{P\Sigma}^{-1}(h(X))}$$

between the reduced H-principal subbundles P^h of P and the global sections h of the quotient fibre bundle $P/H \to X$ (6.5.3). □

Given such a section h, let us consider the restriction h^*P_Σ (2.7.4) of the H-principal bundle P_Σ (6.5.2) to $h(X) \subset \Sigma$. This is a H-principal bundle over X [177], which is equivalent to the reduced subbundle P^h of P.

In general, there are topological obstructions to the reduction of a structure group of a principal bundle to its subgroup. In accordance with Theorem 1.1.2, the structure group G of a principal bundle P is always reducible to its closed subgroup H, if the quotient G/H is homeomorphic to a Euclidean space $\mathbb{R}^k$.

THEOREM 6.5.3. [286]. A structure group G of a principal bundle is always reducible to its maximal compact subgroup H since the quotient space G/H is homeomorphic to a Euclidean space. □

Two H-principal subbundles P^h and $P^{h'}$ of a G-principal bundle P are not isomorphic to each other in general.

PROPOSITION 6.5.4. (i) Every vertical automorphism $\Phi \in \mathrm{Gau}(P)$ of the principal bundle $P \to X$ sends an H-principal subbundle P^h onto an equivalent H-principal subbundle $P^{h'}$. (ii) Conversely, let two reduced subbundles P^h and $P^{h'}$ of a principal fibre bundle P be isomorphic to each other, and $\Phi : P^h \to P^{h'}$ be an isomorphism. Then Φ is extended to a vertical automorphism of P. □

Proof. (i) Let

$$\Psi^h = \{(U_\alpha, z^h_\alpha), \rho^h_{\alpha\beta}\}, \qquad z^h_\alpha(x) = z^h_\beta(x)\rho^h_{\alpha\beta}(x), \qquad x \in U_\alpha \cap U_\beta,$$

be an atlas of the reduced subbundle P^h, where z^h_α are local sections of $P^h \to X$ and $\rho^h_{\alpha\beta}$ are the transition functions. Given a vertical automorphism Φ of P, let us provide the reduced subbundle $P^{h'} = \Phi(P^h)$ with the atlas

$$\Psi^{h'} = \{(U_\alpha, z^{h'}_\alpha), \rho^{h'}_{\alpha\beta}\}$$

determined by the local sections $z^{h'}_\alpha = \Phi \circ z^h_\alpha$ of $P^{h'} \to X$. Then it is readily observed that

$$\rho^{h'}_{\alpha\beta}(x) = \rho^h_{\alpha\beta}(x), \qquad x \in U_\alpha \cap U_\beta.$$

(ii) Any isomorphism Φ of reduced structures P^h and $P^{h'}$ on P determines a G-valued function f on P^h given by the relation

$$pf(p) = \Phi(p), \qquad p \in P^h.$$

Obviously, this function is H-equivariant. Its prolongation to a G-equivariant function on P is defined as

$$f(pg) = g^{-1}f(p)g, \qquad p \in P^h, \qquad g \in G.$$

In accordance with the relation (6.3.4), this function yields a principal automorphism of P whose restriction to P^h coincides with Φ. **QED**

PROPOSITION 6.5.5. If the quotient G/H is homeomorphic to a Euclidean space $\mathbb{R}^k$, all H-principal subbundles of a G-principal bundle P are equivalent to each other [286]. □

Given a reduced subbundle P^h of a principal bundle P, let

$$Y^h = (P^h \times V)/H \tag{6.5.4}$$

be the associated fibre bundle with a typical fibre V. Let $P^{h'}$ be another reduced subbundle of P which is isomorphic to P^h, and

$$Y^{h'} = (P^{h'} \times V)/H.$$

The fibre bundles Y^h and $Y^{h'}$ are isomorphic, but not canonically isomorphic in general.

PROPOSITION 6.5.6. Let P^h be an H-principal subbundle of a G-principal bundle P. Let Y^h be the P^h-associated bundle (6.5.4) with a typical fibre V. If V carries a representation of the whole group G, the fibre bundle Y^h is canonically isomorphic to the P-associated fibre bundle

$$Y = (P \times V)/G.$$

□

Indeed, every element of Y can be represented as $(p,v)\cdot G$, $p \in P^h$. Then the desired isomorphism is

$$Y^h \ni (p,v)\cdot H \quad \Longleftrightarrow \quad (p,v)\cdot G \in Y.$$

It follows that, given a H-principal subbundle P^h of P, any P-associated fibre bundle Y with the structure group G is canonically equipped with a structure of the P^h-associated fibre bundle Y^h with the structure group H. Briefly, we can write

$$Y = (P \times V)/G \simeq (P^h \times V)/H = Y^h.$$

However, if $P^h \neq P^{h'}$, the P^h- and $P^{h'}$-associated bundle structures on Y are not equivalent. Given bundle atlases Ψ^h of P^h and $\Psi^{h'}$ of $P^{h'}$, the union of the associated atlases of Y has necessarily G-valued transition functions between the charts from Ψ^h and $\Psi^{h'}$.

In accordance with Theorem 6.5.2, the set of reduced H-principal subbundles P^h of P is in bijective correspondence with the set of Higgs fields h. Given such a subbundle P^h, let Y^h (6.5.4) be the associated vector bundle with a typical fibre V which admits a representation of the group H of exact symmetries, but not the whole symmetry group G. Its sections s_h describe matter fields in the presence of the Higgs fields h and some principal connection A_h on P^h. In general, the fibre bundle Y^h (6.5.4) is not associated or canonically associated (see Remark 6.1.3) with other H-principal subbundles $P^{h'}$ of P. It follows that, in this case, V-valued matter fields can be represented only by pairs with Higgs fields. The goal is to describe the totality of these pairs (s_h, h) for all Higgs fields h.

For this purpose, let us consider the composite fibre bundle (6.5.1) and the composite fibre bundle

$$Y \xrightarrow{\pi_{Y\Sigma}} \Sigma \xrightarrow{\pi_{\Sigma X}} X \tag{6.5.5}$$

where $Y \to \Sigma$ is a vector bundle

$$Y = (P \times V)/H$$

associated with the corresponding H-principal bundle P_Σ (6.5.2). Given a global section h of the fibre bundle $\Sigma \to X$ (6.5.3) and the P^h-associated fibre bundle (6.5.4), there is the canonical injection

$$i_h : Y^h = (P^h \times V)/H \hookrightarrow Y$$

over X whose image is the restriction

$$h^*Y = (h^*P \times V)/H$$

of the fibre bundle $Y \to \Sigma$ to $h(X) \subset \Sigma$, i.e.,

$$i_h(Y^h) \cong \pi_{Y\Sigma}^{-1}(h(X)) \tag{6.5.6}$$

(see Proposition 2.7.1). Then, by virtue of Proposition 2.7.2, every global section s_h of the fibre bundle Y^h corresponds to the global section $i_h \circ s_h$ of the composite fibre bundle (6.5.5). Conversely, every global section s of the composite fibre bundle (6.5.5) which projects onto a section $h = \pi_{Y\Sigma} \circ s$ of the fibre bundle $\Sigma \to X$ takes its values into the subbundle $i_h(Y^h) \subset Y$ in accordance with the relation (6.5.6). Hence, there is one-to-one correspondence between the sections of the fibre bundle Y^h (6.5.4) and the sections of the composite fibre bundle (6.5.5) which cover h.

Thus, it is precisely the composite fibre bundle (6.5.5) whose sections describe the above-mentioned totality of pairs (s_h, h) of matter fields and Higgs fields in gauge theory with broken symmetries [123, 264, 268].

Turn now to the properties of connections compatible with a reduced structure. Recall the following theorems [177].

THEOREM 6.5.7. Since principal connections, by definition, are equivariant (see (6.1.12)), every principal connection A_h on a reduced H-principal subbundle P^h of a G-principal bundle P gives rise to a principal connection on P. □

THEOREM 6.5.8. A principal connection A on a G-principal bundle P is reducible to a principal connection on a reduced H-principal subbundle P^h of P if and only if the corresponding global section h of the P-associated fibre bundle $P/H \to X$ is an integral section of the associated principal connection A on $P/H \to X$. □

THEOREM 6.5.9. Given the composite fibre bundle (6.5.1), let A_Σ be a principal connection on the H-principal bundle $P \to P/H$. Then, for any reduced H-principal subbundle P^h of P, the pull-back connection $i_h^* A_\Sigma$ (2.7.11) is a principal connection on P^h. □

This theorem is the corollary of Theorem 6.1.2.

As a consequence of Theorem 6.5.9, there is the following feature of the dynamics of field systems with symmetry breaking. Let the composite fibre bundle Y (6.5.5) be provided with coordinates $(x^\lambda, \sigma^m, y^i)$, where (x^λ, σ^m) are fibred coordinates on the fibre bundle $\Sigma \to X$. Let

$$A_\Sigma = dx^\lambda \otimes (\partial_\lambda + A^i_\lambda \partial_i) + d\sigma^m \otimes (\partial_m + A^i_m \partial_i) \tag{6.5.7}$$

be a principal connection on the vector bundle $Y \to \Sigma$. This connection defines the splitting (2.7.13) of the vertical tangent bundle VY and leads to the vertical covariant differential (2.7.15) which reads

$$\boxed{\widetilde{D} = dx^\lambda \otimes (y^i_\lambda - A^i_\lambda - A^i_m \sigma^m_\lambda)\partial_i.} \tag{6.5.8}$$

As was mentioned above, the operator (6.5.8) possesses the following property. Given a global section h of $\Sigma \to X$, its restriction

$$\begin{aligned} \widetilde{D}_h &= \widetilde{D} \circ J^1 i_h : J^1 Y^h \to T^*X \otimes VY^h, \\ \widetilde{D}_h &= dx^\lambda \otimes (y^i_\lambda - A^i_\lambda - A^i_m \partial_\lambda h^m)\partial_i, \end{aligned} \tag{6.5.9}$$

to Y^h is precisely the familiar covariant differential relative to the pull-back principal connection A_h (2.7.11) on the fibre bundle $Y^h \to X$. Thus, one may construct a Lagrangian on the jet manifold $J^1 Y$ of a composite fibre bundle which factorizes through $\widetilde{D}_A$, that is,

$$L : J^1 Y \xrightarrow{\widetilde{D}} T^*X \underset{Y}{\otimes} VY_\Sigma \to \overset{n}{\wedge} T^*X. \tag{6.5.10}$$

In Section 7.5, we will apply the above scheme of symmetry breaking in gauge theory to describing Dirac fermion fields in gauge gravitation theory.

6.6 Effects of flat principal connections

This Section addresses the effects related to flat principal connections, treated as *vacuum gauge fields*. In electromagnetic theory, two such effects are well known. These are an Aharonov–Bohm effect and the quantization of a magnetic flux.

Example 6.6.1. Let $\mathbb{R}^3$ be a 3-dimensional Euclidean space provided with the Cartesian coordinates (x, y, z) or the cylindrical coordinates (ρ, α, z). Let us consider its submanifold $X = \mathbb{R}^3 \setminus \{\rho = 0\}$. This submanifold admits the vacuum electromagnetic field

$$A = \frac{\Phi}{2\pi} d\alpha = \frac{\Phi_y}{x^2 + y^2} dx - \frac{\Phi_x}{x^2 + y^2} dy. \tag{6.6.1}$$

The strength of this field $F = dA$ vanishes everywhere on X, i.e., the 1-form A (6.6.1) is closed. It is readily observed that the form

$$A = df, \quad f = \frac{\Phi\alpha}{2\pi}, \tag{6.6.2}$$

is exact on any contractible open subset of X (e.g., given by the coordinate relations $0 < \varepsilon < \alpha < 2\pi - \varepsilon$, $\varepsilon > 0$), but not everywhere on X. It follows that A (6.6.1) belongs to a non-vanishing element of the De Rham cohomology group $H^1(X) = \mathbb{R}$ of X (see Appendix 6.8). Being represented by the 1-forms (6.6.1), elements of this cohomology group are parametrised by coefficients Φ in the expression (6.6.1). We can denote them by $[\Phi] \subset H^1(X)$ so that the corresponding group operation in $H^1(X)$ reads

$$[\Phi] + [\Phi'] = [\Phi + \Phi'].$$

Note that the field (6.6.1) can be extended to the whole space $\mathbb{R}^3$ in terms of generalized functions

$$A = \frac{\Phi}{2\pi\rho}\theta(\rho)d\alpha, \tag{6.6.3}$$
$$F = \frac{\Phi}{2}\delta(\rho^2)d\rho \wedge d\alpha,$$

where $\theta(\rho)$ is the step function, while $\delta(\rho^2)$ is the Dirac δ-function. The field (6.6.3) satisfies the Stokes formula

$$\int_{\partial S} A = \int_0^{2\pi} A_\alpha \rho d\alpha = \int_S F\rho d\rho d\alpha = \Phi, \tag{6.6.4}$$

where S is a circle in the plane $z = 0$ whose centre is the point $\rho = 0$. One can think of A (6.6.3) as being an electromagnetic field generated by an infinitely thin and infinitely long solenoid along the axis z. This is a well-known example demonstrating the Aharonov–Bohm effect.

Remark 6.6.2. Note that the Berry's phase phenomenon in quantum systems depending on classical parameters is also a kind of the Aharonov–Bohm effect (see Section 10.5). •

Another well-known effect of a vacuum electromagnetic field is the quantization of a magnetic flux in a ringed superconductor. Let us consider the model of an electromagnetic field A and a complex scalar field ϕ of the Cooper pair condensate which satisfy the vacuum field equations

$$F_{\mu\nu} = 0, \quad D_\mu\phi = (\partial_\mu + ieA_\mu)\phi = 0, \quad \phi^2 = \text{const.},$$

in the region $\rho > b > 0$ of $\mathbb{R}^3$. These equations have a solution

$$\phi = a\exp\left(-\frac{i\alpha\Phi}{2\pi e}\right), \quad A = \frac{\Phi}{2\pi\rho}d\alpha, \quad a = \text{const.} \tag{6.6.5}$$

Since the wave function ϕ must obey the periodicity condition $\phi(\alpha) = \phi(\alpha+2\pi)$, the amplitude Φ of the electromagnetic potential A (6.6.5) is not arbitrary, but takes the values

$$\Phi = 2\pi ne, \qquad n \in \mathbb{Z}.$$

It follows that the magnetic flux (6.6.4) also takes the quantized values $\Phi = 2\pi ne$. •

These examples show that effect of vacuum gauge fields takes place on multi-connected spaces with non-trivial homotopy and cohomology groups (see Appendix 6.8). Let us consider vacuum gauge fields A of a group G on a manifold X. We aim to show the following (see also [3]).

PROPOSITION 6.6.1. There is one-to-one correspondence between the set of gauge conjugate vacuum gauge fields A and the set $\mathrm{Hom}\,(\pi_1(X), G)/G$ of conjugate homomorphisms of the homotopy group $\pi_1(X)$ of X to the group G. Recall that two elements $a, b \in G$ are said to be conjugate if there exists an element $g \in G$ such that $ga = bg$. □

Remark 6.6.3. Holonomy groups. Let us recall the notion of a holonomy group [177]. Let $\pi_P : P \to X$ be a G-principal bundle and A a principal connection on P. Let $c : [0,1] \to X$ be a smooth piecewise closed curve through a point $x \in X$, i.e., $c(0) = c(1) = x$. For any point $p \in P_x = \pi_P^{-1}(x)$, there exists the *horizontal lift* c_p of the curve c in P through p, i.e., $c(0) = p$ and every tangent vector $\dot c_p(t)$, $t \in [0,1]$, (if it exists) is the horizontal lift $A\dot c(t)$ of the tangent vector $\dot c(t)$ by means of the connection A. Then the map

$$\gamma_c : P_x \ni p = c_p(0) \mapsto c_p(1) \in P_x \tag{6.6.6}$$

defines an isomorphism g_c of the fibre P_x. This isomorphism can be seen as a *parallel transport* of the point p along the curve c with respect to the connection A. Let us consider the group C_x of all smooth piecewise closed curves through a point $x \in X$ and its subgroup C_x^0 of the contractible curves. Then the set $\mathcal{K}_x = \{\gamma_c, c \in C_x\}$ of isomorphisms (6.6.6) and its subset $\mathcal{K}_x^0 = \{\gamma_c, \, c \in \mathcal{C}_x^0\}$ constitute the groups called the *holonomy group* and the *restricted holonomy group* for the connection A at a point $x \in X$, respectively. If a manifold X is connected, the holonomy groups $\mathcal{K}_x$ for all $x \in X$ are isomorphic to each other, and one can speak about an *abstract holonomy group* $\mathcal{K}$ and its subgroup $\mathcal{K}^0$.

Since $\mathcal{C}_x/\mathcal{C}_x^0 = \pi_1(X,x)$, there is a homomorphism

$$\pi_1(X,x) \to \mathcal{K}_x/\mathcal{K}_x^0 \tag{6.6.7}$$

of the homotopy group $\pi_1(X,x)$ at a point $x \in X$ onto the quotient group $\mathcal{K}_x/\mathcal{K}_x^0$. In particular, if a manifold X is simply connected, then $\mathcal{K}_x = \mathcal{K}_x^0$ for all $x \in X$ and, consequently, $\mathcal{K} = \mathcal{K}^0$.

There exists a monomorphism of a holonomy group into the structure group G, which is however is not canonical. For a point $p \in P$, it is given by the mapping

$$\mathcal{K}_x \ni \gamma_c \mapsto g_c \in \mathcal{K}_p \subset G, \qquad x = \pi_P(p), \tag{6.6.8}$$

where the element g_c is determined by the relation

$$\gamma_c(p) = pg_c.$$

The subgroup $\mathcal{K}_p$ [$\mathcal{K}_p^0$] of the structure group G is also called the holonomy group [restricted holonomy group] at a point $p \in P$. Obviously, the holonomy subgroups $\mathcal{K}_p$ and $\mathcal{K}_{p'}$ for different points $p, p' \in P$ are conjugate in G. Recall the following important properties of the holonomy groups $\mathcal{K}_p$ and $\mathcal{K}_p^0$ [177].

THEOREM 6.6.2. The holonomy group $\mathcal{K}_p$ [restricted holonomy group $\mathcal{K}_p^0$] is a Lie subgroup [connected Lie subgroup] of the structure group G, and the quotient group $\mathcal{K}_p/\mathcal{K}_p^0$ is countable. □

THEOREM 6.6.3. If the holonomy group $\mathcal{K}_p$ of a principal connection A does not coincide with the structure group G, the latter is reducible to $\mathcal{K}_p$. The corresponding reduced subbundle $P(p)$ of the principal bundle P consists of the points of P which can be connected with p by horizontal curves in P. It means that A is reducible to a principal connection on $P(p)$. □

THEOREM 6.6.4. The values of the curvature R (2.3.3) of the principal connection A at points of the reduced subbundle $P(p)$ constitute a subspace of the Lie algebra

$\mathfrak{g}_r$ of the group G which coincides with the Lie algebra of the restricted holonomy group $\mathcal{K}_p^0$. □ •

As follows from Theorem 6.6.4, if a principal connection A on a G-principal bundle $P \to X$ is flat, the restricted holonomy group $\mathcal{K}^0$ of A is trivial, while the holonomy group $\mathcal{K}$ is discrete (countable). In particular, if a principal bundle over a simply connected base X admits a flat connection, this fibre bundle is trivial.

Given a flat principal connection A, the composition of homomorphisms (6.6.7) and (6.6.8) gives the homomorphism

$$\pi_1(X, x) \to G$$

which is not unique, but depends on a point $p \in P$ in the expression (6.6.8). Therefore, every flat principal connection A on a G-principal bundle $P \to X$ defines a class of conjugate homomorphisms of the homotopy group $\pi_1(X)$ to G in accordance with Proposition 6.6.1. A converse assertion is based on the following lemma.

LEMMA 6.6.5. Let K be a discrete group. A connected K-principal bundle over a connected manifold X exists if and only if there is a subgroup $N \subset \pi_1(X)$ such that $\pi_1(X)/N = K$. □

To prove this Lemma, let us consider the exact sequence of homotopy groups of a fibre bundle $\pi : Y \to X$. Given $y \in Y$ and $x = \pi(y)$, this exact sequence reads

$$\begin{aligned} \cdots \to \pi_k(Y_x, y) \to \pi_k(Y, y) \to \pi_k(X, x) \to \pi_{k-1}(Y_x, y) \to \\ \cdots \to \pi_1(X, x) \to \pi_0(Y_x, y) \to \pi_0(Y, y) \to \pi_0(X, x) \to 0. \end{aligned} \tag{6.6.9}$$

If Y is a connected principal bundle with a discrete structure group K, we have

$$\pi_{k>0}(Y_x, y) = \pi_0(Y, y) = \pi_0(X, x) = 0, \qquad \pi_0(Y_x, y) = K.$$

Then the exact sequence (6.6.9) is reduced to the short exact sequences

$$\begin{aligned} &0 \to \pi_k(Y, y) \to \pi_k(X, x) \to 0, \qquad k > 1, \\ &0 \to \pi_1(Y, y) \to \pi_1(X, x) \to K \to 0. \end{aligned}$$

It follows that $\pi_1(Y, y)$ is a subgroup of $\pi_1(X, x)$ and

$$\pi_1(X, x)/\pi_1(Y, y) = K. \tag{6.6.10}$$

This is a necessary condition for a desired fibre bundle over X to exist. One can show that, since a manifold X is a locally contractible space, the condition that $K = \pi_1(X,.)/N$ for some subgroup $N \subset \pi_1(X,.)$ is sufficient.

Let us return to Proposition 6.6.1. Let $\pi_1(X) \to G$ be a homomorphism whose image is a subgroup $K \in G$ and whose kernel is a subgroup $N \in \pi_1(X)$. By virtue of Lemma 6.6.5, there exists a connected K-principal fibre bundle $Y \to X$ and, consequently, a G-principal fibre bundle $P \to X$ which has $Y \to X$ as a subbundle and whose structure group G is reducible to the discrete subgroup K. In accordance with Theorem 6.5.1, there exists an atlas

$$\Psi_P = \{(U_\alpha, \psi_\alpha^P), \rho_{\alpha\beta}\}$$

of the principal bundle P with K-valued constant transition functions $\rho_{\alpha\beta}$. This is an atlas of local constant trivializations. Following Proposition 2.6.2, one can define a flat connection A on $P \to X$ whose local coefficients with respect to the atlas Ψ_P equal zero. This is a principal connection. Certainly, the holonomy group $\mathcal{K}$ of this connection is K.

Remark 6.6.4. In the topological field theory, the space $\mathrm{Hom}(\pi_1(X), G)/G$ is treated as the moduli space of flat connections [30]. This space has a reach geometrical structure which has been extremely studied in the particular case when X is a compact Riemannian surface [12, 129, 158], though it is not a smooth manifold in general (it may not even be Hausdorff if G is non-compact). •

Now, we can extend the description of an Aharonov–Bohm effect in Example 6.6.1 to an arbitrary gauge model on a principal bundle $P \to X$ which admits a flat principal connection A with a non-trivial discrete holonomy group $\mathcal{K}$. Let $Y \to X$ be a P-associated vector bundle (6.1.22). The notion of holonomy group is extended in a straightforward manner to associated principal connections on $Y \to X$. In particular, if A is a flat principal connection with a holonomy group K on $P \to X$, the associated connection A (6.1.24) on $Y \to X$ is a flat connection with the same holonomy group. There exists an atlas

$$\Psi = \{(U_\alpha, \psi_\alpha), \rho_{\alpha\beta}\}$$

of local constant trivializations of the fibre bundle $Y \to X$ such that the connection A on $Y \to X$ takes the form

$$A = dx^\lambda \otimes \partial_\lambda.$$

Let c be a smooth piecewise closed curve through a point $x \in X$ which crosses the charts $U_{\alpha_1}, \ldots, U_{\alpha_k}$ of the atlas Ψ. Then the parallel transport of a vector $v \in Y_x$ along this curve with respect to the flat connection A is

$$v \mapsto (\rho_{\alpha_2\alpha_1} \cdots \rho_{\alpha_1\alpha_k})(v).$$

It depends only on the homotopic class of a curve c.

Another physical effect discussed in Example 6.6.1 is the quantization of a magnetic flux in a ringed superconductor. To describe this phenomenon in geometric terms, one needs the notion of relative homology and cohomology (see Appendix 6.8).

Given a manifold Y and its submanifold X, let us consider an electromagnetic field on Y whose restriction to X is the vacuum one. The strength of this field is represented by a relative cocycle $\widetilde{\sigma}^2 \in \mathfrak{O}^2(Y, X)$. Let S be a surface in Y whose boundary lies in X. It is represented by a relative cycle $\widetilde{b}_2 \in B_2(Y, X)$. Then the flux of the above-mentioned electromagnetic field through the surface S is given by the integral

$$\Phi = \int\limits_{\widetilde{b}_2} \widetilde{\sigma}^2 \tag{6.6.11}$$

which depends only on the relative homology class of $\widetilde{b}_2$ and the relative cohomology class of $\widetilde{\sigma}^2$. In particular, if a manifold Y is contractible, the closed 2-form $\widetilde{\sigma}^2$ is exact on Y, i.e., $\widetilde{\sigma}^2 = d\sigma^1$ where the restriction of σ^1 to X is an exact form. Then the integral (6.6.11) reads

$$\Phi = \int\limits_{\widetilde{b}_2} d\sigma^1 = \int\limits_{\partial\widetilde{b}_2 \subset X} \sigma^1. \tag{6.6.12}$$

In accordance with the equalities (6.8.13) and (6.8.15), this integral depends only on the homology class of the curve $\partial\widetilde{b}_2$ in $H_1(X)$ and on the De Rham cohomology class of the form σ^1 in $H^1(X)$. For instance, if $Y = \mathbb{R}^3$ and its submanifold $X = \mathbb{R}^3 \setminus \{\rho < b\}$ is occupied by a superconductor, the integral (6.6.12) is exactly a magnetic flux in Example 6.6.1.

It should be noted that, in contrast with the Aharonov–Bohm effect, the flux quantization one is described only in the case of an Abelian gauge field. Nevertheless, this effect may also take place in a non-Abelian gauge model on a principal bundle whose structure group is reducible to an Abelian subgroup.

6.7 Characteristic classes

There exist non-equivalent principal bundles with the same structure group G over the same manifold X. Their classification is given by the well-known classification theorem. Our interest to classification of non-equivalent principal bundles is based on the fact that the equivalence classes of $U(n)$- and $O(n)$-principal bundles are associated with the De Rham homology classes of certain characteristic forms expressed into the strength of gauge fields. If X is a compact manifold, the principal bundles over X are classified by the characteristic numbers which are integrals of the above-mentioned characteristic forms over X. One meets characteristic forms and characteristic numbers, treated as topological charges, in many models of classical and quantum gauge theory, e.g., in the description of instantones, monopoles and anomalies, in topological gauge theory. There is the extensive literature on this subject. In our book, we will be mainly concerned with the topological field theory and anomalies (see Chapters 12 and 13). This Section summarizes some of the basic facts on the Chern, Pontryagin and other characteristic classes which we will refer to in the sequel (see, e.g., [98, 157, 223]).

The classification theorem is concerned with topological fibre bundles with a structure group (see Remark 6.1.3). Its application to smooth fibre bundles is based on Propositions 6.8.1 and 6.9.1 below.

Let $S(X,G)$ denote the set of equivalence classes of associated fibre bundles with a topological structure group G over a paracompact topological space X (we follow the terminology of Bourbaki; see also [157] where a paracompact space, by definition, is Hausdorff). It is in bijection with the first cohomology set $H^1(X;G_0)$ (see Remark 6.9.2 below). Basing on this bijection, one can state the following important properties of the set $S(X;G)$.

(i) If H is a subgroup of the group G, then the inclusion $H\to G$ implies the natural inclusion $S(X,H)\to S(X,G)$ whose image consists of the equivalence classes of G-principal bundles reducible to H-principal bundles over X. In particular, if a base X is paracompact, G is a Lie group and H is its maximal compact subgroup, then

$$S(X,G)=S(X,H) \tag{6.7.1}$$

(see Theorem 6.5.3). For instance, the equality (6.7.1) takes place when

$$\begin{array}{ll} G = GL(n,\mathbb{C}), & H = U(n); \\ G = GL(n,\mathbb{R}), & H = O(n); \\ G = GL^+(n,\mathbb{R}), & H = SO(n); \\ G = SO(1,3), & H = SO(3). \end{array}$$

(ii) Let $f : X' \to X$ be a continuous map. Every topological G-principal bundle $P \to X$ yields the pull-back topological principal bundle $f^*P \to X'$ (see (1.1.6)) with the same structure group G. Therefore, the map f induces the morphism

$$f^* : S(X,G) \to S(X',G) \tag{6.7.2}$$

which depends only on the homotopy class of the map f. Hence, there is the morphism

$$\pi(X',X) \to S(X',G).$$

Then we come to the following *classification theorem.*

THEOREM 6.7.1. For every topological group G, there exists a topological space $B(G)$, called the *classifying space*, and a G-principal bundle $P_G \to B(G)$, called the *universal bundle*, which possess the following properties.

- For any G-principal bundle P over a paracompact base X, there exists a continuous map $f : X \to B(G)$ such that $P = f^*P_G$.
- If two maps f_1 and f_2 of X to $B(G)$ are homotopic then the pull-back bundles $f_1^*P_G$ and f_2^* are equivalent, and *vice versa*, i.e., $S(X,G) = \pi(X, B(G))$.

□

This theorem shows that the set $S(X,G)$ depends only on the homotopic class of the space X. In other words, this set is a homotopic invariant. In general, the classifying space is infinite-dimensional, but its choice depends on the category of spaces.

Here we will concentrate our attention to fibre bundles with the structure groups $GL(n,\mathbb{C})$ (reduced to $U(n)$) and $GL(n,\mathbb{R})$ (reduced to $O(n)$). They are most interesting for physical applications. The classifying spaces for these groups are

$$\begin{aligned} B(U(n)) &= \lim_{N\to\infty} \mathfrak{G}(n, N-n; \mathbb{C}), \\ B(O(n)) &= \lim_{N\to\infty} \mathfrak{G}(n, N-n; \mathbb{R}), \end{aligned} \tag{6.7.3}$$

where $\mathfrak{G}(n, N-n; \mathbb{C})$ and $\mathfrak{G}(n, N-n; \mathbb{R})$ are the Grassmann manifolds of n-dimensional vector subspaces of $\mathbb{C}^N$ and $\mathbb{R}^N$, respectively. Then an equivalence classes of $U(n)$- and $O(n)$-principal bundles over a manifold X can be represented by elements of the Čech cohomology groups $H^*(X; \mathbb{Z})$, called the *characteristic classes*. Moreover, due to the homomorphism

$$H^*(X; \mathbb{Z}) \to H^*(X; \mathbb{R}) = H^*(X), \tag{6.7.4}$$

these characteristic classes are represented by elements of the De Rham cohomology groups. They are cohomology classes of the certain exterior forms defined as follows [98, 177, 214].

Let $P \to X$ be a principal bundle with a structure Lie group G, $C \to X$ the corresponding bundle of principal connections (6.1.8), $\mathcal{A}$ the canonical principal connection (6.2.4) on the G-principal bundle $J^1P \to C$ and $F_{\mathcal{A}}$ its curvature. We consider the algebra $I(\mathfrak{g}_r)$ of real G-invariant polynomials on the Lie algebra $\mathfrak{g}_r$ of the group G. Then there is the well-known Weil homomorphism of $I(\mathfrak{g}_r)$ into the De Rham cohomology algebra $H^*(C)$. By virtue of this homomorphism, every k-linear element $r \in I(\mathfrak{g})$ is represented by the cohomology class of the closed $2k$-form $r(F_{\mathcal{A}})$ on C, called the *characteristic form*. If A is a section of $C \to X$, we have

$$A^* r(F_{\mathcal{A}}) = r(F_A), \tag{6.7.5}$$

where F_A is the strength of A and $r(F_A)$ is the corresponding characteristic form on X. The characteristic forms (6.7.5) possess the following important properties:

- $r(F_A)$ is a closed form, i.e., $dr(F_A) = 0$;
- $r(F_A) - r(F_{A'})$ is an exact form, whenever A and A' are two principal connections on the same principal bundle P (see Section 13.1).

It follows that characteristic forms $r(F_A)$ for different principal connections A have the same De Rham cohomology class.

We start from the *Chern classes* $c_i(P) \in H^{2i}(X, \mathbb{Z})$ of $GL(k, \mathbb{C})$-principal bundles which are always $U(k)$-principal bundles.

Remark 6.7.1. Note that Chern classes can be defined without reference to any differentiable structure, and are therefore homotopic invariants. The Pontryagin classes that we consider are expressed into the Chern classes by the formula (6.7.25) and, therefore, are homotopic invariants. •

Let A be a complex $(k \times k)$-matrix and $r(A)$ a $GL(k,\mathbb{C})$-invariant polynomial of components of A, i.e.,

$$r(A) = r(gAg^{-1}), \qquad g \in GL(k,\mathbb{C}).$$

If a matrix A has eigen values $a_1, \ldots, a_k$, an invariant polynomial $r(A)$ takes the form

$$r(A) = b + cS_1(a) + dS_2(a) + \cdots$$

where

$$S_j(a) = \sum_{i_1<\ldots<i_j} a_{i_1} \ldots a_{i_j} \tag{6.7.6}$$

are symmetric polynomials of $a_1, \ldots, a_k$.

Example 6.7.2. An important example of an invariant polynomial is

$$\det(\mathbf{1} + A) = 1 + S_1(a) + S_2(a) + \ldots + S_k(a),$$

where $\mathbf{1}$ denotes the unit matrix. •

Let $P \to X$ be a $U(k)$-principal bundle and E the associated vector bundle with the typical fibre $\mathbb{C}^k$ which performs the natural representation of $U(k)$. For brevity, we will call E the *$U(k)$-bundle.*

Let F be the curvature form (6.1.19) of a principal connection on E. The characteristic form

$$c(F) = \det\left(\mathbf{1} + \frac{i}{2\pi}F\right) = 1 + c_1(F) + c_2(F) + \cdots \tag{6.7.7}$$

is called the *total Chern form*, while its components $c_i(F)$ are called *Chern $2i$-forms.* For instance,

$$c_0(F) = 0,$$

$$c_1(F) = \frac{i}{2\pi}\mathrm{Tr}\, F, \tag{6.7.8}$$

$$c_2(F) = \frac{1}{8\pi^2}[\mathrm{Tr}(F\wedge F) - \mathrm{Tr}\, F \wedge \mathrm{Tr}\, F]. \tag{6.7.9}$$

All Chern forms $c_i(F)$ are closed, and their cohomology classes are identified with the Chern classes $c_i(E) \in H^{2i}(X,\mathbb{Z})$ of the $U(k)$-bundle E under the homomorphism (6.7.4). The total Chern form (6.7.7) corresponds to the *total Chern class*

$$c(E) = c_0(E) + c_1(E) + \cdots.$$

Example 6.7.3. Let us consider a $U(1)$-bundle $L \to X$. It is a *complex line bundle* with the typical fibre $\mathbb{C}$ on which the group $U(1)$ acts by the generator $I = i\mathbf{1}$. The curvature form (6.1.25) on this fibre bundle reads

$$F = \frac{i}{2} F_{\lambda\mu} dx^\lambda \wedge dx^\mu. \tag{6.7.10}$$

Then the total Chern form (6.7.7) of a $U(1)$-bundle is

$$\begin{aligned} &c(F) = 1 + c_1(F), \\ &c_1(F) = \frac{i}{2\pi} \operatorname{Tr} F = -\frac{1}{4\pi} F_{\lambda\mu} dx^\lambda \wedge dx^\mu. \end{aligned} \tag{6.7.11}$$

•

Example 6.7.4. Let us consider a $SU(2)$-bundle $E \to X$. The curvature form (6.1.25) on this fibre bundle reads

$$F = \frac{i\sigma_a}{2} F^a, \tag{6.7.12}$$

where σ_a, $a = 1, 2, 3$, are the Pauli matrices. Then we have

$$\begin{aligned} &c(F) = 1 + c_1(F) + c_2(F), \\ &c_1(F) = 0, \\ &c_2(F) = \frac{1}{8\pi^2} \operatorname{Tr}(F \wedge F). \end{aligned} \tag{6.7.13}$$

•

Using the natural properties of Chern forms, one can obtain easily the properties of Chern classes:

(i) $c_i(E) = 0$ if $2i > n = \dim X$;

(ii) $c_i(E) = 0$ if $i > k$;

(iii) $c(E \oplus E') = c(E)c(E')$;

(iv) $c_1(L \oplus L') = c_1(L) + c_1(L')$ where L and L' are complex line bundles;

(v) if $f^*E \to X'$ is the pull-back bundle generated by the morphism $f : X' \to X$, then

$$c(f^*E) = f^*c(E),$$

where f^* is the induced morphism of the cohomology groups

$$f^* : H^*(X;\mathbb{Z}) \to H^*(X';\mathbb{Z}). \tag{6.7.14}$$

The properties (iii) and (v) of Chern classes are utilized in the following theorem.

THEOREM 6.7.2. For any $U(k)$-bundle $E \to X$, there exists a topological space X' and a continuous morphism $f : X' \to X$ so that:

- the pull-back $f^*E \to X'$ is the Whitney sum of line bundles

$$f^*E = L_1 \oplus \cdots \oplus L_k;$$

- the induced morphism f^* (6.7.14) is an inclusion.

□

It follows that the total Chern class of any $U(k)$-bundle E can be seen as

$$c(E) = f^*c(E) = c(L_1 \oplus \cdots \oplus L_k) = c(L_1)\cdots c(L_k) = \tag{6.7.15}$$
$$(1 + a_1)\cdots(1 + a_k),$$

where $a_i = c_1(L_i)$ denotes the Chern class of the line bundle L_i (see Example 6.7.3). The formula (6.7.15) is called the *splitting principle*. In particular, we have

$$c_1(E) = \sum_i a_i,$$
$$c_2(E) = \sum_{i_1<i_2} a_{i_1}a_{i_2},$$
$$c_j(E) = \sum_{i_1<\cdots<i_j} a_{i_1}\cdots a_{i_j}$$

(cf. (6.7.6)).

Example 6.7.5. Let E^* be a $U(k)$-bundle dual of E. In accordance with the splitting principle, we have

$$c(E^*) = c(L_1^*)\cdots c(L_k^*) = (1 + a_1^*)\cdots(1 + a_k^*).$$

Since the generator of the dual representation of $U(1)$ is $I = -i\mathbf{1}$, then $F^* = -F$ and $a_i^* = -a_i$. It follows that

$$c_i(E^*) = (-1)^i c_i(E). \tag{6.7.16}$$

•

If the base X of a $U(k)$-bundle E is a compact n-dimensional manifold, one can form some exterior n-forms from the Chern forms $c_i(F)$ and can integrate them over X. Such integrals are called *Chern numbers*. Chern numbers are integer since the Chern forms belong to the integer cohomology classes. For instance, if $n = 4$, there are two Chern numbers

$$C_2(E) = \int_X c_2(F), \qquad C_1^2(E) = \int_X c_1(F) \wedge c_1(F). \tag{6.7.17}$$

There are also some other characteristic classes of $U(k)$-bundles which are expressed into the Chern classes. We will mention only two of them.

The *Chern character* $\mathrm{ch}(E)$ is given by the invariant polynomial

$$\mathrm{ch}(A) = \mathrm{Tr}\exp\left(\frac{i}{2\pi}A\right) = \sum_{m=0}^{\infty} \frac{1}{m!}\mathrm{Tr}\left(\frac{i}{2\pi}A\right)^m. \tag{6.7.18}$$

It has the properties

$$\mathrm{ch}(E \oplus E') = \mathrm{ch}(E) + \mathrm{ch}(E'),$$
$$\mathrm{ch}(E \otimes E') = \mathrm{ch}(E) \cdot \mathrm{ch}(E').$$

Using the splitting principle, one can express the Chern character into the Chern classes as follows:

$$\mathrm{ch}(E) = \mathrm{ch}(L_1 \oplus \cdots \oplus L_k) = \mathrm{ch}(L_1) + \cdots + \mathrm{ch}(L_k) =$$
$$\exp a_1 + \cdots + \exp a_k = k + \sum_i a_i + \frac{1}{2}\sum_i a_i^2 + \cdots =$$
$$k + \sum_i a_i + \frac{1}{2}\left[(\sum_i a_i)^2 - 2\sum_{i_1<i_2} a_{i_1}a_{a_2}\right] + \cdots =$$
$$k + c_1(E) + \frac{1}{2}[c_1^2(E) - 2c_2(E)] + \cdots.$$

The *Todd class* is defined as

$$\mathrm{td}(E) = \sum_{i=1}^{k} \frac{a_i}{1-\exp(-a_i)} = 1 + \frac{1}{2}c_1 + \frac{1}{12}(c_1^2 + c_2) + \cdots. \tag{6.7.19}$$

It has the property

$$\mathrm{td}(E \oplus E') = \mathrm{td}(E) \cdot \mathrm{td}(E').$$

Turn now to characteristic classes of a real k-dimensional vector bundle E with the structure group $O(k)$. For brevity, we will call E the *$O(k)$-bundle.*

The *Pontryagin classes* of a $O(k)$-bundle E in the De Rham cohomology algebra $H^*(X)$ are associated with the invariant polynomial

$$p(F) = \det\left(\mathbf{1} - \frac{1}{2\pi}F\right) = 1 + p_1(F) + p_2(F) + \cdots \tag{6.7.20}$$

of the curvature F (6.1.25) which takes its values in the Lie algebra $\mathrm{o}(k)$ of the group $O(k)$. Since the generators of the group $O(k)$ satisfy the condition

$$(I)^a_b = -(I)^b_a,$$

only the components of even degrees in F in the decomposition (6.7.20) are different from zero, i.e., $p_i(E) \in H^{4i}(X)$. Pontryagin classes possess the following properties:

(i) $p_i(E) = 0$ if $4i > n = \dim X$;

(ii) $p_i(E) = 0$ if $2i > k$;

(iii) $p(E \oplus E') = p(E) + p(E')$.

Note that, for the Pontryagin classes taken in the Čech cohomology $H^*(X;\mathbb{Z})$, the property (ii) is true only modulo cyclic elements of order 2.

Remark 6.7.6. It should be emphasized that, though fibre bundles with the structure groups $O(k)$ and $GL(k,\mathbb{R})$ have the same characteristic classes, their characteristic forms are different. For instance, if F is the curvature form (6.1.25) which takes its values into the Lie algebra $\mathrm{gl}(k,\mathbb{R})$, the characteristic form $p(F)$ (6.7.20) contains the terms of odd degrees in F in general. Therefore, to construct the characteristic forms corresponding to Pontryagin classes, one should use only $O(k)$- or $O(k-m,m)$-valued curvature forms R. In particular,

$$p_1(R) = -\frac{1}{8\pi^2}\mathrm{Tr}\, R \wedge R, \tag{6.7.21}$$

$$p_2(R) = \left[\frac{1}{32\pi^2}\varepsilon_{abcd}R^{ab} \wedge R^{cd}\right]. \tag{6.7.22}$$

•

Let us consider the relations between the Pontryagin classes of $O(k)$-bundles and the Chern classes of $U(k)$-bundles. There are the following commutative diagrams of group monomorphisms

$$\begin{array}{ccc} O(k) & \longrightarrow & U(k) \\ \downarrow & & \downarrow \\ GL(k,\mathbb{R}) & \longrightarrow & GL(k,\mathbb{C}) \end{array} \tag{6.7.23}$$

$$\begin{array}{ccc} U(k) & \longrightarrow & O(2k) \\ \downarrow & & \downarrow \\ GL(k,\mathbb{C}) & \longrightarrow & GL(2k,\mathbb{R}) \end{array} \tag{6.7.24}$$

The diagram (6.7.23) implies the inclusion

$$\varphi : S(X,O(k)) \to S(X,U(k)),$$

and one can show that

$$p_i(E) = (-1)^i c_{2i}(\varphi(E)). \tag{6.7.25}$$

The diagram (6.7.24) yields the inclusion

$$\rho : S(X,U(k)) \to S(X,O(2k)).$$

Then we have

$$\varphi\rho : S(X,U(k)) \to S(X,O(2k)) \to S(X,U(2k)).$$

It means that, if E is a $U(k)$-bundle, then $\rho(E)$ is a $O(2k)$-bundle, while $\varphi\rho(E)$ is a $U(2k)$-bundle.

Remark 6.7.7. Let A be an element of $U(k)$. The group monomorphisms

$$U(k) \to O(2k) \to U(2k)$$

define the transformation of matrices

$$A \longrightarrow \begin{pmatrix} \operatorname{Re} A & -\operatorname{Im} A \\ \operatorname{Im} A & \operatorname{Re} A \end{pmatrix} \longrightarrow \begin{pmatrix} A & 0 \\ 0 & A^* \end{pmatrix} \tag{6.7.26}$$

written relative to complex coordinates z^i on the space $\mathbb{C}^k$, real coordinates

$$x^i = \operatorname{Re} z^i, \qquad x^{k+i} = \operatorname{Im} z^i$$

on $\mathbb{R}^{2k}$ and complex coordinates

$$z^i = x^i + ix^{k+i}, \quad z^{k+i} = x^i - ix^{k+i}$$

on the space $\mathbb{C}^{2k}$, respectively. •

A glance at the diagram (6.7.26) shows that the fibre bundle $\varphi\rho(E)$ is the Whitney sum of E and E^* and, consequently,

$$c(\varphi\rho(E)) = c(E)c(E^*).$$

Then combining (6.7.16) and (6.7.25) gives the relation

$$\sum_i (-1)^i p_i(\rho(E)) = c(\varphi\rho(E)) = c(E)c(E^*) = \left[\sum_i c_i(E)\right]\left[\sum_j (-1)^j c_j(E)\right] \tag{6.7.27}$$

between the Chern classes of the $U(k)$-bundle E and the Pontryagin classes of the $O(2k)$-bundle $\rho(E)$.

Example 6.7.8. By *Pontryagin classes* $p_i(X)$ *of a manifold* X are meant those of the tangent bundle $T(X)$. Let a manifold X be oriented and $\dim X = 2m$. One says that a manifold X admits an *almost complex structure* if its structure group $GL(2m, \mathbb{R})$ is reducible to the image of $GL(m, \mathbb{C})$ in $GL(2m, \mathbb{R})$. By *Chern classes* $c_i(X)$ of such a manifold X^{2m} are meant those of the tangent bundle $T(X)$ seen as a $GL(m, \mathbb{C})$-bundle, i.e.,

$$c_i(X) = c_i(\rho(T(X))).$$

Then the formula (6.7.27) provides the relation between Pontryagin and Chern classes of a manifold X admitting an almost complex structure:

$$\sum_i (-1)^i p_i(X) = \left[\sum_i c_i(X)\right]\left[\sum_j (-1)^j c_j(X)\right].$$

In particular, we have

$$p_1(X) = c_1^2(X) - 2c_2(X), \qquad (6.7.28)$$
$$p_2(X) = c_2^2(X) - 2c_1(X)c_3(X) + 2c_4(X).$$

•

If the structure group of a $O(k)$-bundle E is reduced to $SO(k)$, the *Euler class* $e(E)$ of E can be defined as an element of the Čech cohomology group $H^k(X,\mathbb{Z})$ which satisfies the conditions:

(i) $2e(E) = 0$ if k is odd;

(ii) $e(f^*E) = f^*e(E)$;

(iii) $e(E \oplus E') = e(E)e(E')$;

(iv) $e(E) = c_1(E)$ if $k = 2$.

The last condition follows from the isomorphism of groups $SO(2)$ and $U(1)$.

Let us consider the relationship between the Euler class and the Pontryagin classes. Let E be a $U(k)$-bundle and $\rho(E)$ the corresponding $SO(2k)$-bundle. Then, using the splitting principle and the properties (iii), (iv) of the Euler class, we obtain

$$e(\rho(E)) = e(\rho(L_1) \oplus \cdots \oplus \rho(L_k)) = e(\rho(L_1)) \cdots e(\rho(L_k)) = \qquad (6.7.29)$$
$$c_1(L_1) \cdots c_1(L_k) = a_1 \cdots a_k = c_k(E).$$

At the same time, we deduce from (6.7.27) that

$$p_k(\rho(E)) = c_k^2(E). \qquad (6.7.30)$$

Combining (6.7.29) and (6.7.30) gives the desired relation

$$e(E) = [p_k(E)]^{1/2} \qquad (6.7.31)$$

for any $SO(2k)$-bundle.

For instance, let X be a $2k$-dimensional oriented compact manifold. Its tangent bundle TX has the structure group $SO(2k)$. Then the integral

$$e = \int_X e(R)$$

coincides with the *Euler characteristic* of the compact manifold X.

We refer the reader to Remark (12.3.1) below for the notion of a signature of a 4-dimensional topological manifold and to [157] to that of a $4k$-dimensional smooth manifold.

Let us also mention the *Stiefel–Whitney classes* $w_i \in H^i(X, \mathbb{Z}_2)$ of the tangent bundle $T(X)$. In particular, a manifold X is orientable if and only if $w_1 = 0$. If X admits an almost complex structure, then

$$w_{2i+1} = 0, \quad w_{2i} = c_i \bmod 2.$$

In contrast to the above mentioned characteristic classes, the Stiefel–Whitney ones are not represented by De Rham cohomology classes of exterior forms.

6.8 Appendix. Homotopy, homology and cohomology

For the sake of convenience of the reader, we recall the basic notions of algebraic topology. Dealing below with topological spaces, groups and fibre bundles, one should bear in mind that manifolds have the homotopy type of CW-complexes [85]. An application of algebraic topology to gauge theory is based on the following two mathematical facts.

PROPOSITION 6.8.1. [286]. Let $P \to X$ be a principal bundle. Let f_1 and f_2 be two mappings of a manifold X' to X. If these mappings are homotopic, the pull-back bundles $f_1^*P \to X'$ and $f_2^*P \to X'$ are isomorphic. □

PROPOSITION 6.8.2. [157]. The De Rham cohomology groups $H^*(X)$ of a paracompact manifold X are isomorphic to the Čech cohomology groups $H^*(X, \mathbb{R})$ with coefficients in $\mathbb{R}$ (see Remark 8.3.6 below). This isomorphism enables one to represent characteristic classes of principal bundles as the De Rham cohomology classes of characteristic exterior forms expressed into terms of principal connections (see the next Section). □

Let us start from homotopy theory (see, e.g., [139, 305]). Continuous maps f and f' of a topological space X to a topological space X' are said to be *homotopic* if there is a continuous map

$$g : [0, 1] \times X \to X'$$

whose restriction to $\{0\} \times X$ $[\{1\} \times X]$ coincides with f $[f']$. The set of equivalence classes of homotopic maps $X \to X'$ is denoted by $\pi(X, X')$. The topological spaces X and X' are called *homotopically equivalent* or simply *homotopic* if there exist mappings $f : X \to X'$ and $g : X' \to X$ such that $g \circ f$ is homotopic to the identity map $\mathrm{Id}\, X$, and $f \circ g$ is homotopic to $\mathrm{Id}\, X'$. In particular, a topological space is called *contractible* if it is homotopic to any its point. For instance, Euclidean spaces $\mathbb{R}^k$ are contractible.

Let (S^k, a) be a k-dimensional sphere and $a \in S^k$ a point. Let us consider the set of equivalence classes $\pi_k(X, b)$ of homotopic maps of S^k to a topological space X which send a onto a fixed point b of X'. If X is pathwise connected, this set does not depend on the choice of a and b, and one can talk about the set $\pi_k(X)$ of equivalence classes of homotopic maps $S^k \to X$. This set can be provided with a group structure, and is called the *kth homotopy group* of the topological space X, while the first homotopy group $\pi_1(X)$ is also known as the *fundamental group* of X. The homotopy groups $\pi_{k>1}(X)$ are always Abelian. By $\pi_0(X)$ is denoted the set of connected components of X.

A topological space X is said to be *p-connected* if it is pathwise connected and its homotopy groups $\pi_{k\leq p}(X)$ are trivial. A 1-connected space is also called *simply connected*. A contractible topological space is p-connected for any $p \in \mathbb{N}$.

There is the important relation for the homotopy groups of the product of topological spaces:

$$\pi_k(X \times X') = \pi_k(X) \times \pi_k(X').$$

For more complicated constructions of topological spaces we refer the reader to the Van Kampen theorem [71].

Homotopy groups of topological spaces are *homotopic invariants* in the sense that they are the same for homotopic topological spaces. Other homotopic invariants are homology and cohomology groups of topological spaces.

Let us recall briefly the basic notions of homology and cohomology of complexes (see, e.g., [41, 204]). A sequence

$$0 \xleftarrow{\partial_0} B_0 \xleftarrow{\partial_1} B_2 \xleftarrow{\partial_2} \cdots \xleftarrow{\partial_p} B_p \xleftarrow{\partial_{p+1}} \cdots \tag{6.8.1}$$

of Abelian groups B_p and homomorphisms ∂_p is said to be a *chain complex* if

$$\partial_p \circ \partial_{p+1} = 0, \qquad \forall p \in \mathbb{N},$$

i.e., $\operatorname{Im}\partial_{p+1} \subset \operatorname{Ker}\partial_p$. Elements of $\operatorname{Im}\partial_{p+1}$ and $\operatorname{Ker}\partial_p$ are called *p-boundaries* and *p-cycles*, respectively. The quotient

$$H_p(B_*) = \operatorname{Ker}\partial_p/\operatorname{Im}\partial_{p+1}$$

is called the *pth homology group* of the chain complex B_* (6.8.1). The chain complex (6.8.1) is called *exact* at an element B_p if $H_p(B_*) = 0$. B_* is an exact sequence if it is exact at each element.

A sequence

$$0 \hookrightarrow B^0 \xrightarrow{\delta^0} B^1 \xrightarrow{\delta^1} \cdots \xrightarrow{\delta^{p-1}} B^p \xrightarrow{\delta^p} \cdots \tag{6.8.2}$$

of Abelian groups B^p and homomorphisms δ^p is said to be a *cochain complex* if

$$\delta^p \circ \delta^{p-1} = 0, \qquad \forall p \in \mathbb{N},$$

i.e., $\operatorname{Im}\delta^{p-1} \subset \operatorname{Ker}\delta^p$. Elements of $\operatorname{Im}\delta^{p-1}$ and $\operatorname{Ker}\delta^p$ are called *p-coboundaries* and *p-cocycles*, respectively.

The *pth cohomology group* of the cochain complex B^* (6.8.2) is the quotient

$$H^p(B^*) = \operatorname{Ker}\delta^p/\operatorname{Im}\delta^{p-1}.$$

Though there are different homology and cohomology theories (see [304] for a survey), one usually refers to:

- the singular homology $H_*(X;\mathcal{K})$ and the singular cohomology $H^*(X;\mathcal{K})$ with coefficients in a numeral ring $\mathcal{K}$,
- the De Rham cohomology $H^*(X)$,
- the Čech cohomology with coefficients in a ring $\mathcal{K}$

[70, 92, 139]. In this Section, we are concerned only with De Rham cohomology.

The *De Rham complex*

$$0 \hookrightarrow \mathbb{R} \hookrightarrow \mathfrak{O}^0(X) \xrightarrow{d} \mathfrak{O}^1(X) \xrightarrow{d} \cdots \xrightarrow{d} \mathfrak{O}^p(X) \xrightarrow{d} \cdots \tag{6.8.3}$$

of exterior forms on a manifold X is a cochain complex, whose cohomology group $H^p(X)$ is called the pth *De Rham cohomology group*. It is the quotient of the space of closed p-forms by the subspace of exact p-forms. By $H^0(X)$ is meant a vector space whose dimension equals the number of connected components of X. It is clear that $H^p(X) = 0$, $p > \dim X$. If X is contractible, $H^{p>0}(X) = 0$. Let us also recall the *Künneth formula*

$$H^m(X \times X') = \sum_{k+l=m} H^k(X) \otimes H^l(X') \tag{6.8.4}$$

for the De Rham cohomology groups of the product $X \times X'$. Elements of the De Rham cohomology groups constitute an algebra $H^*(X)$ with respect to the *cup-product* induced by the exterior product of exterior forms:

$$[a] \cup [b] = [a \wedge b], \qquad a \in [a], \qquad b \in [b]. \tag{6.8.5}$$

It is called the *De Rham cohomology algebra.*

There are the following important relations between homotopy, homology and cohomology groups.

(i) There exists the *Hurewicz homomorphism*

$$h_k : \pi_k(X) \to H_k(X; \mathbb{Z}) \tag{6.8.6}$$

of homotopy groups $\pi_k(X)$ to singular homology groups $H_k(X;\mathbb{Z})$ with coefficients in $\mathbb{Z}$. In particular, the kernel of the homomorphism h_1 (6.8.6) is the *commutant* of the group $\pi_1(X)$ which is generated by elements $a^{-1}b^{-1}ab \in \pi_1(X)$. If the homotopy group $\pi_1(X)$ is commutative, it is isomorphic to the homology group $H_1(X;\mathbb{Z})$. If $\pi_i(X) = 0$ for all $i < k$, then the homomorphism h_k is an isomorphism, while h_{k+1} is an epimorphism.

(ii) There is the isomorphism

$$H_k(X; \mathbb{R}) = H_k(X; \mathbb{Z}) \otimes \mathbb{R}, \tag{6.8.7}$$

where by $\otimes$ is meant the tensor product of Abelian groups. One can show that, if G is a finite Abelian group, then $G \otimes \mathbb{R} = 0$. It follows that a homology group $H_k(X;\mathbb{R})$ has no a finite subgroup. The isomorphism (6.8.7) defines the homomorphism

$$H_k(X; \mathbb{Z}) \to H_k(X; \mathbb{R}). \tag{6.8.8}$$

(iii) The singular homology $H_*(X;\mathcal{K})$ and the singular cohomology $H^*(X;\mathcal{K})$ make up a dual pair where elements of $H^k(X;\mathcal{K})$ are $\mathcal{K}$-valued characters (linear forms) on the group $H_k(X;\mathcal{K})$.

(iv) For paracompact and second countable manifolds which we deal with in field theory, the singular cohomology groups $H^*(X;\mathbb{R})$, the De Rham cohomology groups

$H^*(X)$ and the Čech cohomology groups with coefficients in $\mathbb{R}$ (see Appendix 6.9) coincide with each other [85, 157].

(v) In accordance with the *De Rham duality theorem*, there is the bilinear form

$$\langle | \rangle : H^p(X) \otimes H_p(X, \mathbb{R}) \to \mathbb{R} \tag{6.8.9}$$

given by the integration

$$\langle \sigma^p | c_p \rangle = \int_{c_p} \sigma^p, \tag{6.8.10}$$

where σ^p are closed p-forms on X and c_p are p-cycles in X. By virtue of the Stokes theorem, the integral (6.8.10) depends only on the cohomology class of σ^p and the homology class of c_p. As a consequence the De Rham cohomology group $H^p(X)$ and the singular homology group $H_p(X, \mathbb{R})$ are isomorphic as vector spaces.

(vi) Let X be a compact oriented manifold. We have the bilinear form

$$\begin{aligned} &(,) : H^p(X) \otimes H^{n-p}(X) \to \mathbb{R}, \\ &(\sigma^p, \sigma^{n-p}) = \int_X \sigma^p \wedge \sigma^{n-p}. \end{aligned} \tag{6.8.11}$$

This bilinear form defines the *Poincaré duality isomorphisms* of vector spaces

$$H^p(X) = H^{n-p}(X), \quad H^p(X) = H_{n-p}(X; \mathbb{R}), \quad H_p(X; \mathbb{R}) = H^{n-p}(X; \mathbb{R}).$$

Relative homology and cohomology

Let Y be a topological space and X its subspace. Then the group $B_k(X)$ of singular chains in X belongs to the group $B_k(Y)$ of singular chains in X. The quotient

$$B_k(Y, X) = B_k(Y)/B_k(X).$$

is called the group of *relative k-chains* [92]. Its elements $\widetilde{b}_k$ are singular k-chains b_k in Y modulo singular k-chains in X. The boundary operator

$$\partial : B_k(Y) \to B_{k-1}(Y)$$

yields the boundary operator

$$\partial : B_k(Y, X) \to B_{k-1}(Y, X)$$

on relative chains. By a *relative cycle* $\widetilde{b}_k$ is meant a class of singular k-chains b_k in Y such that $\partial b_k \subset X$, i.e. $\partial\widetilde{b}_k = 0$. Accordingly, a *relative boundary* $\widetilde{b}_k = \partial\widetilde{b}_{k+1}$ is defined as a singular k-boundary $b_k = \partial b_{k+1}$ in Y modulo k-chains in X. The quotient of the group of relative k-cycles by the group of relative k-boundaries is called the kth *relative homology group* $H_k(Y,X)$. We have the exact sequence

$$\cdots \to H_k(X) \to H_k(Y) \to H_k(Y,X) \to H_{k-1}(X) \to \cdots. \tag{6.8.12}$$

Example 6.8.1. If Y is a contractible space, $H_k(Y) = 0$, $k > 0$. Then the exact sequence (6.8.12) reduces to the short exact sequences

$$0 \to H_{k+1}(Y,X) \to H_k(X) \to 0, \qquad k > 0.$$

It follows that

$$H_{k+1}(Y,X) = H_k(X), \qquad k > 0. \tag{6.8.13}$$

•

Similarly, relative cohomology groups are introduced. Let X be a submanifold of a manifold Y. Let us consider the homomorphism $\mathfrak{O}^k(Y) \to \mathfrak{O}^k(X)$ which restricts k-forms on Y to k-forms on X. Its kernel is called the group of *relative k-cochains* $\mathfrak{O}^k(Y,X)$ of the pair (Y,X). Elements $\widetilde{\sigma}^k$ of this group can be represented by k-forms σ^k on Y which vanish on $X \subset Y$. Therefore, the exterior differential d acts on relative cochains $\widetilde{\sigma}^k$ in an ordinary way. Accordingly, the *relative cocycles* $\widetilde{\sigma}^k$ such that $d\widetilde{\sigma}^k = 0$ and the *relative coboundaries* $\widetilde{\sigma}^k = d\widetilde{\sigma}^{k-1}$ are introduced. It is readily observed that a relative cocycle is a closed form, while a relative cocycle which is an exact form $\widetilde{\sigma}^k = d\sigma^{k-1}$ on Y fails to be a relative coboundary in general since σ^{k-1} does not necessarily vanish on X. The quotient of the additive group of relative k-cocycles by the additive group of relative k-coboundaries is called the kth *relative cohomology group* $H^k(Y,X)$. There is the exact sequence

$$\ldots H^k(Y,X) \to H^k(Y) \to H^k(X) \to H^{k+1}(Y,X) \to \cdots. \tag{6.8.14}$$

Example 6.8.2. If Y is a contractible manifold, $H^k(Y) = 0$, $k > 0$. Then the exact sequence (6.8.14) reduces to the short exact sequences

$$0 \to H^k(X) \to H^{k+1}(Y,X) \to 0, \qquad k > 0.$$

It follows that

$$H^{k+1}(Y,X) = H^k(X), \qquad k > 0. \tag{6.8.15}$$

•

6.9 Appendix. Čech cohomology

Recall briefly the notion of Čech cohomology [85, 157]. Let $\mathfrak{U} = \{U_i\}_{i\in I}$ be an open covering of a paracompact topological space X. Let us consider functions ϕ which associate an element of an Abelian group K to each $(p+1)$-tuple $(i_0,\ldots,i_p)$ of indices in I such that $U_{i_0}\cap\cdots\cap U_{i_p}\neq\emptyset$. One can think of ϕ as being constant K-valued functions on the set $U_{i_0}\cap\cdots\cap U_{i_p}$. These functions form an Abelian group $B^p(\mathfrak{U},K)$. Let us consider the cochain morphism

$$\delta^p : B^p(\mathfrak{U},K)\to B^{p+1}(\mathfrak{U},K),$$
$$(\delta^p\phi)(i_0,\ldots,i_{p+1}) = \sum_{k=0}^{p+1}(-1)^k\phi(i_0,\ldots,\widehat{i}_k,\ldots i_{p+1}),$$

where $\widehat{i}_k$ means that the index i_k is omitted. One can check that

$$\delta^{p+1}\circ\delta^p = 0.$$

Hence, we have the cochain complex

$$\cdots\longrightarrow B^p(\mathfrak{U},K)\xrightarrow{\delta^p} B^{p+1}(\mathfrak{U},K)\longrightarrow\cdots,$$

and its cohomology groups

$$H^p(\mathfrak{U};K) = \mathrm{Ker}\,\delta^p/\mathrm{Im}\,\delta^{p-1} \tag{6.9.1}$$

can be defined.

Example 6.9.1. For instance,

$$(\delta^0\phi)(i_0,i_1) = \phi(i_1)-\phi(i_0),$$
$$(\delta^1\phi)(i_0,i_1,i_2) = \phi(i_1,i_2)-\phi(i_0,i_2)+\phi(i_0,i_1).$$

The cohomology group $H^1(\mathfrak{U};K)$ (6.9.1) consists of the class of functions $\phi(i_k,i_j)$ which satisfies the cocycle condition

$$\phi(i_k,i_j)-\phi(i_p,i_j)+\phi(i_p,i_k) = 0, \tag{6.9.2}$$

module functions $\phi(i_k,i_j)=\phi(i_k)-\phi(i_j)$. •

Of course, the cohomology groups (6.9.1) depend on an open covering $\mathfrak{U}$ of the topological space X. Let $\mathfrak{U}'$ be a refinement of the covering $\mathfrak{U}$. Then there exists a homomorphism

$$H^p(\mathfrak{U};K)\to H^p(\mathfrak{U}';K).$$

One can take the direct limit of the groups $H^p(\mathfrak{U}, K)$ with respect to these homomorphisms, where $\mathfrak{U}$ runs through all open coverings of X [157]. This limit is the *pth Čech cohomology group* $H^p(X; K)$ of X with coefficients in K. By definition, $H^0(X; K) = K$.

Remark 6.9.2. Čech cohomology groups introduced above are particular cohomology groups with coefficients in a sheaf, namely, in the sheaf of locally constant K-valued functions or simply in the constant sheaf K (see Example 8.3.1 below). Here we consider again a particular case of such cohomology groups when $\phi(i_0, \ldots, i_p)$ are local continuous functions $\phi_{i_0,\ldots,i_p}$ on $U_{i_0} \cap \cdots \cap U_{i_p}$ with values in an Abelian group K. Let us write the group operation in K in a multiplicative form. One says that the set of functions $\{(U_i \cap U_j; \phi_{ij})\}$ is a *cocycle* if these functions obey the cocycle condition (6.9.2), i.e.,

$$\phi_{ij}\phi_{jk} \mid_{U_i \cap U_j \cap U_k} = \phi_{ik} \mid_{U_i \cap U_j \cap U_k} . \tag{6.9.3}$$

In particular, $f_{ii} = 1$ and $f_{ij} = f_{ji}^{-1}$. Two cocycles $\{\phi_{ij}\}$ and $\{\phi'_{ij}\}$ are said to be equivalent if there exists a set of K-valued functions $\{U_i; f_i\}$ such that

$$\phi'_{ij} = f_i \phi_{ij} f_j^{-1}. \tag{6.9.4}$$

Then the set of equivalent cocycles on an open covering $\mathfrak{U}$ of a paracompact topological space X constitute a cohomology group $H^1(\mathfrak{U}; K_0)$. The direct limit $H^1(X; K_0)$ of the groups $H^1(\mathfrak{U}, K_0)$, where $\mathfrak{U}$ runs all open coverings of X is said to be the first *cohomology group with coefficients in the sheaf* K_0 of continuous K-valued functions on X. Similarly, the higher cohomology groups with coefficients in the sheaf K_0 are introduced.

Note that, given an arbitrary group G, the first cohomology set $H^1(X; G_0)$ with coefficients in the sheaf of continuous G-valued functions can be defined in the same manner [157]. This set fails to be a group, unless G is Abelian. If G is a Lie group, one can consider the cohomology set $H^1(X; G_\infty)$ with coefficients in the sheaf G_∞ of smooth G-valued functions on X.

PROPOSITION 6.9.1. [157]. There is a bijection $H^1(X; G_\infty) \to H^1(X; G_0)$. □ •

Chapter 7

Space-time connections

The geometry of gravitation theory is a geometry of the tangent bundle TX of a world manifold X and associated fibre bundles. The connections which one deals with in gravitation theory are linear and affine connections on the tangent bundle TX, spinor connections, the associated principal connections on the fibre bundle LX of linear frames in TX and some others. They are not necessarily expressed into a pseudo-Riemannian metric on X, and play the role of independent dynamic variables in metric-affine and gauge gravitation theories.

7.1 Linear world connections

In the absence of fermion fields, gravitation theory is formulated on fibre bundles $T \to X$, called *natural bundles*, which admit the canonical lift of any diffeomorphism of its base X. The reader is referred to [179] for a detailed exposition of the category of natural bundles. Tensor bundles (1.1.12) exemplify natural bundles.

Throughout this Chapter, by X is meant a 4-dimensional orientable manifold, called a *world manifold*. Let an orientation of X be chosen. Unless otherwise stated, a coordinate atlas Ψ_X of X and the corresponding holonomic bundle atlas Ψ (1.1.11) of the tangent bundle TX is assumed to be fixed.

The tangent bundle TX of a world manifold X has the structure group

$$GL_4 = GL^+(4, \mathbb{R}).$$

It is associated with the GL_4-principal bundle

$$\pi_{LX} : LX \to X$$

of oriented linear frames in the tangent spaces to a world manifold X. For the sake of brevity, we will call LX the *frame bundle*. Its (local) sections are termed *frame fields*. Given holonomic frames $\{\partial_\mu\}$ in the tangent bundle TX associated with the holonomic atlas (1.1.11), every element $\{H_a\}$ of the frame bundle LX takes the form $H_a = H^\mu_a\partial_\mu$, where H^μ_a is a matrix element of the natural representation of the group GL_4. These matrix elements constitute the bundle coordinates

$$(x^\lambda, H^\mu_a), \qquad H'^\mu_a = \frac{\partial x'^\mu}{\partial x^\lambda} H^\lambda_a,$$

on LX. In these coordinates, the canonical action (6.1.1) of GL_4 on LX reads

$$R_g : H^\mu_a \mapsto H^\mu_b g^b{}_a, \qquad g \in GL_4.$$

The frame bundle LX is equipped with the canonical $\mathbb{R}^4$-valued 1-form

$$\boxed{\theta_{LX} = H^a_\mu dx^\mu \otimes t_a,} \tag{7.1.1}$$

where $\{t_a\}$ is a fixed basis for $\mathbb{R}^4$ and H^a_μ is the inverse matrix of H^μ_a.

The frame bundle $LX \to X$ belongs to the category of natural bundles. Every diffeomorphism f of X gives rise canonically to the general principal automorphism

$$\widetilde{f} : (x^\lambda, H^\lambda_a) \mapsto (f^\lambda(x), \partial_\mu f^\lambda H^\mu_a) \tag{7.1.2}$$

of LX and, consequently, to the corresponding automorphisms (6.3.2) of the associated bundles T. These automorphisms are called *general covariant transformations* or *holonomic automorphisms*. In particular, if $T = TX$ is the tangent bundle, $\widetilde{f} = Tf$ is the familiar tangent map to the diffeomorphism f. We will denote the group of holonomic automorphisms by $\mathrm{HOL}(X) \subset \mathrm{AUT}(LX)$. This is isomorphic to $\mathrm{Diff}(X)$. Note that the gauge group $\mathrm{Gau}(LX) \subset \mathrm{AUT}(LX)$ of vertical automorphisms of LX does not contain any holonomic automorphism, except the identity morphism.

Remark 7.1.1. By this reason, gauge gravitation theory cannot follow gauge theory of internal symmetries in a straightforward manner [161, 263]. •

The lift (7.1.2) leads to the canonical lift $\widetilde{\tau}$ of every vector field τ on X onto the principal bundle LX and the associated fibre bundles. The canonical lift of τ onto LX is defined by the relation

$$\mathbf{L}_{\widetilde{\tau}}\theta_{LX} = 0.$$

The corresponding canonical lift $\widetilde{\tau}$ on a tensor bundle T (1.1.12) is given by the expression (1.2.2). Let us introduce the collective index A for the tensor bundle coordinates

$$y^A = \dot{x}^{\alpha_1\cdots\alpha_m}_{\beta_1\cdots\beta_k}. \tag{7.1.3}$$

In this notation, the canonical lift $\widetilde{\tau}$ (1.2.2) onto T of a vector field τ on X reads

$$\boxed{\widetilde{\tau} = \tau^\lambda\partial_\lambda + u^{A\beta}_\alpha\partial_\beta\tau^\alpha\partial_A.} \tag{7.1.4}$$

The expression (7.1.4) is the general form of the canonical lift of a vector field τ on X onto a natural bundle T, when this lift depends only on the first partial derivatives of the components of τ. One can think of such a canonical lift as being the generator of a local 1-parameter group of general covariant transformations of a natural bundle T.

Since the tangent bundle TX is associated with the frame bundle LX, every world connection K (2.4.7) on a world manifold X is associated with a principal connection on LX whose connection form $\overline{K}$ (6.1.15) reads

$$\overline{K} = H^b_\nu(dH^\nu_a - K_\beta{}^\nu{}_\alpha H^\alpha_a dx^\beta)\otimes\epsilon^a_b, \tag{7.1.5}$$

where ϵ^a_b are the basis elements of the left Lie algebra $\mathrm{gl}(4,\mathbb{R})$. It follows that there is one-to-one correspondence between the world connections and the sections of the quotient fibre bundle

$$\boxed{C_K = J^1LX/GL_4,} \tag{7.1.6}$$

called the *bundle of world connections.*

With respect to the holonomic frames in TX, the fibre bundle C_K (7.1.6) is provided with the coordinates $(x^\lambda, k_\lambda{}^\nu{}_\alpha)$ so that, for any section K of $C_K\to X$,

$$k_\lambda{}^\nu{}_\alpha\circ K = K_\lambda{}^\nu{}_\alpha$$

are the coefficients of the world connection K (2.4.7).

Though the bundle of world connections $C_K\to X$ (7.1.6) is not a LX-associated bundle, it is a natural bundle, and admits the canonical lift

$$\widetilde{f}_C : J^1LX/GL_4 \to J^1\widetilde{f}(J^1LX)/GL_4$$

of any diffeomorphism f of X and, consequently, the canonical lift

$$\widetilde{\tau}=\tau^\mu\partial_\mu+[\partial_\nu\tau^\alpha k_\mu{}^\nu{}_\beta-\partial_\beta\tau^\nu k_\mu{}^\alpha{}_\nu-\partial_\mu\tau^\nu k_\nu{}^\alpha{}_\beta+\partial_{\mu\beta}\tau^\alpha]\frac{\partial}{\partial k_\mu{}^\alpha{}_\beta} \tag{7.1.7}$$

of any vector field τ on X.

The jet manifold J^1C_K has the canonical splitting (6.2.21). We will denote the corresponding 2-form $\mathcal{F}$ by R. It has the coordinate expression

$$\boxed{R_{\lambda\mu}{}^\alpha{}_\beta=k_{\lambda\mu}{}^\alpha{}_\beta-k_{\mu\lambda}{}^\alpha{}_\beta+k_\mu{}^\alpha{}_\varepsilon k_\lambda{}^\varepsilon{}_\beta-k_\lambda{}^\alpha{}_\varepsilon k_\mu{}^\varepsilon{}_\beta.} \tag{7.1.8}$$

It is readily observed that, if K is a section of $C_K\to X$, then $R\circ J^1K$ is the curvature (2.4.9) of the world connection K.

Recall that the torsion of a world connection is said to be the vertical-valued 2-form T (2.4.10) on TX. Due to the canonical vertical splitting (1.2.34), the torsion (2.4.10) defines the tangent-valued 2-form (2.4.11) on X and the soldering form

$$\boxed{T=T_\mu{}^\nu{}_\lambda\dot{x}^\lambda dx^\mu\otimes\dot{\partial}_\nu} \tag{7.1.9}$$

on TX. It is clear that, given a world connection K (2.4.7) and its torsion form T (7.1.9), the sum $K+\lambda T$, $\lambda\in\mathbb{R}$, is a world connection. In particular, every world connection K defines a unique symmetric world connection

$$K'=K-\frac{1}{2}T. \tag{7.1.10}$$

This is a corollary of a more general result. If K and K' are world connections, so is $\lambda K+(1-\lambda)K'$.

Remark 7.1.2. Note that, a world connection K is both torsionless and curvature-free if and only if there exists a holonomic atlas of constant local trivializations for K, i.e., all components $K_\mu{}^\nu{}_\lambda$ of K vanish with respect to this atlas. •

Every world connection K yields the horizontal lift

$$\boxed{K\tau=\tau^\lambda(\partial_\lambda+K_\lambda{}^\beta{}_\alpha\dot{x}^\alpha\dot{\partial}_\beta)} \tag{7.1.11}$$

of a vector field τ on X onto TX and the associated bundles. One can think of this lift as being the generator of a local 1-parameter group of non-holonomic automorphisms of these bundles. Note that, in the pioneer gauge gravitation models, the canonical lift (1.2.3) and the horizontal lift (7.1.11) were treated as generators of the gauge group of translations (see [149, 161] and references therein).

Remark 7.1.3. We refer the reader to [203] for some other lifts onto TX of vector fields on X, which are combinations of the horizontal lift (7.1.11) and the vertical lift (1.2.8). •

The horizontal lift of a vector field τ on X onto the frame bundle LX by means of a world connection K reads

$$K\tau = \tau^\lambda \left(\partial_\lambda + \Gamma_\lambda{}^\nu{}_\alpha H^\alpha_a \frac{\partial}{\partial H^\nu_a}\right). \tag{7.1.12}$$

A horizontal vector field u on the frame bundle LX is called *standard* if the morphism

$$u\rfloor\theta_{LX} : LX \to \mathbb{R}^4$$

is constant on LX. It is readily observed that every standard horizontal vector field on LX takes the form

$$u_v = H^\lambda_b v^b \left(\partial_\lambda + K_\lambda{}^\nu{}_\alpha H^\alpha_a \frac{\partial}{\partial H^\nu_a}\right) \tag{7.1.13}$$

where $v = v^b t_b \in \mathbb{R}^4$. A glance at this expression shows that a standard horizontal vector field is not projectable.

Remark 7.1.4. A world connection K defines a parallelization of the frame bundle LX [66]. Standard horizontal vector fields u_{t_a} (7.1.13) and fundamental vector fields $\alpha^{-1}(\epsilon^a_b)$ form a basis of the tangent space T_pLX at any point $p \in LX$. •

Since TX is an LX-associated fibre bundle, we have the canonical morphism

$$LX \times \mathbb{R}^4 \to TX,$$
$$(H^\mu_a, v^a) \mapsto \dot{x}^\mu = H^\mu_a v^a.$$

The tangent map to this morphism sends every standard horizontal vector field (7.1.13) on LX to the horizontal vector field

$$u = \dot{x}^\lambda(\partial_\lambda + K_\lambda{}^\nu{}_\alpha \dot{x}^\alpha \dot{\partial}_\nu) \tag{7.1.14}$$

on the tangent bundle TX. This is a holonomic vector field (see Definition 5.2.1). Then the vector field (7.1.14) defines the second order dynamic equation (5.2.2)

$$\ddot{x}^\nu = K_\lambda{}^\nu{}_\alpha \dot{x}^\lambda \dot{x}^\alpha \tag{7.1.15}$$

on X which is the geodesic equation (5.2.4) with respect to the world connection K. Solutions of the geodesic equation (7.1.15) (i.e., the geodesic lines of the connection K) are the projection of integral curves of the vector field (7.1.14) onto X. Moreover, there is the following theorem [177].

THEOREM 7.1.1. The projection of an integral curve of any standard horizontal vector field (7.1.13) on LX onto X is a geodesic in X. Conversely, any geodesic in X can be built in this way. □

A world connection K on a manifold X is said to be *complete* if the holonomic horizontal vector field (7.1.14) on the tangent bundle TX is complete. It takes place if and only if any standard horizontal vector field (7.1.13) on the frame bundle LX is complete.

It is readily observed that, if world connections K and K' differ from each other only in the torsion, they define the same holonomic vector field (7.1.14) and the same geodesic equation (7.1.15).

7.2 Lorentz connections

Gravitation theory is the theory with different types of symmetry breaking. We refer the reader, e.g., to [66, 131] for the general theory of reduced structures on the frame bundle LX. This Section is devoted to the reduced Lorentz structures and the Lorentz connections compatible with these structures.

The geometric formulation of the equivalence principle requires that the structure group GL_4 of the frame bundle LX over a world manifold X must be reducible to the Lorentz group $SO(1,3)$, while the condition of existence of fermion fields implies that GL_4 is reducible to the proper Lorentz group $\mathrm{L} = SO^0(1,3)$ [161, 263]. Recall that L is homeomorphic to $\mathbf{RP}^3 \times \mathbb{R}^3$, where $\mathbf{RP}^3$ is a real 3-dimensional projective space.

Unless otherwise stated, by a *Lorentz structure* we will mean a reduced L-principal subbundle L^hX, called the *Lorentz subbundle*, of the frame bundle LX.

Remark 7.2.1. There is the topological obstruction for a reduced Lorentz structure to exist on a world manifold X. All non-compact manifolds and compact manifolds whose Euler characteristic equals zero admit a reduced $SO(1,3)$-structure and, as a consequence, a pseudo-Riemannian metric [83]. A reduced L-structure exists if X

is additionally time-orientable. In gravitational models some conditions of causality should be also satisfied (see [147]). A compact space-time does not possess this property. At the same time, a non-compact world manifold X has a spin structure if and only if it is parallelizable (i.e., the tangent bundle $TX \to X$ is trivial) [119, 306]. It should be noted that also paracompactness of manifolds has a physical reason. A manifold is paracompact if and only if it admits a Riemannian structure [177]. •

From now on, we will assume that a Lorentz structure on a world manifold exists. Then one can show that different Lorentz subbundles L^hX and $L^{h'}X$ of the frame bundle LX are equivalent as L-principal bundles [154]. It means that there exists a vertical automorphism of the frame bundle LX which sends isomorphically L^hX onto $L^{h'}X$ (see Proposition 6.5.4).

By virtue of Theorem 6.5.2, there is one-to-one correspondence between the L-principal subbundles L^hX of the frame bundle LX and the global sections h of the quotient fibre bundle

$$\boxed{\Sigma_{\rm T} = LX/{\rm L},} \tag{7.2.1}$$

called the *tetrad bundle*. This is an LX-associated fibre bundle with the typical fibre $GL_4/{\rm L}$, homeomorphic to $S^3 \times \mathbb{R}^7$. Its global sections are called the *tetrad fields*.

The fibre bundle (7.2.1) is the two-fold covering of the *metric bundle*

$$\boxed{\Sigma_{\rm PR} = LX/SO(1,3),} \tag{7.2.2}$$

whose typical fibre is homeomorphic to the topological space $\mathbf{RP}^3 \times \mathbb{R}^7$, and whose global sections are pseudo-Riemannian metrics g on X. In particular, every tetrad field h defines uniquely a pseudo-Riemannian metric g. For the sake of simplicity, we will often identify the metric bundle with an open subbundle of the tensor bundle

$$\Sigma_{\rm PR} \subset \overset{2}{\vee} TX.$$

Therefore, we can equip $\Sigma_{\rm PR}$ with the coordinates $(x^\lambda, \sigma^{\mu\nu})$.

Remark 7.2.2. In General Relativity, a pseudo-Riemannian metric (a tetrad field) describes a gravitational field. Then, following the general scheme of gauge theory, we can treat a gravitational field as a Higgs field associated with a Lorentz structure [161, 263]. •

Every tetrad field h defines an associated *Lorentz atlas* $\Psi^h = \{(U_\zeta, z_\zeta^h)\}$ of the frame bundle LX such that the corresponding local sections z_ζ^h of LX take their values into the Lorentz subbundle L^hX. In accordance with Theorem (6.5.1) the transition functions of the Lorentz atlases of the frame bundle LX and associated bundles are L-valued.

Given a Lorentz atlas Ψ^h, the pull-back

$$h^a \otimes t_a = z_\zeta^{h*}\theta_{LX} = h_\lambda^a dx^\lambda \otimes t_a \tag{7.2.3}$$

of the canonical form θ_{LX} (7.1.1) by a local section z_ζ^h is said to be a (local) *tetrad form*. The tetrad form (7.2.3) determines the *tetrad coframes*

$$h^a = h_\mu^a(x)dx^\mu, \qquad x \in U_\zeta, \tag{7.2.4}$$

in the cotangent bundle T^*X. These coframes are associated with the Lorentz atlas Ψ^h. The coefficients h_μ^a of the tetrad form and the inverse matrix elements

$$h_a^\mu = H^\mu{}_a \circ z_\zeta^h \tag{7.2.5}$$

are called the *tetrad functions*. Given a Lorentz atlas Ψ^h, the tetrad field h can be represented by the family of tetrad functions $\{h_a^\mu\}$. We have the well-known relation

$$g = \eta_{ab}h^a \otimes h^b, \qquad g_{\mu\nu} = h_\mu^a h_\nu^b \eta^{ab}, \tag{7.2.6}$$

between the tetrad functions and the metric functions of the corresponding pseudo-Riemannian metric $g : X \to \Sigma_{\rm PR}$. A glance at the expressions (7.2.6) shows that this pseudo-Riemannian metric g has the Minkowski metric functions with respect to any Lorentz atlas Ψ^h.

In particular, given the Minkowski space $M = \mathbb{R}_4$ equipped with the Minkowski metric η, let us consider the L^hX-associated fibre bundle of Minkowski spaces

$$M^hX = (L^hX \times M)/{\rm L}. \tag{7.2.7}$$

By virtue of Proposition 6.5.6, the fibre bundle M^hX (7.2.7) is isomorphic to the cotangent bundle

$$T^*X = (LX \times \mathbb{R}_4)/GL_4 = (L^hX \times M)/{\rm L} = M^hX. \tag{7.2.8}$$

Given the isomorphism (7.2.8), we say that the cotangent bundle T^*X is provided with a *Minkowski structure*. Note that different Minkowski structures M^hX and $M^{h'}X$ on T^*X are not equivalent.

If the frame bundle LX over a world manifold admits a Lorentz subbundle L^hX, its structure group L is always reducible to the maximal compact subgroup $SO(3)$. It means that there exists an $SO(3)$-principal subbundle $L_0^hX \subset L^hX \subset LX$, called a *space-time structure*. The corresponding global section of the quotient fibre bundle $L^hX/SO(3) \to X$ with the typical fibre $\mathbb{R}^3$ is a 3-dimensional *spatial distribution* $FX \subset TX$ on X. Its generating 1-form written relative to a Lorentz atlas is the global tetrad form h^0 [263]. We have the corresponding *space-time decomposition*

$$TX = FX \oplus NF, \tag{7.2.9}$$

where NF is the 1-dimensional fibre bundle defined by the time-like vector field $h_0 = h_0^\mu \partial_\mu$. In particular, if the generating form h^0 is exact, the space-time decomposition (7.2.9) obeys Hawking's condition of stable causality [147].

There is the commutative diagram

$$\begin{array}{ccc} GL_4 & \longrightarrow & SO(4) \\ \downarrow & & \downarrow \\ \mathrm{L} & \longrightarrow & SO(3) \end{array} \tag{7.2.10}$$

of the reduction of structure groups of the frame bundle LX in gravitation theory [263]. This diagram leads us to the well-known theorem [147].

THEOREM 7.2.1. For any pseudo-Riemannian metric g on a world manifold X, there exists a space-time decomposition (7.2.9) with the generating tetrad form h^0 and a Riemannian metric g^R (associated with a reduced $GL_4{\downarrow}SO(4)$ structure) such that

$$g^R = 2h^0 \otimes h^0 - g. \tag{7.2.11}$$

□

Turn now to Lorentz connections. A principal connection on a Lorentz subbundle L^hX of the frame bundle LX is called the *Lorentz connection*. It reads

$$A_h = dx^\lambda \otimes (\partial_\lambda + \frac{1}{2}A_\lambda{}^{ab}\varepsilon_{ab}), \tag{7.2.12}$$

where $\varepsilon_{ab} = -\varepsilon_{ba}$ are generators of the Lorentz group. Recall that the Lorentz group L acts on $\mathbb{R}^4$ by the generators

$$\varepsilon_{ab}{}^c{}_d = \eta_{ad}\delta^c_b - \eta_{bd}\delta^c_a. \tag{7.2.13}$$

By virtue of Theorem 6.5.7, every Lorentz connection (7.2.12) is extended to a principal connection on the frame bundle LX which is given by the same expression (7.2.13) and, thereby, it defines a world connection K whose coefficients are

$$\boxed{K_\lambda{}^\mu{}_\nu = h^k_\nu \partial_\lambda h^\mu_k + \eta_{ka} h^\mu_b h^k_\nu A_\lambda{}^{ab}.} \tag{7.2.14}$$

This world connection is also called the Lorentz connection. Its holonomy group is a subgroup of the Lorentz group L. Conversely, let K be a world connection whose holonomy group is a subgroup of L. By virtue of Theorem 6.6.3, it defines a Lorentz subbundle of the frame bundle LX, and is a Lorentz connection on this subbundle (see also [276]).

Now let K be an arbitrary world connection. Given a pseudo-Riemannian metric g corresponding to a Lorentz structure L^hX, every world connection K admits the decomposition

$$K_{\mu\nu\alpha} = \{_{\mu\nu\alpha}\} + C_{\mu\nu\alpha} + \frac{1}{2} Q_{\mu\nu\alpha} \tag{7.2.15}$$

in the Christoffel symbols $\{_{\mu\nu\alpha}\}$ (2.4.13), the *non-metricity tensor*

$$Q_{\mu\nu\alpha} = Q_{\mu\alpha\nu} = \nabla^K_\mu g_{\nu\alpha} = \partial_\mu g_{\nu\alpha} + K_{\mu\nu\alpha} + K_{\mu\alpha\nu} \tag{7.2.16}$$

and the *contorsion*

$$C_{\mu\nu\alpha} = -C_{\mu\alpha\nu} = \frac{1}{2}(T_{\nu\mu\alpha} + T_{\nu\alpha\mu} + T_{\mu\nu\alpha} + Q_{\alpha\nu\mu} - Q_{\nu\alpha\mu}), \tag{7.2.17}$$

where $T_{\mu\nu\alpha} = -T_{\alpha\nu\mu}$ is the torsion of K. The tensor fields T and Q, in turn, are decomposed into three and four irreducible pieces, respectively (we refer the reader to [150, 220, 241] for details and outcomes of this decomposition).

By virtue of Theorem 6.5.8, a world connection K is reducible to a principal connection on the Lorentz subbundle L^hX if and only if it satisfies the *metricity condition*

$$\nabla^K_\mu g_{\nu\alpha} = 0. \tag{7.2.18}$$

Then K is called a *metric connection* for g. Obviously, a metric connection is the Lorentz one. Conversely, every Lorentz connection obeys the metricity condition (7.2.18) for some pseudo-Riemannian metric g (which is not necessarily unique [294]). A metric connection reads

$$K_{\mu\nu\alpha} = \{_{\mu\nu\alpha}\} + \frac{1}{2}(T_{\nu\mu\alpha} + T_{\nu\alpha\mu} + T_{\mu\nu\alpha}). \tag{7.2.19}$$

The *Levi–Civita connection*, by definition, is a torsion-free metric connection $K_{\mu\nu\alpha} = \{_{\mu\nu\alpha}\}$.

Remark 7.2.3. There is a classification, e.g., of vacuum pseudo-Riemannian geometries by the Lie algebras of a constrained holonomy group which are subalgebras of the Lorentz Lie algebra [143]. •

Remark 7.2.4. It is easily seen that different metric connections (7.2.19) for the same world metric g lead to different geodesic equations (7.1.15). Note on a problem if a spinless test particle in the presence of a non-symmetric metric connection K moves along the geodesics of the Levi-Civita connection as in General Relativity (a geodesic motion) or along the geodesics of the total connection K (an *autoparallel* motion) (see, e.g., [107] for a discussion). •

Though a world connection is not a Lorentz connection in general, any world connection K defines a Lorentz connection K_h on each L-principal subbundle L^hX of the frame bundle as follows.

It is readily observed that the Lie algebra of the general linear group GL_4 is the direct sum

$$\mathfrak{g}(GL_4) = \mathfrak{g}(\mathrm{L}) \oplus \mathfrak{m}$$

of the Lie algebra $\mathfrak{g}(\mathrm{L})$ of the Lorentz group and a subspace $\mathfrak{m}$ such that

$$ad(l)(\mathfrak{m}) \subset \mathfrak{m}, \qquad \forall l \in \mathrm{L}.$$

Let $\overline{K}$ be the connection form (6.1.15) of a world connection K on LX. Then, by virtue of the well-known theorem [177], the pull-back onto L^hX of the $\mathfrak{g}(\mathrm{L})$-valued component $\overline{K}_L$ of $\overline{K}$ is the connection form of a principal connection K_h on the Lorentz subbundle L^hX. To obtain the connection parameters of K_h, let us consider the local connection 1-form (6.1.16) of the connection K with respect to a Lorentz atlas Ψ^h of LX given by the tetrad forms h^a. This reads

$$z_\zeta^{h*}\overline{K} = -K_\lambda{}^b{}_a dx^\lambda \otimes \varepsilon_b^a,$$
$$K_\lambda{}^b{}_a = -h_\mu^b \partial_\lambda h_a^\mu + K_\lambda{}^\mu{}_\nu h_\mu^b h_a^\nu,$$

where $\{e_b^a\}$ is the basis for the right Lie algebra of the group GL_4. Then the Lorentz part of this form is precisely the local connection 1-form (6.1.16) of the connection

K_h on L^hX. We have

$$z_\zeta^{h*}\overline{K}_L = -\frac{1}{2}A_\lambda{}^{ab}dx^\lambda \otimes \varepsilon_{ab}, \tag{7.2.20}$$
$$A_\lambda{}^{ab} = \frac{1}{2}(\eta^{kb}h^a_\mu - \eta^{ka}h^b_\mu)(\partial_\lambda h^\mu_k - h^\nu_k K_\lambda{}^\mu{}_\nu).$$

If K is a Lorentz connection extended from L^hX, then obviously $K_h = K$.

Lorentz non-symmetric connections play a prominent role in describing fermion fields on a world manifold (see Section 7.5). At the same time, the torsion of a world connection does not make a contribution in the strength (6.2.25) of gauge potentials in gauge theory because a world connection K in the expression (6.2.23) for connections Γ_A on the bundle of principal connections C is necessarily symmetric.

7.3 Relativistic mechanics

In accordance with Theorem (5.4.2), non-relativistic dynamic equations on a configuration bundle $Q \to \mathbb{R}$ are equivalent to some particular geodesic equations (5.4.7) on the tangent bundle TQ. Let us compare them with relativistic geodesic equations.

In contrast with non-relativistic mechanics, a configuration space Q of relativistic mechanics has no preferable fibration over $\mathbb{R}$ [213, 271]. Therefore, one should use formalism of jets of 1-dimensional submanifolds of Q. In these terms, relativistic mechanics can be formulated on an arbitrary configuration space Q, dim$Q > 1$ [213]. Here we restrict our consideration to relativistic mechanics of a test particle on a 4-dimensional configuration space Q.

Let us provide Q with a coordinate atlas $\{U; (q^0, q^{1,2,3})\}$ (1.3.20) together with the transition functions

$$q^0 \to \tilde{q}^0(q^0, q^j), \qquad q^i \to \tilde{q}^i(q^0, q^j)$$

(1.3.21). Note that, given a coordinate chart $(U; q^0, q^i)$, we have a local fibre bundle

$$U \ni (q^0, q^i) \mapsto q^0 \in \mathbb{R} \tag{7.3.1}$$

which can be treated as a configuration space of a local non-relativistic mechanical system.

The velocity phase space of relativistic mechanics on the configuration space Q is the first order jet manifold J_1^1Q of 1-dimensional subbundles of Q. It is endowed with the coordinates (q^0, q^i, q_0^i) (1.3.22). Their transition functions (1.3.23) read

$$\widetilde{q}_0^i = \left(\frac{\partial\widetilde{q}^i}{\partial q^0} + q_0^j\frac{\partial\widetilde{q}^i}{\partial q^j}\right) / \left(\frac{\partial\widetilde{q}^0}{\partial q^0} + q_0^k\frac{\partial\widetilde{q}^0}{\partial q^k}\right). \tag{7.3.2}$$

A glance at this expression shows that the fibre coordinates q_0^i on $J_1^1Q \to Q$ are the standard coordinates on the projective space $\mathbf{RP}^4$, i.e., the velocity phase space $J_1^1Q \to Q$ of relativistic mechanics is a projective bundle.

Example 7.3.1. Let consider the configuration space $Q = \mathbb{R}^4$, provided with the Cartesian coordinates (q^0, q^i). Let

$$\begin{aligned}
\widetilde{q}^0 &= q^0\mathrm{ch}\alpha - q^1\mathrm{sh}\alpha, \\
\widetilde{q}^1 &= -q^0\mathrm{sh}\alpha + q^1\mathrm{ch}\alpha, \\
\widetilde{q}^{2,3} &= q^{2,3}.
\end{aligned} \tag{7.3.3}$$

be a Lorentz transformation of the plane (q^0, q^1). Substituting these expressions in the formula (7.3.2), we obtain

$$\widetilde{q}_0^1 = \frac{-\mathrm{sh}\alpha + q_0^1\mathrm{ch}\alpha}{\mathrm{ch}\alpha - q_0^1\mathrm{sh}\alpha}, \qquad \widetilde{q}_0^{2,3} = \frac{q_0^{2,3}}{\mathrm{ch}\alpha - q_0^1\mathrm{sh}\alpha}.$$

This is precisely the transformation law of 3-velocities in Special Relativity if

$$\mathrm{ch}\alpha = \frac{1}{\sqrt{1-v^2}}, \qquad \mathrm{sh}\alpha = \frac{v}{\sqrt{1-v^2}},$$

where v is the velocity of a reference frame, moving along the axis q^1. •

Thus, one can think of the velocity phase space J_1^1Q as being the space of *3-velocities* v of a relativistic system. We will call J_1^1Q the *3-velocity phase space.* Given a coordinate chart $(U; q^0, q^i)$, 3-velocities $v = (q_0^i)$ of a relativistic system can be seen as velocities of a local non-relativistic system (7.3.1) with respect to the corresponding local reference frame $\Gamma = \partial_t$ on U. However, the notion of a reference frame in non-relativistic mechanics fails to be extended to relativistic mechanics since the relativistic transformations $q_0^i \to q_0'^i$ and $\Gamma^i \to \Gamma'^i$ are not affine and the relative velocity $q_0^i - \Gamma^i$ is not maintained under these transformations.

To introduce relativistic velocities, let us take the tangent bundle TQ of the configurations space Q, equipped with the holonomic coordinates $(q^0, q^i, \dot{q}^0, \dot{q}^i)$. In accordance with (1.3.24), there is the multivalued morphism $\widetilde{\lambda}$ from the 3-velocity phase space J^1_1Q to the tangent bundle TQ when a point $(q^0, q^i, q^i_0) \in J^1_1Q$ corresponds to a line

$$\widetilde{\lambda} : q^i_0 \mapsto (\dot{q}^0, \dot{q}^i = \dot{q}^0 q^i_0) \subset TQ \tag{7.3.4}$$

in the tangent space to Q at the point (q^0, q^i). Conversely, there is the morphism

$$\varrho : TQ \underset{Q}{\to} J^1_1 Q, \qquad q^i_0 \circ \varrho = \frac{\dot{q}^i}{\dot{q}^0}, \tag{7.3.5}$$

such that

$$\varrho \circ \widetilde{\lambda} = \mathrm{Id}\, J^1_1 Q.$$

Indeed, it is readily observed that q^i_0 and $\dot{q}^i/\dot{q}^0$ have the same transformation laws. It should be emphasized that, though the expressions (7.3.4) and (7.3.5) are singular at $\dot{q}^0 = 0$, this point belongs to another coordinate chart, and the morphisms $\widetilde{\lambda}$ and ϱ are well defined. Thus, one can think of the tangent bundle TQ as being the space of *relativistic velocities* or *4-velocities* of a relativistic system. It is called the *4-velocity phase space*.

Since the morphism $\widetilde{\lambda}$ of J^1_1Q onto TQ is multivalued and the converse morphism (7.3.5) is a surjection, one may try to find a subbundle W of the tangent bundle TQ such that $\varrho : W \to J^1_Q$ is an injection. Let us assume that Q is oriented and endowed with a pseudo-Riemannian metric g. The pair (Q, g) is called a *relativistic system*. The metric g defines the subbundle of *velocity hyperboloids*

$$W_g = \{\dot{q}^\lambda \in TQ \; : \; g_{\mu\nu}(q)\dot{q}^\mu \dot{q}^\nu = 1\} \tag{7.3.6}$$

of TQ. Of course, W_g is neither vector nor affine subbundle of TQ. Let Q be time-oriented with respect to the pseudo-Riemannian metric g. This means that W_g is a disjoint union of two connected subbundles W^+ and W^-. Then it is readily observed that the restriction of the morphism ϱ (7.3.5) to each of these subbundles is an injection into J^1_1Q.

Let us consider the image of this injection in a fibre of J^1_1Q at a point $q \in Q$. There are local coordinates (q^0, q^i) on a neighbourhood of $q \in Q$ such that the pseudo-Riemannian metric $g(q)$ at q comes to the Minkowski metric $g(q) = \eta$. With respect to these coordinates, the velocity hyperboloid $W_q \subset T_qQ$ is given by the equation

$$(\dot{q}^0)^2 - \sum_i (\dot{q}^i)^2 = 1.$$

This is the union of the subset W_q^+, where $\dot{q}^0 > 0$, and W_q^-, where $\dot{q}^0 < 0$. Restricted to W^+, the morphism (7.3.5) takes the familiar form of the relations between 3- and 4-velocities

$$\dot{q}^0 = \frac{1}{\sqrt{1 - \sum(q_0^i)^2}}, \qquad \dot{q}^i = \frac{q_0^i}{\sqrt{1 - \sum(q_0^i)^2}}$$

in Special Relativity. The image of each of the hyperboloids W^+ and W^- in the 3-velocity phase space J_1^1Q by the morphism (7.3.5) is the open ball

$$\sum_i (q_0^i)^2 < 1, \tag{7.3.7}$$

i.e, 3-velocities of a relativistic system (Q, g) are bounded.

Turn now to a dynamics of relativistic mechanics.

In a straightforward manner, Lagrangian formalism fails to be appropriate to relativistic mechanics because a Lagrangian $L = \mathcal{L}dq^0$, by very definition, is defined only locally on a coordinate chart (1.3.22) of the 3-velocity phase space J_1^1Q. For instance, the Lagrangian

$$L_m = -m\sqrt{1 - \sum_i (q_0^i)^2 dq^0} \tag{7.3.8}$$

of a free relativistic point mass m in Special Relativity can be introduced on each coordinate chart of J_1^1Q, but it is not maintained in a straightforward manner by the Lorentz transformations (7.3.3) because of the term dq^0. At the same time, given a motion $q^i = c^i(q^0)$ with respect to a coordinate chart (q^0, q^i), its Lorentz transformation (7.3.3) reads

$$\begin{aligned}
&\widetilde{q}^0 = q^0 \mathrm{ch}\alpha - c^1(q^0)\mathrm{sh}\alpha, \\
&\widetilde{c}^1(\widetilde{q}^0) = -q^0(\widetilde{q}^0)\mathrm{sh}\alpha + c^1(q^0(\widetilde{q}^0))\mathrm{ch}\alpha, \\
&\widetilde{c}^{2,3}(\widetilde{q}^0) = c^{2,3}(q^0(\widetilde{q}^0)).
\end{aligned}$$

Then we have the Lorentz invariance of the pull-back c^*L_m of the Lagrangian (7.3.8), i.e., $c^*L_m = \widetilde{c}^*\widetilde{L}_m$, where $d\widetilde{q}^0 = d_{q^0}(\widetilde{q}^0)dq^0$ and d_{q^0} is the total derivative.

Therefore, let us start from Hamiltonian relativistic mechanics. Then we will come to dynamic equations on the 4-velocity phase space TQ.

Given a coordinate chart of the relativistic configuration space Q, the homogeneous Legendre bundle corresponding to the local non-relativistic system (7.3.1) is the cotangent bundle T^*Q. This fact motivates us to choose T^*Q as the *relativistic momentum phase space*, equipped with the holonomic coordinates (q^λ, p_λ). The cotangent bundle T^*Q is endowed with the canonical symplectic form

$$\Omega = dp_\mu \wedge dq^\mu. \tag{7.3.9}$$

Let us describe a relativistic system (Q, g) as a Hamiltonian system of conservative mechanics on the symplectic manifold T^*Q, characterized by a *relativistic Hamiltonian*

$$\mathbf{H} : T^*Q \to \mathbb{R}. \tag{7.3.10}$$

Any relativistic Hamiltonian $\mathbf{H}$ (7.3.10) defines the Hamiltonian map

$$\widehat{\mathbf{H}} : T^*Q \underset{Q}{\to} TQ, \qquad \dot q^\mu = \partial^\mu \mathbf{H}, \tag{7.3.11}$$

from the relativistic momentum phase space T^*Q to the 4-velocity phase space TQ (see [213] for details). Since the 4-velocities of a relativistic system live in the velocity hyperboloids (7.3.6), we have the constraint subspace

$$\begin{aligned} &N = \widehat{\mathbf{H}}^* W_g, \\ &g_{\mu\nu}\partial^\mu \mathbf{H} \partial^\nu \mathbf{H} = 1, \end{aligned} \tag{7.3.12}$$

of the relativistic momentum phase space T^*Q. It follows that the relativistic system (Q, g) can be described as an autonomous Dirac constraint system on the primary constraint space N (7.3.12) (see also [259]). Its solutions are integral curves of the Hamiltonian vector field $u = u^\mu \partial_\mu + u_\mu \partial^\mu$ on $N \subset T^*Q$ which obeys the Hamilton equation

$$\boxed{u \rfloor i_N^* \Omega = -i_N^* d\mathbf{H}.} \tag{7.3.13}$$

Example 7.3.2. Let us consider a point electric charge e in the Minkowski space in the presence of an electromagnetic potential A_λ. Its relativistic Hamiltonian reads

$$\mathbf{H} = -\frac{1}{2m}\eta^{\mu\nu}(p_\mu - eA_\mu)(p_\nu - eA_\nu),$$

while the constraint space N (7.3.12) is

$$\eta^{\mu\nu}(p_\mu - eA_\mu)(p_\nu - eA_\nu) = m^2.$$

Then the Hamilton equation (7.3.13) takes the form

$$\boxed{u\rfloor i_N^*\Omega = 0.} \tag{7.3.14}$$

Its solution is

$$\dot{p}_k = u_k = u^\mu \partial_k A_\mu, \tag{7.3.15}$$

$$\dot{q}^k = u^k = \frac{u^0\eta^{ik}(p_i - eA_i)}{\sqrt{m^2 - \eta^{ij}(p_i - eA_i)(p_j - eA_j)}}. \tag{7.3.16}$$

The equality (7.3.16) leads to the usual expression for the 3-velocities

$$p_k = \frac{m\eta_{ki}q_t^i}{\sqrt{1+\eta_{ij}q_t^iq_t^j}} + A_k.$$

Substituting this expression in the equality (7.3.15), we obtain the familiar equation of motion of a relativistic charge in an electromagnetic field. •

Example 7.3.3. The relativistic Hamiltonian for a point mass m in a gravitational field g on a 4-dimensional manifold Q reads

$$\mathbf{H} = -\frac{1}{2m} g^{\mu\nu}(q)p_\mu p_\nu,$$

while the constraint space N (7.3.12) is

$$g^{\mu\nu}p_\mu p_\nu = m^2.$$

As in previous Examples, the equation (7.3.13) takes the form (7.3.14). •

Given a relativistic Hamiltonian $\mathbf{H}$ on the relativistic momentum phase space T^*Q, let

$$\xi_H = \partial^i\mathbf{H}\partial_i - \partial_i\mathbf{H}\partial^i \tag{7.3.17}$$

be the corresponding Hamiltonian vector field whose integral curves are solutions of the Hamilton equation

$$\xi_H\rfloor\Omega = -d\mathbf{H}.$$

If the Hamiltonian map $\widehat{\mathbf{H}}$ (7.3.11) is a diffeomorphism, the Hamiltonian vector field (7.3.17) yields the holonomic vector field

$$\begin{aligned} &\xi = T\widehat{\mathbf{H}} \circ \xi_H \circ \widehat{\mathbf{H}}^{-1} = \dot{q}^\mu \partial_\mu + \xi^\mu \dot{\partial}_\mu, \\ &\xi^\mu = (\partial_\alpha \partial^\mu \mathbf{H} \partial^\alpha \mathbf{H} - \partial^\alpha \partial^\mu \mathbf{H} \partial_\alpha \mathbf{H}) \circ \widehat{\mathbf{H}}^{-1}, \end{aligned} \tag{7.3.18}$$

on the 4-velocity phase space TQ, where

$$(q^\mu, \dot{q}^\mu, \dot{\mathrm{q}}^\mu, \ddot{q}^\mu) \circ T\widehat{\mathbf{H}} = (q^\mu, \partial^\mu \mathbf{H}, \dot{q}^\mu, \partial_\alpha \partial^\mu \mathbf{H} \dot{q}^\alpha + \partial^\alpha \partial^\mu \mathbf{H} \dot{p}_\alpha)$$

is the tangent morphism to $\widehat{\mathbf{H}}$. The holonomic vector field (7.3.18) defines an autonomous second order dynamic equation

$$\ddot{q}^\mu = (\partial_\alpha \partial^\mu \mathbf{H} \partial^\alpha \mathbf{H} - \partial^\alpha \partial^\mu \mathbf{H} \partial_\alpha \mathbf{H}) \circ \widehat{\mathbf{H}}^{-1}, \tag{7.3.19}$$

called the *relativistic dynamic equation*, on the relativistic configuration space Q.

For instance, the relativistic dynamic equation (7.3.19) for a point electric charge in Example 7.3.2 takes the well-known form

$$\ddot{q}^\mu = \eta^{\mu\nu} \dot{q}^\lambda F_{\nu\lambda}, \tag{7.3.20}$$

where F is the electromagnetic strength. The relativistic dynamic equation (7.3.19) for a point mass m in a gravitational field g in Example 7.3.3 reads

$$\ddot{q}^\mu = \{_\lambda{}^\mu{}_\nu\} \dot{q}^\lambda \dot{q}^\nu, \tag{7.3.21}$$

where $\{_\lambda{}^\mu{}_\nu\}$ are the Christoffel symbols of the pseudo-Riemannian metric g.

The equations (7.3.20) and (7.3.21) exemplify relativistic dynamic equations which are geodesic equations

$$\ddot{q}^\mu = K^\mu_\lambda(q^\alpha, \dot{q}^\alpha) \dot{q}^\lambda \tag{7.3.22}$$

with respect to a connection K on the tangent bundle $TQ \to Q$ (see Definition 5.2.4). We call (7.3.22) the *relativistic geodesic equation*. For instance, a connection K in the equation (7.3.21), is the Levi–Civita connection of the pseudo-Riemannian metric g. In the equation (7.3.20), K is the zero Levi–Civita connection of the Minkowski metric plus the soldering form

$$\sigma = \eta^{\mu\nu} F_{\nu\lambda} dq^\lambda \otimes \dot{\partial}_\mu. \tag{7.3.23}$$

We say that a relativistic geodesic equation (7.3.22) on the 4-velocity phase space TQ describes a relativistic system (Q, g) if its geodesic vector field does not leave the subbundle of velocity hyperboloids (7.3.6). It suffices to require that the condition

$$\boxed{(\partial_\lambda g_{\mu\nu}\dot{q}^\mu + 2g_{\mu\nu}K^\mu_\lambda)\dot{q}^\lambda\dot{q}^\nu = 0} \qquad (7.3.24)$$

holds for all tangent vectors which belong to W_g (7.3.6). Obviously, the Levi–Civita connection $\{_\lambda{}^\mu{}_\nu\}$ of the metric g fulfills the condition (7.3.24). Any connection K on $TQ \to Q$ can be written as

$$K^\mu_\lambda = \{_\lambda{}^\mu{}_\nu\}\dot{q}^\nu + \sigma^\mu_\lambda(q^\alpha, \dot{q}^\alpha),$$

where $\sigma = \sigma^\mu_\lambda dq^\lambda \otimes \dot{\partial}_\lambda$ is a soldering form. Then the condition (7.3.24) takes the form

$$\boxed{g_{\mu\nu}\sigma^\mu_\lambda\dot{q}^\lambda\dot{q}^\nu = 0.} \qquad (7.3.25)$$

It is readily observed that the soldering form (7.3.23) in the equation (7.3.20) obeys this condition for the Minkowski metric η.

Now let us compare relativistic and non-relativistic geodesic equations [213, 127, 128]. In physical applications, one usually thinks of non-relativistic mechanics as being an approximation of small velocities of relativistic theory. At the same time, the 3-velocities in mathematical formalism of non-relativistic mechanics are not bounded.

Let a relativistic configuration space Q admit a fibration $Q \to \mathbb{R}$, where $\mathbb{R}$ is a time axis. One can think of the fibre bundle $Q \to \mathbb{R}$ as being a configuration space of a non-relativistic mechanical system. In order to compare relativistic and non-relativistic dynamics, one should consider pseudo-Riemannian metric on TQ, compatible with the fibration $Q \to \mathbb{R}$. Note that $\mathbb{R}$ is a time of non-relativistic mechanics. It is the same for all non-relativistic observers. In the framework of a relativistic theory, this time can be seen as a cosmological time. Given a fibration $Q \to \mathbb{R}$, a pseudo-Riemannian metric on the tangent bundle TQ is said to be *admissible* if it is defined by a pair (g^R, Γ) of a Riemannian metric on Q and a non-relativistic reference frame Γ, i.e.,

$$g = \frac{2\Gamma \otimes \Gamma}{\mid \Gamma \mid^2} - g^R, \qquad (7.3.26)$$

$$\mid \Gamma \mid^2 = g^R_{\mu\nu}\Gamma^\mu\Gamma^\nu = g_{\mu\nu}\Gamma^\mu\Gamma^\nu,$$

in accordance with Theorem 7.2.1. The vector field Γ is time-like relative to the pseudo-Riemannian metric g (7.3.26), but not with respect to other admissible pseudo-Riemannian metrics in general. There is the canonical imbedding (5.1.3) of the velocity phase space J^1Q of non-relativistic mechanics into the affine subbundle

$$\dot{q}^0 = 1, \qquad \dot{q}^i = q^i_0 \tag{7.3.27}$$

of the 4-velocity phase space TQ. Then one can think of (7.3.27) as the 4-velocities of a non-relativistic system. The relation (7.3.27) differs from the familiar relation (7.3.5) between 4- and 3-velocities of a relativistic system. In particular, the temporal component $\dot{q}^0$ of 4-velocities of a non-relativistic system equals 1. It follows that the 4-velocities of relativistic and non-relativistic systems occupy different subbundles of the 4-velocity space TQ. Moreover, Theorem 5.4.2 shows that both relativistic and non-relativistic equations of motion can be seen as the geodesic equations on the same tangent bundle TQ. The difference between them lies in the fact that their solutions live in the different subbundles (7.3.6) and (7.3.27) of TQ. At the same time, relativistic equations, expressed in the 3-velocities $\dot{q}^i/\dot{q}^0$ of a relativistic system, tend exactly to the non-relativistic equations on the subbundle (7.3.27) when $\dot{q}^0 \to 1$, $g_{00} \to 1$, i.e., where non-relativistic mechanics and the non-relativistic approximation of a relativistic theory only mutually coincide.

Let (q^0, q^i) be a non-relativistic reference frame on Q compatible with the fibration of $Q \to \mathbb{R}$. Given a non-relativistic geodesic equation (5.4.7), we will say that the relativistic geodesic equation (7.3.20) is the *relativization* of (5.4.7) if the spatial parts of these equations are the same. In accordance with Lemma (5.5.4), any relativistic geodesic equation with respect to a connection K is a relativization of a non-relativistic geodesic equation with respect to the connection

$$\widetilde{K} = dq^\lambda \otimes (\partial_\lambda + (K^i_\lambda - \Gamma^i K^0_\lambda)\partial_i),$$

where $\Gamma^i = 0$ is the connection corresponding to the reference frame (q^0, q^i). Of course, for different reference frames, we have different non-relativistic limits of the same relativistic equation. The converse procedure is more intricate.

Following Section 5.9, a generic quadratic dynamic equation (5.4.8) can be written in the form

$$q^i_{00} = -(m^{-1})^{ik}\{_{\lambda k\mu}\}q^\lambda_0 q^\mu_0 + b^i_\mu(q^\nu)q^\mu_0, \qquad q^0_0 = 1, \tag{7.3.28}$$

where $\{_{\lambda k \mu}\}$ are the Christoffel symbols of some pseudo-Riemannian metric g, whose spatial part is the mass tensor $(-m_{ik})$, while

$$b^i_k(q^\mu)q^k_0 + b^i_0(q^\mu) \tag{7.3.29}$$

is an external force. In view of Proposition 5.4.5, the decomposition of the right-hand side of the equation (7.3.28) into two parts is not unique. With respect to the coordinates where $g_{0i} = 0$, one may construct the relativistic geodesic equation

$$\ddot{q}^\mu = \{_\lambda{}^\mu{}_\nu\}\dot{q}^\lambda \dot{q}^\nu + \sigma^\mu_\lambda \dot{q}^\lambda, \tag{7.3.30}$$

where the soldering form σ must fulfill the condition (7.3.25). It takes place only if

$$g_{ik}b^i_j + g_{ij}b^i_k = 0,$$

i.e., the external force (7.3.29) is the Lorentz-type force plus some potential one. Then we have

$$\sigma^0_0 = 0, \qquad \sigma^0_k = -g^{00}g_{kj}b^j_0, \qquad \sigma^j_k = b^j_k.$$

However, the relativization (7.3.30) fails to exhausts all examples. Let a non-relativistic acceleration $\xi^i(x^\mu)$ be a spatial part of a 4-vector ξ^λ in the Minkowski space. Then one can write the relativistic equation

$$\ddot{q}^\lambda = \xi^\lambda - \eta_{\alpha\beta}\xi^\beta \dot{q}^\alpha \dot{q}^\lambda. \tag{7.3.31}$$

This is the case, e.g., for the relativistic hydrodynamics that we usually meet in the literature on gravitation theory. However, the non-relativistic limit $\dot{q}^0 = 1$ of (7.3.31) does not coincide with the initial non-relativistic equation.

7.4 Metric-affine gravitation theory

Metric-affine gravitation theory deals with a pseudo-Riemannian metric g and a world connection K considered as independent dynamic variables. We refer the reader to [150, 220, 222, 241] for a general formulation of this gravitation theory and to [77, 150, 296] for the study of its solutions.

Affine-metric gravitation theory is formulated on natural bundles. Due to Proposition 6.5.6 and Theorem 6.5.7, this theory includes in a natural way the gauge theory of Lorentz connections on natural bundles. Moreover, the gauge theory of

Lorentz connections on spinor bundles is also reduced to metric-affine theory of world connections (see the next Section).

The gauge transformations of metric-affine gravitation theory on natural bundles are general covariant transformations. All field Lagrangians in gravitation theory, by construction, are invariant under these transformations. As was mentioned above, the generator of a 1-parameter group of general covariant transformations of a natural bundle $T \to X$ is the canonical lift $\widetilde{\tau}$ onto T of a vector field τ on X. Following the general procedure in Section 3.4, let us examine the conservation laws along such canonical lifts of vector fields on X. These are the energy-momentum conservation laws [122, 123, 269]. There are several approaches to discover an energy-momentum conservation law in gravitation theory. Here we treat this conservation law as a particular gauge conservation law. Accordingly, the energy-momentum of gravity is seen as a peculiar Noether current (see, e.g., [16, 33, 100, 150, 156, 289]). A glance at the expression (7.1.4) shows that the generators $\widetilde{\tau}$ of general covariant transformations (as well as the generators (6.3.10) of principal automorphisms in gauge theory) depend on the derivatives of components τ^λ of a vector field τ, which play the role of gauge parameters. Therefore, the main peculiarity of the energy-momentum conservation laws along these generators is that the corresponding energy-momentum currents reduce to a superpotential and that this superpotential depends on components of a vector field τ as gauge parameters.

We have seen in Section 6.3 that these properties are common for gauge conservation laws. In particular, the fact that the energy-momentum current depend on a vector field τ provides the maintenance of the gravitational energy-momentum conservation law under general covariant transformations.

We will start from the tensor field model which clearly illustrates these phenomena.

Let $T \to X$ be the tensor bundle (1.1.12) with fibred coordinates (x^λ, y^A) (7.1.3). The canonical lift onto T of a vector field τ on X is given by the expression $\widetilde{\tau}$ (7.1.4). Let L be a first order Lagrangian on J^1T which is invariant under general covariant transformations, i.e., L satisfies the strong equality

$$\mathbf{L}_{J^1\widetilde{\tau}}L = \partial_\alpha(\tau^\alpha\mathcal{L}) + u^{A\beta}{}_\alpha\partial_\beta\tau^\alpha\partial_A\mathcal{L} + d_\mu(u^{A\beta}{}_\alpha\partial_\beta\tau^\alpha)\pi^\mu_A - y^A_\alpha\partial_\beta\tau^\alpha\pi^\beta_A = 0. \quad (7.4.1)$$

The corresponding weak identity (3.4.2) takes the form

$$0 \approx -d_\lambda[\pi^\lambda_A(y^A_\alpha\tau^\alpha - u^{A\beta}{}_\alpha\partial_\beta\tau^\alpha) - \tau^\lambda\mathcal{L}]. \quad (7.4.2)$$

Due to the arbitrariness of the gauge parameters τ^α, the equality (7.4.1) is equivalent to the system of strong equalities

$$\partial_\lambda \mathcal{L} = 0, \tag{7.4.3a}$$

$$\delta^\beta_\alpha \mathcal{L} + u^{A\beta}{}_\alpha \delta_A \mathcal{L} + d_\mu(u^{A\beta}{}_\alpha \pi^\mu_A) = y^A_\alpha \pi^\beta_A, \tag{7.4.3b}$$

$$u^{A\beta}{}_\alpha \pi^\mu_A + u^{A\mu}{}_\alpha \pi^\beta_A = 0, \tag{7.4.3c}$$

where $\delta_A \mathcal{L}$ are variational derivatives. Substituting the relations (7.4.3b) and (7.4.3c) in the weak identity (7.4.2), we obtain the energy-momentum conservation law

$$0 \approx -d_\lambda[u^{A\lambda}{}_\alpha \delta_A \mathcal{L} \tau^\alpha + d_\mu(u^{A\lambda}{}_\alpha \pi^\mu_A \tau^\alpha)]. \tag{7.4.4}$$

A glance at the expression (7.4.4) shows that, on-shell, the corresponding energy-momentum current leads to the superpotential form (3.4.6), i.e.,

$$\boxed{\mathfrak{T}^\lambda_T = u^{A\lambda}{}_\alpha \delta_A \mathcal{L} \tau^\alpha + d_\mu(u^{A\lambda}{}_\alpha \pi^\mu_A \tau^\alpha),}$$

where

$$\boxed{U_T{}^{\mu\lambda} = u^{A\lambda}{}_\alpha \pi^\mu_A \tau^\alpha} \tag{7.4.5}$$

is the *energy-momentum superpotential of tensor fields*.

It is readily seen that the energy-momentum superpotential (7.4.5) emerges from the dependence of the canonical lift $\widetilde{\tau}$ (7.1.4) on the derivatives of the components of the vector field τ. This dependence guarantees that the energy-momentum conservation law (7.4.4) is maintained under general covariant transformations.

Let us now consider tensor fields, treated as matter fields, in the presence of a metric gravitational field described by the second order Hilbert–Einstein Lagrangian. The first variational formula and the associated procedure of constructing Lagrangian conservation laws can also be extended to this case [123, 238]. As a result, we obtain that the corresponding energy-momentum current reduces to the superpotential which is the sum of the energy-momentum superpotential (7.4.5) of tensor fields and the Komar superpotential

$$\boxed{U_K{}^{\mu\lambda} = \tfrac{1}{2\kappa}\sqrt{|\sigma|}(\sigma^{\lambda\nu}\nabla_\nu\tau^\mu - \sigma^{\mu\nu}\nabla_\nu\tau^\lambda)} \tag{7.4.6}$$

of a metric fields, where $\sigma^{\lambda\nu}$ are coordinates on the metric bundle $\Sigma_{\rm PR}$ (7.2.2) and

$$\nabla_\nu \tau^\mu = \partial_\nu \tau^\mu - \{_\nu{}^\mu{}_\alpha\}\tau^\alpha,$$

$$\{_\nu{}^\mu{}_\alpha\} = -\frac{1}{2}\sigma^{\mu\beta}(\sigma_{\nu\beta\alpha} + \sigma_{\alpha\beta\nu} - \sigma_{\beta\alpha\nu}).$$

Example 7.4.1. Let us consider *Proca fields* as an example of tensor matter fields. They are represented by sections of the cotangent bundle $T = T^*X$, and their configuration space is the jet manifold J^1T^*X coordinated by $(x^\lambda, k_\mu, k_{\mu\lambda})$. Given a world connection K, the configuration space J^1T^*X admits the splitting

$$J^1T^*X = S_+ \underset{T^*X}{\oplus} (T^*X \underset{X}{\times} \overset{2}{\wedge} T^*X), \tag{7.4.7}$$

where S_+ is an affine bundle modelled over the vector bundle

$$T^*X \underset{X}{\times} (\overset{2}{\vee} T^*X) \to T^*X.$$

In coordinates, this splitting reads

$$k_{\lambda\mu} = \frac{1}{2}(\mathcal{S}_{\lambda\mu} + \mathcal{F}_{\lambda\mu}),$$

$$\mathcal{F}_{\mu\nu} = k_{\mu\nu} - k_{\nu\mu} + T_\mu{}^\alpha{}_\nu k_\alpha, \tag{7.4.8}$$

$$\mathcal{S}_{\mu\nu} = k_{\mu\nu} + k_{\nu\mu} - T_\mu{}^\alpha{}_\nu k_\alpha, \tag{7.4.9}$$

where $T_\mu{}^\alpha{}_\nu$ is the torsion of the world connection K.

Let us consider the relation (7.4.3c). In coordinates $(\mathcal{F}_{\mu\nu}, \mathcal{S}_{\mu\nu})$ associated with the splitting (7.4.7), it takes the form

$$\frac{\partial \mathcal{L}_{\rm P}}{\partial \mathcal{S}_{\lambda\nu}} = 0.$$

It follows that, in order to be invariant under general covariant transformations, the Lagrangian $L_{\rm P}$ of Proca fields must factorize through the morphism

$$\mathcal{F} : J^1T^*X \underset{T^*X}{\longrightarrow} T^*X \underset{X}{\times} \overset{2}{\wedge} T^*X.$$

Indeed, the standard Lagrangian of Proca fields takes the form

$$L_{\rm P} = [-\frac{1}{4\gamma}\sigma^{\mu\alpha}\sigma^{\nu\beta}\mathcal{F}_{\alpha\beta}\mathcal{F}_{\mu\nu} - \frac{1}{2}m^2\sigma^{\mu\lambda}k_\mu k_\lambda]\sqrt{\mid g\mid}\omega. \tag{7.4.10}$$

Let us consider Proca fields in the presence of a symmetric world connection K. In this case, the strength $\mathcal{F}$ (7.4.8) is independent of K, and we come to the above tensor field model in the presence of a metric gravitational field. The canonical lift $\widetilde{\tau}$ (7.1.4) onto T^*X is

$$\widetilde{\tau} = \tau^\mu\partial_\mu - \partial_\alpha\tau^\nu k_\nu \frac{\partial}{\partial k_\alpha}.$$

Substituting it in the expression (7.4.5), we obtain the energy-momentum superpotential

$$\boxed{U_{\rm P}{}^{\mu\lambda} = -k_\alpha \pi^{\mu\lambda}\tau^\alpha} \tag{7.4.11}$$

of Proca fields, where

$$\pi^{\mu\lambda} = -\frac{1}{\gamma} g^{\mu\alpha} g^{\lambda\beta} \mathcal{F}_{\alpha\beta}\sqrt{\mid g\mid}.$$

•

Let us turn now to the energy-momentum conservation laws in metric-affine gravitation theory.

Recall that world connections are represented by sections of the fibre bundle C_K (7.1.6). Therefore, the configuration space of the metric-affine gravity is the first order jet manifold of the fibred product

$$\Sigma_{\rm PR} \underset{X}{\times} C_K. \tag{7.4.12}$$

Let $L_{\rm MA}$ be a Lagrangian on this configuration space equipped with the adapted coordinates $(x^\lambda, \sigma^{\alpha\beta}, k_\mu{}^\alpha{}_\beta, \sigma_\lambda{}^{\alpha\beta}, k_{\lambda\mu}{}^\alpha{}_\beta)$. The problem lies in the fact that a Lagrangian $L_{\rm MA}$ which factorizes through the curvature $R_{\lambda\mu}{}^\alpha{}_\beta$ (7.1.8) is invariant under the transformations

$$k_\mu{}^\alpha{}_\beta \mapsto k_\mu{}^\alpha{}_\beta + V_\mu \delta^\alpha_\beta, \tag{7.4.13}$$

called the *projective freedom*. One can think of (7.4.13) as being the gauge transformations associated with the dilatation subgroup of the group GL_4. The projective invariance of $L_{\rm MA}$ implies that of a matter Lagrangian which imposes rigorous constraints on matter sources. In fact, these matter sources are only fermion fields (see the next Section). One also suggests to include in a metric-affine Lagrangian $L_{\rm MA}$ different terms expressed into the irreducible parts of the torsion T and the non-metricity Q (see [241] and references therein). The projective symmetry of such a Lagrangian is broken, and one utilizes a Proca field [77] and the hypermomentum fluid [15, 241] as hypothetical matter sources of the metric-affine gravity. Here we restrict our consideration to the case when a metric-affine Lagrangian $L_{\rm MA}$ factorizes through the curvature is independent of the derivative coordinates $\sigma_\lambda{}^{\alpha\beta}$ of a world metric, while matter sources are absent. The following relations take place:

$$\pi^{\lambda\nu}{}_\alpha{}^\beta = -\pi^{\nu\lambda}{}_\alpha{}^\beta, \tag{7.4.14}$$

$$\frac{\partial \mathcal{L}_{\rm MA}}{\partial k_\nu{}^\alpha{}_\beta} = \pi^{\lambda\nu}{}_\alpha{}^\sigma k_\lambda{}^\beta{}_\sigma - \pi^{\lambda\nu}{}_\sigma{}^\beta k_\lambda{}^\sigma{}_\alpha. \tag{7.4.15}$$

We also have the equalities

$$\pi^\lambda_A u^{A\beta\mu}_{\ \alpha} = \pi^{\lambda\mu}{}_\alpha{}^\beta,$$

$$\pi^\varepsilon_A u^{A\beta}_{\ \alpha} = -\partial^\varepsilon{}_\alpha{}^\beta \mathcal{L}_{\rm MA} - \pi^{\varepsilon\beta}{}_\sigma{}^\gamma k_\alpha{}^\sigma{}_\gamma.$$

Given a vector field τ on a world manifold X, its canonical lift onto the product (7.4.12) reads

$$\widetilde{\tau} = \tau^\lambda\partial_\lambda + (\sigma^{\nu\beta}\partial_\nu\tau^\alpha + \sigma^{\alpha\nu}\partial_\nu\tau^\beta)\partial_{\alpha\beta} + (u^{A\beta}_{\ \alpha}\partial_\beta\tau^\alpha + u^{A\beta\mu}_{\ \alpha}\partial_{\beta\mu}\tau^\alpha)\partial_A,$$

where we use the compact notation

$$(u^{A\beta}_{\ \alpha}\partial_\beta\tau^\alpha + u^{A\beta\mu}_{\ \alpha}\partial_{\beta\mu}\tau^\alpha)\partial_A,$$

$$y^A = k_\mu{}^\alpha{}_\beta,$$

$$u_\mu{}^\alpha{}_{\beta\gamma}^{\varepsilon\sigma} = \delta^\varepsilon_\mu\delta^\sigma_\beta\delta^\alpha_\gamma,$$

$$u_\mu{}^\alpha{}_{\beta\gamma}^{\varepsilon} = k_\mu{}^\varepsilon{}_\beta\delta^\alpha_\gamma - k_\mu{}^\alpha{}_\gamma\delta^\varepsilon_\beta - k_\gamma{}^\alpha{}_\beta\delta^\varepsilon_\mu,$$

for the vertical part of the vector field (7.1.7)

Let the Lagrangian $L_{\rm MA}$ be invariant under general covariant transformations, i.e.,

$$\mathbf{L}_{J^1\widetilde{\tau}} L_{\rm MA} = 0. \tag{7.4.16}$$

Then, on-shell, the first variational formula (3.2.2) leads to the weak conservation law

$$0 \approx -d_\lambda[\pi^\lambda_A(y^A_\alpha\tau^\alpha - u^{A\beta}_{\ \alpha}\partial_\beta\tau^\alpha - u^{A\varepsilon\beta}_{\ \alpha}\partial_{\varepsilon\beta}\tau^\alpha) - \tau^\lambda\mathcal{L}_{\rm MA}], \tag{7.4.17}$$

where

$$\mathfrak{T}_{\rm MA}{}^\lambda = \pi^\lambda_A(y^A_\alpha\tau^\alpha - u^{A\beta}_{\ \alpha}\partial_\beta\tau^\alpha - u^{A\varepsilon\beta}_{\ \alpha}\partial_{\varepsilon\beta}\tau^\alpha) - \tau^\lambda\mathcal{L}_{\rm MA} \tag{7.4.18}$$

is the energy-momentum current of the metric-affine gravity.

Remark 7.4.2. It is readily observed that, in the local gauge where the vector field τ is constant, the energy-momentum current (7.4.18) leads to the canonical energy-momentum tensor

$$\mathfrak{T}_{\mathrm{MA}}{}^{\lambda} = (\pi^{\lambda\mu}{}_{\beta}{}^{\nu} k_{\alpha\mu}{}^{\beta}{}_{\nu} - \delta^{\lambda}_{\alpha} L_{\mathrm{MA}})\tau^{\alpha}.$$

This tensor was suggested in order to describe the energy-momentum complex in the Palatini model [80, 230, 239]. •

Due to the arbitrariness of the gauge parameters τ^{λ}, the equality (7.4.16) is equivalent to the system of strong equalities

$$\partial_{\lambda}\mathcal{L}_{\mathrm{MA}} = 0,$$

$$\delta^{\beta}_{\alpha}\mathcal{L}_{\mathrm{MA}} + 2\sigma^{\beta\mu}\delta_{\alpha\mu}\mathcal{L}_{\mathrm{MA}} + u^{A\beta}{}_{\alpha}\delta_{A}\mathcal{L}_{\mathrm{MA}} + d_{\mu}(\pi^{\mu}_{A}u^{A\beta}{}_{\alpha}) - y^{A}_{\alpha}\pi^{\beta}_{A} = 0, \tag{7.4.19}$$

$$[(u^{A\varepsilon\sigma}{}_{\gamma}\partial_{A} + u^{A\varepsilon}{}_{\gamma}\partial^{\sigma}_{A})\mathcal{L}_{\mathrm{MA}}]\partial_{\sigma\varepsilon}\tau^{\gamma} = 0, \tag{7.4.20}$$

$$\pi^{(\lambda\varepsilon}{}_{\gamma}{}^{\sigma)} = 0, \tag{7.4.21}$$

where $\delta_{\alpha\mu}\mathcal{L}$, $\delta_{A}\mathcal{L}$ and $\delta^{\beta}\mathcal{L}$ are the corresponding variational derivatives.

Remark 7.4.3. It is readily observed that the equalities (7.4.20) and (7.4.21) hold due to the relations (7.4.15) and (7.4.14), respectively. •

Substituting the term $y^{A}_{\alpha}\pi^{\beta}_{A}$ from the expression (7.4.19) in the energy-momentum conservation law (7.4.17), we bring it into the form

$$\begin{aligned} 0 \approx -d_{\lambda}[&2\sigma^{\lambda\mu}\tau^{\alpha}\delta_{\alpha\mu}\mathcal{L}_{\mathrm{MA}} + u^{A\lambda}{}_{\alpha}\tau^{\alpha}\delta_{A}\mathcal{L}_{\mathrm{MA}} - \pi^{\lambda}_{A}u^{A\beta}{}_{\alpha}\partial_{\beta}\tau^{\alpha} + \\ &d_{\mu}(\pi^{\lambda\mu}{}_{\alpha}{}^{\beta})\partial_{\beta}\tau^{\alpha} + d_{\mu}(\pi^{\mu}_{A}u^{A\lambda}{}_{\alpha})\tau^{\alpha} - d_{\mu}(\pi^{\lambda\mu}{}_{\alpha}{}^{\beta}\partial_{\beta}\tau^{\alpha})]. \end{aligned} \tag{7.4.22}$$

After separating the variational derivatives, the energy-momentum conservation law (7.4.22) of the metric-affine gravity and Proca fields leads to the superpotential form

$$\begin{aligned} 0 \approx -d_{\lambda}[&2\sigma^{\lambda\mu}\tau^{\alpha}\delta_{\alpha\mu}\mathcal{L}_{\mathrm{MA}} + (k_{\mu}{}^{\lambda}{}_{\gamma}\delta^{\mu}{}_{\alpha}{}^{\gamma}\mathcal{L}_{\mathrm{MA}} - k_{\mu}{}^{\sigma}{}_{\alpha}\delta^{\mu}{}_{\sigma}{}^{\lambda}\mathcal{L}_{\mathrm{MA}} - k_{\alpha}{}^{\sigma}{}_{\gamma}\delta^{\lambda}{}_{\sigma}{}^{\gamma}\mathcal{L}_{\mathrm{MA}})\tau^{\alpha} \\ &+\delta^{\lambda}{}_{\alpha}{}^{\mu}\mathcal{L}\partial_{\mu}\tau^{\alpha} - d_{\mu}(\delta^{\mu}{}_{\alpha}{}^{\lambda}\mathcal{L})\tau^{\alpha} + d_{\mu}(\pi^{\mu\lambda}{}_{\alpha}{}^{\nu}\partial_{\nu}\tau^{\alpha})], \end{aligned}$$

where the energy-momentum current on-shell reduces to the *generalized Komar superpotential*

$$\boxed{U_{\mathrm{MA}}{}^{\mu\lambda} = \pi^{\mu\lambda}{}_{\alpha}{}^{\nu}(\partial_{\nu}\tau^{\alpha} - k_{\sigma}{}^{\alpha}{}_{\nu}\tau^{\sigma})} \tag{7.4.23}$$

[122, 123, 269]. We can rewrite this superpotential as

$$U_{\rm MA}{}^{\mu\lambda} = 2\frac{\partial \mathcal{L}_{\rm MA}}{\partial R_{\mu\lambda}{}^{\alpha}{}_{\nu}}(D_\nu \tau^\alpha + T_\nu{}^\alpha{}_\sigma \tau^\sigma),$$

where D_ν is the covariant derivative relative to the connection $k_\nu{}^\alpha{}_\sigma$ and $T_\nu{}^\alpha{}_\sigma$ is the torsion of this connection.

Example 7.4.4. Let us consider the Hilbert–Einstein Lagrangian

$$L_{\rm HE} = -\frac{1}{2\kappa}R\sqrt{|\,\sigma\,|}\omega,$$
$$R = \sigma^{\lambda\nu}R_{\lambda\alpha}{}^{\alpha}{}_{\nu},$$

in the metric-affine gravitation model. Then the generalized Komar superpotential (7.4.23) comes to the Komar superpotential (7.4.6) if we substitute the Levi–Civita connection $k_\nu{}^\alpha{}_\sigma = \{_\nu{}^\alpha{}_\sigma\}$. One may generalize this example by considering the Lagrangian

$$L = f(R)\sqrt{-g}\omega,$$

where $f(R)$ is a polynomial of the scalar curvature R. In the case of a symmetric connection, we restate the superpotential

$$U^{\mu\lambda} = -\frac{\partial f}{\partial R}\sqrt{-g}(g^{\lambda\nu}D_\nu\tau^\mu - g^{\mu\nu}D_\nu\tau^\lambda)$$

of the Palatini model [33] just as the superpotential when a Lagrangian of the Palatini model factorizes through the product $R^{\alpha\beta}R_{\alpha\beta}$ [34]. •

7.5 Spin connections

We will restrict our consideration to Dirac spinor fields since all fermion matter, observable till now, is described by these fields. Let us consider gauge theory on spinor bundles over a world manifold whose sections describe Dirac fermion fields. The key point is that these spinor bundles are not preserved under general covariant transformations. From the physical viewpoint, a Dirac fermion matter is responsible for a spontaneous symmetry breaking in gravitation theory. From the mathematical viewpoint, every Dirac spin structure on a world manifold X is associated with a certain tetrad gravitational field h on X, i.e., with a certain reduced Lorentz structure

$P^h \subset LX$. The problem of a spin connection for Dirac spinor fields in the presence of a background gravitational field has been solved by V.Fock and D.Ivanenko in 1929, and their spin Levi–Civita connection (see (7.5.13) below) is naturally extended to an arbitrary Lorentz connection A_h on P^h. Here we study a spin connection associated with an arbitrary linear world connection on X. Therefore, we should consider a spinor structure which is not associated with a certain tetrad field h and which is subject to general covariant transformations (see also [74, 101, 243]). To solve this problem, one can follow the general scheme of describing symmetry breaking in gauge theories in terms of composite bundles in Section 6.5. We will construct a composite bundle $S \to \Sigma_{\rm T} \to X$, where $S \to \Sigma_{\rm T}$ is a spinor bundle over the tetrad bundle (7.2.1) [123, 124, 272, 273]. Given a section h of the tetrad bundle $\Sigma_{\rm T}$, the restriction of this spinor bundle to $h(X) \subset \Sigma_{\rm T}$ describes the familiar Dirac spin structure in the presence of h. Conversely, every Dirac spin structure S^h on a world manifold can be seen in this way. Note that $S \to X$ is not a spin bundle, and is subject to general covariant transformations. A desired spin connection is a connection on the composite bundle $S \to X$.

Remark 7.5.1. Dirac spinors. We describe Dirac spinors in terms of Clifford algebras (see, e.g., [69, 123, 240, 255] and [47, 194] for a general Clifford algebra technique).

Let M be the Minkowski space equipped with the Minkowski metric η, and let $\{e^a\}$ be a fixed basis for M. By $\mathbb{C}_{1,3}$ is denoted the complex *Clifford algebra* generated by elements of M. This is the complexified quotient of the tensor algebra $\otimes M$ of M by the two-sided ideal generated by elements

$$e\otimes e' + e'\otimes e - 2\eta(e,e') \in \otimes M, \qquad e, e' \in M.$$

The complex Clifford algebra $\mathbb{C}_{1,3}$ is isomorphic to the real Clifford algebra $\mathbb{R}_{2,3}$, whose generating space is $\mathbb{R}^5$ equipped with the metric

$$\mathrm{diag}(1,-1,-1,-1,\ 1).$$

Its subalgebra generated by elements of $M \subset \mathbb{R}^5$ is the real Clifford algebra $\mathbb{R}_{1,3}$.

A *spinor space* V is defined as a minimal left ideal of $\mathbb{C}_{1,3}$ on which this algebra acts on the left. We have the representation

$$\gamma : M\otimes V \to V, \qquad \gamma(e^a) = \gamma^a, \tag{7.5.1}$$

of elements of the Minkowski space $M \subset \mathbb{C}_{1,3}$ by the Dirac γ-matrices on V. Different ideals V lead to equivalent representations (7.5.1). The spinor space V is provided with the *spinor metric*

$$a(v,v') = \frac{1}{2}(\overline{v}v' + \overline{v}'v) = \frac{1}{2}(v^+\gamma^0 v' + v'^+\gamma^0 v) \tag{7.5.2}$$

since the element $e^0 \in M$ satisfies the conditions

$$(e^0)^+ = e^0, \qquad (e^0 e)^+ = e^0 e, \qquad \forall e \in M.$$

By definition, the *Clifford group* $G_{1,3}$ consists of the invertible elements l_s of the real Clifford algebra $\mathbb{R}_{1,3}$ such that the inner automorphisms defined by these elements preserve the Minkowski space $M \subset \mathbb{R}_{1,3}$, i.e.,

$$l_s e l_s^{-1} = l(e), \qquad e \in M, \tag{7.5.3}$$

where l is a Lorentz transformation of M. Hence, we have an epimorphism of the Clifford group $G_{1,3}$ onto the Lorentz group $O(1,3)$. Since the action (7.5.3) of the Clifford group on the Minkowski space M is not effective, one usually consider its *pin* and *spin subgroups*. The subgroup $Pin(1,3)$ of $G_{1,3}$ is generated by elements $e \in M$ such that $\eta(e,e) = \pm 1$. The even part of $\mathrm{Pin}(1,3)$ is the spin group $Spin(1,3)$. Its component of the unity

$$L_s = Spin^0(1,3) \simeq SL(2,\mathbb{C})$$

is the well-known two-fold universal covering group

$$z_L : L_s \to \mathrm{L} = L_s/\mathbb{Z}_2, \qquad \mathbb{Z}_2 = \{1,-1\}, \tag{7.5.4}$$

of the proper Lorentz group L.

The Clifford group $G_{1,3}$ acts on the spinor space V by left multiplications

$$G_{1,3} \ni l_s : v \mapsto l_s v, \qquad v \in V.$$

This action preserves the representation (7.5.1), i.e.,

$$\gamma(lM \otimes l_s V) = l_s \gamma(M \otimes V).$$

The spin group L_s acts on the spinor space V by means of the generators

$$L_{ab} = \frac{1}{4}[\gamma_a, \gamma_b]. \tag{7.5.5}$$

Since $L^+_{ab}\gamma^0 = -\gamma^0 L_{ab}$, this action preserves the spinor metric (7.5.2). •

A *Dirac spin structure* on a world manifold X is said to be a pair (P^h, z_s) of an L_s-principal bundle $P^h \to X$ and a principal bundle morphism

$$z_h : P^h \underset{X}{\to} LX \tag{7.5.6}$$

from P^h to the frame bundle LX [14, 27, 194]. More generally, one can define a spin structure on any vector bundle $E \to X$ [194]. Since the homomorphism $L_s \to GL_4$ factorizes through the epimorphism (7.5.4), every bundle morphism (7.5.6) factorizes through a morphism

$$\begin{aligned} &z_h : P^h \to L^h X, \\ &z_h \circ R_g = R_{z_L(g)}, \qquad \forall g \in L_s, \end{aligned} \tag{7.5.7}$$

of P^h to some L-principal subbundle $L^h X$ of the frame bundle LX.

It follows that the necessary condition of the existence of a Dirac spin structure on X is the existence of a Lorentz structure. From the physics viewpoint, it means that the existence of Dirac's fermion matter implies the existence of a gravitational field.

Fermion fields in the presence of a tetrad field h are described by sections of the P^h-associated spinor bundle

$$S^h = (P^h \times V)/L_s \to X \tag{7.5.8}$$

whose typical fibre V carriers the spinor representation (7.5.5) of the spin group L_s. To describe Dirac fermions and, in particular, to construct the Dirac operator, the spinor bundle S^h (7.5.8) must be represented as a subbundle of the bundle of Clifford algebras, i.e., as a spinor structure on the cotangent bundle T^*X.

Every fibre bundle of Minkowski spaces $M^h X$ (7.2.7) over a world manifold X is extended to the fibre bundle of Clifford algebras $C^h X$ with the fibres generated by the fibres of $M^h X$ [27]. This fibre bundle $C^h X$ has the structure group $\mathrm{Aut}(\mathbb{C}_{1,3})$ of inner automorphisms of the Clifford algebra $\mathbb{C}_{1,3}$. In general, $C^h X$ does not contain a spinor subbundle because a spinor subspace V is not stable under inner automorphisms of $\mathbb{C}_{1,3}$. As was shown [27], a spinor subbundle of $C^h X$ exists if the transition functions of $C^h X$ can be lifted from $\mathrm{Aut}(\mathbb{C}_{1,3})$ to the Clifford group $G_{1,3}$. This agrees with the usual condition of the existence of a spin structure which holds for a world manifold X. Such a spinor subbundle is the bundle S^h (7.5.8) associated with the universal two-fold covering (7.5.7) of the Lorentz bundle L^hX. We will call P^h (and S^h) the h-associated Dirac spin structure on a world manifold.

Remark 7.5.2. All spin structures on a manifold X which are related to the two-fold universal covering groups possess the following two properties [140]. Let $P \to X$ be a principal bundle whose structure group G has the fundamental group $\pi_1(G) = \mathbb{Z}_2$. Let $\widetilde{G}$ be the universal covering group of G.

- The topological obstruction to the existence of a $\widetilde{G}$-principal bundle $\widetilde{P} \to X$ covering the bundle $P \to X$ is given by the Čech cohomology group $H^2(X;\mathbb{Z}_2)$ of X with coefficients in $\mathbb{Z}_2$. Roughly speaking, the principal bundle P defines an element of $H^2(X;\mathbb{Z}_2)$ which must be zero so that $P \to X$ can give rise to $\widetilde{P} \to X$.

- Non-equivalent lifts of $P \to X$ to $\widetilde{G}$-principal bundles are classified by elements of the Čech cohomology group $H^1(X;\mathbb{Z}_2)$.

In particular, the well-known topological obstruction to the existence of a Dirac spin structure is the second Stiefel–Whitney class $w_2(X) \in H^2(X;\mathbb{Z}_2)$ of X [194]. In the case of 4-dimensional non-compact manifolds, all Riemannian and pseudo-Riemannian spin structures are equivalent [14, 119]. •

There exists the bundle morphism

$$\boxed{\gamma_h : T^*X \otimes S^h = (P^h \times (M \otimes V))/L_s \to (P^h \times \gamma(M \otimes V))/L_s = S^h,} \tag{7.5.9}$$

where by γ is meant the left action (7.5.1) of $M \subset \mathbb{C}_{1,3}$ on $V \subset \mathbb{C}_{1,3}$. One can think of (7.5.9) as being the representation of covectors to X^4 by the Dirac γ-matrices on elements of the spinor bundle S^h. Relative to an atlas $\{z_\zeta\}$ of P^h and to the associated Lorentz atlas $\{z_h \circ z_\zeta\}$ of LX, the representation (7.5.9) reads

$$y^A(\gamma_h(h^a(x) \otimes v)) = \gamma^{aA}{}_B y^B(v), \qquad v \in S^h_x,$$

where y^A are the corresponding bundle coordinates of S^h, and h^a are the tetrad coframes (7.2.4). For brevity, we will write

$$\widehat{h}^a = \gamma_h(h^a) = \gamma^a, \qquad \widehat{dx}^\lambda = \gamma_h(dx^\lambda) = h^\lambda_a(x)\gamma^a.$$

Let A_h be a principal connection on S^h and let

$$D : J^1S^h \to T^*X \underset{S^h}{\otimes} S^h,$$
$$D = (y^A_\lambda - A^{ab}{}_\lambda L_{ab}{}^A{}_B y^B)dx^\lambda \otimes \partial_A,$$

be the corresponding covariant differential (2.2.7), where

$$VS^h = S^h \underset{X}{\times} S^h.$$

The first order differential *Dirac operator* is defined on S^h by the composition

$$\mathcal{D}_h = \gamma_h \circ D : J^1S^h \to T^*X \otimes S^h \to S^h, \tag{7.5.10}$$
$$y^A \circ \mathcal{D}_h = h_a^\lambda \gamma^{aA}{}_B(y_\lambda^B - \frac{1}{2}A^{ab}{}_\lambda L_{ab}{}^A{}_B y^B).$$

Remark 7.5.3. The spinor bundle S^h is a complex fibre bundle with a real structure group over a real manifold. Of course, one can regard such a fibre bundle as the real one. In particular, the jet manifold J^1S^h with coordinates $(x^\lambda, y^A, y_\lambda^A)$ is defined as usual. •

The h-associated spinor bundle S^h is equipped with the fibre spinor metric

$$a_h : S^h \underset{X}{\times} S^h \to \mathbb{R},$$
$$a_h(v,v') = \frac{1}{2}(v^+\gamma^0 v' + v'^+\gamma^0 v), \qquad v,v' \in S^h.$$

Using this metric and the Dirac operator (7.5.10), one can define *Dirac's Lagrangian* on J^1S^h in the presence of a background tetrad field h and a background connection A_h on S^h as

$$L_h : J^1S^h \to \overset{4}{\wedge} T^*X,$$
$$L_h = [a_h(i\mathcal{D}_h(w), w) - m a_h(w,w)]h^0 \wedge \cdots \wedge h^3, \qquad w \in J^1S^h.$$

Its coordinate expression is

$$\mathcal{L}_h = \{\frac{i}{2}h_q^\lambda[y_A^+(\gamma^0\gamma^q)^A{}_B(y_\lambda^B - \frac{1}{2}A_\lambda{}^{ab}L_{ab}{}^B{}_C y^C) - \tag{7.5.11}$$
$$(y_{\lambda A}^+ - \frac{1}{2}A_\lambda{}^{ab} y_C^+ L_{ab}^+)(\gamma^0\gamma^q)^A{}_B y^B] - m y_A^+(\gamma^0)^A{}_B y^B\} \det(h_\mu^a).$$

Note that there is one-to-one correspondence between the principal connections, called *spin connections*, on the h-associated principal spinor bundle P^h and the Lorentz connections on the L-principal bundle L^hX. Indeed, it follows from Theorem 6.1.3 that every principal connection

$$A_h = dx^\lambda \otimes (\partial_\lambda + \frac{1}{2}A_\lambda{}^{ab}\varepsilon_{ab}) \tag{7.5.12}$$

on P^h defines a principal connection on L^hX which is given by the same expression (7.5.12). Conversely, the pull-back $z_h^*\overline{A}_h$ on P^h of the connection form $\overline{A}_h$ of a Lorentz connection A_h on L^hX is equivariant under the action of the group L_s on P^h and, consequently, it is a connection form of a spin connection on P^h.

In particular, the Levi–Civita connection of a pseudo-Riemannian metric g gives rise to the spin connection with the components

$$\boxed{A_\lambda{}^{ab} = \eta^{kb}h^a_\mu(\partial_\lambda h^\mu_k - h^\nu_k\{_\lambda{}^\mu{}_\nu\})} \tag{7.5.13}$$

on the g-associated spinor bundle S^g.

We consider the general case of a spin connection generated on P^h by a world connection K. The Lorentz connection K_h induced by K on L^hX is given by the local connection 1-form (7.2.20), and it defines the corresponding spin connection on S^h

$$\boxed{K_h = dx^\lambda \otimes [\partial_\lambda + \tfrac{1}{4}(\eta^{kb}h^a_\mu - \eta^{ka}h^b_\mu)(\partial_\lambda h^\mu_k - h^\nu_k K_\lambda{}^\mu{}_\nu)L_{ab}{}^A{}_B y^B\partial_A],} \tag{7.5.14}$$

where L_{ab} are the generators (7.5.5) [123, 248, 270].

Substituting the spin connection (7.5.14) in the Dirac operator (7.5.10) and Dirac's Lagrangian (7.5.11), we obtain a description of Dirac fermion fields in the presence of an arbitrary world connection on a world manifold, not only of the Lorentz type.

Motivated by the connection (7.5.14) (see Remark 2.4.2), one can obtain the canonical lift

$$\widetilde{\tau} = \tau^\lambda\partial_\lambda + \frac{1}{4}(\eta^{kb}h^a_\mu - \eta^{ka}h^b_\mu)(\tau^\lambda\partial_\lambda h^\mu_k - h^\nu_k\partial_\nu\tau^\mu)L_{ab}{}^A{}_B y^B\partial_A \tag{7.5.15}$$

of vector fields τ on X onto the spinor bundle S^h [123, 269]. The lift (7.5.15) can be brought into the form

$$\widetilde{\tau} = \tau_{\{\}} - \frac{1}{4}(\eta^{kb}h^a_\mu - \eta^{ka}h^b_\mu)h^\nu_k\nabla_\nu\tau^\mu L_{ab}{}^A{}_B y^B\partial_A,$$

where $\tau_{\{\}}$ is the horizontal lift of τ by means of the spin Levi–Civita connection for the tetrad field h, and $\nabla_\nu\tau^\mu$ are the covariant derivatives of τ relative to the Levi–Civita connection [101, 181].

From now on, we will assume that a world manifold X is non-compact and parallelizable in accordance with Remark 7.2.1. In this case, all Dirac spin structures are equivalent, i.e., the principal spinor bundles P^h and $P^{h'}$ are isomorphic [14,

119]. This property remains true for all spin structures on X which are generated by the two-fold universal covering groups (see Remark 7.5.2). Nevertheless, the associated structures of the bundles of Minkowski spaces M^hX and $M^{h'}X$ (7.2.7) on the cotangent bundle T^*X are not equivalent, and so are the representations γ_h and $\gamma_{h'}$ (7.5.9) [123, 263]. It follows that every Diràc fermion field must be described in a pair (s_h, h) with a certain tetrad field h, and Dirac fermion fields in the presence of different tetrad fields fail to be given by sections of the same spinor bundle. This fact exhibits the physical nature of gravity as a Higgs field. The goal is to construct a bundle over X whose sections exhaust the whole totality of fermion-gravitation pairs [123, 272, 273]. Following the general scheme of describing symmetry breaking in Section 6.5, we will use the fact that the frame bundle LX is the principal bundle $LX \to \Sigma_{\rm T}$ over the tetrad bundle $\Sigma_{\rm T}$ (7.2.1) with the structure Lorentz group L.

The group GL_4 has the first homotopy group $\mathbb{Z}_2$. Therefore, GL_4 admits the universal two-fold covering group $\widetilde{GL}_4$ such that the diagram

$$\begin{array}{ccc} \widetilde{GL}_4 & \longrightarrow & GL_4 \\ \uparrow & & \uparrow \\ L_s & \xrightarrow{z_L} & L \end{array} \tag{7.5.16}$$

commutes. Let us consider the corresponding two-fold covering bundle $\widetilde{LX} \to X$ of the frame bundle LX [74, 194, 243, 288]. However, the spinor representation of the group $\widetilde{GL}_4$ is infinite dimensional. Therefore, the $\widetilde{LX}$-associated spinor bundle over X describes infinite-dimensional "world" spinor fields, but not the Dirac ones (see [150] for details). At the same time, since the fibre bundle

$$\boxed{\widetilde{LX} \to \Sigma_{\rm T}} \tag{7.5.17}$$

is an $L_{\rm s}$-principal bundle over the tetrad bundle $\Sigma_{\rm T} = \widetilde{LX}/L_{\rm s}$, the commutative diagram

$$\begin{array}{ccc} \widetilde{LX} & \xrightarrow{\widetilde{z}} & LX \\ & \searrow \swarrow & \\ & \Sigma_{\rm T} & \end{array} \tag{7.5.18}$$

provides a Dirac spin structure on the tetrad bundle $\Sigma_{\rm T}$. One can show that the spin structure (7.5.18) is unique [123, 124, 272, 273]. This spin structure, called the *universal spin structure*, possesses the following property.

Owing to the commutative diagram (7.5.16), we have the commutative diagram

$$\begin{array}{ccc} \widetilde{LX} & \xrightarrow{\widetilde{z}} & LX \\ \uparrow & & \uparrow \\ P^h & \xrightarrow{z_h} & L^hX \end{array} \tag{7.5.19}$$

for any tetrad field h [114, 123, 272, 273]. This means that, given a tetrad field h, the restriction $h^*\widetilde{LX}$ of the L_s-principal bundle (7.5.17) to $h(X) \subset \Sigma_T$ is isomorphic to the L_s-principal subbundle P^h of the fibre bundle $\widetilde{LX} \to X$ which is the h-associated Dirac spin structure.

Let us consider the spinor bundle

$$S = (\widetilde{LX} \times V)/L_s \to \Sigma_T \tag{7.5.20}$$

associated with the L_s-principal bundle (7.5.17), and the corresponding composite spinor bundle

$$\boxed{S \to \Sigma_T \to X,} \tag{7.5.21}$$

which however is not a spinor bundle over X.

Given a tetrad field h, there is the canonical isomorphism

$$i_h : S^h = (P^h \times V)/L_s \to (h^*\widetilde{LX} \times V)/L_s$$

of the h-associated spinor bundle S^h (7.5.8) onto the restriction h^*S of the spinor bundle $S \to \Sigma_T$ to $h(X) \subset \Sigma_T$ (see Proposition 2.7.1). Thence, every global section s_h of the spinor bundle S^h corresponds to the global section $i_h \circ s_h$ of the composite spinor bundle (7.5.21). Conversely, every global section s of the composite spinor bundle (7.5.21), which projects onto a tetrad field h, takes its values into the subbundle $i_h(S^h) \subset S$ (see Proposition 2.7.2).

Let the frame bundle $LX \to X$ be provided with a holonomic atlas $\{U_\zeta, T\phi_\zeta\}$, and let the principal bundles $\widetilde{LX} \to \Sigma_T$ and $LX \to \Sigma_T$ have the associated atlases $\{U_\epsilon, z_\epsilon^s\}$ and $\{U_\epsilon, z_\epsilon = \widetilde{z} \circ z_\epsilon^s\}$, respectively. With these atlases, the composite spinor bundle S (7.5.21) is equipped with the bundle coordinates $(x^\lambda, \sigma_a^\mu, y^A)$, where $(x^\lambda, \sigma_a^\mu)$ are coordinates of the tetrad bundle Σ_T such that σ_a^μ are the matrix components of the group element $(T\phi_\zeta \circ z_\epsilon)(\sigma)$, $\sigma \in U_\epsilon$, $\pi_{\Sigma X}(\sigma) \in U_\zeta$. For any tetrad field h, we have $(\sigma_a^\lambda \circ h)(x) = h_a^\lambda(x)$ where $h_a^\lambda(x) = H_a^\lambda \circ z_\epsilon \circ h$ are the tetrad functions (7.2.5) with respect to the Lorentz atlas $\{z_\epsilon \circ h\}$ of L^hX.

The spinor bundle $S \to \Sigma_T$ is the subbundle of the bundle of Clifford algebras which is generated by the bundle of Minkowski spaces

$$E_M = (LX \times M)/\mathrm{L} \to \Sigma_T \tag{7.5.22}$$

associated with the L-principal bundle $LX \to \Sigma$. Since the fibre bundles $LX \to X$ and $GL_4 \to GL_4/\mathrm{L}$ are trivial, so is the fibre bundle (7.5.22). Hence, it is isomorphic to the product $\Sigma_T \underset{X}{\times} T^*X$. Then there exists the representation

$$\gamma_\Sigma : T^*X \underset{\Sigma_T}{\otimes} S = (\widetilde{LX} \times (M \otimes V))/L_s \to (\widetilde{LX} \times \gamma(M \otimes V))/L_s = S \tag{7.5.23}$$

given by the coordinate expression

$$\boxed{\widehat{dx}^\lambda = \gamma_\Sigma(dx^\lambda) = \sigma^\lambda_a \gamma^a.}$$

Restricted to $h(X) \subset \Sigma_T$, this representation recovers the morphism γ_h (7.5.9).

Using the representation γ_Σ (7.5.23), one can construct the total Dirac operator on the composite spinor bundle S as follows. Since the bundles $\widetilde{LX} \to \Sigma_T$ and $\Sigma_T \to X$ are trivial, let us consider a principal connection A (6.5.7) on the L_s-principal bundle $\widetilde{LX} \to \Sigma_T$ given by the local connection form

$$A = (A_\lambda{}^{ab} dx^\lambda + A^{kab}_\mu d\sigma^\mu_k) \otimes L_{ab}, \tag{7.5.24}$$

$$A_\lambda{}^{ab} = -\frac{1}{2}(\eta^{kb}\sigma^a_\mu - \eta^{ka}\sigma^b_\mu)\sigma^\nu_k K_\lambda{}^\mu{}_\nu,$$

$$A^{kab}_\mu = \frac{1}{2}(\eta^{kb}\sigma^a_\mu - \eta^{ka}\sigma^b_\mu), \tag{7.5.25}$$

where K is a world connection on X. This connection defines the associated spin connection

$$A_\Sigma = dx^\lambda \otimes (\partial_\lambda + \frac{1}{2}A_\lambda{}^{ab} L_{ab}{}^A{}_B y^B \partial_A) + \tag{7.5.26}$$

$$d\sigma^\mu_k \otimes (\partial^k_\mu + \frac{1}{2}A^{kab}_\mu L_{ab}{}^A{}_B y^B \partial_A)$$

on the spinor bundle $S \to \Sigma_T$. The choice of the connection (7.5.24) is motivated by the fact that, given a tetrad field h, the restriction of the spin connection (7.5.26) to S^h is exactly the spin connection (7.5.14).

The connection (7.5.26) yields the first order differential operator $\widetilde{D}$ (2.7.15) on the composite spinor bundle $S \to X$ which reads

$$\boxed{\widetilde{D} : J^1S \to T^*X \underset{\Sigma_T}{\otimes} S,}$$

$$\widetilde{D} = dx^\lambda \otimes [y^A_\lambda - \frac{1}{2}(A_\lambda{}^{ab} + A^{kab}_\mu \sigma^\mu_{\lambda k})L_{ab}{}^A{}_B y^B]\partial_A = \tag{7.5.27}$$
$$dx^\lambda \otimes [y^A_\lambda - \frac{1}{4}(\eta^{kb}\sigma^a_\mu - \eta^{ka}\sigma^b_\mu)(\sigma^\mu_{\lambda k} - \sigma^\nu_k K_\lambda{}^\mu{}_\nu)L_{ab}{}^A{}_B y^B]\partial_A.$$

The corresponding restriction $\widetilde{D}_h$ (6.5.9) of the operator $\widetilde{D}$ (7.5.27) to $J^1S^h \subset J^1S$ recovers the familiar covariant differential on the h-associated spinor bundle $S^h \to X^4$ relative to the spin connection (7.5.15).

Combining (7.5.23) and (7.5.27) gives the first order differential operator

$$\boxed{\mathcal{D} = \gamma_{\Sigma_T} \circ \widetilde{D} : J^1S \to T^*X \underset{\Sigma_T}{\otimes} S \to S,} \tag{7.5.28}$$

$$y^B \circ \mathcal{D} = \sigma^\lambda_a \gamma^{aB}{}_A [y^A_\lambda - \frac{1}{4}(\eta^{kb}\sigma^a_\mu - \eta^{ka}\sigma^b_\mu)(\sigma^\mu_{\lambda k} - \sigma^\nu_k K_\lambda{}^\mu{}_\nu)L_{ab}{}^A{}_B y^B],$$

on the composite spinor bundle $S \to X$. One can think of $\mathcal{D}$ as being the *total Dirac operator* on S since, for every tetrad field h, the restriction of $\mathcal{D}$ to $J^1S^h \subset J^1S$ is exactly the Dirac operator $\mathcal{D}_h$ (7.5.10) on the spinor bundle S^h in the presence of the background tetrad field h and the spin connection (7.5.14).

Thus, we come to the model of the metric-affine gravity and Dirac fermion fields. The total configuration space of this model is the jet manifold J^1Y of the bundle product

$$Y = C_K \underset{\Sigma_T}{\times} S \tag{7.5.29}$$

where C_K is the bundle of world connections (7.1.6). This product is coordinated by $(x^\mu, \sigma^\mu_a, k_\mu{}^\alpha{}_\beta, y^A)$.

Let $J^1_{\Sigma_T}Y$ denote the first order jet manifold of the fibre bundle $Y \to \Sigma_T$. This fibre bundle can be endowed with the spin connection

$$A_Y : Y \longrightarrow J^1_{\Sigma_T}Y \xrightarrow{\text{pr}_2} J^1_{\Sigma_T}S,$$
$$A_Y = dx^\lambda \otimes (\partial_\lambda + \widetilde{A}_\lambda{}^{ab} L_{ab}{}^A{}_B y^B \partial_A) + d\sigma^\mu_k \otimes (\partial^k_\mu + A^{kab}_\mu L_{ab}{}^A{}_B y^B \partial_A), \tag{7.5.30}$$

where A^{kab}_μ is given by the expression (7.5.25), and

$$\widetilde{A}_\lambda{}^{ab} = -\frac{1}{2}(\eta^{kb}\sigma^a_\mu - \eta^{ka}\sigma^b_\mu)\sigma^\nu_k k_\lambda{}^\mu{}_\nu.$$

Using the connection (7.5.30), we obtain the first order differential operator

$$\widetilde{D}_Y : J^1Y \to T^*X \underset{\Sigma}{\otimes} S,$$

$$\widetilde{D}_Y = dx^\lambda \otimes [y^A_\lambda - \frac{1}{4}(\eta^{kb}\sigma^a_\mu - \eta^{ka}\sigma^b_\mu)(\sigma^\mu_{\lambda k} - \sigma^\nu_k k_\lambda{}^\mu{}_\nu)L_{ab}{}^A{}_B y^B]\partial_A, \tag{7.5.31}$$

and the total Dirac operator

$$\mathcal{D}_Y = \gamma_\Sigma \circ \widetilde{D} : J^1Y \to T^*X \underset{\Sigma}{\otimes} S \to S,$$

$$y^B \circ \mathcal{D}_Y = \sigma^\lambda_a \gamma^{aB}{}_A [y^A_\lambda - \frac{1}{4}(\eta^{kb}\sigma^a_\mu - \eta^{ka}\sigma^b_\mu)(\sigma^\mu_{\lambda k} - \sigma^\nu_k k_\lambda{}^\mu{}_\nu)L_{ab}{}^A{}_B y^B], \tag{7.5.32}$$

on the fibre bundle $Y \to X$. Given a world connection $K : X \to C_K$, the restrictions of the spin connection A_Y (7.5.30), the operator $\widetilde{D}_Y$ (7.5.31) and the Dirac operator $\mathcal{D}_Y$ (7.5.32) to K^*Y are exactly the spin connection (7.5.26) and the operators (7.5.27) and (7.5.28), respectively.

The total Lagrangian on the configuration space J^1Y of the metric-affine gravity and fermion fields is the sum

$$L = L_{\rm MA} + L_{\rm D} \tag{7.5.33}$$

of a metric-affine Lagrangian

$$L_{\rm MA}(R_{\mu\lambda}{}^\alpha{}_\beta, \sigma^{\mu\nu}), \qquad \sigma^{\mu\nu} = \sigma^\mu_a \sigma^\nu_b \eta^{ab},$$

and Dirac's Lagrangian

$$L_{\rm D} = [a_Y(i\mathcal{D}(w), w) - m a_S(w,w)]\sigma^0 \wedge \cdots \wedge \sigma^3, \qquad w \in J^1S,$$

where $\sigma^a = \sigma^a_\mu dx^\mu$ and

$$a_Y(v, v') = \frac{1}{2}(v^+\gamma^0 v' + v'^+\gamma^0 v)$$

is the fibre spinor metric onto the fibre bundle $Y \to (C_K \times \Sigma_{\rm T})$. Its coordinate expression is

$$\begin{aligned} L_{\rm D} = \{&\frac{i}{2}\sigma^\lambda_q [y^+_A(\gamma^0\gamma^q)^A{}_B(y^B_\lambda - \frac{1}{4}(\eta^{kb}\sigma^a_\mu - \eta^{ka}\sigma^b_\mu)(\sigma^\mu_{\lambda k} - \sigma^\nu_k k_\lambda{}^\mu{}_\nu)L_{ab}{}^B{}_C y^C) - \\ &(y^+_{\lambda A} - \frac{1}{4}(\eta^{kb}\sigma^a_\mu - \eta^{ka}\sigma^b_\mu)(\sigma^\mu_{\lambda k} - \sigma^\nu_k k_\lambda{}^\mu{}_\nu)y^+_C L^{+C}_{ab}{}_A(\gamma^0\gamma^q)^A{}_B y^B] - \\ &m y^+_A(\gamma^0)^A{}_B y^B\}\sqrt{|\sigma|}, \qquad \sigma = \det(\sigma_{\mu\nu}). \end{aligned} \tag{7.5.34}$$

It is readily observed that

$$\frac{\partial\mathcal{L}_{\rm D}}{\partial k_\lambda{}^\mu{}_\nu} + \frac{\partial\mathcal{L}_{\rm D}}{\partial k_\nu{}^\mu{}_\lambda} = 0, \tag{7.5.35}$$

i.e., Dirac's Lagrangian (7.5.34) depends only on the torsion of a world connection.

Note that, since the universal spin structure is unique, the $\widetilde{GL}_4$-principal bundle $\widetilde{LX} \to X^4$ as well as the frame bundle LX admits the canonical lift of any diffeomorphism f of the base X [74, 123, 154]. This lift yields the general covariant transformation of the associated spinor bundle $S \to \Sigma_{\rm T}$ over the general covariant transformations of the tetrad bundle $\Sigma_{\rm T}$. The corresponding canonical lift onto S of a vector field on X can be constructed (see [123, 272, 273] for details). The goal is the energy-momentum conservation law in gauge theory on spinor bundles. One can show that the corresponding energy-momentum current reduces to the generalize Komar superpotential (7.4.23) [123, 270]. It should be emphasized that Dirac fermion fields do not contribute to this superpotential because of the relation (7.5.35).

7.6 Affine world connections

Being a vector bundle, the tangent bundle TX of a world manifold X has a natural structure of an affine bundle. Therefore, one can consider affine connections on TX, called *affine world connections*. Here we will study them as principal connections.

Let $Y \to X$ be an affine bundle with a k-dimensional typical fibre V. It is associated with a principal bundle AY of affine frames in Y, whose structure group is the general affine group $GA(k,\mathbb{R})$. Then any affine connection on $Y \to X$ is associated with a principal connection on $AY \to X$. These connections are represented by global sections of the affine bundle $J^1AY/GA(k,\mathbb{R}) \to X$, and they always exist.

As was mentioned in Section 2.5, every affine connection Γ (2.5.1) on $Y \to X$ defines a linear connection $\overline{\Gamma}$ (2.5.2) on the underlying vector bundle $\overline{Y} \to X$. This connection $\overline{\Gamma}$ is associated with a linear principal connection on the principal bundle $L\overline{Y}$ of linear frames in $\overline{Y}$, whose structure group is the general linear group $GL(k,\mathbb{R})$. We have the exact sequence of groups

$$0 \to T_k \to GA(k,\mathbb{R}) \to GL(k,\mathbb{R}) \to \mathbf{1}, \tag{7.6.1}$$

where T_k is the group of translations in $\mathbb{R}^k$. It is readily observed that there is the corresponding principal bundle morphism $AY \to L\overline{Y}$ over X such that the principal connection $\overline{\Gamma}$ on $L\overline{Y}$ is the image of the principal connection Γ on $AY \to X$ under this morphism in accordance with Theorem 6.1.3.

The exact sequence (7.6.1) admits a splitting $GL(k, \mathbb{R}) \hookrightarrow GA(k, \mathbb{R})$, but this splitting fails to be canonical (see, e.g., [177]). It depends on the morphism

$$V \ni v \mapsto v - v_0 \in \overline{V},$$

i.e., on the choice of an origin v_0 of the affine space V. Given v_0, the image of the corresponding monomorphism $GL(k, \mathbb{R}) \hookrightarrow GA(k, \mathbb{R})$ is the stabilizer $G(v_0) \subset GA(k, \mathbb{R})$ of v_0. Different subgroups $G(v_0)$ and $G(v'_0)$ are related with each other as follows:

$$G(v'_0) = T(v'_0 - v_0)G(v_0)T^{-1}(v'_0 - v_0),$$

where $T(v'_0 - v_0)$ is the translation along the vector $(v'_0 - v_0) \in \overline{V}$.

Remark 7.6.1. Accordingly, the well-known morphism of a k-dimensional affine space V onto a hypersurface $\overline{y}^{k+1} = 1$ in $\mathbb{R}^{k+1}$ and the corresponding representation of elements of $GA(k, \mathbb{R})$ by particular $(k + 1) \times (k + 1)$-matrices also fail to be canonical. They depend on a point $v_0 \in V$ sent to vector $(0, \ldots, 0, 1) \in \mathbb{R}^{k+1}$. •

One can say something more if $Y \to X$ is a vector bundle provided with the natural structure of an affine bundle whose origin is the canonical zero section $\widehat{0}$. In this case, we have the canonical splitting of the exact sequence (7.6.1) such that $GL(k, \mathbb{R})$ is a subgroup of $GA(k, \mathbb{R})$ and $GA(k, \mathbb{R})$ is the semidirect product of $GL(k, \mathbb{R})$ and the group $T(k, \mathbb{R})$ of translations in $\mathbb{R}^k$. Given a $GA(k, \mathbb{R})$-principal bundle $AY \to X$, its affine structure group $GA(k, \mathbb{R})$ is always reducible to the linear subgroup $GL(k, \mathbb{R})$ since the quotient $GA(k, \mathbb{R})/GL(k, \mathbb{R})$ is a vector space $\mathbb{R}^k$ provided with the natural affine structure. The corresponding quotient bundle is isomorphic to the vector bundle $Y \to X$. There is the canonical injection of the linear frame bundle $LY \hookrightarrow AY$ onto the reduced $GL(k, \mathbb{R})$-principal subbundle of AY which corresponds to the zero section $\widehat{0}$ of $Y \to X$. In this case, every principal connection on the linear frame bundle LY gives rise to a principal connection on the affine frame bundle in accordance with Theorem 6.5.7. This is equivalent to the fact that any affine connection Γ on a vector bundle $Y \to X$ defines a linear connection $\overline{\Gamma}$ on $Y \to X$ and that every linear connection on $Y \to X$ can be seen as an affine one (see Section 2.5). Hence, any affine connection Γ on the vector bundle $Y \to X$ is represented by the sum of the associated linear connection $\overline{\Gamma}$ and a basic soldering form σ on $Y \to X$. Due to the vertical splitting (1.1.15), this soldering form is given by a global section of the tensor product $T^*X \otimes Y$.

Let now $Y \to X$ be the tangent bundle $TX \to X$ considered as an affine bundle. Then the relationship between affine and linear world connections on TX is the repetition of that we have said above. In particular, any affine world connection

$$K = dx^\lambda \otimes [K_\lambda{}^\alpha{}_\mu(x)\dot{x}^\mu + \sigma^\alpha_\lambda(x)]\partial_\alpha \tag{7.6.2}$$

on $TX \to X$ is represented by the sum of the associated linear world connection

$$\overline{K} = K_\lambda{}^\alpha{}_\mu(x)\dot{x}^\mu dx^\lambda \otimes \partial_\alpha \tag{7.6.3}$$

on $TX \to X$ and a basic soldering form

$$\sigma = \sigma^\alpha_\lambda(x) dx^\lambda \otimes \partial_\alpha \tag{7.6.4}$$

on $Y \to X$, which is the $(1,1)$-tensor field on X. For instance, if $\sigma = \theta_X$, we have the Cartan connection (2.5.4).

It is readily observed that the soldered curvature (2.3.8) of any soldering form (7.6.4) equals zero. Then we obtain from (2.3.11) that the torsion (2.5.3) of the affine connection K (7.6.2) with respect to σ (7.6.4) coincides with that of the associated linear connection $\overline{K}$ (7.6.3) and reads

$$\begin{aligned} &T = \frac{1}{2} S_\lambda{}^\alpha{}_\mu dx^\mu \wedge dx^\lambda \otimes \dot{\partial}_\alpha, \\ &T_\lambda{}^\alpha{}_\mu = K_\lambda{}^\alpha{}_\nu \sigma^\nu_\mu - K_\mu{}^\alpha{}_\nu \sigma^\nu_\lambda. \end{aligned} \tag{7.6.5}$$

The relation between the curvatures of the affine world connection K (7.6.2) and the associated linear connection $\overline{K}$ (7.6.3) is given by the general expression (2.3.12) where $\rho = 0$ and T is (7.6.5).

Remark 7.6.2. On may think on the physical meaning of the tensor field σ (7.6.4). Sometimes, it is mistakenly identified with a tetrad field, but, as we have seen, these fields have the different mathematical nature (see [161, 263] for a discussion). One can use $\sigma^\alpha_\lambda dx^\lambda$ as a non-holonomic coframes in the metric-affine gauge theory with non-holonomic GL_4 gauge transformations (see, e.g., [150]). In the gauge theory of dislocations in continuous media [164] and the analogous gauge model of the fifth force, the field σ is treated as an elastic distortion [261, 262, 263]). •

Chapter 8

Algebraic connections

In quantum theory, we almost never deal with bundles in their traditional geometric description as fibrations of manifolds, but with modules and sheaves of their sections. By this reason, jets and connections should be described in the same algebraic terms. This Chapter is devoted to the notion of connections on modules and sheaves over commutative algebras and rings (see [184]). Such an algebraic notion of connections is equivalent to the geometric one in the case of smooth vector bundles. Generalizing this construction to modules and sheaves over graded commutative algebras, we come to graded connections and superconnections in Chapter 9. The further generalization is connections on modules over non-commutative algebras. These are connections in non-commutative geometry studied in Chapter 14.

8.1 Jets of modules

We start from some basic elements of the differential calculus in modules [123, 184, 185] (the reader is referred to [193, 204] for algebraic theory of rings and modules; see also Section 14.1).

Let $\mathcal{K}$ be a commutative ring (i.e., a commutative associative unital $\mathbb{Z}$-algebra) and $\mathcal{A}$ a commutative unital $\mathcal{K}$-algebra, i.e., $\mathcal{A}$ is both a $\mathcal{K}$-module and a commutative ring (called sometimes a $\mathcal{K}$-ring). For the sake of simplicity, the reader can think of $\mathcal{A}$ as being a ring of real smooth functions on a manifold. Let P and Q be left $\mathcal{A}$-modules. Right modules are studied in a similar way. The set $\operatorname{Hom}_{\mathcal{K}}(P, Q)$ of $\mathcal{K}$-module homomorphisms of P into Q is endowed with the $\mathcal{A}-\mathcal{A}$-bimodule structure by the left and right multiplications

$$(a\phi)(p) = a\phi(p), \qquad (\phi \star a)(p) = \phi(ap), \qquad a \in \mathcal{A}, \qquad p \in P. \tag{8.1.1}$$

However, this is not a central $\mathcal{A}$-bimodule because $a\phi \neq \phi \star a$ in general. Let us denote

$$\delta_a \phi = a\phi - \phi \star a. \tag{8.1.2}$$

DEFINITION 8.1.1. An element $\Delta \in \mathrm{Hom}_{\mathcal{K}}(P, Q)$ is called an s-order *linear differential operator* from the $\mathcal{A}$-module P to the $\mathcal{A}$-module Q if

$$\delta_{a_0} \circ \cdots \circ \delta_{a_s} \Delta = 0$$

for arbitrary collections of $s+1$ elements of $\mathcal{A}$. It is also called a Q-valued differential operator on P. Throughout this Chapter, by differential operators are meant linear differential operators. □

Example 8.1.1. By virtue of Definition 8.1.1, a first order linear differential operator Δ obeys the condition

$$\boxed{\delta_a \circ \delta_b \Delta)(p) = \Delta(abp) - a\Delta(bp) - b\Delta(ap) + ab\Delta(p) = 0} \tag{8.1.3}$$

for all $p \in P$, $b, c \in \mathcal{A}$. •

The set $\mathrm{Diff}_s(P, Q) \subset \mathrm{Hom}_{\mathcal{K}}(P, Q)$ of s-order Q-valued differential operators on P is endowed with the $\mathcal{A} - \mathcal{A}$-bimodule structure (8.1.1). It is clear that

$$\mathrm{Diff}_s(P, Q) \subset \mathrm{Diff}_k(P, Q), \qquad k \geq s.$$

At the same time, one must distinguish the $\mathcal{A}$-bimodule $\mathrm{Diff}_s(P, Q)$ from the same set provided separately with the left $\mathcal{A}$-module structure and the right $\mathcal{A}$-module structure. We denote these modules by $\mathrm{Diff}_s^{\rightarrow}(P, Q)$ and $\mathrm{Diff}_s^{\leftarrow}(P, Q)$, respectively. For example, put $P = \mathcal{A}$ and consider the morphism

$$\mathfrak{D}_s : \mathrm{Diff}_s^{\leftarrow}(\mathcal{A}, Q) \to Q,$$
$$\mathfrak{D}_s(\Delta) \stackrel{\mathrm{def}}{=} \Delta(\mathbf{1}), \qquad \mathbf{1} \in \mathcal{A}.$$

This morphism is an s-order differential operator on the right module $\mathrm{Diff}_s^{\leftarrow}(\mathcal{A}, Q)$, and a 0-order differential operator on the left module $\mathrm{Diff}_s^{\rightarrow}(\mathcal{A}, Q)$.

THEOREM 8.1.2. For any differential operator $\Delta \in \mathrm{Diff}\,_s^{\leftarrow}(P,Q)$, there exists a unique homomorphism

$$\mathfrak{f}_\Delta : P \to \mathrm{Diff}\,_s^{\leftarrow}(\mathcal{A},Q),$$
$$[\mathfrak{f}_\Delta(p)](a) \stackrel{\mathrm{def}}{=} \Delta(ap), \qquad \forall a \in \mathcal{A},$$

such that the following diagram commutes:

$$\begin{array}{ccc} & \mathfrak{f}_\Delta & \\ P & \longrightarrow & \mathrm{Diff}\,_s^{\leftarrow}(\mathcal{A},Q) \\ & \Delta \searrow \swarrow \mathfrak{D}_s & \\ & Q & \end{array}$$

□

The correspondence $\Delta \mapsto \mathfrak{f}_\Delta$ defines the isomorphism

$$\boxed{\mathrm{Hom}\,_{\mathcal{A}}(P, \mathrm{Diff}\,_s^{\leftarrow}(\mathcal{A},Q)) = \mathrm{Diff}\,_s^{\leftarrow}(P,Q).} \tag{8.1.4}$$

In other words, every differential operator from an $\mathcal{A}$-module P to an $\mathcal{A}$-module Q is represented by a morphism of P to the module of differential operators from $\mathcal{A}$ to Q. One says that the $\mathcal{A}$-module $\mathrm{Diff}\,_s^{\leftarrow}(\mathcal{A},Q)$ is the *representative object* of the functor $P \to \mathrm{Diff}\,_s^{\leftarrow}(P,Q)$. Therefore, we can concentrate our attention mainly to Q-valued differential operators on the algebra $\mathcal{A}$.

DEFINITION 8.1.3. A first order differential operator ∂ from $\mathcal{A}$ to an $\mathcal{A}$-module Q is called the Q-valued *derivation* of the algebra $\mathcal{A}$ if it obeys the *Leibniz rule*

$$\boxed{\partial(aa') = a\partial(a') + a'\partial(a),} \qquad \forall a, a' \in \mathcal{A}. \tag{8.1.5}$$

This is a particular condition (8.1.3). □

Since $a\partial$, $\forall a \in \mathcal{A}$, is also a derivation, derivations constitute the submodule $\mathfrak{d}(\mathcal{A},Q)$ of the left $\mathcal{A}$-module $\mathrm{Diff}\,_1^{\rightarrow}(\mathcal{A},Q)$. At the same time, $\partial{\star}a$ is not a derivation in general. Therefore, $\mathfrak{d}(\mathcal{A},Q)$ is endowed with the structure of a right $\mathcal{K}$-module only. There exists the right $\mathcal{K}$-module monomorphism

$$i : \mathfrak{d}(\mathcal{A},Q) \to \mathrm{Diff}\,_1^{\leftarrow}(\mathcal{A},Q). \tag{8.1.6}$$

It is easily seen that a first order differential operator Δ belongs to $\mathfrak{d}(\mathcal{A},Q)$ if and only if $\Delta(\mathbf{1}) = 0$. Hence, we have the exact sequence of $\mathcal{K}$-modules

$$0 \longrightarrow \mathfrak{d}(\mathcal{A},Q) \stackrel{i}{\longrightarrow} \mathrm{Diff}\,_1^{\leftarrow}(\mathcal{A},Q) \longrightarrow Q \longrightarrow 0.$$

Remark 8.1.2. Let $i : P \to Q$ be an $\mathcal{A}$-submodule of the $\mathcal{A}$-module Q. Any P-valued derivation ∂ of $\mathcal{A}$ yields the Q-valued derivation $i \circ \partial$ of $\mathcal{A}$, and we obtain the homomorphism of the left $\mathcal{A}$-modules

$$\partial^1 i : \mathfrak{d}(\mathcal{A}, P) \to \mathfrak{d}(\mathcal{A}, Q). \tag{8.1.7}$$

A difficulty arises if P is not an $\mathcal{A}$-submodule of Q just as in the case of the injection (8.1.6). •

Let us apply the derivation functor (8.1.7) to the injection (8.1.6). The module $\mathfrak{d}(\mathcal{A}, \mathrm{Diff}_1^{\leftarrow}(\mathcal{A}, Q))$ consists of the derivations of $\mathcal{A}$ with values into the right $\mathcal{A}$-module $\mathrm{Diff}_1^{\leftarrow}(\mathcal{A}, Q)$. If $\partial \in \mathfrak{d}(\mathcal{A}, \mathrm{Diff}_1^{\leftarrow}(\mathcal{A}, Q))$, then $\partial(a) \in \mathrm{Diff}_1^{\leftarrow}(\mathcal{A}, Q)$, $\forall a \in \mathcal{A}$, such that

$$\partial(aa') = \partial(a) \star a' + \partial(a') \star a, \qquad \forall a, a' \in \mathcal{A}.$$

Let us consider the module $\mathfrak{d}(\mathcal{A}, \mathfrak{d}(\mathcal{A}, Q))$, where $\mathfrak{d}(\mathcal{A}, Q)$ is regarded as a left $\mathcal{A}$-module. Elements $\partial \in \mathfrak{d}(\mathcal{A}, \mathfrak{d}(\mathcal{A}, Q))$ satisfy the condition

$$\partial(aa') = a'\partial(a) + a\partial(a'), \qquad \forall a, a' \in \mathcal{A}.$$

Then it is easily verified that the intersection

$$\mathfrak{d}_2(\mathcal{A}, Q) = \mathfrak{d}(\mathcal{A}, \mathfrak{d}(\mathcal{A}, Q)) \cap \mathfrak{d}(\mathcal{A}, \mathrm{Diff}_1^{\leftarrow}(\mathcal{A}, Q))$$

consists of those elements of $\mathfrak{d}(\mathcal{A}, \mathrm{Diff}_1^{\leftarrow}(\mathcal{A}, Q))$ which obey the relation

$$(\partial(a))(a') = (\partial(a'))(a).$$

This is a left $\mathcal{A}$-module. We have the obvious monomorphism

$$\mathfrak{d}_2(\mathcal{A}, Q) \to \mathfrak{d}(\mathcal{A}, \mathrm{Diff}_1^{\leftarrow}(\mathcal{A}, Q)). \tag{8.1.8}$$

Set up inductively

$$\mathfrak{d}_{n+1}(\mathcal{A}, Q) \stackrel{\mathrm{def}}{=} \mathfrak{d}(\mathcal{A}, \mathfrak{d}_n(\mathcal{A}, Q)) \cap \mathfrak{d}(\mathcal{A}, (\mathrm{Diff}_1^{\leftarrow}(\mathcal{A}, P))^n),$$

where

$$(\mathrm{Diff}_1^{\leftarrow}(Q))^k \stackrel{\mathrm{def}}{=} \mathrm{Diff}_1^{\leftarrow}(\mathcal{A}, \cdots, \mathrm{Diff}_1^{\leftarrow}(\mathcal{A}, \mathrm{Diff}_1^{\leftarrow}(\mathcal{A}, Q)) \cdots).$$

The monomorphism (8.1.8) is generalized to higher order derivations as

$$\boxed{\mathfrak{d}_k(\mathcal{A},Q)\to\mathfrak{d}_{k-1}(\mathcal{A},\mathrm{Diff}\,\overleftarrow{_1}(\mathcal{A},Q)).} \tag{8.1.9}$$

Turn now to the modules of jets. Given an $\mathcal{A}$-module P, let us consider the tensor product $\mathcal{A}\underset{\mathcal{K}}{\otimes}P$ of $\mathcal{K}$-modules provided with the left $\mathcal{A}$-module structure

$$b(a\otimes p)\overset{\mathrm{def}}{=}(ba)\otimes p, \qquad \forall b\in\mathcal{A}. \tag{8.1.10}$$

For any $b\in\mathcal{A}$, we introduce the left $\mathcal{A}$-module morphism

$$\delta^b(a\otimes p)=(ba)\otimes p-a\otimes(bp). \tag{8.1.11}$$

Let μ^{k+1} be the submodule of the left $\mathcal{A}$-module $\mathcal{A}\underset{\mathcal{K}}{\otimes}P$ generated by all elements of the type

$$\delta^{b_0}\circ\cdots\circ\delta^{b_k}(\mathbf{1}\otimes p).$$

DEFINITION 8.1.4. The *k-order jet module* of the $\mathcal{A}$-module P is defined to be the quotient $\mathfrak{J}^k(P)$ of $\mathcal{A}\otimes P$ by μ^{k+1}. It is a left $\mathcal{A}$-module with respect to the multiplication

$$b(a\otimes p \bmod \mu^{k+1})=ba\otimes p \bmod \mu^{k+1}. \tag{8.1.12}$$

□

Besides the left $\mathcal{A}$-module structure induced by (8.1.10), the k-order jet module $\mathfrak{J}^k(P)$ also admits the left $\mathcal{A}$-module structure given by the multiplication

$$b\star(a\otimes p \bmod \mu^{k+1})=a\otimes(bp)\bmod \mu^{k+1}. \tag{8.1.13}$$

It is called the $\star$-*left module* structure. There is the $\star$-left $\mathcal{A}$-module homomorphism

$$\boxed{J^k: P\to\mathfrak{J}^k(P), \qquad J^kp=\mathbf{1}\otimes p \bmod \mu^{k+1},} \tag{8.1.14}$$

such that $\mathfrak{J}^k(P)$ as a left $\mathcal{A}$-module is generated by the elements J^kp, $p\in P$. It is readily observed that the homomorphism $\mathfrak{J}^k$ (8.1.14) is a k-order differential operator (compare the relation (8.1.3) and the relation (8.1.15) below).

Remark 8.1.3. If P is a $\mathcal{A}-\mathcal{A}$-bimodule, the tensor product $\mathcal{A}\underset{\mathcal{K}}{\otimes} P$ is also provided with the right $\mathcal{A}$-module structure

$$(a\otimes p)b \overset{\text{def}}{=} a\otimes pb, \qquad \forall b\in\mathcal{A},$$

and so is the jet module $\mathfrak{J}^k(P)$:

$$(a\otimes p \operatorname{mod}\mu^{k+1})b = a\otimes (pb)\operatorname{mod}\mu^{k+1}.$$

If P is a central bimodule, i.e.,

$$ap = pa, \quad \forall a\in\mathcal{A}, \quad p\in P,$$

the $\star$-left $\mathcal{A}$-module structure (8.1.13) is equivalent to the right $\mathcal{A}$-module structure (8.1.15). •

The jet modules possess the properties similar to those of jet manifolds. In particular, since $\mu^r\subset\mu^s$, $r>s$, there is the the inverse system of epimorphisms

$$\mathfrak{J}^s(P) \overset{\pi^s_{s-1}}{\longrightarrow} \mathfrak{J}^{s-1}(P) \longrightarrow\cdots \overset{\pi^1_0}{\longrightarrow} P.$$

Given the repeated jet module $\mathfrak{J}^s(\mathfrak{J}^k(P))$, there exists the monomorphism $\mathfrak{J}^{s+k}(P)\to\mathfrak{J}^s(\mathfrak{J}^k(P))$.

Example 8.1.4. The first order jet module $\mathfrak{J}^1(P)$ consists of elements $a\otimes p\operatorname{mod}\mu^2$, i.e., elements $a\otimes p$ modulo the relations

$$\begin{aligned}&\delta^a\circ\delta^b(\mathbf{1}\otimes p) = \\ &\qquad (\delta_a\circ\delta_b\mathfrak{J}^1)(p) = \mathbf{1}\otimes(abp) - a\otimes(bp) - b\otimes(ap) + ab\otimes p = 0.\end{aligned} \tag{8.1.15}$$

The morphism $\pi^1_0 : \mathfrak{J}^1(P)\to P$ reads

$$\boxed{\pi^1_0 : a\otimes p\operatorname{mod}\mu^2 \to ap.} \tag{8.1.16}$$

•

THEOREM 8.1.5. For any differential operator $\Delta\in\operatorname{Diff}^{\rightarrow}_s(P,Q)$ there is a unique homomorphism $\mathfrak{f}^\Delta : \mathfrak{J}^s(P)\to Q$ such that the diagram

$$\begin{array}{ccc} P & \overset{J^k}{\longrightarrow} & \mathfrak{J}^s(P) \\ & \Delta\searrow\;\swarrow\mathfrak{f}^\Delta & \\ & Q & \end{array}$$

is commutative. □

Proof. The proof is based on the following fact [185]. Let $h \in \mathrm{Hom}_{\mathcal{A}}(\mathcal{A}\otimes P, Q)$ and

$$\widehat{a} : P \ni p \to a\otimes p \in \mathcal{A}\otimes P,$$

then

$$\delta_b(h\circ\widehat{a})(p) = h(\delta^b(a\otimes p)).$$

QED

The correspondence $\Delta \mapsto \mathfrak{f}^\Delta$ defines the isomorphism

$$\boxed{\mathrm{Hom}_{\mathcal{A}}(\mathfrak{J}^s(P), Q) = \mathrm{Diff}_s(P,Q),} \tag{8.1.17}$$

which shows that the jet module $\mathfrak{J}^s(P)$ is the representative object of the functor $Q \to \mathrm{Diff}^{\to}_s(P,Q)$.

Let us consider the particular jet modules $\mathfrak{J}^s(\mathcal{A})$ of the algebra $\mathcal{A}$, denoted simply by $\mathfrak{J}^s$. The module $\mathfrak{J}^s$ can be provided with the structure of a commutative algebra with respect to the multiplication

$$(aJ^sb)\cdot(a'J^sb) = aa'J^s(bb').$$

In particular, the algebra $\mathfrak{J}^1$ consists of the elements $a\otimes b$ modulo the relations

$$\boxed{a\otimes b + b\otimes a = ab\otimes\mathbf{1} + \mathbf{1}\otimes ab.} \tag{8.1.18}$$

It has the left $\mathcal{A}$-module structure

$$c((a\otimes b) \bmod \mu^2) = (ca)\otimes b \bmod \mu^2 \tag{8.1.19}$$

(8.1.12) and the $\star$-left $\mathcal{A}$-module structure

$$c\star((a\otimes b) \bmod \mu^2) = a\otimes (cb) \bmod \mu^2 \tag{8.1.20}$$

(8.1.13) which coincides with the right $\mathcal{A}$-module structure (8.1.15) (see Remark 8.1.3). We have the canonical monomorphism of left $\mathcal{A}$-modules

$$i_1 : \mathcal{A} \to \mathfrak{J}^1, \qquad i_1 : a \mapsto a\otimes\mathbf{1} \bmod \mu^2, \tag{8.1.21}$$

and the corresponding projection

$$\boxed{\mathfrak{J}^1 \to \mathfrak{J}^1/\mathrm{Im}\, i_1 = (\mathrm{Ker}\,\mu^1)\mathrm{mod}\,\mu^2 = \mathfrak{O}^1,} \tag{8.1.22}$$
$$a \otimes b \,\mathrm{mod}\,\mu^2 \to (a \otimes b - ab \otimes \mathbf{1})\,\mathrm{mod}\,\mu^2.$$

The quotient $\mathfrak{O}^1$ (8.1.22) consists of the elements

$$(a \otimes b - ab \otimes \mathbf{1})\,\mathrm{mod}\,\mu^2, \qquad \forall a, b \in \mathcal{A}.$$

It is provided both with the central $\mathcal{A}$-bimodule structure

$$c((a \otimes b - ab \otimes \mathbf{1})\,\mathrm{mod}\,\mu^2) = (ca \otimes b - cab \otimes \mathbf{1})\,\mathrm{mod}\,\mu^2, \tag{8.1.23}$$
$$((\mathbf{1} \otimes ab - b \otimes a)\,\mathrm{mod}\,\mu^2)c = (\mathbf{1} \otimes abc - b \otimes ac)\,\mathrm{mod}\,\mu^2 \tag{8.1.24}$$

and the $\star$-left $\mathcal{A}$-module structure

$$c \star ((a \otimes b - ab \otimes \mathbf{1})\,\mathrm{mod}\,\mu^2) = (a \otimes cb - acb \otimes \mathbf{1})\,\mathrm{mod}\,\mu^2. \tag{8.1.25}$$

It is readily observed that the projection (8.1.22) is both the left and $\star$-left module morphisms. Then we have the $\star$-left module morphism

$$d^1 : \mathcal{A} \xrightarrow{J^1} \mathfrak{J}^1 \to \mathfrak{O}^1, \tag{8.1.26}$$
$$\boxed{d^1 : b \to \mathbf{1} \otimes b \,\mathrm{mod}\,\mu^2 \to (\mathbf{1} \otimes b - b \otimes \mathbf{1})\,\mathrm{mod}\,\mu^2,}$$

such that the central $\mathcal{A}$-bimodule $\mathfrak{O}^1$ is generated by the elements $d^1(b)$, $b \in \mathcal{A}$, in accordance with the law

$$ad^1b = (a \otimes b - ab \otimes \mathbf{1})\,\mathrm{mod}\,\mu^2 = (\mathbf{1} \otimes ab) - b \otimes a)\,\mathrm{mod}\,\mu^2 = (d^1b)a. \tag{8.1.27}$$

PROPOSITION 8.1.6. The morphism d^1 (8.1.26) is a derivation from $\mathcal{A}$ to $\mathfrak{O}^1$ seen both as a left $\mathcal{A}$-module and $\mathcal{A}$-bimodule. □

Proof. Using the relations (8.1.18), one obtains in an explicit form that

$$d^1(ba) = (\mathbf{1} \otimes ba - ba \otimes \mathbf{1})\,\mathrm{mod}\,\mu^2 =$$
$$(b \otimes a + a \otimes b - ba \otimes \mathbf{1} - ab \otimes \mathbf{1})\,\mathrm{mod}\,\mu^2 = bd^1a + ad^1b. \tag{8.1.28}$$

This is a $\mathfrak{O}^1$-valued first order differential operator. At the same time,

$$d^1(ba) = (\mathbf{1} \otimes ba - ba \otimes \mathbf{1} + b \otimes a - b \otimes a)\,\mathrm{mod}\,\mu^2 = (d^1b)a + bd^1a.$$

QED

With the derivation d^1 (8.1.26), we get the left and $\star$-left module splitting

$$\mathfrak{J}^1 = \mathcal{A} \oplus \mathfrak{O}^1, \tag{8.1.29}$$
$$a\mathfrak{J}^1(cb) = ai_1(cb) + ad^1(cb). \tag{8.1.30}$$

Accordingly, there is the exact sequence

$$\boxed{0 \to \mathfrak{O}^1 \to \mathfrak{J}^1 \to \mathcal{A} \to 0} \tag{8.1.31}$$

which is split by the monomorphism (8.1.21).

PROPOSITION 8.1.7. There is the isomorphism

$$\boxed{\mathfrak{J}^1(P) = \mathfrak{J}^1 \otimes P,} \tag{8.1.32}$$

where by $\mathfrak{J}^1 \otimes P$ is meant the tensor product of the right ($\star$-left) $\mathcal{A}$-module $\mathfrak{J}^1$ (8.1.20) and the left $\mathcal{A}$-module P, i.e.,

$$[a \otimes b \bmod \mu^2] \otimes p = [a \otimes \mathbf{1} \bmod \mu^2] \otimes bp.$$

□

Proof. The isomorphism (8.1.32) is given by the assignment

$$(a \otimes bp) \bmod \mu^2 \leftrightarrow [a \otimes b \bmod \mu^2] \otimes p. \tag{8.1.33}$$

QED

The isomorphism (8.1.29) leads to the isomorphism

$$\boxed{\mathfrak{J}^1(P) = (\mathcal{A} \oplus \mathfrak{O}^1) \otimes P,}$$
$$(a \otimes bp) \bmod \mu^2 \leftrightarrow [(ab + ad^1(b)) \bmod \mu^2] \otimes p,$$

and to the splitting of left and $\star$-left $\mathcal{A}$-modules

$$\mathfrak{J}^1(P) = (\mathcal{A} \otimes P) \oplus (\mathfrak{O}^1 \otimes P), \tag{8.1.34}$$

Applying the projection π_0^1 (8.1.16) to the splitting (8.1.34), we obtain the exact sequence of left and $\star$-left $\mathcal{A}$-modules

$$\boxed{0 \longrightarrow \mathfrak{O}^1 \otimes P \to \mathfrak{J}^1(P) \xrightarrow{\pi_0^1} P \longrightarrow 0,} \tag{8.1.35}$$
$$0 \to [(a \otimes b - ab \otimes \mathbf{1}) \bmod \mu^2] \otimes p \to [(c \otimes \mathbf{1} + a \otimes b - ab \otimes \mathbf{1}) \bmod \mu^2] \otimes p$$
$$= (c \otimes p + a \otimes bp - ab \otimes p) \bmod \mu^2 \to cp,$$

similar to the exact sequence (8.1.31). This exact sequence has the canonical splitting by the $\star$-left $\mathcal{A}$-module morphism

$$P \ni ap \mapsto a \otimes p + d^1(a) \otimes p.$$

However, the exact sequence (8.1.35) needs not be split by a left $\mathcal{A}$-module morphism. Its splitting by a left $\mathcal{A}$-module morphism (see (8.2.1) below) implies a connection. On can treat the canonical splitting (8.1.21) of the exact sequence (8.1.31) as being the canonical connection on the algebra $\mathcal{A}$.

In the case of $\mathfrak{J}^s$, the isomorphism (8.1.17) takes the form

$$\mathrm{Hom}_{\mathcal{A}}(\mathfrak{J}^s, Q) = \mathrm{Diff}_s(\mathcal{A}, Q). \tag{8.1.36}$$

Then Theorem 8.1.5 and Proposition 8.1.6 lead to the isomorphism

$$\boxed{\mathrm{Hom}_{\mathcal{A}}(\mathfrak{O}^1, Q) = \mathfrak{d}(\mathcal{A}, Q).} \tag{8.1.37}$$

In other words, any Q-valued derivation of $\mathcal{A}$ is represented by the composition $h \circ d^1$, $h \in \mathrm{Hom}_{\mathcal{A}}(\mathfrak{O}^1, Q)$, due to the property $d^1(\mathbf{1}) = 0$.

Example 8.1.5. If $Q = \mathcal{A}$, the isomorphism (8.1.37) reduces to the duality relation

$$\boxed{\mathrm{Hom}_{\mathcal{A}}(\mathfrak{O}^1, \mathcal{A}) = \mathfrak{d}(\mathcal{A}),} \tag{8.1.38}$$
$$u(a) = u(d^1 a), \qquad a \in \mathcal{A},$$

i.e., the module $\mathfrak{d}\mathcal{A}$ coincides with the left $\mathcal{A}$-dual $\mathfrak{O}^{1*}$ of $\mathfrak{O}^1$. •

Let us define the modules $\mathfrak{O}^k$ as the skew tensor products of the $\mathcal{K}$-modules $\mathfrak{O}^1$.

PROPOSITION 8.1.8. [185]. There are the isomorphisms

$$\mathrm{Hom}_{\mathcal{A}}(\mathfrak{O}^k, Q) = \mathfrak{d}_k(\mathcal{A}, Q), \tag{8.1.39}$$
$$\mathrm{Hom}_{\mathcal{A}}(\mathfrak{J}^1(\mathfrak{O}^k), Q) = \mathfrak{d}_k(\mathrm{Diff}_1^{\leftarrow}(Q)). \tag{8.1.40}$$

□

The isomorphism (8.1.39) is the higher order extension of the isomorphism (8.1.37). It shows that the module $\mathfrak{O}^k$ is a representative object of the derivation functor $Q \to \mathfrak{d}_k(\mathcal{A}, Q)$.

The monomorphism (8.1.9) and the isomorphism (8.1.40) imply the homomorphism

$$h^k : \mathfrak{J}^1(\mathfrak{O}^{k-1}) \to \mathfrak{O}^k$$

and define the operators of exterior differentiation

$$d^k = h^k \circ J^1 : \mathfrak{O}^{k-1} \to \mathfrak{O}^k. \tag{8.1.41}$$

These operators constitute the De Rham complex

$$0 \longrightarrow \mathcal{A} \xrightarrow{d^1} \mathfrak{O}^1 \xrightarrow{d^2} \cdots \mathfrak{O}^k \xrightarrow{d^{k+1}} \cdots. \tag{8.1.42}$$

Remark 8.1.6. Let an $\mathcal{A}$-module P be a $\mathcal{K}$-ring such that there is the monomorphism $\mathcal{A} \to P$. Point out the essential difference between the jet modules $\mathfrak{J}^k(P)$ of P as an $\mathcal{A}$-module and the jet modules $\mathfrak{J}^k$ of P as a $\mathcal{K}$-ring. In particular, we have the canonical monomorphism (8.1.21) of P to $\mathfrak{J}^1$, but not to $\mathfrak{J}^k(P)$. •

Let us turn now to the case when $\mathcal{A}$ is the ring $C^\infty(X)$ of smooth functions on a manifold X.

Remark 8.1.7. Whenever referring to a topology on the ring $C^\infty(X)$, we will mean the topology of compact convergence for all derivatives. $C^\infty(X)$ is a *Fréchet ring* with respect to this topology. Recall that a Fréchet space is a complete locally convex metrizable topological real vector space (see, e.g., [254]). •

To obtain a geometric realization of the modules over the ring $C^\infty(X)$, one should consider the subcategory of the geometric modules.

DEFINITION 8.1.9. The $\mathfrak{O}^0(X)$-module P is called a *geometric module* if

$$\bigcap_{x\in X} \mu_x P = 0,$$

where by μ_x is meant the maximal ideal of functions vanishing at a point $x \in X$. □

For the sake of brevity, one can say that elements of the geometric modules over $C^\infty(X)$ are defined only by their values at points of the manifold X. Every such a $C^\infty(X)$-module P is identified with the module $Y(X)$ of global sections of some vector bundle $Y \to X$ (which is not necessarily finite-dimensional); P is called the *structure module* of the vector bundle Y. The fibre of this bundle over x is the quotient $\mathbb{R}$-module $P/\mu_x P$.

Remark 8.1.8. A differentiable manifold can be reconstructed as the real spectrum of its ring of smooth functions. Let Z be a manifold and $\mu_z \subset C^\infty(Z)$ the maximal ideal of functions vanishing at a point $z \in Z$. We have $C^\infty(Z)/\mu_z = \mathbb{R}$. The *real spectrum* $\mathrm{Spec}_R C^\infty(Z)$ of $C^\infty(Z)$ is called the set of all maximal ideals μ of $C^\infty(Z)$ such that

$$\mathbb{R} \hookrightarrow C^\infty(Z) \to C^\infty(Z)/\mu$$

is an isomorphism (see, e.g., [11]). If the real spectrum $\mathrm{Spec}_R C^\infty(Z)$ is provided with the Zariski topology (which coincides with the Gelfand topology), then the map

$$Z \ni z \mapsto \mu_z \in \mathrm{Spec}_R C^\infty(Z)$$

is a homeomorphism. The spectrum and the real spectrum of a graded commutative rings can also be defined and provided with the corresponding Zariski topology [20].

It should be emphasized that, if X and X' are differentiable manifolds, the natural morphism

$$C^\infty(X) \otimes C^\infty(X') \to C^\infty(X \times X'),$$
$$f(x) \otimes f'(x') \mapsto f(x) f'(x'),$$

induces an isomorphism of Fréchet $\mathbb{R}$-algebras

$$C^\infty(X) \widehat{\otimes} C^\infty(X') \cong C^\infty(X \times X'), \tag{8.1.43}$$

where the left-hand side is the completion of $C^\infty(X) \otimes C^\infty(X')$ with respect to Grothendieck's topology. Recall that, if E and E' are locally convex vector spaces, there is a unique locally convex topology on $E \otimes E'$, called *Grothendieck's topology* such that, for any locally convex vector space F, the continuous linear maps $E \otimes E' \to F$ are in natural one-to-one correspondence with the continuous bilinear maps $E \times E \to F$. This is the finest topology under which the canonical mapping of $E \times E'$

to $E \otimes E'$ is continuous [254]. In particular, if E is a topological algebra, one can write the multiplication operation as the morphism $E\widehat{\otimes}E \to E$. •

Let us restrict our consideration to the subcategory of *locally free finite $C^\infty(X)$-modules*, which is equivalent to the category of smooth finite-dimensional vector bundles on X [184, 303]. In this case, we have the following identifications.

- The module $\mathfrak{d}(C^\infty(X))$ is identified with the $C^\infty(X)$-module $\mathcal{T}(X)$ of vector fields on the manifold X.
- The module $\mathfrak{O}^1$ coincides with the module $\mathfrak{O}^1(X)$ of 1-forms on X.
- The operator d^k (8.1.41) is the familiar exterior differential of exterior forms on X.
- If $Y(X)$ is the structure module of sections of a vector bundle $Y \to X$, the modules of jets $\mathfrak{J}^k(Y(X))$ are identified with the modules $J^kY(X)$ of sections of the jet bundles $J^kY \to X$.
- The jet functor J^k (8.1.14) is exactly the ordinary functor of the k-order jet prolongation. Namely, if $Y(X)$ is a structure module of a vector bundle $Y \to X$ and $s \in P$, then J^ks is the k-order jet prolongation of s.

If X is compact, the Serre–Swan theorem states the equivalence between the category of projective $C^\infty(X)$-modules of finite rank and smooth complex vector bundles over X [168, 287, 300] (see Theorem 14.1.1 below).

8.2 Connections on modules

To introduce the notion of a connection in the category of $\mathcal{A}$-modules, let us return to the exact sequence (8.1.35). It has no canonical splitting. Moreover, it needs not be split in general.

DEFINITION 8.2.1. By a *connection* on a $\mathcal{A}$-module P is called a left $\mathcal{A}$-module morphism

$$\boxed{\Gamma : P \to \mathfrak{J}^1(P),} \tag{8.2.1}$$

$$\Gamma(ap) = a\Gamma(p), \tag{8.2.2}$$

which splits the exact sequence (8.1.35). □

This splitting reads

$$J^1p = \Gamma(p) + \nabla^\Gamma(p), \tag{8.2.3}$$

where ∇^Γ is the complementary morphism

$$\boxed{\nabla^\Gamma : P \to \mathfrak{O}^1 \otimes P,} \tag{8.2.4}$$
$$\nabla^\Gamma(p) = \mathbf{1} \otimes p \operatorname{mod} \mu^2 - \Gamma(p).$$

This complementary morphism makes the sense of a *covariant differential* on the module P, but we will follow the tradition to use the terms "covariant differential" and "connection" on modules and sheaves synonymously. With the relation (8.2.2), we find that ∇^Γ obeys the *Leibniz rule*

$$\boxed{\nabla^\Gamma(ap) = da \otimes p + a\nabla^\Gamma(p).} \tag{8.2.5}$$

DEFINITION 8.2.2. By a *connection* on a $\mathcal{A}$-module P is meant any morphism ∇ (8.2.4) which obeys the Leibniz rule (8.2.5), i.e., ∇ is a $(\mathfrak{O}^1 \otimes P)$-valued first order differential operator on P. □

In view of Definition (8.2.2) and of the isomorphism (8.1.34), it is more convenient to rewrite the exact sequence (8.1.35) into the form

$$\boxed{0 \to \mathfrak{O}^1 \otimes P \to (\mathcal{A} \oplus \mathfrak{O}^1) \otimes P \to P \to 0.} \tag{8.2.6}$$

Then a connection ∇ on P can be defined as a left $\mathcal{A}$-module splitting of this exact sequence.

Let us show that, in the case of structure modules of vector bundles, the above notion of a connection on modules is adequate to a familiar connection on a vector bundle.

PROPOSITION 8.2.3. If $Y \to X$ is a vector bundle, there exists the exact sequence of vector bundles over X

$$\boxed{0 \to T^*X \underset{X}{\otimes} Y \xrightarrow{\epsilon} J^1Y \to Y \to 0,} \tag{8.2.7}$$

where the morphism ϵ is given by the coordinate expression

$$y^i \circ \epsilon = 0, \qquad y^i_\lambda \circ \epsilon = \overline{y}^i_\lambda$$

with respect to the coordinates $(x^\lambda, \overline{y}^i_\lambda)$ on $T^*X \otimes Y$. □

The proof is based on the inspection of coordinate transformation laws.

Due to the canonical vertical splitting (1.1.15) of VY, the exact sequence (8.2.7) gives rise to the exact sequence of fibre bundles over Y

$$0 \to T^*X \underset{Y}{\otimes} VY \xrightarrow{\epsilon} J^1Y \to VY \to 0, \tag{8.2.8}$$

where $y^i_\lambda \circ \epsilon = \overline{y}^i_\lambda$ with respect to the coordinates $(x^\lambda, y^i, \overline{y}^i_\lambda)$ on $T^*X \otimes VY$. It is readily observed that any splitting over Y of the exact sequence (8.2.8) yields a splitting of the exact sequence (8.2.7), and *vice versa*. It is clear that any linear connection Γ on the vector bundle $Y \to X$ yields a desired splitting of the exact sequence (8.2.8)

$$\Gamma : VY = Y \times Y \underset{Y}{\to} J^1Y,$$
$$J^1Y = \Gamma(VY) \oplus D_\Gamma(J^1Y), \tag{8.2.9}$$

where D_Γ is the covariant differential (2.2.7), and *vice versa*. Since J^1Y is both an affine subbundle of the tensor bundle $T^*X \otimes TY$ and a vector bundle over X, its elements over $x \in X$ are vectors

$$y^i e_i + dx^\lambda(\partial_\lambda + y^i_\lambda e_i),$$

written with respect to fibre bases $\{e_i\}$ for the vector bundle $Y \to X$ and to the holonomic bases $\{\partial_i = e_i\}$ for VY. Then the corresponding splitting of the exact sequence (8.2.7) reads

$$i_Y + \Gamma : y^i e_i \mapsto y^i e_i \oplus \Gamma,$$
$$y^i e_i + dx^\lambda(\partial_\lambda + y^i_\lambda e_i) = y^i e_i \oplus \Gamma \oplus D_\Gamma. \tag{8.2.10}$$

The exact sequence of vector bundles (8.2.7) implies the exact sequence of the modules of their sections

$$\boxed{0 \to \mathfrak{O}^1(X) \otimes Y(X) \to J^1Y(X) \to Y(X) \to 0.} \tag{8.2.11}$$

One can derive this result from Theorem 1.1.4. Moreover, every splitting of the exact sequence (8.2.7) define a splitting the exact sequence (8.2.11), and *vice versa*. Given the splitting (8.2.10) of the exact sequence (8.2.7) of vector bundles by means of a linear connection Γ, the corresponding splitting of the exact sequence (8.2.11) is

$$Y(X) \ni s \mapsto s \oplus \Gamma \circ s \in J^1Y(X),$$
$$s + J^1 s = s \oplus \Gamma \circ s \oplus \nabla^\Gamma s,$$

where ∇^Γ is the covariant differential (2.2.8) with respect to the connection Γ. It is a $C^\infty(X)$-module morphism

$$\nabla^\Gamma : Y(X) \to \mathfrak{O}^1(X) \otimes Y(X) \tag{8.2.12}$$

which satisfies the Leibniz rule

$$\nabla(fs) = df \otimes s + f\nabla(s), \quad f \in C^\infty(X), \quad s \in Y(X),$$

and splits the exact sequence

$$\boxed{0 \to \mathfrak{O}^1(X) \otimes Y(X) \to (C^\infty(X) \oplus \mathfrak{O}^1(X)) \otimes Y(X) \to Y(X) \to 0} \tag{8.2.13}$$

(cf. (8.2.6)).

It should be emphasized that, in contrast with the case of vector bundles and their structure modules, an arbitrary exact sequence of modules need not admit a splitting, and it may happen that a connection on a module fails to exist.

The morphism (8.2.4) can be extended naturally to the morphism

$$\nabla : \mathfrak{O}^1 \otimes P \to \mathfrak{O}^2 \otimes P.$$

Then it is readily observed that, in the case of a structure module $P = Y(X)$ of a vector bundle $Y \to X$, the morphism

$$\boxed{R = \nabla^2 : P \to \mathfrak{O}^2 \otimes P} \tag{8.2.14}$$

restarts the curvature (2.4.3) of a linear connection.

DEFINITION 8.2.4. The morphism R (8.2.14) is called the *curvature* of a connection ∇ on a module P. □

In the case of the ring $C^\infty(X)$ and a locally free $C^\infty(X)$-module $\mathcal{S}$ of finite rank, there exist the isomorphisms

$$\begin{aligned} &\mathfrak{O}^1(X) = \mathrm{Hom}_{C^\infty(X)}(\mathfrak{d}(C^\infty(X)), C^\infty(X)), \\ &\mathrm{Hom}_{C^\infty(X)}(\mathfrak{d}(C^\infty(X)), \mathcal{S}) = \mathfrak{O}^1(X) \otimes \mathcal{S}. \end{aligned} \tag{8.2.15}$$

With these isomorphisms, we come to other equivalent definitions of a connection on modules.

DEFINITION 8.2.5. Any morphism

$$\boxed{\nabla : \mathcal{S} \to \mathrm{Hom}_{C^\infty(X)}(\mathfrak{d}(C^\infty(X)), \mathcal{S})} \tag{8.2.16}$$

satisfying the Leibniz rule (8.2.5) is called a connection on a $C^\infty(X)$-module $\mathcal{S}$. □

DEFINITION 8.2.6. By a connection on a $C^\infty(X)$-module $\mathcal{S}$ is meant a $C^\infty(X)$-module morphism

$$\boxed{\mathfrak{d}(C^\infty(X)) \ni \tau \mapsto \nabla_\tau \in \mathrm{Diff}\,_1^{\to}(\mathcal{S}, \mathcal{S})} \tag{8.2.17}$$

such that the first order differential operators ∇_τ obey the rule

$$\boxed{\nabla_\tau(fs) = (\tau\rfloor df)s + f\nabla_\tau s.} \tag{8.2.18}$$

□

Using the expression (2.4.4) for the curvature of a linear connection, one can define the *curvature* of the connection (8.2.17) as a 0-order differential operator on the module $\mathcal{S}$

$$\boxed{R(\tau, \tau') = [\nabla_\tau, \nabla_{\tau'}] - \nabla_{[\tau,\tau']}} \tag{8.2.19}$$

for any two vector fields $\tau, \tau' \in \mathfrak{d}(C^\infty(X))$. In the case of a structure module of a vector bundle, we restart the curvature (2.4.4).

If a $\mathcal{S}$ is a commutative $C^\infty(X)$-ring, Definition 8.2.6 can be modified as follows.

DEFINITION 8.2.7. By a *connection* on $C^\infty(X)$-ring $\mathcal{S}$ is meant any $C^\infty(X)$-module morphism

$$\boxed{\mathfrak{d}(C^\infty(X)) \ni \tau \mapsto \nabla_\tau \in \mathfrak{d}\mathcal{S}} \tag{8.2.20}$$

which is a connection on $\mathcal{S}$ as a $C^\infty(X)$-module, i.e., obeys the Leinbniz rule (8.2.18). □

In Definition 8.2.7, we require additionally of ∇_τ to be a derivation of $\mathcal{S}$ in order to maintain its algebra structure. Two such connections ∇_τ and ∇'_τ differ from each other in a derivation of the ring $\mathcal{S}$ which vanishes on $C^\infty(X) \subset \mathcal{S}$. The curvature of the connection (8.2.20) is given by the formula (8.2.19).

8.3 Connections on sheaves

There are several equivalent definitions of sheaves [40, 157, 292]. We will start from the following one. A *sheaf* on a topological space X is a topological fibre bundle $S \to X$ whose fibres, called the *stalks*, are Abelian groups S_x provided with the discrete topology.

A *presheaf* on a topological space X is defined if an Abelian group S_U corresponds to every open subset $U \subset X$ ($S_\emptyset = 0$) and, for any pair of open subsets $V \subset U$, there is the *restriction homomorphism*

$$r^U_V : S_U \to S_V$$

such that

$$r^U_U = \mathrm{Id}\, S_U, \qquad r^U_W = r^V_W r^U_V, \qquad W \subset V \subset U.$$

Every presheaf $\{S_U, r^U_V\}$ on a topological space X yields a sheaf on X whose stalk S_x at a point $x \in X$ is the *direct limit* of the Abelian groups S_U, $x \in U$, with respect to the restriction homomorphisms r^U_V. We refer the reader to [217] and also to Section 11.1 for the general notion of a direct limit. Here, by a direct limit is meant that, for each open neighbourhood U of a point x, every element $s \in S_U$ determines an element $s_x \in S_x$, called the *germ* of s at x. Two elements $s \in S_U$ and $s' \in S'_V$ define the same germ at x if and only if there is an open neighbourhood $W \ni x$ such that

$$r^U_W s = r^V_W s'.$$

Example 8.3.1. For instance, let X be a topological space, $C^0(U)$ the additive Abelian group of all continuous real functions on $U \subset X$, while the homomorphism

$$r^U_V : C^0(U) \to C^0(V)$$

is the restriction of these functions to $V \subset U$. Then $\{C^0(U), r^U_V\}$ is a presheaf. Two real functions s and s' on X define the same germ s_x if they coincide on an open neighbourhood of x. The sheaf C^0_X generated by the presheaf $\{C^0(U), r^U_V\}$ is called the *sheaf of continuous functions*. The *sheaf* C^∞_X *of smooth functions* on a manifold X is defined in a similar way. Let us also mention the presheaf of constant real functions on open subsets of X. The corresponding sheaf is called a *constant sheaf*.

•

Two different presheaves may generate the same sheaf. Conversely, every sheaf S defines a presheaf of Abelian groups $S(U)$ of its local sections. This presheaf is called the *canonical presheaf* of the sheaf S. It is easily seen that the sheaf generated by the canonical presheaf $\{S(U), r^U_V\}$ of the sheaf S coincides with S. Therefore, we will identify sometimes sheaves and canonical presheaves. Note that, if a sheaf S is constructed from a presheaf $\{S_U, r^U_V\}$, there is the natural homomorphism $S_U \to S(U)$ which however is neither a monomorphism nor an epimorphism.

The direct sum, the tensor product and homomorphisms of sheaves on the same topological space are defined in a natural way.

Let S and S' be sheaves on the same topological space X and $\mathrm{Hom}\,(S\mid_U, S'\mid_U)$ the Abelian group of sheaf homomorphisms $S\mid_U \to S'\mid_U$ for any open subset $U \subset X$. These groups define the sheaf $Hom(S, S')$ on X. It should be emphasized that

$$Hom(S, S')(U) \neq \mathrm{Hom}\,(S(U), S'(U)). \tag{8.3.1}$$

Let $\varphi : X \to X'$ be a continuous map. Given a sheaf S on X, the *direct image* $\varphi_* S$ on X' of the sheaf S is given by the assignment $X' \supset U \mapsto S(f^{-1}(U))$ since $\varphi^{-1}(U)$ is an open set in X. To define the *inverse image* $\varphi^* S'$ on X of a sheaf S' on X', we correspond to any open set $V \subset X$ the limit of $S'(U)$ over all open subsets $U \subset X'$, $V \subset \varphi^{-1}(U)$, with respect to the inclusion (recall that $\varphi(V)$ is not open in Y in general). In particular, if $S' = C^\infty_{X'}$ is the sheaf of smooth functions on X in Example 8.3.1, its inverse image $\varphi^* C^\infty_{X'}$ is the sheaf of pull-back functions on X.

The notion of a sheaf is extended to sheaves of modules, commutative rings and graded commutative algebras [146].

Given a sheaf $\mathcal{A}$ on a topological space X, the pair $(X, \mathcal{A})$ is called a *locally ringed space* if every stalk $\mathcal{A}_x$, $x \in X$, is a *local* commutative ring, i.e., has a unique maximal ideal. By a morphism of locally ringed spaces $(X, \mathcal{A}) \to (X', \mathcal{A}')$ is meant a pair (φ, Φ) of a continuous map $\varphi : X \to X'$ and a sheaf morphism $\Phi : \mathcal{A}' \to \varphi_* \mathcal{A}$. A morphism (φ, Φ) is said to be:

- a monomorphism if φ is an injection and Φ is a surjection,
- an epimorphism if φ is a surjection, while Φ is an injection.

By a *sheaf* $\mathfrak{d}\mathcal{A}$ *of derivations* of $\mathcal{A}$ is meant a subsheaf of endomorphisms of $\mathcal{A}$ such that any section u of $\mathfrak{d}\mathcal{A}$ over an open subset $U \subset X$ is a derivation of the ring $\mathcal{A}(U)$. It should be emphasized that, because of the inequality (8.3.1), the converse property is not true in general. A derivation of the ring $\mathcal{A}(U)$ needs not be a section of the sheaf $\mathfrak{d}\mathcal{A}|_U$ because it may happen that, given open sets $U' \subset U$, there is no restriction morphism $\mathfrak{d}(\mathcal{A}(U)) \to \mathfrak{d}(\mathcal{A}(U'))$.

Let $(X, \mathcal{A})$ be a locally ringed space. A sheaf P on X is called a *sheaf of $\mathcal{A}$-modules* if every stalk P_x, $x \in X$, is an $\mathcal{A}_x$-module or, equivalently, if $P(U)$ is an $\mathcal{A}(U)$-module for any open subset $U \subset X$. A sheaf of $\mathcal{A}$-modules P is said to be *locally free*, if there exists an open neighbourhood U of every point $x \in X$ such that $P(U)$ is a free $\mathcal{A}(U)$-module. If all these free modules are of the same rank, one says that P is a sheaf of locally free modules of *constant rank*.

Example 8.3.2. The sheaf C^∞_X of smooth functions on a manifold X in Example 8.3.1 is a sheaf of commutative rings. The stalk of germs of these functions C^∞_{Xx} at a point x is a local ring, and the pair (X, C^∞_X) is a locally ringed space. In particular, every manifold morphism $\varphi : X \to X'$ yields the pull-back morphism (φ, Φ) of locally ringed spaces $(X, C^\infty_X) \to (X', C^\infty_{X'})$, where

$$\Phi(C^\infty_{X'}) = (\varphi_* \circ \varphi^*)(C^\infty_{X'}) \subset \varphi_*(C^\infty_X). \qquad (8.3.2)$$

We come to the following alternative definition of a smooth manifold [20].

PROPOSITION 8.3.1. Let X be a topological space, $\{U_\zeta\}$ an open covering of X, and S_ζ a sheaf on U_ζ for every U_ζ. Let us assume that:

- if $U_\zeta \cap U_\xi \neq \emptyset$, there is a sheaf isomorphism

$$\rho_{\zeta\xi} : S_\xi\, |_{U_\zeta \cap U_\xi} \to S_\zeta\, |_{U_\zeta \cap U_\xi};$$

- these sheaf isomorphisms fulfill the cocycle condition

$$\rho_{\xi\zeta} \circ \rho_{\zeta\iota}(S_\iota\, |_{U_\zeta \cap U_\xi \cap U_\iota}) = \rho_{\xi\iota}(S_\iota\, |_{U_\zeta \cap U_\xi \cap U_\iota})$$

for every triple U_ζ, U_ξ, U_ι.

Then there exist a sheaf S on X and sheaf isomorphisms $\phi_\zeta : S|_{U_\zeta} \to S_\zeta$ such that

$$\phi_\zeta|_{U_\zeta \cap U_\xi} = \rho_{\zeta\xi} \circ \phi_\xi|_{U_\zeta \cap U_\xi} .$$

□

PROPOSITION 8.3.2. Let X be a paracompact topological space, and let $(X, \mathcal{A})$ be a locally ringed space which is locally isomorphic to $(\mathbb{R}^n, C^\infty_{R^n})$. Then X is an n-dimensional differentiable manifold, and there is a natural isomorphism of locally ringed spaces $(X, \mathcal{A})$ and (X, C^∞_X). □

Let $X \times X'$ be the product of two paracompact topological spaces. For any two open subsets $U \subset X$ and $U' \subset X'$, let us consider the topological tensor product of rings $C^\infty(U)\widehat{\otimes}C^\infty(U')$ (see Remark 8.1.8). These tensor products define a sheaf of rings on $X \times X$ which we denote by $C^\infty_X\widehat{\otimes}C^\infty_{X'}$. Due to the metric isomorphism (8.1.43) written for all $U \subset X$ and $U' \subset X'$, we obtain the sheaf isomorphism

$$\boxed{C^\infty_X\widehat{\otimes}C^\infty_{X'} = C^\infty_{X\times X'}.} \tag{8.3.3}$$

•

Example 8.3.3. Let $Y \to X$ be a vector bundle. The germs of its sections make up the sheaf S_Y of sections of $Y \to X$. The stalk S_{Yx} of this sheaf at a point $x \in X$ consists of the germs at x of sections of $Y \to X$ in a neighbourhood of a point x. The stalk S_{Yx} is a module over the ring C^∞_{Xx} of the germs at $x \in X$ of smooth functions on X. Hence, S_Y is a sheaf of modules over the sheaf C^∞_X of rings with respect to pointwise operations. The canonical presheaf of S_Y is isomorphic to the presheaf of local sections of the vector bundle $Y \to X$. It is called the *structure sheaf* of the vector bundle $Y \to X$. Similarly to manifolds, a vector bundle can be characterized by its structure sheaf, locally isomorphic to the sheaf $C^\infty_U \otimes \mathbb{R}^m$. In order to glue these local sheaves and to yield a globally defined sheaf in accordance with Proposition 8.3.1, we need a cocycle of transition functions which is an element of the first cohomology set $H^1(X, GL(m,\mathbb{R})_\infty)$ with coefficients in the sheaf $GL(m,\mathbb{R})_\infty$ of smooth mapping from X into $GL(m,\mathbb{R})$ (see Remark 6.9.2). Given a structure sheaf S, the fibre Y_x of a vector bundle at a point $x \in X$ is the quotient $S_x/\mathcal{M}_x$ of the stalk S_x by the submodule $\mathcal{M}_x$ of germs in S_x whose evaluation vanishes.

Similarly, the sheaf of sections of any fibre bundle $Y \to X$ can be defined. It is a *sheaf of sets*, which have no any algebraic structure. •

Let $(X, \mathcal{A})$ be a locally ringed space and P a sheaf of $\mathcal{A}$-modules on X. For any open subset $U \subset X$, let us consider the jet module $\mathfrak{J}^1(P(U))$ of the module $P(U)$. It consists of the elements of $\mathcal{A}(U) \otimes P(U)$ modulo the pointwise relations (8.1.15). Hence, there is the restriction morphism $\mathfrak{J}^1(P(U)) \to \mathfrak{J}^1(P(V))$ for any open subsets $V \subset U$, and the jet modules $\mathfrak{J}^1(P(U))$ constitute a presheaf. This presheaf defines the *sheaf* $\mathfrak{J}^1 P$ *of jets* of P (or simply the *jet sheaf*). The jet sheaf $\mathfrak{J}^1\mathcal{A}$ of the sheaf $\mathcal{A}$ of local rings is introduced in a similar way. Since the relations (8.1.15) and (8.1.18) on the ring $\mathcal{A}(U)$ and modules $P(U)$, $\mathfrak{J}^1(P(U))$, $\mathfrak{J}^1(\mathcal{A}(U))$ are pointwise relations for any open subset $U \subset X$, they commute with the restriction morphisms. Therefore, the direct limits of the quotients modulo these relations exist [217]. Then we have the sheaf $\mathfrak{O}^1\mathcal{A}$ of 1-forms of the sheaf $\mathcal{A}$, the sheaf isomorphism

$$\mathfrak{J}^1(P) = (\mathcal{A} \oplus \mathfrak{O}^1\mathcal{A}) \otimes P,$$

and the exact sequences of sheaves

$$\boxed{0 \to \mathfrak{O}^1\mathcal{A} \otimes P \to \mathfrak{J}^1(P) \to P \to 0,} \tag{8.3.4}$$

$$\boxed{0 \to \mathfrak{O}^1\mathcal{A} \otimes P \to (\mathcal{A} \oplus \mathfrak{O}^1\mathcal{A}) \otimes P \to P \to 0.} \tag{8.3.5}$$

They reflect the quotient (8.1.22), the isomorphism (8.1.34) and the exact sequences of modules (8.1.35), (8.2.6), respectively.

Remark 8.3.4. It should be emphasized that, because of the inequality (8.3.1), the duality relation (8.1.38) is not extended to the sheaves $\mathfrak{d}\mathcal{A}$ and $\mathfrak{O}^1\mathcal{A}$ in general, unless $\mathfrak{d}\mathcal{A}$ and $\mathfrak{O}^1$ are locally free sheaves of finite rank. If P is a locally free sheaf of finite rank, so is $\mathfrak{J}^1 P$ (see Example 8.1.4). •

Following Definitions 8.2.1, 8.2.2 of a connection on modules, we come to the following notion of a connection on sheaves.

DEFINITION 8.3.3. Given a locally ringed space $(X, \mathcal{A})$ and a sheaf P of $\mathcal{A}$-modules on X, a *connection* on a sheaf P is defined as a splitting of the exact sequence (8.3.4) or, equivalently, the exact sequence (8.3.5). □

To state the relationship between connections on modules and connections on sheaves, let us recall some basic facts concerning exact sequences of sheaves.

Referring to the properties of direct systems [217], we have used that an exact sequence of presheaves defines an exact sequence of sheaves constructed from these presheaves. Moreover, if an exact sequence of presheaves is split, we have the corresponding splitting of the above mentioned exact sequence of sheaves. The converse relation is more intricate.

Let us consider an exact sequence of sheaves

$$0 \to S' \to S \to S'' \to 0 \tag{8.3.6}$$

on a topological space X. As in the case of vector bundles, the sheaf S'' in this exact sequence is necessarily the quotient S/S'. Given an open subset $U \subset X$, the exact sequence (8.3.6) yields the following two exact sequences of Abelian groups:

$$0 \to S'(U) \to S(U) \to S''(U) \tag{8.3.7}$$

and

$$0 \to S'(U) \to S(U) \to S''_U \to 0, \tag{8.3.8}$$

where $S''_U = S(U)/S'(U)$ does not coincide with the group of sections $S''(U)$ of the quotient S/S' over U in general. A sheaf S on X is called *flabby* if, for every pair $U \subset U'$ of open subsets of X, the restriction morphism $S(U') \to S(U)$ is surjective. This is equivalent to the condition that every section $s \in S(U)$ can be extended to a global section $s \in S(X)$. It states the following.

PROPOSITION 8.3.4. If the sheaf S' in the exact sequence (8.3.6) is flabby, then $S''_U = S''(U)$, and we have the exact sequence

$$0 \to S'(U) \to S(U) \to S''(U) \to 0 \tag{8.3.9}$$

for every open subset $U \subset X$, i.e., there exists the exact sequence of the canonical presheaves

$$0 \to \{S'(U)\} \to \{S(U)\} \to \{S''(U)\} \to 0 \tag{8.3.10}$$

of sheaves in the exact sequence (8.3.6). □

Let the exact sequence of sheaves (8.3.6) admit a splitting, i.e., $S = S' \oplus S''$, then $\{S(U)\} = \{S'(U)\} \oplus \{S''(U)\}$, and the canonical presheaves form the exact sequence (8.3.10). Moreover, this exact sequence is split in a suitable manner.

Thus, we come to the following compatibility of the notion of a connection on sheaves with that of a connection on modules.

PROPOSITION 8.3.5. If there exists a connection on a sheaf P in Definition 8.3.3, then there exists a connection on a module $P(U)$ for any open subset $U \subset X$. Conversely, if for any open subsets $V \subset U \subset X$ there are connections on the modules $P(U)$ and $P(V)$ related by the restriction morphism, then the sheaf P admits a connection. □

Example 8.3.5. Let $Y \to X$ be a vector bundle. Every linear connection Γ on $Y \to X$ defines a connection on the structure module $Y(X)$ such that the restriction $\Gamma|_U$ is a connection on the module $Y(U)$ for any open subset $U \subset X$. Then we have a connection on the structure sheaf Y_X. Conversely, a connection on the structure sheaf Y_X defines a connection on the module $Y(X)$ and, consequently, a connection on the vector bundle $Y \to X$. •

As an immediate consequence of Proposition 8.3.5, we find that the exact sequence of sheaves (8.3.5) is split if and only if there exists a sheaf morphism

$$\boxed{\nabla : P \to \mathfrak{O}^1\mathcal{A} \otimes P,} \tag{8.3.11}$$

satisfying the Leibniz rule

$$\boxed{\nabla(fs) = df \otimes s + f\nabla(s),} \qquad f \in \mathcal{A}(U), \qquad s \in P(U),$$

for any open subset $U \in X$. It leads to the following equivalent definition of a connection on sheaves in the spirit of Definition 8.2.2.

DEFINITION 8.3.6. The sheaf morphism (8.3.11) is a *connection* on the sheaf P. □

Similarly to the case of connections on modules, the *curvature* of the connection (8.3.11) on a sheaf P is given by the expression

$$\boxed{R = \nabla^2 : P \to \mathfrak{O}^2_X \otimes P.} \tag{8.3.12}$$

The exact sequence (8.3.5) needs not be split in general. One can obtain the criteria of the existence of a connection on a sheaf in terms of cohomology groups.

Remark 8.3.6. Given the exact sequence of sheaves (8.3.6) on a paracompact topological space, we have the following exact sequence of the cohomology groups

$$\cdots \to H^{k-1}(X;S'') \to H^k(X;S') \to H^k(X;S) \to H^k(X;S'') \to \cdots \qquad (8.3.13)$$

with coefficients in the sheaves S', S and S'' [157].

We omit the definition of cohomology groups with coefficients in a sheaf (see, e.g., [157]), which is the repetition of the notions of the Čech cohomology groups with coefficients in a constant sheaf and the cohomology groups with coefficients in a sheaf of G-valued functions in Section 6.9. Nevertheless, we will need some basic properties of cohomology groups with coefficients in a sheaf in the sequel.

Note that, given a sheaf S on a topological space X, the cohomology group $H^0(X;S)$ is isomorphic to the group $S(X)$ of sections of the sheaf S over X. If S is a sheaf of $\mathcal{K}$-modules, then the cohomology groups $H^k(X;S)$ are also $\mathcal{K}$-modules.

A sheaf S on a topological space X is called *acyclic* if the cohomology groups $H^{k>0}(X;S)$ vanish. A flabby sheaf is acyclic.

A sheaf S on a topological space X is said to be *soft* if every section of S on a closed subset of X is the restriction of some global section of S. Any soft sheaf on a paracompact space is acyclic. On a paracompact space, any flabby sheaf is soft.

A sheaf S on a paracompact space is said to be *fine* if, for each *locally finite open covering* $\mathfrak{U} = \{U_i\}_{i\in I}$ of X (i.e., every point of X has a neighbourhood which intersects only with a finite number of elements of this covering), there exists a system $\{h_i\}$ of endomorphisms $h_i : S \to S$ such that:

- there is a closed subset $V_i \subset U_i$ and $h_i(S_x) = 0$ if $x \notin V_i$;
- $\sum\limits_{i\in I} h_i$ is the identity map.

A fine sheaf is soft and acyclic.

In particular, let S be a sheaf of modules over the sheaf C^0_X of continuous functions on a paracompact space X. Let $\mathfrak{U} = \{U_i\}_{i\in I}$ be a locally finite open covering of X. Then $\mathfrak{U}$ has an associated *partition of unity* $\{\phi_i\}$, i.e.:

(i) ϕ_i are real non-negative continuous functions on X,

(ii) $\mathrm{supp}\phi_i \subset U_i$,

(iii) $\sum\limits_{i\in I} \phi_i(x) = 1$ for all $x \in X$.

The functions ϕ_i can be used in order to define the homomorphisms $h_i : S \to S$ as follows. For any open subset $U \subset X$, let us put $h_i(f) = \phi_i f$, $f \in S(U)$. This defines an endomorphism h_i of the canonical presheaf $\{S(U)\}$ and, consequently, an endomorphism of the sheaf S. It is readily observed that these endomorphisms h_i satisfy the definition of a fine sheaf. If X is a smooth manifold and $\mathfrak{U}$ is an open covering of X, there exists an associated partition of unity performed by smooth functions. It follows that the sheaf C^∞_X of smooth functions on a manifold X is fine and acyclic, and so are the sheaves of sections of smooth vector bundles over a manifold X.

PROPOSITION 8.3.7. [20]. Let $f : X \to X'$ be a continuous map and S a sheaf on X. If either:

- f is a closed immersion or
- every point $x' \in X'$ has a base of open neighbourhoods $\{U\}$ such that the sheaves $S\mid_{f^{-1}(U)}$ are acyclic,

then the cohomology groups $H^*(X;S)$ and $H^*(X';f_*S)$ are isomorphic. □

Let us consider the exact sequence of sheaves

$$0 \to S \xrightarrow{h} S_0 \xrightarrow{h^0} \to S_1 \xrightarrow{h^1} \cdots \qquad (8.3.14)$$

on a paracompact topological space X. This sequence is called a *resolution* of the sheaf S if the cohomology groups $H^q(X;S_p)$ vanish for $q \geq 1$ and $p \geq 0$. For instance, it takes place if the sheaves $S_{p\geq 0}$ are fine. Then the exact sequence (8.3.14) is said to be the *fine resolution* of the sheaf S. The exact sequence (8.3.14) yields the cochain complex

$$0 \to S(X) \xrightarrow{h_*} S_0(X) \xrightarrow{h^0_*} \to S_1(X) \xrightarrow{h^1_*} \cdots \qquad (8.3.15)$$

which is exact only at $S(X)$. There is the well-known De Rham theorem (see, e.g., [157]).

THEOREM 8.3.8. Given the resolution (8.3.14) of a sheaf S on a paracompact space X, the q-cohomology group of the cochain complex (8.3.15) is isomorphic to the cohomology group $H^q(X;S)$ of X with coefficients in the sheaf S, i.e.,

$$H^{q>0}(X;S) = \mathrm{Ker}\, h^q_*/\mathrm{Im}\, h^{q-1}_*, \qquad H^0(X;S) = \mathrm{Ker}\, h^0_*. \qquad (8.3.16)$$

□

For instance, let X be a connected smooth manifold, $S = \mathbb{R}$ the constant sheaf of $\mathbb{R}$-valued functions on X and $S_p = \mathfrak{O}^p_X$ the sheaves of exterior p-forms on X. There is the sequence of the fine sheaves

$$0 \to \mathbb{R} \to \mathfrak{O}^0_X \to \mathfrak{O}^1_X \to \cdots \tag{8.3.17}$$

which is exact in accordance with the Poincaré lemma, and so is the fine resolution of the sheaf $\mathbb{R}$ on X. The corresponding sequence (8.3.15) is the De Rham complex (6.8.3). By virtue of Theorem 8.3.8, we have the above mentioned isomorphism of the Čech and De Rham cohomology groups $H^q(X;\mathbb{R}) = H^q(X)$. •

Turn now to the exact sequence (8.3.5). Let P be a locally free sheaf of $\mathcal{A}$-modules. Then we have the exact sequence of sheaves

$$0 \to Hom(P, \mathfrak{O}^1\mathcal{A} \otimes P) \to Hom(P, (\mathcal{A} \oplus \mathfrak{O}^1\mathcal{A}) \otimes P) \to Hom(P, P) \to 0$$

and the corresponding exact sequence (8.3.13) of the cohomology groups

$$0 \to H^0(X; Hom(P, \mathfrak{O}^1\mathcal{A} \otimes P)) \to H^0(X; Hom(P, (\mathcal{A} \oplus \mathfrak{O}^1\mathcal{A}) \otimes P)) \to$$
$$H^0(X; Hom(P, P)) \to H^1(X; Hom(P, \mathfrak{O}^1\mathcal{A} \otimes P)) \to \cdots.$$

The identity morphism $\mathrm{Id} : P \to P$ belongs obviously to $H^0(X; Hom(P, P))$. Its image in $H^1(X; Hom(P, \mathfrak{O}^1\mathcal{A} \otimes P))$ is called sometimes the *Atiyah class*. If this class vanishes, there exists an element of $Hom(P, (\mathcal{A} \oplus \mathfrak{O}^1\mathcal{A}) \otimes P))$ whose image is $\mathrm{Id}\, P$, i.e., a splitting of the exact sequence (8.3.5).

In particular, let X be a manifold and $\mathcal{A} = C^\infty_X$ the sheaf of smooth functions on X. The sheaf $\mathfrak{d}C^\infty_X$ of its derivations is isomorphic to the sheaf of vector fields on a manifold X. It follows that:

- there is the restriction morphism $\mathfrak{d}(C^\infty(U)) \to \mathfrak{d}(C^\infty(V))$ for any open sets $V \subset U$,
- $\mathfrak{d}C^\infty_X$ is a locally free sheaf of C^∞_X-modules of finite rank,
- the sheaves $\mathfrak{d}C^\infty_X$ and $\mathfrak{O}^1_X$ are mutually dual.

Let P be a locally free sheaf of C^∞_X-modules. In this case, $Hom(P,\mathfrak{O}^1_X\otimes P)$ is a locally free sheaf of C^∞_X-modules. It is fine and acyclic. Its cohomology group $H^1(X;Hom(P,\mathfrak{O}^1_X\otimes P))$ vanishes, and the exact sequence

$$\boxed{0\to\mathfrak{O}^1_X\otimes P\to(C^\infty_X\oplus\mathfrak{O}^1_X)\otimes P\to P\to 0} \tag{8.3.18}$$

admits a splitting.

In conclusion, let us consider a sheaf S of commutative C^∞_X-rings on a manifold X. Building on Definition 8.2.7, we come to the following notion of a connection on a sheaf S of commutative C^∞_X-rings.

DEFINITION 8.3.9. Any morphism

$$\mathfrak{d}C^\infty_X\ni\tau\to\nabla_\tau\in\mathfrak{d}S,$$

which is a connection on S as a sheaf of C^∞_X-modules, is called a connection on the sheaf S of rings. □

Its curvature is given by the expression

$$\boxed{R(\tau,\tau')=[\nabla_\tau,\nabla_{\tau'}]-\nabla_{[\tau,\tau']},} \tag{8.3.19}$$

similar to the expression (8.2.19) for the curvature of a connection on modules.

Chapter 9

Superconnections

Elements of the graded calculus and supergeometry are present in many quantum field models. By this reason, we start our exposition of connections in quantum field theory from superconnections. Superconnections exemplify the algebraic connections phrased in terms of graded modules and sheaves of graded commutative algebras. We will refer to the properties of modules and sheaves in the previous Chapter which are extended to the graded ones [146].

With respect to mathematical prerequisites, the reader is expected to be familiar with the basics of theory of supersymmetries and supermanifolds (see, e.g., [20, 59, 79]). Nevertheless, Section 9.1 aims to recall some element of the graded tensor calculus which we will refer to in the sequel. We omit the categorial aspects of the constructions below, which are not essential for applications.

9.1 Graded tensor calculus

Unless otherwise stated, by a graded structure throughout this Chapter is meant a $\mathbb{Z}_2$-graded structure. We use the notation [.] for the $\mathbb{Z}_2$-graded parity, in contrast with the above stated notation $|\,.\,|$ for the $\mathbb{Z}$-graded one.

By a *graded commutative ring* $\mathcal{K}$ is meant a ring which has two subgroups $\mathcal{K}_0$ and $\mathcal{K}_1$, called respectively *even* and *odd*, such that $\mathcal{K} = \mathcal{K}_0 \oplus \mathcal{K}_1$ and

$$a_r a_s = (-1)^{rs} a_s a_r \in \mathcal{K}_{r+s}, \qquad a_r \in \mathcal{K}_r, \qquad a_s \in \mathcal{K}_s, \qquad s, r = 0, 1.$$

A commutative ring is a particular graded commutative ring where $\mathcal{K}_1 = 0$.

A *graded $\mathcal{K}$-module* Q is a $\mathcal{K}$-bimodule which has two subgroups Q_0 and Q_1 such that $Q = Q_0 \oplus Q_1$ and $\mathcal{K}_r Q_s \subset Q_{r+s}$. A graded $\mathcal{K}$-module is said to be *free* if it has a basis generated by *homogeneous elements*, i.e., elements which belong to either Q_0 or Q_1. Each $\mathcal{K}$-graded module is obviously a $\mathcal{K}_0$-module. A basis for Q of finite cardinality is said to be of type (n,m) if it is formed by n even and m odd elements.

In particular, by a *graded vector space* $B = B_0 \oplus B_1$ is meant a graded $\mathbb{R}$-module. A graded vector space is said to be (n,m)-dimensional if $\dim B_0 = n$ and $\dim B_1 = m$.

A *graded commutative $\mathcal{K}$-algebra* A with a unit is a graded commutative ring which is also a graded $\mathcal{K}$-module. The *graded tensor product* $A_1 \otimes A_2$ of two graded commutative $\mathcal{K}$-algebras A_1 and A_2 is defined as the tensor product of underlying $\mathcal{K}$-modules equipped with the multiplication

$$(a_1 \otimes a_2) \cdot (a'_1 \otimes a'_2) = (-1)^{[a_2][a'_1]}(a_1 a'_1 \otimes a_2 a'_2).$$

By a *graded commutative algebra* A throughout is meant a unital graded commutative $\mathbb{R}$-algebra. A graded commutative algebra is said to be of odd rank m (or simply of rank m) if it is a free algebra generated over $\mathbb{R}$ by m odd elements. A *graded commutative Banach algebra* is a graded commutative algebra if it is a Banach algebra and the condition

$$\|a_0 + a_1\| = \|a_0\| + \|a_1\|$$

is fulfilled.

Let V be a vector space, and let

$$\Lambda = \wedge V = \mathbb{R} \bigoplus_{k=1} \overset{k}{\wedge} V \tag{9.1.1}$$

be its *exterior algebra*. This is a $\mathbb{Z}$-graded commutative algebra provided with the $\mathbb{Z}_2$-graded structure

$$\Lambda_0 = \mathbb{R} \bigoplus_{k=1} \overset{2k}{\wedge} V, \quad \Lambda_1 = \bigoplus_{k=1} \overset{2k-1}{\wedge} V.$$

It is called the *Grassmann algebra*. Given a basis $\{c^i, i \in I\}$ for the vector space V, the elements of the Grassmann algebra take the form

$$a = \sum_{k=0} \sum_{(i_1 \cdots i_k)} a_{i_1 \cdots i_k} c^{i_1} \cdots c^{i_k}, \tag{9.1.2}$$

where the sum is over all the collections $(i_1 \cdots i_k)$ of indices such that no two of them are the permutations of each other. For the sake of simplicity, we will omit the symbol of the exterior product of elements of a Grassmann algebra. By definition, a Grassmann algebra admits the splitting

$$\boxed{\Lambda = \mathbb{R} \oplus \mathcal{R} = \mathbb{R} \oplus \mathcal{R}_0 \oplus \mathcal{R}_1} \tag{9.1.3}$$

where $\mathcal{R} = \mathcal{R}_0 \oplus \mathcal{R}_1$ is the ideal of nilpotents of Λ. The corresponding projections $\sigma : \Lambda \to \mathbb{R}$ and $s : \Lambda \to \mathcal{R}$ are called the *body* and *soul* maps, respectively

Remark 9.1.1. Note that there is a different definition of a Grassmann algebra [162], which is equivalent to the above one only in the case of an infinite-dimensional vector space V [59] (see [45] for the Arens–Michael algebras of Grassmann origin, which are most general graded commutative algebras suitable for superanalysis, and Remark 9.3.4 below for supermanifolds over these algebras). The Grassmann algebra Λ (9.1.1) of a vector space V is a particular *exterior algebra* $\wedge_\mathcal{K} Q$ of a graded $\mathcal{K}$-module Q. The exterior algebra $\wedge_\mathcal{K} Q$ is the graded commutative $\mathcal{K}$-algebra defined as the tensor algebra $\otimes Q$ modulo the relations

$$q \otimes q' = (-1)^{[q][q']} q' \otimes q.$$

If $\mathcal{K} = \mathbb{R}$ and $Q_0 = 0$, $Q_1 = V$, we come to the Grassmann algebra (9.1.1). Hereafter, we will restrict our consideration to Grassmann $\mathbb{R}$-algebras of finite rank. •

A Grassmann algebra of finite rank becomes a graded commutative Banach algebra if its elements (9.1.2) are endowed with the norm

$$\|a\| = \sum_{k=0} \sum_{(i_1 \cdots i_k)} | a_{i_1 \cdots i_k} | .$$

Let B be a graded vector space. Given a Grassmann algebra Λ of rank N, the graded vector space B can be extended to a graded Λ-module

$$\Lambda B = (\Lambda B)_0 \oplus (\Lambda B)_1 = (\Lambda_0 \otimes B_0 \oplus \Lambda_1 \otimes B_1) \oplus (\Lambda_1 \otimes B_0 \oplus \Lambda_0 \otimes B_1),$$

called the *graded Λ-envelope*. The graded envelope

$$\boxed{B^{n|m} = (\Lambda_0^n \oplus \Lambda_1^m) \oplus (\Lambda_1^n \oplus \Lambda_0^m)} \tag{9.1.4}$$

of the (n, m)-dimensional graded vector space $\mathbb{R}^n \oplus \mathbb{R}^m$ is a free Λ-module of type (n, m). The Λ_0-module

$$\boxed{B^{n,m} = (\Lambda_0^n \oplus \Lambda_1^m)}$$

is called an (n,m)-dimensional *supervector space*. Note that Λ_0-modules $B^{n|m}$ and $B^{n+m,n+m}$ are isomorphic.

Unless otherwise stated (see the DeWitt topology below), whenever referring to a topology on a supervector space $B^{n,m}$, we will mean the Euclidean topology of $(2^{N-1}[n+m])$-dimensional $\mathbb{R}$-vector space.

Given a superspace $B^{n|m}$ over a Grassmann algebra Λ, a Λ-module endomorphism of $B^{n|m}$ is represented by a square $(n+m)\times(n+m)$-matrix

$$L = \begin{pmatrix} L_1 & L_2 \\ L_3 & L_4 \end{pmatrix} \tag{9.1.5}$$

with entries in Λ. It is called a *supermatrix*. One says that a supermatrix L is:

- *even* if L_1 and L_4 have even entries, while L_2 and L_3 have odd entries,
- *odd* if L_1 and L_4 have odd entries, while L_2 and L_3 have the even ones.

Endowed with this gradation, the set of supermatrices (9.1.5) is a Λ-graded algebra. Unless otherwise stated, by a supermatrix will be meant a homogeneous supermatrix.

The familiar notion of a trace is extended to supermatrices (9.1.5) as a *supertrace*

$$\boxed{\operatorname{Str} L = \operatorname{Tr} L_1 - (-1)^{[L]}\operatorname{Tr} L_4.}$$

For instance, if $\mathbf{1}_{n|m}$ is a unit matrix, we have $\operatorname{Str}(\mathbf{1}_{n|m}) = n-m$. A *supertransposition* L^{st} of a supermatrix L is the matrix

$$L^{st} = \begin{pmatrix} L_1^t & (-1)^{[L]}L_3^t \\ -(-1)^{[L]}L_2^t & L_4^t \end{pmatrix},$$

where L^t denotes the ordinary transposition. There are the relations

$$\operatorname{Str}(L^{st}) = \operatorname{Str} L,$$

$$(LL')^{st} = (-1)^{[L][L']}L'^{st}L^{st}, \tag{9.1.6}$$

$$\operatorname{Str}(LL') = (-1)^{[L][L']}\operatorname{Str}(L'L) \quad \text{or} \quad \operatorname{Str}([L,L']) = 0. \tag{9.1.7}$$

In order to extend the notion of a determinant to supermatrices, let us consider invertible supermatrices L (9.1.5) corresponding to even isomorphisms of the superspace $B^{n,m}$. Recall that a supermatrix L is invertible if and only if:

- L_1 and L_4 are invertible;
- $\sigma(L)$ is invertible, where σ is the body morphism.

Invertible supermatrices constitute a group $GL(n|m;\Lambda)$, called a *general linear graded group*. Then a *superdeterminant* of $L \in GL(n|m;\Lambda)$ is defined as

$$\boxed{\mathrm{Sdet}\, L = \det(L_1 - L_2 L_4^{-1} L_3)(\det L_4^{-1}).}$$

It satisfies the relations

$$\mathrm{Sdet}(LL') = (\mathrm{Sdet}\, L)(\mathrm{Sdet}\, L'),$$
$$\mathrm{Sdet}(L^{st}) = \mathrm{Sdet}\, L,$$
$$\mathrm{Sdet}(\exp(L)) = \exp(\mathrm{Sdet}\,(L)).$$

9.2 Connections on graded manifolds

Graded manifolds are not supermanifolds in a strict sense (see Axioms 1-4 in Remark 9.3.4 below). At the same time, every graded manifold defines a DeWitt H^∞-supermanifold, and *vice versa* (see Theorem 9.3.8 below). Principal graded bundles and connections on these bundles are described similarly to that on principal superbundles (see Section 9.5). We refer the reader to [20, 28, 183, 285] for a general theory of graded manifolds. This Section is devoted to connections which can be introduced on a graded manifold itself due to the fact that graded functions, unlike superfunctions, can be represented by sections of some smooth exterior bundle.

DEFINITION 9.2.1. By a *graded manifold* of dimension (n, m) is meant the pair $(Z, \mathcal{A})$ of an n-dimensional smooth manifold Z and a sheaf $\mathcal{A} = \mathcal{A}_0 \oplus \mathcal{A}_1$ of graded commutative $\mathbb{R}$-algebras of rank m such that [20]:

(i) There is the exact sequence of sheaves

$$0 \to \mathcal{R} \to \mathcal{A} \xrightarrow{\sigma} C^\infty_Z \to 0, \qquad \mathcal{R} = \mathcal{A}_1 + (\mathcal{A}_1)^2, \tag{9.2.1}$$

where C^∞_Z is the sheaf of smooth functions on Z.

(ii) $\mathcal{R}/\mathcal{R}^2$ is a locally free C^∞_Z-module of finite rank, and $\mathcal{A}$ is locally isomorphic to the exterior bundle $\wedge_{C^\infty_Z}(\mathcal{R}/\mathcal{R}^2)$. □

The sheaf $\mathcal{A}$ is called a *structure sheaf* of the graded manifold $(Z,\mathcal{A})$, while the manifold Z is a *body* of $(Z,\mathcal{A})$. This terminology is motivated by the above mentioned correspondence between the graded manifolds and the DeWitt supermanifolds. Sections of $\mathcal{A}$ are called *graded functions*.

A graded manifold is a graded locally ringed space. By a *morphism of graded manifolds* $(Z,\mathcal{A}) \to (Z'\mathcal{A}')$ is meant their morphism $\varphi : Z \to Z'$, $\Phi : \mathcal{A}' \to \varphi_*\mathcal{A}$ as locally ringed spaces, where Φ is an even graded morphism.

By very definition, a graded manifold $(Z,\mathcal{A})$ has the following local structure. Given a point $z \in Z$, there exists its open neighbourhood U such that

$$\boxed{\mathcal{A}(U) \cong C^\infty(U) \otimes \wedge\mathbb{R}^m.} \tag{9.2.2}$$

It means that the restriction $\mathcal{A}\mid_U$ of the structure sheaf $\mathcal{A}$ to U is isomorphic to the sheaf $C^\infty_U \otimes \wedge\mathbb{R}^m$ of sections of some exterior bundle $\wedge E^*_U = U \times \wedge\mathbb{R}^m \to U$. Then U is called a *splitting domain* of the graded manifold $(Z,\mathcal{A})$.

The well-known Batchelor's theorem [20, 22] shows that such a structure of graded manifolds is global. The global structure sheaf $\mathcal{A}$ of the graded manifold $(Z,\mathcal{A})$ is glued of the local sheaves $C^\infty_U \otimes \wedge\mathbb{R}^m$ by means of transition functions in Proposition 8.3.1, which constitute a cocycle of the sheaf $\mathrm{Aut}\,(\wedge\mathbb{R}^m)_\infty$ of smooth mappings from Z to $\mathrm{Aut}\,(\wedge\mathbb{R}^m)$. Batchelor's theorem is based on the bijection between the cohomology sets $H^1(Z;\mathrm{Aut}\,(\wedge\mathbb{R}^m)_\infty)$ and $H^1(Z;GL(m,\mathbb{R}^m)_\infty)$.

THEOREM 9.2.2. Let $(Z,\mathcal{A})$ be a graded manifold. There exists a vector bundle $E \to Z$ with an m-dimensional typical fibre V such that the structure sheaf $\mathcal{A}$ is isomorphic to the sheaf

$$\boxed{\mathcal{A}_E = C^\infty_Z \otimes \wedge V^*} \tag{9.2.3}$$

of sections of the exterior bundle

$$\wedge E^* = \mathbb{R} \underset{Z}{\oplus} (\overset{m}{\underset{k=1}{\oplus}} \overset{k}{\wedge} E^*), \tag{9.2.4}$$

whose typical fibre is the Grassmann algebra $\wedge V^*$ □

It should be emphasized that Batchelor's isomorphism in Theorem 9.2.2 is not canonical. One can speak only on one-to-one correspondence between the classes of isomorphic graded manifolds of odd rank m and the classes of equivalent m-dimensional vector bundles over the same smooth manifold Z. At the same time, this isomorphism enables one to obtain some properties of a graded manifold.

COROLLARY 9.2.3. The structure sheaf $\mathcal{A}$ of the graded manifold $(Z,\mathcal{A})$ is isomorphic to the sheaf of sections of a 2^N-dimensional real vector bundle $\Upsilon \to Z$, called the *characteristic vector bundle*, with the typical fibre $\wedge\mathbb{R}^m$ and the structure group $\mathrm{Aut}\,(\wedge\mathbb{R}^m)$ such that every splitting domain U of the graded manifold $(Z,\mathcal{A})$ is a trivialization chart of the vector bundle $\Upsilon \to Z$ with the corresponding $\mathrm{Aut}\,(\wedge\mathbb{R}^m)$-valued transition functions. The structure group $\mathrm{Aut}\,(\wedge\mathbb{R}^m)$ of the vector bundle $\Upsilon \to Z$ is reducible to the group $GL(m,\mathbb{R})$, and $\Upsilon \to Z$ is isomorphic to the exterior vector bundle $\wedge E^* \to Z$ in accordance with Theorem 9.2.2. □

COROLLARY 9.2.4. The structure sheaf $\mathcal{A}$ of the graded manifold $(Z,\mathcal{A})$ is fine and, consequently, acyclic (see Remark 8.3.6). □

COROLLARY 9.2.5. The direct product of two graded manifolds $(Z,\mathcal{A})$ and $(Z',\mathcal{A}')$ is the graded manifold whose body is $Z\times Z$, while the structure sheaf $\mathcal{A}\widehat{\otimes}\mathcal{A}'$ is isomorphic to the tensor product

$$(C^\infty_Z \widehat{\otimes} C^\infty_{Z'})\otimes\wedge(V\oplus V')^* \tag{9.2.5}$$

(see the isomorphisms (8.3.3) and (9.2.3)). □

Given a splitting domain U, graded functions on U read

$$f=\sum_{k=0}^{m}\frac{1}{k!}f_{a_1\ldots a_k}(z)c^{a_1}\cdots c^{a_k}, \tag{9.2.6}$$

where $f_{a_1\cdots a_k}(z)$ are smooth functions on U, $\{c^a\}$ are the fibre basis for E^*, and we omit the symbol of the exterior product of elements c. In particular, the sheaf epimorphism σ in (9.2.1) can be seen as the body morphism. We will call $\{c^a\}$ the local *basis* of a graded manifold.

If U' is another splitting domain and $U\cap U'\neq\emptyset$, we have the transition functions

$$c'^a=\rho^a(z^A,c^b) \tag{9.2.7}$$

where $\rho^a(z^A,c^b)$ are graded functions on $U\cap U'$. These are transition functions of the characteristic bundle $\Upsilon\to Z$ in Corollary 9.2.3. The corresponding coordinate transformation law of graded functions (9.2.6) is obvious. If U and U' are trivializations charts of the same exterior bundle in Theorem 9.2.2, the transition functions take the form

$$\boxed{c'^a=\rho^a_b(z^A)c^b} \tag{9.2.8}$$

where $\rho^a_b(z)$ are smooth functions on $U \cap U'$.

Given a graded manifold $(Z, \mathcal{A})$, by the sheaf $\mathfrak{d}\mathcal{A}$ of graded derivations of $\mathcal{A}$ is meant a subsheaf of endomorphisms of the structure sheaf $\mathcal{A}$ such that any section u of $\mathfrak{d}\mathcal{A}$ over an open subset $U \subset Z$ is a *graded derivation* of the graded algebra $\mathcal{A}(U)$, i.e.

$$\boxed{u(ff') = u(f)f' + (-1)^{[u][f]} f u(f')} \tag{9.2.9}$$

for the homogeneous elements $u \in (\mathfrak{d}\mathcal{A})(U)$ and $f, f' \in \mathcal{A}(U)$.

LEMMA 9.2.6. [20]. If $U' \subset U$ are open sets, there is a surjection $\mathfrak{d}(\mathcal{A}(U)) \to \mathfrak{d}(\mathcal{A}(U'))$. $\Box$

It follows that $(\mathfrak{d}\mathcal{A})(U) = \mathfrak{d}(\mathcal{A}(U))$, i.e., the canonical presheaf of the sheaf of graded derivations $\mathfrak{d}\mathcal{A}$ is isomorphic to the presheaf of derivations of graded modules $\mathcal{A}(U)$. Sections of $\mathfrak{d}\mathcal{A}$ are called *graded vector fields* on the graded manifold $(Z, \mathcal{A})$ (or simply on Z if there is no danger of confusion). One can show that, given a splitting domain U and the corresponding trivial vector bundle $E_U = U \times V$, graded vector fields u on U are represented by sections of the vector bundle

$$\wedge E^*_U \underset{U}{\otimes} (E_U \underset{U}{\oplus} TU) \to U$$

[20, 183]. They take the form

$$\boxed{u = u^A \partial_A + u^a \partial_a,} \tag{9.2.10}$$

where u^A, u^a are local graded functions, $\{\partial_a\}$ are the dual bases of $\{c^a\}$, and (z^A) are coordinates on $U \subset Z$. The derivations (9.2.10) act on graded functions $f \in \mathcal{A}_E(U)$ (9.2.6) by the rule

$$\boxed{u(f_{a\ldots b} c^a \cdots c^b) = u^A \partial_A (f_{a\ldots b}) c^a \cdots c^b + u^k f_{a\ldots b} \partial_k \rfloor (c^a \cdots c^b).} \tag{9.2.11}$$

Remark 9.2.1. With derivations (9.2.11), one can provide the ring $\mathcal{A}(U)$ with the Fréchet topology such that there is the metric isomorphism (9.2.2). •

Let U' be another splitting domain together with the transition functions (9.2.7). Then the action rule (9.2.11) implies the corresponding coordinate transformation law

$$u'^A = u^A, \qquad u'^a = u^j \partial_j \rho^a + u^A \partial_A \rho^a$$

of graded vector fields (where we leave $z^A = z'^A$ for the sake of simplicity). If U and U' are trivialization charts of the same vector bundle E in Theorem 9.2.2 together with the transition functions (9.2.8), we have

$$u'^A = u^A, \qquad u'^a = u^j \rho^a_j + u^A \partial_A(\rho^a_j) c^j.$$

It follows that, given Batchelor's isomorphism in Theorem 9.2.2 and the corresponding vector bundle $E \to Z$, graded vector fields on Z can be represented by sections of the vector bundle $\mathcal{V}_E \to Z$ which is locally isomorphic to the vector bundle

$$\mathcal{V}_E \mid_U \approx \wedge E^* \underset{Z}{\otimes} (\mathrm{pr}_2 VE \underset{Z}{\oplus} TZ) \mid_U, \tag{9.2.12}$$

and has the transition functions

$$\begin{aligned} z'^A_{i_1\ldots i_k} &= \rho^{-1a_1}_{i_1} \cdots \rho^{-1a_k}_{i_k} z^A_{a_1\ldots a_k}, \\ v'^i_{j_1\ldots j_k} &= \rho^{-1b_1}_{j_1} \cdots \rho^{-1b_k}_{j_k} \left[\rho^i_j v^j_{b_1\ldots b_k} + \frac{k!}{(k-1)!} z^A_{b_1\ldots b_{k-1}} \partial_A \rho^i_{b_k} \right] \end{aligned} \tag{9.2.13}$$

of the bundle coordinates $(z^A_{a_1\ldots a_k}, v^i_{b_1\ldots b_k})$, $k = 0, \ldots, m$, with respect to the fibre bases $\{c^a\}$ for $E^* \to Z$ and the dual holonomic fibre bases $\{\partial_a\}$ for the vertical tangent bundle $VE \to E$ (recall the canonical splitting $VE = E \times E$). These transition functions fulfill the cocycle relations (1.1.4). There is the exact sequence over Z of vector bundles

$$\boxed{0 \to \wedge E^* \underset{Z}{\otimes} \mathrm{pr}_2 VE \to \mathcal{V}_E \to \wedge E^* \underset{Z}{\otimes} TZ \to 0.} \tag{9.2.14}$$

Remark 9.2.2. In view of the local isomorphism (9.2.12), one can think of $\mathcal{V}_E$ as a local Ne'eman–Quillen superbundle (see Section 9.7). •

By virtue of Lemma 9.2.6 and Proposition 8.3.1, the sheaf of sections of the vector bundle $\mathcal{V}_E \to Z$ is isomorphic to the sheaf $\mathfrak{d}\mathcal{A}$. Global sections of $\mathcal{V}_E \to Z$ constitute the $\mathcal{A}(Z)$-module of graded vector fields on Z, which is also a *Lie superalgebra* with respect to the bracket

$$\boxed{[u, u'] = uu' + (-1)^{[u][u']+1} u'u.} \tag{9.2.15}$$

In particular,

$$[\partial_A, \partial_B] = [\partial_A, \partial_a] = [\partial_a, \partial_b] = 0. \tag{9.2.16}$$

Remark 9.2.3. As was mentioned above, derivations of the sheaf C^∞_X of smooth function on a manifold X are sections of the tangent bundle $TX \to X$. By analogy, one assumes that graded derivations of the sheaf $\mathcal{A}$ of graded functions on a manifold Z can be represented by sections of some graded tangent bundle $(STZ, ST\mathcal{A}) \to (Z, \mathcal{A})$ [54]. Nevertheless, the transformation law (9.2.13) shows that the projection

$$\mathcal{V}_E\mid_U \to \mathrm{pr}_2 VE \underset{Z}{\oplus} TZ$$

is not global, i.e., $\mathcal{V}_E$ is not an exterior bundle. It means that the sheaf of derivations $\mathfrak{d}\mathcal{A}$ is not a structure sheaf of a graded manifold (cf. Example 9.3.6 below). •

There are many physical models where a vector bundle E is introduced from the beginning (see Sections 9.6, 11.3). In this case, we can restrict our consideration to the sheaf $\mathcal{A}_E$ (9.2.3) of sections of the exterior bundle (9.2.4) [117]. Accordingly, its automorphisms reduce to the bundle isomorphisms of $E \to Z$. We will call the pair $(Z, \mathcal{A}_E)$ a *simple graded manifold*. This is not the terminology of [54] where this term is applied to all graded manifolds of finite rank in connection with Batchelor's theorem. The simple graded manifold $(Z, \mathcal{A}_E)$ is characterized entirely by the structure module $\mathcal{A}_E(Z) = \wedge E^*(Z)$ of the exterior bundle $\wedge E^*$. Therefore, $\mathcal{A}_E(Z)$ is also called the *structure module of a simple graded manifold.*

Let $(Z, \mathcal{A}_E)$ and $(Z', \mathcal{A}_{E'}$ be simple graded manifolds and $\zeta : E \to E'$ a linear bundle morphism over a morphism $\varphi : Z \to Z'$. Then every section s^* of the dual bundle $E'^* \to Z'$ defines the pull-back section $\zeta^* s^*$ of the dual bundle $E^* \to Z$ by the law

$$v_z \rfloor \zeta^* s^*(z) = \zeta(v_z) \rfloor s^*(\varphi(z)), \qquad \forall v_z \in E_z.$$

As a consequence, we obtain the pull-back $\zeta^*\mathcal{A}_{E'}$ of the sheaf $\mathcal{A}_{E'}$ onto Z, which is a subsheaf of $\mathcal{A}_E$. Note that the pair $(Z, \zeta^*\mathcal{A}_{E'}$ is not a graded manifold in general. Then the associated morphism of simple graded manifolds

$$S\zeta = (\varphi, \varphi_* \circ \zeta^*) : (Z, \mathcal{A}_E) \to (Z', \mathcal{A}_{E'}) \tag{9.2.17}$$

can be defined (cf. (8.3.2)). With respect to the local basis $\{c^a\}$ and $\{c'^a\}$ for $\mathcal{A}_E$ and $\mathcal{A}_{E'}$, the morphism (9.2.17 reads $S\zeta(c'^a) = \zeta^a_b(z)c^b$.

Turn now to the exact sequence (9.2.14). Its splitting

$$\widetilde{\gamma} : \dot{z}^A \partial_A \mapsto \dot{z}^A(\partial_A + \widetilde{\gamma}^a_A \partial_a) \tag{9.2.18}$$

is represented by a section

$$\boxed{\widetilde{\gamma} = dz^A \otimes (\partial_A + \widetilde{\gamma}_A^a \partial_a)} \tag{9.2.19}$$

of the vector bundle $T^*Z \underset{Z}{\otimes} \mathcal{V}_E \to Z$ such that the composition

$$Z \xrightarrow{\widetilde{\gamma}} T^*Z \underset{Z}{\otimes} \mathcal{V}_E \to T^*Z \underset{Z}{\otimes} (\wedge E^* \underset{Z}{\otimes} TZ) \xrightarrow{\sigma} T^*Z \underset{Z}{\otimes} TZ$$

is the canonical form $dz^A \otimes \partial_A$ on Z. Note that the coefficients $\widetilde{\gamma}$ are odd. The splitting (9.2.18) transforms every vector field τ on Z into a graded vector field

$$\boxed{\tau = \tau^A \partial_A \mapsto \nabla_\tau = \tau^A(\partial_A + \widetilde{\gamma}_A^a \partial_a),} \tag{9.2.20}$$

which is a graded derivation of $\mathcal{A}_E$ such that

$$\boxed{\nabla_\tau(sf) = (\tau\rfloor ds)f + s\nabla_\tau(f),} \quad f \in \mathcal{A}_E(U), \quad s \in C^\infty(U), \qquad \forall U \subset Z.$$

Then, in accordance with Definition 8.3.9 extended to graded commutative rings, one can think of the graded derivation ∇_τ (9.2.20) and, consequently, of the splitting (9.2.18) as being a *graded connection* on the simple graded manifold $(Z, \mathcal{A}_E)$. In particular, this connection provides the corresponding decomposition

$$\boxed{u = u^A \partial_A + u^a \partial_a = u^A(\partial_A + \widetilde{\gamma}_A^a \partial_a) + (u^a - u^A \widetilde{\gamma}_A^a)\partial_a}$$

of graded vector fields on Z.

Remark 9.2.4. By virtue of the isomorphism (9.2.2), any connection $\widetilde{\gamma}$ on the graded manifold $(Z, \mathcal{A})$, restricted to a splitting domain U, takes the form (9.2.18). Given two splitting domains U and U' of $(Z, \mathcal{A})$ with the transition functions (9.2.7), the connection components $\widetilde{\gamma}_A^a$ obey the transformation law

$$\boxed{\widetilde{\gamma}'^a_A = \widetilde{\gamma}_A^b \partial_b \rho^a + \partial_A \rho^a.} \tag{9.2.21}$$

If U and U' are the trivialization charts of the same vector bundle E in Theorem 9.2.2 together with the transition functions (9.2.8), the transformation law (9.2.21) takes the form

$$\widetilde{\gamma}'^a_A = \rho^a_b(z)\widetilde{\gamma}^b + \partial_A \rho^a_b(z) c^b. \tag{9.2.22}$$

If not refer to the particular transformation law (9.2.22), one can think of the graded connections (9.2.19) on the simple graded manifold $(Z, \mathcal{A}_E)$ as being connections on the simple graded manifold $(Z, \mathcal{A}_E)$. In view of Remark 9.2.4, γ_S is also a graded connection on the graded manifold $(Z, \mathcal{A}) \cong (Z, \mathcal{A}_E)$, but its linear form (9.2.23) is not maintained under the transformation law (9.2.21). •

By virtue of Theorem 1.1.4, graded connections (9.2.18) always exist. For instance, every linear connection

$$\gamma = dz^A \otimes (\partial_A + \gamma_A{}^a{}_b v^b \partial_a)$$

on the vector bundle $E \to Z$ yields the graded connection

$$\boxed{\gamma_S = dz^A \otimes (\partial_A + \gamma_A{}^a{}_b c^b \partial_a)} \qquad (9.2.23)$$

on $(Z, \mathcal{A}_E)$ such that, for any vector field τ on Z and any graded function f, the graded derivation $\nabla_\tau(f)$ with respect to the connection (9.2.23) is exactly the covariant derivative of f relative to the linear connection γ. Its is not surprising because graded connections (9.2.19), in fact, are particular connections on the exterior bundle $\wedge E^*$ (9.2.4) which provide derivations of its structure module $\wedge E^*(Z)$.

Graded connections $\widetilde{\gamma}$ (9.2.19) on the simple graded manifold $(Z, \mathcal{A}_E)$ constitute an affine space modelled over the linear space of sections $\varphi = \varphi^a_A dz^A \otimes \partial_a$ of the vector bundle

$$T^*Z \underset{Z}{\otimes} \wedge E^* \underset{Z}{\otimes} E \to Z.$$

In particular, any graded connection can be represented by a sum of a graded connection preserving the $\mathbb{Z}$-gradation, e.g., a linear connection (9.2.23) and an above mentioned field φ.

The *curvature* of the graded connection ∇_τ (9.2.20) is defined by the expression (8.3.19):

$$R(\tau, \tau') = [\nabla_\tau, \nabla_{\tau'}] - \nabla_{[\tau,\tau']},$$
$$R(\tau, \tau') = \tau^A \tau'^B R^a_{AB} \partial_a : \mathcal{A}_E \to \mathcal{A}_E,$$
$$\boxed{R^a_{AB} = \partial_A \widetilde{\gamma}^a_B - \partial_B \widetilde{\gamma}^a_A + \widetilde{\gamma}^k_A \partial_k (\widetilde{\gamma}^a_B) - \widetilde{\gamma}^k_B \partial_k (\widetilde{\gamma}^a_A).} \qquad (9.2.24)$$

It can also be written in the form (8.3.12):

$$R : \mathcal{A}_E \to \mathfrak{O}^2_X \otimes \mathcal{A}_E,$$
$$R = \frac{1}{2} R^a_{AB} dz^A \wedge dz^B \otimes \partial_a. \qquad (9.2.25)$$

Remark 9.2.5. Let $Z \to X$ be a fibre bundle coordinated by (x^λ, z^i) and

$$\gamma = \Gamma + \gamma_\lambda{}^a{}_b v^b dx^\lambda \otimes \partial_a$$

a connection on the composite fibre bundle $E \to Z \to X$ which is a linear morphism over a connection Γ on $Z \to X$. Then we have the bundle monomorphism

$$\gamma_S : \wedge E^* \underset{Z}{\otimes} TX \ni u^\lambda \partial_\lambda \mapsto u^\lambda(\partial_\lambda + \Gamma^i_\lambda \partial_i + \gamma_\lambda{}^a{}_b c^b \partial_a) \in \mathcal{V}_E$$

over Z, called a *composite graded connection* on $Z \to X$. It is represented by a section

$$\boxed{\gamma_S = \Gamma + \gamma_\lambda{}^a{}_b c^b dx^\lambda \otimes \partial_a} \tag{9.2.26}$$

of the fibre bundle $T^*X \underset{Z}{\otimes} \mathcal{V}_E \to Z$ such that the composition

$$Z \overset{\gamma_S}{\to} T^*X \underset{Z}{\otimes} \mathcal{V}_E \to T^*X \underset{Z}{\otimes} (\wedge E^* \underset{Z}{\otimes} TZ) \overset{\sigma}{\to} T^*X \underset{Z}{\otimes} TZ \to T^*X \underset{Z}{\otimes} TX$$

is the pull-back onto Z of the canonical form $dx^\lambda \otimes \partial_\lambda$ on X. •

Given a graded manifold $(Z, \mathcal{A})$, the dual of the sheaf $\mathfrak{d}\mathcal{A}$ is the sheaf $\mathfrak{d}^*\mathcal{A}$ generated by the $\mathcal{A}$-module morphisms

$$\phi : \mathfrak{d}(\mathcal{A}(U)) \to \mathcal{A}(U). \tag{9.2.27}$$

One can think of its sections as being *graded exterior 1-forms* on the graded manifold $(Z, \mathcal{A})$.

In the case of a simple graded manifold $(Z, \mathcal{A}_E)$, graded 1-forms can be represented by sections of the vector bundle $\mathcal{V}^*_E \to Z$ which is the $\wedge E^*$-dual of $\mathcal{V}_E$. This vector bundle is locally isomorphic to the vector bundle

$$\mathcal{V}^*_E \mid_U \approx \wedge E^* \underset{Z}{\otimes} (\mathrm{pr}_2 V E^* \underset{Z}{\oplus} T^*Z) \mid_U, \tag{9.2.28}$$

and is characterized by the transition functions

$$v'_{j_1\dots j_k j} = \rho^{-1}{}^{a_1}_{j_1} \cdots \rho^{-1}{}^{a_k}_{j_k} \rho^{-1}{}^{a}_{j} v_{a_1\dots a_k a},$$
$$z'_{i_1\dots i_k A} = \rho^{-1}{}^{b_1}_{i_1} \cdots \rho^{-1}{}^{b_k}_{i_k} \left[z_{b_1\dots b_k A} + \frac{k!}{(k-1)!} v_{b_1\dots b_{k-1} j} \partial_A \rho^j_{b_k} \right]$$

of the bundle coordinates $(z_{a_1\dots a_k A}, v_{b_1\dots b_k j})$, $k = 0, \dots, m$, with respect to the dual bases $\{dz^A\}$ for T^*Z and $\{dc^b\}$ for $\mathrm{pr}_2 V^*E = E^*$. In view of the local isomorphism

(9.2.28), $\mathcal{V}^*_E$ as like as $\mathcal{V}_E$ can be regarded as a local Ne'eman–Quillen superbundle. The sheaf of sections of $\mathcal{V}^*_E \to Z$ is isomorphic to the sheaf $\mathfrak{d}^*\mathcal{A}_E$. Global sections of the vector bundle $\mathcal{V}^*_E \to Z$ constitute the $\mathcal{A}_E(Z)$-module of graded exterior 1-forms

$$\boxed{\phi = \phi_A dz^A + \phi_a dc^a} \tag{9.2.29}$$

on $(Z, \mathcal{A}_E)$. Then the morphism (9.2.27) can be seen as the interior product

$$\boxed{u\rfloor\phi = u^A\phi_A + (-1)^{[\phi]}u^a\phi_a.} \tag{9.2.30}$$

Given a splitting domain U of the graded manifold $(Z, \mathcal{A})$, sections of the sheaf $\mathfrak{d}^*\mathcal{A}\mid_U$ takes the form (9.2.29). If U' is another splitting domain together with the transition functions (9.2.7), the graded forms obey the transformation law

$$\phi'_a = \frac{\partial c^b(z^B, c'^j)}{\partial c'^a}\phi_b, \qquad \phi'_A = \phi_A + \partial_A c^b(z^B, c'^j)\phi_b.$$

If U and U' are trivialization charts of the same vector bundle E in Theorem 9.2.2 with the transition functions (9.2.8), we have

$$\boxed{\phi'_a = \rho^{-1}{}^b_a\phi_b, \qquad \phi'_A = \phi_A + \rho^{-1}{}^b_a\partial_A(\rho^a_j)\phi_b c^j.}$$

There is the exact sequence

$$\boxed{0 \to \wedge E^* \underset{Z}{\otimes} T^*Z \to \mathcal{V}^*_E \to \wedge E^* \underset{Z}{\otimes} \mathrm{pr}_2 VE^* \to 0.} \tag{9.2.31}$$

Any graded connection $\widetilde{\gamma}$ (9.2.19) yields the splitting of the exact sequence (9.2.31), and defines the corresponding decomposition of graded 1-forms

$$\boxed{\phi = \phi_A dz^A + \phi_a dc^a = (\phi_A + \phi_a\widetilde{\gamma}^a_A)dz^A + \phi_a(dc^a - \widetilde{\gamma}^a_A dz^A).}$$

In conclusion, let us recall the basic elements of graded exterior differential calculus [117, 155, 183].

Graded k-forms ϕ are defined as sections of the graded exterior products $\overset{k}{\wedge}\mathfrak{d}^*\mathcal{A}$ of the sheaf $\mathfrak{d}^*\mathcal{A}$ such that

$$\phi\wedge\sigma = (-1)^{|\phi||\sigma| + [\phi][\sigma]}\sigma\wedge\phi. \tag{9.2.32}$$

In particular,

$$dx^\lambda\wedge dc^i = -dc^i\wedge dx^\lambda, \qquad dc^i\wedge dc^j = dc^j\wedge dc^i. \tag{9.2.33}$$

The interior product (9.2.30) is extended to higher degree graded exterior forms by the rule

$$u\rfloor(\phi\wedge\sigma)=(u\rfloor\phi)\wedge\sigma+(-1)^{|\phi|+[\phi][u]}\phi\wedge(u\rfloor\sigma). \tag{9.2.34}$$

The *graded exterior differential* d of graded functions is introduced by the condition $u\rfloor df = u(f)$ for an arbitrary graded vector field u, and is extended uniquely to higher degree graded forms by the rules

$$d(\phi\wedge\sigma)=(d\phi)\wedge\sigma+(-1)^{|\phi|}\phi\wedge(d\sigma), \qquad d\circ d=0. \tag{9.2.35}$$

It takes the coordinate form

$$d\phi=dz^A\wedge\partial_A(\phi)+dc^a\wedge\partial_a(\phi), \tag{9.2.36}$$

where the left derivatives ∂_A, ∂_a act on the coefficients of graded forms by the rule (9.2.11), and they are graded commutative with the forms dz^A, dc^a. The Poincaré lemma is also extended to graded exterior forms [20, 183]. The Lie derivative of a graded form ϕ along a graded vector field u is given by the familiar formula

$$\mathbf{L}_u\phi=u\rfloor d\phi+d(u\rfloor\phi), \tag{9.2.37}$$

and possesses the property

$$\mathbf{L}_u(\phi\wedge\phi')=\mathbf{L}_u(\phi)\wedge\phi'+(-1)^{[u][\phi]}\phi\wedge\mathbf{L}_u(\phi').$$

Remark 9.2.6. Graded exterior k-forms can be seen as sections of the exterior products of the vector bundle $\mathcal{V}_E^*\to Z$. Therefore, the sheaves of graded exterior forms $\overset{k}{\wedge}\mathfrak{d}^*\mathcal{A}$ are fine. They constitute the fine resolution

$$0\to\mathbb{R}\to\mathcal{A}\to\mathfrak{d}^*\mathcal{A}\to\overset{2}{\wedge}\mathfrak{d}^*\mathcal{A}\to\cdots$$

of the constant sheaf $\mathbb{R}$ of real functions on the manifold Z, and define the corresponding cochain complex

$$0\to\mathbb{R}\to\mathcal{A}(Z)\overset{d}{\longrightarrow}\mathfrak{d}^*\mathcal{A}(Z)\overset{d}{\longrightarrow}(\overset{2}{\wedge}\mathfrak{d}^*\mathcal{A})(Z)\overset{d}{\longrightarrow}\cdots \tag{9.2.38}$$

of graded exterior forms on Z, called the *graded De Rham complex*. Then, by virtue of Theorem 8.3.8, there is an isomorphism

$$\boxed{H^q(Z)=H^q(Z;\mathbb{R})=H^q_{GR}(Z)} \tag{9.2.39}$$

between the De Rham cohomology groups $H^*(Z)$ of smooth exterior forms on Z and the cohomology groups $H^q_{GR}(Z)$ of the complex (9.2.38), called *graded De Rham cohomology groups* [183]. •

9.3 Connections on supervector bundles

Superconnections are introduced on supervector bundles over G-supermanifolds. There are two reasons which motivate this choice. Firstly, the category of these supervector bundles is equivalent to the category of locally free sheaves of finite rank just as it takes place in the case of smooth vector bundles (e.g., this is not the case of GH^∞-supermanifolds). It enables one to extend the familiar differential geometric notions to supervector bundles. Secondly, derivations of the structure sheaf of a G-supermanifold constitute a locally free sheaf. It is important from the differential geometric point of view (this is not the case of G^∞-supermanifolds). Moreover, this sheaf is a structure sheaf of some G-superbundle (in contrast, with graded manifolds (see Remark (9.2.3)).

We will start from the notions of a superfunction, a supermanifold and a supervector bundle (see [20] for a detailed exposition).

Superfunctions

By analogy with manifolds, supermanifolds are constructed by gluing of open subsets of supervector spaces $B^{n,m}$ by means of transition superfunctions. There are different classes of superfunctions. Nevertheless, they can be introduced in a unified manner as follows.

Let $B^{n,m} = \Lambda_0^n \oplus \Lambda_1^m$ be a supervector space, where Λ is an N-dimensional Grassmann algebra and $N \geq m$. In accordance with the decomposition (9.1.3), any element $q \in B^{n,m}$ is split uniquely as

$$q = x + y = (\sigma(x^i) + s(x^i))e_i^0 + y^j e_j^1, \tag{9.3.1}$$

where $\{e_i^0, e_j^1\}$ is a basis for $B^{n,m}$ and $\sigma(x^i) \in \mathbb{R}$, $s(x^i) \in \mathcal{R}_0$, $y^j \in \mathcal{R}_1$. By

$$\sigma^{n,m} : B^{n,m} \to \mathbb{R}^n, \qquad s^{n,m} : B^{n,m} \to \mathcal{R}^{n,m}$$

are denoted the corresponding body and soul morphisms.

Let Λ' be an N'-dimensional Grassmann algebra $(0 \leq N' \leq N)$, which is treated as a subalgebra of Λ, i.e., the basis $\{c^a\}$ for Λ' can be regarded as a subset of the basis $\{c^a, c^b\}$ for Λ. Given an open subset $U \subset \mathbb{R}^n$, let us consider a Λ'-valued graded function

$$f(z) = \sum_{k=0}^{N'} \frac{1}{k!} f_{a_1 \ldots a_k}(z) c^{a_1} \cdots c^{a_k} \tag{9.3.2}$$

with smooth coefficients $f_{a_1\cdots a_k}(z)$, $z \in \mathbb{R}^n$. The corresponding graded function $f(x)$ at a point $x \in (\sigma^{n,0})^{-1}(U) \subset B^{n,0}$ is defined as the formal Taylor series

$$f(\sigma(x)+s(x)) = f(\sigma(x)) + \sum_{k=0}^{N'} \frac{1}{k!}\left[\sum_{p=1}^{N} \frac{1}{p!}\frac{\partial^q f_{a_1\ldots a_k}(\sigma(x))}{\partial r^{i_1}\cdots\partial r^{i_p}} s(x^{i_1})\cdots s(x^{i_p})\right] c^{a_1}\cdots c^{a_k}. \tag{9.3.3}$$

Then a graded function $F(q)$ at a point $q \in (\sigma^{n,m})^{-1}(U) \subset B^{n,m}$, by definition, is given by the sum

$$F(x+y) = \sum_{r=0}^{N} \frac{1}{r!} f_{j_1\cdots j_r}(x) y^{j_1}\cdots y^{j_r}, \tag{9.3.4}$$

where $f_{j_1\cdots j_r}(x)$ are graded functions (9.3.3). Graded functions (9.3.4) are called *superfunctions* on the supervector space $B^{n,m}$. They define the sheaf $S_{N'}$ of graded commutative algebras of rank N' on $B^{n,m}$. Let $S^0_{N'}$ be its subsheaf whose sections are superfunctions $f(x+y) = f(x)$ (9.3.3) independent of the odd variables y^j. The expression (9.3.4) implies that, for any open subset U of $B^{n,m}$, there exists the epimorphism

$$\begin{gathered}\lambda : S^0_{N'}(U) \otimes \wedge\mathbb{R}^m \to S_{N'}(U), \\ \lambda : \sum_{r=0}^{N} \frac{1}{r!} f_{j_1\cdots j_r}(x) \otimes (y^{j_1}\cdots y^{j_r}) \to \sum_{r=0}^{N} \frac{1}{r!} f_{j_1\cdots j_r}(x) y^{j_1}\cdots y^{j_r},\end{gathered} \tag{9.3.5}$$

having identified $\wedge\mathbb{R}^m$ with the exterior algebra generated by the $(y^1,\ldots,y^m)$. Then we have the corresponding sheaf epimorphism

$$\lambda : S^0_{N'} \otimes \wedge\mathbb{R}^m \to S_{N'}, \tag{9.3.6}$$

where $\wedge\mathbb{R}^m$ is the constant sheaf on $B^{n,m}$.

PROPOSITION 9.3.1. The sheaf morphism (9.3.6) is injective and, consequently, an isomorphism if and only if

$$N - N' \geq m. \tag{9.3.7}$$

□

In this case, the representation of a superfunction $F(x+y)$ by the sum (9.3.4) is unique.

Using the representation (9.3.4), one can define derivatives of superfunctions. Let $f(x) \in S_{N'}(U)$ be a graded function on $U \subset B^{n,0}$. Since f, by definition, is the Taylor series (9.3.3), its partial derivative along an even coordinate x^i is defined in a natural way as

$$\partial_i f(x) = (\partial_i f)(\sigma(x) + s(x)) = \frac{\partial f}{\partial r^i}(\sigma(x)) + \tag{9.3.8}$$
$$\sum_{k=0}^{N'} \frac{1}{k!} \left[\sum_{p=1}^{N} \frac{1}{p!} \frac{\partial^{p+1} f_{a_1 \ldots a_k}(\sigma(x))}{\partial r^i \partial r^{i_1} \cdots \partial r^{i_p}} s(x^{i_1}) \cdots s(x^{i_p}) \right] c^{a_1} \cdots c^{a_k}.$$

This notion of an even derivative is extended to superfunctions F on $B^{n,m}$ when the representation (9.3.4) is not necessarily unique.

The definition of an odd derivative of superfunctions however meets difficulties. An odd derivative is defined as an image of a $\mathbb{Z}$-graded derivation of order -1 of the exterior algebra $\wedge\mathbb{R}^m$ by the morphism (9.3.5), i.e.,

$$\frac{\partial}{\partial y^j}(\lambda(f \otimes y)) = \lambda(f \otimes \partial_j(y)), \qquad y \in \wedge\mathbb{R}^m.$$

This definition is consistent only if λ is an isomorphism, i.e., the relation (9.3.7) holds. If otherwise, there exists a non-vanishing element $f \otimes y$ such that $\lambda(f \otimes y) = 0$, while $\lambda(f \otimes \partial_j(y)) \neq 0$. For example, this is an element $f \otimes (y^1 \cdots y^m)$ if $N - N' = m - 1$.

Example 9.3.1. We will follow below the terminology of [20].

If $N' = 0$, the sheaf $S_{N'}$ coincides with the sheaf $\mathcal{H}^\infty$ of H^∞-superfunctions (first considered by M.Batchelor [23] and B. DeWitt [79]). In this case, superfunctions (9.3.4) read

$$F(x+y) = \sum_{r=0}^{N} \frac{1}{r!} \left[\sum_{p=0}^{N} \frac{1}{p!} \frac{\partial^p f_{j_1 \ldots j_r}(\sigma(x))}{\partial r^{i_1} \cdots \partial r^{i_p}} s(x^{i_1}) \cdots s(x^{i_p}) \right] y^{j_1} \cdots y^{j_r}. \tag{9.3.9}$$

If $N' = N$, we deal with G^∞-superfunctions, introduced by A. Rogers [257]. In this case, the inequality (9.3.7) does not hold, unless $m = 0$.

If the condition (9.3.7) is fulfilled, superfunctions are called GH^∞-superfunctions. They include H^∞-superfunctions as a particular case.

Superfunctions of the above mentioned types are called *smooth superfunctions*.

•

Let $\mathcal{GH}_{N'}$ denote the sheaf of GH^∞-superfunctions on a supervector space $B^{n,m}$. Let us define the sheaf of graded commutative Λ-algebras

$$\mathcal{G}_{N'} = \mathcal{GH}_{N'} \otimes \Lambda, \tag{9.3.10}$$

where Λ is regarded as a graded algebra over Λ'. The sheaf $\mathcal{G}_{N'}$ (9.3.10) possesses the following important properties [20].

- There is the *evaluation morphism*

$$\begin{aligned} &\delta : \mathcal{G}_{N'} \to C^\Lambda_{B^{n,m}}, \\ &\delta : F \otimes a \mapsto Fa, \qquad F \in \mathcal{GH}_{N'}, \qquad a \in \Lambda, \end{aligned} \tag{9.3.11}$$

  where $C^\Lambda_{B^{n,m}} \cong C^0_{B^{n,m}} \otimes \Lambda$ is the sheaf of continuous Λ-valued functions on $B^{n,m}$. This morphism enables one to evaluate germs of sections of $\mathcal{G}_{N'}$. Its image is isomorphic to the sheaf $\mathcal{G}^\infty$ of G^∞-superfunctions on $B^{n,m}$.

- For any two integers N' and N'' satisfying the condition (9.3.7), there is the canonical isomorphism of sheaves of graded commutative Λ-algebras $\mathcal{G}_{N'}$ and $\mathcal{G}_{N''}$. Therefore, we can define a canonical sheaf $\mathcal{G}_{n,m}$ of graded commutative Λ-algebras on the supervector space $B^{n,m}$. With no loss of generality, one can think of sections of $\mathcal{G}_{n,m}$ as being the tensor products $F \otimes a$ of H^∞-superfunctions F (9.3.9) and elements $a \in \Lambda$. We call these sections the *G-functions.*

- The sheaf $\mathfrak{d}\mathcal{G}_{n,m}$ of graded derivations of $\mathcal{G}_{n,m}$ is locally free over $\mathcal{G}_{n,m}$, of rank (n, m). On every open set $U \subset B^{n,m}$, the $\mathcal{G}_{n,m}(U)$-module $\mathfrak{d}\mathcal{G}_{n,m}(U)$ is generated by the derivations $\partial/\partial x^i$, $\partial/\partial y^j$ which act on $\mathcal{G}_{n,m}(U)$ by the rule.

$$\frac{\partial}{\partial x^i}(F \otimes a) = \frac{\partial F}{\partial x^i} \otimes a, \qquad \frac{\partial}{\partial y^j}(F \otimes a) = \frac{\partial F}{\partial y^j} \otimes a. \tag{9.3.12}$$

Supermanifolds

Turn now to the notion of a supermanifold.

DEFINITION 9.3.2. A paracompact topological space M is said to be an (n, m)-dimensional *smooth supermanifold* if it admits an atlas

$$\Psi = \{U_\zeta, \phi_\zeta\}, \qquad \phi_\zeta : U_\zeta \to B^{n,m},$$

such that the transition functions $\phi_\zeta \circ \phi_\xi^{-1}$ are supersmooth morphisms. □

Obviously, a smooth supermanifold of dimension (n, m) is also a real smooth manifold of dimension $2^{N-1}(n+m)$.

If transition superfunctions are H^∞, G^∞ or GH^∞-superfunctions, one deals with H^∞, G^∞ or GH^∞ supermanifolds, respectively. By virtue of Proposition 8.3.2 extended to graded locally ringed spaces, the preceding definition is equivalent to the following one.

DEFINITION 9.3.3. A smooth supermanifold is a locally ringed space (M, S) which is locally isomorphic to $(B^{n,m}, \mathcal{S})$, where $\mathcal{S}$ is one of the sheaves of smooth superfunctions on $B^{n,m}$ under consideration. The sheaf S is called the *structure sheaf* of a supermanifold. □

By a morphism of smooth supermanifolds is meant their morphism (φ, Φ) as locally ringed spaces, where Φ is an even graded morphism. In particular, every morphism $\varphi : M \to M'$ yields the smooth supermanifold morphism $(\varphi, \Phi = \varphi^*)$. Thus, a choice of a class of superfunctions on a smooth supermanifold determines a class of morphisms of this supermanifold.

Note that smooth supermanifolds are effected by serious inconsistencies as follows.

Since it is impossible to define derivatives of G^∞-superfunctions with respect to odd variables, the sheaf of derivations of the sheaf of G^∞-superfunctions is not locally free and the transition functions of the G^∞-tangent bundle are not the Jacobian matrices (see Example 9.3.6 below). Nevertheless, any G-supermanifold considered below has an underlying G^∞-supermanifold.

In the case of GH^∞-supermanifolds, one meets the phenomenon that the space of values of GH^∞-superfunctions changes from point to point since the Grassmann algebra Λ of rank N is not a free module with respect to the subalgebra Λ' of rank N'. This fact leads to difficulties in the definition of GH^∞ vector bundles if one follows the construction of smooth vector bundles in Example 8.3.3.

By these reasons, supervector bundles in the category of G-supermanifolds usually are considered. With the canonical sheaf $\mathcal{G}_{n,m}$ of graded commutative Λ-algebras on the supervector space $B^{n,m}$, one gives the following definition of G-supermanifolds.

DEFINITION 9.3.4. An (n, m) dimensional *G-supermanifold* is a graded locally Λ-ringed space (M, G_M) satisfying the following conditions:

- M is a paracompact topological space;
- (M, G_M) is locally isomorphic to $(B^{n,m}, \mathcal{G}_{n,m})$;
- there exists a morphism of sheaves of Λ-algebras $\delta : G_M \to C_M^\Lambda$, where $C_M^\Lambda \cong C_M^0 \otimes \Lambda$ is sheaf of continuous Λ-valued functions on M, and δ is locally isomorphic to the evaluation morphism (9.3.11).

□

Example 9.3.2. The triple $(B^{n,m}, \mathcal{G}_{n,m}, \delta)$, where δ is the evaluation morphism (9.3.11), is called a *standard supermanifold.* •

Remark 9.3.3. Any GH^∞-supermanifold (M, GH_M^∞) with the structure sheaf GH_M^∞ is naturally extended to the G-supermanifold $(M, GH_M^\infty \otimes \Lambda)$. Every G-supermanifold defines an *underlying G^∞-supermanifold* $(M, \delta(G_M))$, where $\delta(G_M) = G_M^\infty$ is the sheaf of G^∞-superfunctions on M. •

Morphisms of G-supermanifolds are morphisms of graded locally ringed spaces. In particular, every morphism (φ, φ^*) of GH^∞-supermanifolds

$$(M, GH_M^\infty) \to (M', GH_{M'}^\infty)$$

is extended trivially to the morphism (φ, Φ) of G-supermanifolds

$$(M, GH_M^\infty) \otimes \Lambda \to (M', GH_{M'}^\infty \otimes \Lambda)$$

where $\Phi(F \otimes \lambda) = \varphi^*(F) \otimes \lambda$.

As in the case of smooth supermanifolds, the underlying space M of a G-supermanifold (M, G_M) is provided with the structure of a real smooth manifold of dimension $2^{N-1}(n+m)$, and morphisms of G-supermanifolds are smooth morphisms of the underlying smooth manifolds. Nevertheless, it may happen that non-isomorphic G-supermanifold have isomorphic underlying smooth manifolds.

Similarly to the properties of the sheaf $\mathfrak{d}\mathcal{G}_{n,m}$, the sheaf $\mathfrak{d}G_M$ of graded derivations of G_M is locally free, with the local bases $\{\partial/\partial x^i, \partial/\partial y^j\}$. The *supertangent space*

$T_q(M, G_M)$ to the G-supermanifold (M, G_M) at a point $q \in M$ is the graded Λ-module of graded derivations $G_{Mq} \to \Lambda$. It is isomorphic to the quotient $\mathfrak{d}G_{Mq}/(\mathcal{M}_q \cdot \mathfrak{d}G_{Mq})$, and its basis is given by the elements

$$\left(\frac{\partial}{\partial x^i}\right)_q (s_q) = \frac{\partial s}{\partial x^i}(q), \qquad \left(\frac{\partial}{\partial y^j}\right)_q (s_q) = \frac{\partial s}{\partial y^j}(q), \qquad \forall s \in G_{Mq}.$$

For any open subset $U \subset B^{n,m}$, the space $\mathcal{G}_{n,m}(U)$ can be provided with the topology such that it is a graded Fréchet algebra. There are isometrical isomorphisms

$$\mathcal{G}_{n,m}(U) \cong \mathcal{H}^\infty(U) \otimes \Lambda \cong C^\infty(\sigma^{n,m}(U)) \otimes \Lambda \otimes \wedge\mathbb{R}^m \cong C^\infty(\sigma^{n,m}(U)) \otimes \wedge\mathbb{R}^{N+m}. \tag{9.3.13}$$

Remark 9.3.4. We present briefly the axiomatic approach to supermanifolds which enables one to obtain, for different choice of a graded commutative algebra, all the previously known types of supermanifolds. These are R^∞-supermanifolds [20, 21, 45]. This approach to supermanifolds develops that of M.Rothstein [258] (see [20, 21] for a discussion). The R^∞-supermanifolds are introduced over the above mentioned Arens–Michael algebras of Grassmann origin [45], but we omit here the topological side of their definition though just the topological properties differ R^∞-supermanifolds from *R-supermanifolds* of M.Rothstein.

Let Λ be a real graded commutative algebra of the above mentioned type (for the sake of simplicity, the reader can think of Λ as being a Grassmann algebra). A *superspace* over Λ is a triple $(M, \mathcal{A}, \delta)$, where M is a paracompact topological space, $\mathcal{A}$ is a sheaf of Λ-algebras, and $\delta : \mathcal{A} \to C^\Lambda_M$ is an evaluation morphism to the sheaf C^Λ_M of continuous Λ-valued functions on M. Sections of $\mathcal{A}$ are called R^∞-*superfunctions.* We define the graded ideal $\mathcal{M}_q$ of the stalk $\mathcal{A}_q$, $q \in M$, formed by the germs of R^∞-superfunctions f vanishing at a point q, i.e., such that $\delta(f)(q) = 0$.

An R^∞-*supermanifold* of dimension (n, m) is a superspace $(M, \mathcal{A}, \delta)$ satisfying the following four axioms [45].

Axiom 1. The graded $\mathcal{A}$-dual $\mathfrak{d}^*\mathcal{A}$ of the sheaf of derivations is a locally free graded $\mathcal{A}$-module of rank (n, m). Every point $q \in M$ has an open neighbourhood U with sections $x^1, \cdots, x^n \in \mathcal{A}(U)_0$, $y^1, \cdots, y^m \in \mathcal{A}(U)_1$ such that $\{dx^i, dy^j\}$ is a graded basis of $\mathfrak{d}^*\mathcal{A}(U)$ over $\mathcal{A}(U)$.

Axiom 2. Given the above mentioned coordinate chart, the assignment

$$q \to (\delta(x^i), \delta(y^j))$$

defines a homeomorphism of U onto an open subset in $B^{n,m}$
Axiom 3. For every $q \in M$, the ideal $\mathcal{M}_q$ is finitely generated.
Axiom 4. For every open subset $U \subset M$, the topological algebra $\mathcal{A}(U)$ is Hausdorff and complete.

An R-supermanifold over a graded commutative Banach algebra, satisfying Axiom 4, is an R^∞-supermanifold.

A standard supermanifold in Example 9.3.2 is an R^∞-supermanifold. Moreover, in the case of a finite Grassmann algebra Λ, the category of R^∞ supermanifolds and the category of G-supermanifolds are equivalent. •

Let (M, G_M) be a G-supermanifold. As was mentioned above, it satisfies Axioms 1-4. Sections u of the sheaf $\mathfrak{d}G_M$ of graded derivations are called *supervector fields* on the G-supermanifold (M, G_M), while sections ϕ of the dual sheaf $\mathfrak{d}^*G_M$ are *1-superforms* on (M, G_M). Given a coordinate chart $(q^i) = (x^i, y^j)$ on $U \subset M$, supervector fields and 1-superforms read

$$u = u^i\partial_i, \qquad \phi = \phi_i dq^i,$$

where coefficients u^i and ϕ_i are G-functions on U. The graded exterior differential calculus on supervector fields and superforms obeys the same formulas (9.2.15), (9.2.30) – (9.2.37) as that for graded vector fields and graded forms.

Let us consider the cohomology of G-supermanifolds. Given a G-supermanifold (M, G_M), let $\mathfrak{O}^*_{\Lambda M} = \mathfrak{O}^*_M \otimes \Lambda$ be the sheaves of smooth Λ-valued exterior forms on M. These sheaves are fine, and constitute the fine resolution

$$0 \to \Lambda \to C^\infty_M \otimes \Lambda \to \mathfrak{O}^1_M \otimes \Lambda \to \cdots$$

of the constant sheaf Λ on M. We have the corresponding De Rham complex of Λ-valued exterior forms on M

$$0 \to \Lambda \to C^\infty_\Lambda(M) \to \mathfrak{O}^1_\Lambda(M) \to \cdots .$$

By virtue of Theorem 8.3.8, the cohomology groups $H^*_\Lambda(M)$ of this complex are isomorphic to the cohomology groups $H^*(M;\Lambda)$ with coefficients in the constant sheaf Λ on M and, consequently, are related to the De Rham cohomology as follows:

$$\boxed{H^*_\Lambda(M) = H^*(M;\Lambda) = H^*(M) \otimes \Lambda.} \tag{9.3.14}$$

Thus, the cohomology groups of Λ-valued forms do not provide us with information on the G-supermanifold structure of M.

Let us turn to cohomology of exterior superforms. The sheaves $\overset{k}{\wedge}\mathfrak{d}^*G_M$ of exterior superforms constitute the sequence

$$0 \to \Lambda \to G_M \to \mathfrak{d}^*G_M \to \cdots. \tag{9.3.15}$$

The Poincaré lemma is extended to superforms [44], and this sequence is exact. However, the structure sheaf G_M is not acyclic in general. It follows that the exact sequence (9.3.15) fails to be a resolution of the constant sheaf Λ on M (this is not the terminology of [20]), and the homology groups $H^*_S(M)$ of the De Rham complex of exterior superforms are not equal to the cohomology groups $H^*(M;\Lambda)$ and the De Rham cohomology groups $H^*(M)$ of the manifold M. In particular, the cohomology groups $H^*_S(M)$ are not topological invariants, but they are invariant under G-isomorphisms of G-supermanifolds.

PROPOSITION 9.3.5. The structure sheaf $\mathcal{G}_{n,m}$ of the standard G-supermanifold $(B^{n,m}, \mathcal{G}_{n,m})$ is acyclic, i.e.,

$$\boxed{H^{k>0}(B^{n,m}; \mathcal{G}_{n,m}) = 0.}$$

□

The proof is based on the isomorphism (9.3.13) and some cohomological constructions [20, 45].

DeWitt supermanifolds

There exists a particular class of supermanifolds, called DeWitt supermanifolds. Their notion implies introducing in $B^{n,m}$ a topology, called the *DeWitt topology*, which is coarser than the Euclidean one. This is the coarsest topology such that the body projection $\sigma^{n,m} : B^{n,m} \to \mathbb{R}^n$ is continuous. The open sets in the DeWitt topology have the form $V \times \mathcal{R}^{n,m}$, where V are open sets in $\mathbb{R}^n$. Obviously, this topology is not Hausdorff.

DEFINITION 9.3.6. A smooth supermanifold [G-supermanifold, R^∞-supermanifold] is said to be a *DeWitt supermanifold* if it admits an atlas such that the local morphisms $\phi_\zeta : U_\zeta \to B^{n,m}$ in Definition 9.3.2 [Definition 9.3.4, Axiom 2] are continuous in the DeWitt topology, i.e., $\phi_\zeta(U_\zeta) \subset B^{n,m}$ are open in the DeWitt topology. □

In particular, the G^∞-supermanifold underlying a DeWitt G-supermanifold is also a DeWitt supermanifold, and so is the G-extension of a DeWitt GH^∞- supermanifold.

Given an atlas (U_ζ, ϕ_ζ) of a DeWitt supermanifold in accordance with Definition 9.3.6, it is readily observed that its transition functions $\phi_\zeta \circ \phi_\xi^{-1}$ must preserve the fibration $\sigma^{n,m} : B^{n,m} \to \mathbb{R}^n$ whose fibre $(\sigma^{n,m})^{-1}(z)$ over $z \in \mathbb{R}^n$ is equipped with the coarsest topology where $\emptyset$ and $(\sigma^{n,m})^{-1}(z)$ only are open sets. It states the following fact.

PROPOSITION 9.3.7. Every DeWitt supermanifold is a locally trivial topological fibre bundle

$$\sigma_M : M \to Z_M \tag{9.3.16}$$

over an n-dimensional smooth manifold Z_M with the typical fibre $\mathcal{R}^{n,m}$. □

The base Z_M of the fibre bundle (9.3.16) is said to be a *body manifold* of M, while the projection σ_M is called a *body map* of a DeWitt supermanifold.

There is the important correspondence between the DeWitt H^∞-supermanifolds and the above studied graded manifolds. This correspondence is based on the following facts.

- Given a graded manifold $(Z, \mathcal{A})$, its structure sheaf $\mathcal{A}$, by definition, is locally isomorphic to the sheaf $C_U^\infty \otimes \wedge\mathbb{R}^m$ for any suitable $U \subset Z$.

- Given a DeWitt H^∞-supermanifold (M, H_M^∞) and the body map (9.3.16), Proposition 9.3.1 implies that the direct image $\sigma_*(H_M^\infty)$ on Z_M of the sheaf H_M^∞ is locally isomorphic to the sheaf $C_{Z_M}^\infty \otimes \wedge\mathbb{R}^m$. The expression (9.3.9) shows this isomorphism in an explicit form.

- Moreover, the spaces (M, H_M^∞) and $(Z_M, \sigma(H_M^\infty))$ determine the same element of $H^1(Z_M; \mathrm{Aut}\,(\wedge\mathbb{R}^m)_\infty)$.

It states the following Theorem [20, 23].

THEOREM 9.3.8. Given a DeWitt H^∞-supermanifold (M, H_m^∞), the associated pair $(Z_M, \sigma_*(H_M^\infty))$ is a graded manifold. Conversely, for any graded manifold $(Z, \mathcal{A})$, there exists a DeWitt H^∞-supermanifold whose body manifold is Z and $\mathcal{A}$ is isomorphic to $\sigma_*(H_M^\infty)$. □

COROLLARY 9.3.9. By virtue of Batchelor's Theorem 9.2.2 and Theorem 9.3.8, there is one-to-one correspondence between the classes of isomorphic DeWitt H^∞-supermanifolds of odd rank m, with a body manifold Z, and the classes of equivalent m-dimensional vector bundles over Z. □

This result is extended to DeWitt GH^∞-, G^∞- and G-supermanifolds.

Let us say something more on DeWitt G-supermanifolds, which can be utilized as base of supervector bundles.

PROPOSITION 9.3.10. The structure sheaf G_M of a DeWitt G-supermanifold is acyclic, and so is any locally free sheaves of G_M-modules [20, 45]. □

PROPOSITION 9.3.11. [251]. The cohomology groups $H^*_\Lambda(M)$ of the De Rham complex of exterior superforms on a DeWitt G-supermanifold are isomorphic with the De Rham cohomology groups (9.3.14) of Λ-valued exterior forms on the body manifold Z_M, i.e.,

$$\boxed{H^*_\Lambda(M) = H^*(Z_M) \otimes \Lambda.} \tag{9.3.17}$$

□

These results are based on the fact that the structure sheaf G_M on M provided with the DeWitt topology is fine. However, this does not imply automatically that G_M is acyclic since the DeWitt topology is not paracompact. Nevertheless, it follows that the image $\sigma_*(G_M)$ on the body manifold Z_M is fine and acyclic. Then combining Propositions 8.3.7 and 9.3.5 leads to Proposition 9.3.10. In particular, the sheaves of exterior superforms on a DeWitt G-supermanifold are acyclic. Then they constitute the resolution of the constant sheaf Λ on M, and one gains the isomorphisms

$$H^*_\Lambda(M) = H^*(M;\Lambda) = H^*(M) \otimes \Lambda.$$

Since the typical fibre of the fibre bundle $M \to Z_M$ is contractible, then $H^*(M) = H^*(Z_M)$ such that the isomorphisms (9.3.17) take place.

Supervector bundles

As was manifested above, let us consider vector bundles in the category of G-supermanifolds (see [20] for a detailed exposition).

We will start from the definition of the product of two G-supermanifolds. Let $(B^{n,m}, \mathcal{G}_{n,m})$ and $(B^{r,s}, \mathcal{G}_{r,s})$ be two standard supermanifolds in Example 9.3.4. Given open sets $U \subset B^{n,m}$ and $V \subset B^{r,s}$, we consider the presheaf

$$U \times V \to \mathcal{G}_{n,m}(U) \widehat{\otimes} \mathcal{G}_{r,s}(V), \tag{9.3.18}$$

where $\widehat{\otimes}$ denotes the tensor product of modules completed in Grothendieck's topology (see Remark 8.1.8). Due to the isomorphism (9.3.13), it is readily observed that the structure sheaf $\mathcal{G}_{n+r,m+s}$ of the standard supermanifold on $B^{n+r,m+s}$ is isomorphic to that defined by the presheaf (9.3.18). This construction is generalized to arbitrary G-supermanifolds as follows.

PROPOSITION 9.3.12. Let (M, G_M) and $(M', G_{M'})$ be two G-supermanifolds of dimensions (n, m) and (r, s), respectively. Their *product* $(M, G_M) \times (M', G_{M'})$ is defined as the graded locally ringed space $(M \times M', G_M \widehat{\otimes} G_{M'})$, where $G_M \widehat{\otimes} G_{M'}$ is the sheaf constructed from the presheaf

$$U \times U' \to G_M(U) \widehat{\otimes} G_{M'}(U'),$$
$$\delta : G_M(U) \widehat{\otimes} G_{M'}(U') \to C^\infty_{\sigma(U)} \widehat{\otimes} C^\infty_{\sigma(U')} = C^\infty_{\sigma_M(U) \times \sigma_M(U')},$$

for any open subsets $U \subset M$ and $U' \subset M'$. This product is a G-supermanifold of dimension $(n + r, m + s)$. □

Moreover, there are the epimorphisms

$$\mathrm{pr}_1 : (M, G_M) \times (M', G_{M'}) \to (M, G_M),$$
$$\mathrm{pr}_2 : (M, G_M) \times (M', G_{M'}) \to (M', G_{M'}).$$

One may define a section, e.g., of the fibration pr_1 over an open subset $U \subset M$ as the G-supermanifold morphism

$$s_U : (U, G_M|_U) \to (M, G_M) \times (M', G_{M'})$$

such that $\mathrm{pr}_1 \circ s_U$ is the identity morphism of $(U, G_M|_U)$. Sections s_U for all open subsets $U \subset M$ define a sheaf on M. We are interested in providing this sheaf with a suitable graded G_M-structure. Recall that, in the case of a smooth vector bundle over a manifold X, the sheaf of its section is a sheaf of C^∞_X-modules (see Example 8.3.3).

For this purpose, let us consider the product

$$(M, G_M) \times (B^{r|s}, \mathcal{G}_{r|s}), \tag{9.3.19}$$

where $B^{r|s}$ is the graded envelope (9.1.4). It is called a *product G-supermanifold.* Since the Λ_0-modules $B^{r|s}$ and $B^{r+s,r+s}$ are isomorphic, $B^{r|s}$ has a natural structure of an $(r+s, r+s)$-dimensional G-supermanifold. Because $B^{r|s}$ is a free graded Λ-module of the type (r, s), the sheaf $S_M^{r|s}$ of sections of the fibration

$$(M, G_M) \times (B^{r|s}, \mathcal{G}_{r|s}) \to (M, G_M) \tag{9.3.20}$$

has the structure of the sheaf of free graded G_M-modules of rank (r, s). Conversely, given a G-supermanifold (M, G_M) and a sheaf S of free graded G_M-modules of rank (r, s) on M, there exists a product G-supermanifold (9.3.19) such that S is isomorphic to the sheaf of sections of the fibration (9.3.20).

Turn now to the notion of a supervector bundle over G-supermanifolds. Similarly to smooth vector bundles (see Example 8.3.3), one can require that the category of supervector bundles over G-supermanifolds to be equivalent to the category of locally free graded sheaves on G-supermanifolds. Then we can restrict ourselves to locally trivial *G-superbundles* with the standard fibre $B^{r|s}$.

DEFINITION 9.3.13. A *supervector bundle* over a G-supermanifold (M, G_M) with the standard fibre $(B^{r|s}, \mathcal{G}_{r|s})$ is a pair $((Y, G_Y), \pi)$ of a G-supermanifold (Y, G_Y) and a G-epimorphism

$$\boxed{\pi : (Y, G_Y) \to (M, G_M)} \tag{9.3.21}$$

such that M admits an open covering $\{U_\varsigma\}$ with a set of local G-isomorphisms

$$\psi_\varsigma : (\pi^{-1}(U_\varsigma), G_Y|_{\pi^{-1}(U_\varsigma)}) \to (U_\varsigma, G_M|_{U_\varsigma}) \times (B^{r|s}, \mathcal{G}_{r|s}).$$

□

It is clear that sections of the supervector bundle (9.3.21) constitute a sheaf of locally free graded G_M-modules. A converse of this fact is the following [20].

PROPOSITION 9.3.14. For any sheaf S of locally free graded G_M-modules of rank (r, s) on a G-manifold (M, G_M), there exists a supervector bundle over (M, G_M) such that S is isomorphic to the sheaf of its sections. □

Given a sheaf S as in Proposition 9.3.14, called the *structure sheaf*, the fibre Y_q, $q \in M$, of the above mentioned supervector bundle is the quotient

$$S_q/\mathcal{M}_q \cong S^{r|s}_{Mq}/(\mathcal{M}_q \cdot S^{r|s}_{Mq}) \cong B^{r|s}$$

of the stalk S_q by the submodule $\mathcal{M}_q$ of the germs $s \in S_q$ whose evaluation $\delta(f)(q)$ vanishes. This fibre is a graded Λ-module isomorphic to $B^{r|s}$. It is provided with the structure of a standard supermanifold.

Remark 9.3.5. The proof of Proposition 9.3.14 uses the fact that, given the transition functions $\rho_{\varsigma\xi}$ of the sheaf S, its evaluations

$$g_{\varsigma\xi} = \delta(\rho_{\varsigma\xi}) \tag{9.3.22}$$

define the morphisms $U_\varsigma \cap U_\xi \to GL(r|s;\Lambda)$ and constitute the cocycle of the sheaf G^∞-morphisms from M to the general linear graded group $GL(r|s;\Lambda)$. Thus, we come to the notion of a G^∞-*vector bundle*. Its definition is the repetition of Definition 9.3.13 if one replaces G-supermanifolds and G-morphisms with the G^∞-ones. Moreover, the G^∞-supermanifold underlying a supervector bundle (see Remark 9.3.3) is a G^∞-supervector bundle whose transition functions $g_{\varsigma\xi}$ are related to those of the supervector bundle by the evaluation morphisms (9.3.22), and are $GL(r|s;\Lambda)$-valued transition functions. •

Since the category of supervector bundles over a G-supermanifold (M, G_M) is equivalent to the category of locally free sheaves of G_M-modules, one can define the usual operations of direct sum, tensor product, etc. of supervector bundles.

Note that any supervector bundle admits the canonical global zero section. Any section of the supervector bundle π (9.3.21), restricted to its a trivialization chart

$$(U, G_M\mid_U) \times (B^{r|s}, \mathcal{G}_{r|s}), \tag{9.3.23}$$

is represented by a sum $s = s^a(q)\epsilon_a$, where $\{\epsilon_a\}$ is the basis for the Λ-module $B^{r|s}$, while $s^a(q)$ are G-functions on U. Given another trivialization chart U' of π, a transition function

$$s'^b(q)\epsilon'_b = s^a(q)h^b{}_a(q)\epsilon_b, \qquad q \in U \cap U', \tag{9.3.24}$$

is given by the $(r+s)\times(r+s)$ matrix h whose entries $h^b{}_a(q)$ are G-functions on $U \cap U'$. One can think of this matrix as being a section of the supervector bundle over $U \cap U$ with the above mentioned group $GL(r|s;\Lambda)$ as a typical fibre.

Example 9.3.6. Given a G-supermanifold (M, G_M), let us consider the locally free graded sheaf $\mathfrak{d}G_M$ of graded derivations of G_M. In accordance with Proposition 9.3.14, there is a supervector bundle $T(M, G_M)$, called *supertangent bundle*, whose sheaf of sections is isomorphic to $\mathfrak{d}G_M$. If $(q^1, \ldots, q^{m+n})$ and $(q'^1, \ldots, q'^{m+n})$ are two coordinate charts on M, the Jacobian matrix

$$h^i_j = \frac{\partial q'^i}{\partial q^j}, \qquad i, j = 1, \ldots, n+m,$$

(see the prescription (9.3.12)) provides the transition morphisms for $T(M, G_M)$. Of course, the fibre of the supertangent bundle $T(M, G_M)$ at a point $q \in M$ is the above mentioned supertangent space $T_q(M, G_M)$.

It should be emphasized that the underlying G^∞-vector bundle of the supertangent bundle $T(M, G_M)$, called G^∞*-supertangent bundle*, has the transition functions $\delta(h^i_j)$ which cannot be written as the Jacobian matrices since the derivatives of G^∞-superfunctions with respect to odd variables are ill-defined and the sheaf $\mathfrak{d}G^\infty_M$ is not locally free. •

Superconnections

Given a supervector bundle π (9.3.21) with the structure sheaf S of its sections, a connection on this supervector bundle is defined as in Definition 8.3.6. The difference is only that S is a sheaf of graded locally free G_M-modules.

As in the case of the exact sequence (8.3.5) in Section 8.3, one can obtain the exact sequence of sheaves

$$\boxed{0 \to \mathfrak{d}^* G_M \otimes S \to (G_M \oplus \mathfrak{d}^* G_M) \otimes S \to S \to 0} \qquad (9.3.25)$$

as a direct limit of the exact sequence (8.1.35) for the graded $G_M(U)$-modules $S(U)$, where by $\mathrm{mod}\mu^2$ is meant the quotient by the graded relations

$$\delta^a \circ \delta^b(\mathbf{1} \otimes p) = \mathbf{1} \otimes (abp) - a \otimes (bp) - (-1)^{[a][b]} b \otimes (ap) + ab \otimes p = 0,$$
$$a \otimes b + (-1)^{[a][b]} b \otimes a = ab \otimes \mathbf{1} + \mathbf{1} \otimes ab,$$

where

$$\delta^b(a \otimes p) = (-1)^{[a][b]} a \otimes (bp) - (ba) \otimes p \qquad (9.3.26)$$

(cf. (8.1.11), (8.1.15) and (8.1.18)). The exact sequence (9.3.25) needs not be split in general. It admits a splitting if and only if there exists an even sheaf morphism

$$\boxed{\nabla : S \to \mathfrak{d}^* G_M \otimes S} \tag{9.3.27}$$

satisfying the Leibniz rule

$$\boxed{\nabla(fs) = df \otimes s + f\nabla(s),} \qquad f \in G_M(U), \qquad s \in S(U), \tag{9.3.28}$$

for any open subset $U \in M$.

DEFINITION 9.3.15. The sheaf morphism (9.3.27) is called a *superconnection* on the supervector bundle π (9.3.21). □

The *curvature* of the superconnection (9.3.27) is given by the expression

$$\boxed{R = \nabla^2 : S \to \overset{2}{\wedge} \mathfrak{d}^*_M \otimes S,} \tag{9.3.29}$$

similar to the expression (8.3.12).

As in the case of sheaves of C^∞_X-modules, the exact sequence (9.3.25) leads to the exact sequence of sheaves

$$0 \to Hom(S, \mathfrak{d}^* G_M \otimes S) \to Hom(S, (G_M \oplus \mathfrak{d}^* G_M) \otimes S) \to Hom(S, S) \to 0$$

and to the corresponding exact sequence of the cohomology groups

$$0 \to H^0(M; Hom(S, \mathfrak{d}^* G_M \otimes S)) \to H^0(M; Hom(S, (G_M \oplus \mathfrak{d}^* G_M) \otimes S)) \to$$
$$H^0(M; Hom(S, S)) \to H^1(M; Hom(S, \mathfrak{d}^* G_M \otimes S)) \to \cdots .$$

The exact sequence (9.3.25) defines the Atiyah class $\mathrm{At}(\pi) \in H^1(M; Hom(S, \mathfrak{d}^* G_M \otimes S))$ of the supervector bundle π (9.3.21). If $\mathrm{At}(\pi)$ vanishes, a superconnection on this supervector bundle exists (see Section 8.3). Of course, a superconnection exists if the cohomology group $H^1(M; Hom(S, \mathfrak{d}^* G_M \otimes S))$ is trivial. In contrast with smooth vector bundles, the structure sheaf G_M of a G-supermanifold is not acyclic in general, the sheaf $Hom(S, \mathfrak{d}^* G_M \otimes S)$ has non-trivial cohomology, and a supervector bundle does not admit necessarily a superconnection.

Example 9.3.7. For instance, the structure sheaf of the standard supermanifold $(B^{n,m}, \mathcal{G}_{n,m})$ is acyclic (see Proposition 9.3.5), and a supervector bundle

$$(B^{n,m}, \mathcal{G}_{n,m}) \times (B^{r|s}, \mathcal{G}_{r|s}) \to (B^{n,m}, \mathcal{G}_{n,m}) \tag{9.3.30}$$

has obviously a superconnection, e.g., the trivial superconnection. •

Example 9.3.8. If (M, G_M) is a DeWitt G-supermanifold, its structure sheaf G_M is acyclic, and so is the locally free sheaf $Hom(S, \mathfrak{d}^* G_M \otimes S)$ (see Proposition 9.3.10). It follows that supervector bundles over DeWitt G-supermanifolds admit superconnections. •

Example 9.3.7 enables one to gain a local coordinate expression of a superconnection on a supervector bundle π (9.3.21) with the typical fibre $B^{r|s}$ and with the base G-supermanifold locally isomorphic to the standard supermanifold $(B^{n,m}, \mathcal{G}_{n,m})$. Let $U \subset M$ (9.3.23) be a trivialization chart of this supervector bundle such that every section s of $\pi|_U$ is represented by a sum $s^a(q)\epsilon_a$, while the sheaf of superforms $\mathfrak{d}^* G_M|_U$ has a local basis $\{dq^i\}$. Then a superconnection ∇ (9.3.27) restricted to this trivialization chart can be given by a collection of coefficients

$$\nabla(\epsilon_a) = dq^i \otimes (\nabla_i{}^b{}_a \epsilon_b) \tag{9.3.31}$$

where $\nabla_i{}^a{}_b$ are G-functions on U. Bearing in mind the Leibniz rule (9.3.28), one can compute the coefficients of the curvature form (9.3.29) of the superconnection (9.3.31). We have

$$\begin{aligned} &R(\epsilon_a) = \frac{1}{2} dq^i \wedge dq^j \otimes R_{ij}{}^b{}_a \epsilon_b, \\ &R_{ij}{}^a{}_b = (-1)^{[i][j]} \partial_i \nabla_j{}^a{}_b - \partial_j \nabla_i{}^a{}_b + (-1)^{[i]([j]+[a]+[k])} \nabla_j{}^a{}_k \nabla_i{}^k{}_b - \\ &\qquad (-1)^{[j]([a]+[k])} \nabla_i{}^a{}_k \nabla_j{}^k{}_b. \end{aligned}$$

In a similar way, one can obtain the transformation law of the superconnection coefficients (9.3.31) under the transition morphisms (9.3.24). In particular, any trivial supervector bundle admits the trivial superconnection $\nabla_i{}^b{}_a = 0$.

9.4 Principal superconnections

In contrast with a supervector bundle, the structure sheaf G_P of a principal superbundle $(P, G_P) \to (M, G_M)$ is not a sheaf of locally free G_M-modules in general. Therefore, the above technique of connections on modules and sheaves is not applied to principal superconnections in a straightforward way. Principal superconnections are introduced on principal superbundles by analogy with principal connections on smooth principal bundles [20] (compare, e.g., the exact sequence (6.1.11) with the exact sequence (9.4.6) below). For the sake of simplicity, let us denote G-supermanifolds (M, G_M) and their morphisms $(\varphi : M \to N, \Phi : G_N \to \varphi_*(M))$ by $\widehat{M}$ and $\widehat{\varphi}$, respectively. Given a point $q \in M$, by $\widehat{q} = (q, \Lambda)$ is meant the trivial G-supermanifold of dimension $(0,0)$. We will start from the notion of a G-Lie supergroup $\widehat{H}$. The relations between G- GH^∞- and G^∞-Lie supergroups follow the relations between the corresponding classes of superfunctions.

DEFINITION 9.4.1. A G-supermanifold $\widehat{H} = (H, \mathcal{H})$ is said to be a *G-Lie supergroup* if there exist the following G-supermanifold morphisms:

- a multiplication $\widehat{m} : \widehat{H} \times \widehat{H} \to \widehat{H}$,
- a unit $\widehat{\varepsilon} : \widehat{e} \to \widehat{H}$,
- an inverse $\widehat{S} : \widehat{H} \to \widehat{H}$,

together with the natural identifications

$$\widehat{e} \times \widehat{H} = \widehat{H} \times \widehat{e} = \widehat{H},$$

which satisfy

- the associativity

$$\widehat{m} \circ (\mathrm{Id} \times \widehat{m}) = \widehat{m} \circ (\widehat{m} \times \mathrm{Id}) : \widehat{H} \times \widehat{H} \times \widehat{H} \to \widehat{H} \times \widehat{H} \to \widehat{H},$$

- the unit property

$$(\widehat{m} \circ (\widehat{\varepsilon} \times \mathrm{Id}))(\widehat{e} \times \widehat{H}) = (\widehat{m} \circ (\mathrm{Id} \times \widehat{\varepsilon}))(\widehat{H} \times \widehat{e}) = \mathrm{Id}\, H,$$

- the inverse property

$$(\widehat{m} \circ (\widehat{S}, \mathrm{Id}))(\widehat{H}) = (\widehat{m} \circ (\mathrm{Id}, \widehat{S}))(\widehat{H}) = \widehat{\varepsilon}(\widehat{e}).$$

□

Given a point $g \in H$, let us denote by $\widehat{g} : \widehat{e} \to \widehat{H}$ the G-supermanifold morphism whose image in H is g. Then one can introduce the notions of the left translation $\widehat{L}_g$ and the right translation $\widehat{R}_g$ as the G-supermanifold morphisms

$$\widehat{L}_g : \widehat{H} = \widehat{e} \times \widehat{H} \xrightarrow{\widehat{g}\times\mathrm{Id}} \widehat{H} \times \widehat{H} \xrightarrow{\widehat{m}} \widehat{H},$$
$$\widehat{R}_g : \widehat{H} = \widehat{H} \times \widehat{e} \xrightarrow{\mathrm{Id}\times\widehat{g}} \widehat{H} \times \widehat{H} \xrightarrow{\widehat{m}} \widehat{H}.$$

Remark 9.4.1. Given a G-Lie supergroup $\widehat{H}$, the underlying smooth manifold H is provided with the structure of a real Lie group of dimension $2^{N-1}(n+m)$, called the *underlying Lie group*. In particular, the actions on the underlying Lie group H, corresponding to the left and right translations by $\widehat{g}$, are ordinary left and right translations by g. •

Let us reformulate the group axioms in Definition 9.4.1 in terms of the structure sheaf $\mathcal{H}$ of the G-Lie supergroup $(H, \mathcal{H})$. We will observe that $\mathcal{H}$ has properties of a sheaf of graded Hopf algebras.

Remark 9.4.2. A real (or complex) vector space A is called a *coalgebra* if there exist the morphisms:

- a *comultiplication* $\Delta : A \to A \otimes A$,
- a counit $\epsilon : A \to \mathbb{R}$

which satisfy the relations

$$(\Delta \otimes \mathrm{Id})\Delta(a) = (\mathrm{Id} \otimes \Delta)\Delta(a),$$
$$(\epsilon \otimes \mathrm{Id})\Delta(a) = (\mathrm{Id} \otimes \epsilon)\Delta(a) = a, \qquad a \in \mathcal{A}.$$

Let A be an associative $\mathbb{R}$-algebra with a unit e, i.e., an $\mathbb{R}$-ring, where the multiplication m is written as

$$m : A \otimes A \ni a \otimes b \mapsto ab \in A, \qquad a, b \in A.$$

Recall that $A \otimes A$ is also a $\mathbb{R}$-ring with respect to the operations

$$(a \otimes b) \otimes (a' \otimes b') \mapsto (aa') \otimes (bb'),$$
$$\lambda(a \otimes b) = (\lambda a) \otimes b = a \otimes (\lambda b), \qquad a, b \in \mathcal{A}, \qquad \lambda \in \mathbb{R}.$$

A *bialgebra* (A, m, Δ, ϵ) is a coalgebra A which is also an $\mathbb{R}$-ring such that

$$\Delta(e) = e \otimes e, \qquad \epsilon(e) = 1.$$

A *Hopf algebra* $(A, m, \Delta, \epsilon, S)$ is a bialgebra, endowed with a *coinverse* $S : A \to A$ such that

$$m((S \otimes \mathrm{Id})\Delta(a)) = m((\mathrm{Id} \otimes S)\Delta(a)) = \epsilon(a)e$$

(see [1] for a survey). •

If $(H, \mathcal{H})$ is a G-Lie supergroup, the structure sheaf $\mathcal{H}$ is provided with the sheaf morphisms:

- a comultiplication $\widehat{m}^* : \mathcal{H} \to m_*(\mathcal{H}\widehat{\otimes}\mathcal{H})$,
- a counit $\widehat{\varepsilon}^* : \mathcal{H} \to e_*(\Lambda)$,
- a coinverse $\widehat{S} : \mathcal{H} \to s_*\mathcal{H}$.

Let us denote

$$k = m \circ (\mathrm{Id} \times m) = m \circ (m \times \mathrm{Id}) : H \times H \times H \to H.$$

Then the group axioms in Definition 9.4.1 are equivalent to the relations

$$\begin{aligned}
&((\mathrm{Id} \otimes \widehat{m}^*) \circ \widehat{m}^*)(\mathcal{H}) = ((\widehat{m}^* \otimes \mathrm{Id}) \circ \widehat{m}^*)(\mathcal{H}) = k_*(\mathcal{H}\widehat{\otimes}\mathcal{H}\widehat{\otimes}\mathcal{H}),\\
&(\widehat{m}^* \circ (\mathrm{Id} \otimes \widehat{\varepsilon}^*))(\mathcal{H}\widehat{\otimes}e_*(\Lambda)) = (\widehat{m}^* \circ (\widehat{\varepsilon}^* \otimes \mathrm{Id}))(e_*(\Lambda)\widehat{\otimes}\mathcal{H}) = \mathrm{Id}\,\mathcal{H},\\
&(\mathrm{Id} \cdot \widehat{S}^*) \circ \widehat{m}^* = (\widehat{S}^* \cdot \mathrm{Id}) \circ \widehat{m}^* = \widehat{\varepsilon}^*.
\end{aligned}$$

Comparing these relations with the axioms of a Hopf algebra in Remark 9.4.2, one can think of the structure sheaf of a G-Lie group as a sheaf of graded topological Hopf algebras.

Example 9.4.3. The general linear graded group $GL(n|m;\Lambda)$ is endowed with the natural structure of H^∞-supermanifold of dimension $(n^2 + m^2, 2nm)$. The matrix multiplication gives the H^∞-morphism

$$m : GL(n|m;\Lambda) \times GL(n|m;\Lambda) \to GL(n|m;\Lambda)$$

such that $F(g, g') \mapsto F(gg')$. It follows that $GL(n|m;\Lambda)$ is an H^∞-Lie supergroup. It is trivially extended to the G-Lie supergroup $\widehat{GL}(n \mid m;\Lambda)$, called the *general linear supergroup*. •

A *Lie superalgebra* $\mathfrak{h}$ of a G-Lie supergroup $\widehat{H}$ is defined as an algebra of left-invariant supervector fields on $\widehat{H}$. Recall that a supervector field u on a G-supermanifold $\widehat{H}$ is a derivation of its structure sheaf $\mathcal{H}$. It is called *left-invariant* if

$$(\mathrm{Id} \otimes u) \circ \widehat{m}^* = \widehat{m}^* \circ u.$$

If u and u' are left-invariant supervector fields, so are $[u, u']$ and $au + a'u'$, $a, a' \in \Lambda$. Hence, left-invariant supervector fields constitute a Lie superalgebra. The Lie superalgebra $\mathfrak{h}$ can be identified with the supertangent space $T_e(\widehat{H})$. Moreover, there is the sheaf isomorphism

$$\boxed{\mathcal{H} \otimes \mathfrak{h} = \mathfrak{d}\mathcal{H},} \tag{9.4.1}$$

i.e., the sheaf of supervector fields on a G-Lie supergroup $\widehat{H}$ is the globally free sheaf of graded $\mathcal{H}$-modules of rank (n, m), generated by left-invariant supervector fields. The Lie superalgebra of right-invariant supervector fields on $\widehat{H}$ is introduced in a similar way.

Let us consider the right action of a G-Lie supergroup $\widehat{H}$ on a G-supermanifold $\widehat{P}$. This is a G-morphism

$$\widehat{\rho} : \widehat{P} \times \widehat{H} \to \widehat{P}$$

such that

$$\widehat{\rho} \circ (\widehat{\rho} \times \mathrm{Id}) = \widehat{\rho} \circ (\mathrm{Id} \times \widehat{m}) : \widehat{P} \times \widehat{H} \times \widehat{H} \to \widehat{P},$$
$$\widehat{\rho} \circ (\mathrm{Id} \times \widehat{\varepsilon})(\widehat{P} \times \widehat{e}) = \mathrm{Id}\, \widehat{P}.$$

The left action of $\widehat{H}$ on $\widehat{P}$ is defined similarly.

Example 9.4.4. Obviously, a G-Lie supergroup acts on itself both on the left and on the right by the multiplication morphism $\widehat{m}$.

The general linear supergroup $\widehat{GL}(n|m; \Lambda)$ acts linearly on the standard supermanifold $B^{n|n}$ on the left by the matrix multiplication which is a G-morphism. •

Let $\widehat{P}$ and $\widehat{P}'$ be G-supermanifolds that are acted on by the same G-Lie supergroup $\widehat{H}$. A G-supermanifold morphism $\widehat{\varphi} : \widehat{P} \to \widehat{P}'$ is said to be $\widehat{H}$-invariant if

$$\widehat{\varphi} \circ \widehat{\rho} = \widehat{\rho}' \circ (\widehat{\varphi} \times \mathrm{Id}) : \widehat{P} \times \widehat{H} \to \widehat{P}'.$$

DEFINITION 9.4.2. A quotient of an action of a G-Lie supergroup on a G-submanifold $\widehat{P}$ is a pair $(\widehat{M},\widehat{\pi})$ of a G-supermanifold $\widehat{M}$ and a G-supermanifold morphism $\widehat{\pi}:\widehat{P}\to\widehat{M}$ such that:

(i) there is the equality

$$\widehat{\pi}\circ\widehat{\rho}=\widehat{\pi}\circ\widehat{\mathrm{pr}}_1:\widehat{P}\times\widehat{H}\to\widehat{M}, \tag{9.4.2}$$

(ii) for every morphism $\widehat{\varphi}:\widehat{P}\to\widehat{M}'$ such that $\widehat{\varphi}\circ\widehat{\rho}=\widehat{\varphi}\circ\widehat{\mathrm{pr}}_1$, there is a unique morphism $\widehat{g}:\widehat{M}\to\widehat{M}'$ with $\widehat{\varphi}=\widehat{g}\circ\widehat{\pi}$.

□

The quotient $(\widehat{M},\widehat{\pi})$ does not necessarily exists. If it exists, there is a monomorphism of the structure sheaf G_M of $\widehat{M}$ into the direct image π_*G_P. Since the G-Lie group $\widehat{H}$ acts trivially on $\widehat{M}$, the image of this monomorphism is a subsheaf of π_*G_P, invariant under the action of $\widehat{H}$. Moreover, there is an isomorphism

$$\boxed{G_M\cong(\pi_*G_P)^H} \tag{9.4.3}$$

between G_M and the subsheaf of G_P of $\widehat{H}$-invariant sections. The latter is generated by sections of G_P on $\pi^{-1}(U)$, $U\subset M$, which are $\widehat{H}$-invariant as G-morphisms $\widehat{U}\to\Lambda$, where one takes the trivial action of $\widehat{H}$ on Λ.

Let us denote the morphism in the equality (9.4.2) by ϑ. It is readily observed, that the invariant sections of $G_P(\pi^{-1}(U))$ are exactly the elements which have the same image under the morphisms

$$\widehat{\rho}^*:G_P(\pi^{-1}(U))\to(\mathcal{H}\widehat{\otimes}G_P)(\vartheta^{-1}(U)),$$
$$\widehat{\mathrm{pr}}_1^*:G_P(\pi^{-1}(U))\to(\mathcal{H}\widehat{\otimes}G_P)(\vartheta^{-1}(U)).$$

Then the isomorphism (9.4.3) leads to the exact sequence of sheaves of Λ-modules on M

$$\boxed{0\longrightarrow G_M\overset{\widehat{\pi}^*}{\longrightarrow}\pi_*G_P\overset{\widehat{\rho}^*-\widehat{\mathrm{pr}}_1^*}{\longrightarrow}\vartheta_*(G_M\widehat{\otimes}\mathcal{H}).} \tag{9.4.4}$$

DEFINITION 9.4.3. A *principal superbundle* of a G-Lie supergroup $\widehat{H}$ is defined as a locally trivial quotient $\pi:\widehat{P}\to\widehat{M}$, i.e., there exists an open covering $\{U_\varsigma\}$ of M together with $\widehat{H}$-invariant isomorphisms

$$\widehat{\psi}_\varsigma:\widehat{P}\mid_{\widehat{U}_\varsigma}\to\widehat{U}_\varsigma\times\widehat{H},$$

where $\widehat{H}$ acts on

$$\widehat{U}_\zeta \times \widehat{H} \to \widehat{U}_\zeta \tag{9.4.5}$$

by the right multiplication. □

Remark 9.4.5. In fact, we need only the condition (i) in Definition 9.4.2 of the action of $\widehat{H}$ on $\widehat{P}$ and the condition of local triviality of $\widehat{P}$. •

In an equivalent way, one can think of a principal superbundle as being glued of trivial principal superbundles (9.4.5) by $\widehat{H}$-invariant transition functions

$$\widehat{\phi}_{\zeta\xi} : \widehat{U}_{\zeta\xi} \times \widehat{H} \to \widehat{U}_{\zeta\xi} \times \widehat{H}, \qquad U_{\zeta\xi} = U_\zeta \cap U_\xi,$$

which fulfill the cocycle condition.

As in the case of smooth principal bundles, the following two types of supervector fields on a principal superbundle are introduced.

DEFINITION 9.4.4. A supervector field u on a principal superbundle $\widehat{P}$ is said to be *invariant* if

$$\widehat{\rho}^* \circ u = (u \otimes \mathrm{Id}) \circ u : G_P \to \rho_*(G_P \widehat{\otimes} \mathcal{H}).$$

□

One can associate with every open subset $V \subset M$ the $G_M(V)$-module of all $\widehat{H}$-invariant supervector fields on $\pi^{-1}(V)$, thus defining the sheaf $\mathfrak{d}^H(\pi_* G_P)$ of G_M-modules.

DEFINITION 9.4.5. A *fundamental* supervector field $\widetilde{v}$ associated with an element $v \in \mathfrak{h}$ is defined by the condition

$$\widetilde{v} = (\mathrm{Id} \otimes v) \circ \widehat{\rho}^* : G_P \to G_P \widehat{\otimes} e_*(\Lambda) = G_P.$$

□

Fundamental supervector fields generate the sheaf $\mathcal{V}G_P$ of G_P-modules of vertical supervector field on the principal superbundle $\widehat{P}$, i.e., $u \circ \pi^* = 0$. Moreover, there is an isomorphism of sheaves of G_P-modules

$$\boxed{G_P \otimes \mathfrak{h} \ni F \otimes v \mapsto F\widetilde{v} \in \mathcal{V}G_P,}$$

which is similar to the isomorphism (9.4.1).

Let us consider the sheaf

$$(\pi_*\mathcal{V}G_P)^H = \pi_*(\mathcal{V}G_P) \cap \mathfrak{d}^H(\pi_*G_P)$$

on M whose sections are vertical $\widehat{H}$-invariant supervector fields.

PROPOSITION 9.4.6. [20]. There is the exact sequence of sheaves of G_M-modules

$$\boxed{0 \to (\pi_*\mathcal{V}G_P)^H \to \mathfrak{d}^H(\pi_*G_P) \to \mathfrak{d}G_M \to 0.} \tag{9.4.6}$$

□

The exact sequence (9.4.6) is similar to the exact sequence (6.1.11) and the corresponding exact sequence of sheaves of C^∞-modules

$$0 \to (V_GP)_X \to (T_GP)_X \to \mathfrak{d}C^\infty_X \to 0$$

in the case of smooth principal bundles. Accordingly, we come to the following definition of a superconnection on a principal superbundle.

DEFINITION 9.4.7. A *superconnection on a principal superbundle* $\pi : \widehat{P} \to \widehat{M}$ (or simply a *principal superconnection*) is defined as a splitting

$$\boxed{\nabla : \mathfrak{d}G_M \to \mathfrak{d}^H(\pi_*G_P)} \tag{9.4.7}$$

of the exact sequence (9.4.6). □

In contrast with principal connections on smooth principal bundles, principal superconnections on a $\widehat{H}$-principal superbundle need not exist.

A principal superconnection can be described in terms of a $\mathfrak{h}$-valued 1-superform on $\widehat{P}$

$$\omega : \mathfrak{d}G_P \to G_P \otimes \mathfrak{h} \cong \mathcal{V}G_P,$$

called a *superconnection form* (cf. (6.1.14)). Indeed, every splitting ∇ (9.4.7) defines the morphism of G_P-modules

$$\widehat{\pi}^*(\mathfrak{d}G_M) \to \widehat{\pi}^*(\mathfrak{d}^H(\pi_*G_P)) \cong \mathfrak{d}G_P$$

which splits the exact sequence

$$0 \to \mathcal{V}G_P \to \mathfrak{d}G_P \to \widehat{\pi}^*(\mathfrak{d}G_M) \to 0.$$

Therefore, there exists the exact sequence

$$0 \to \widehat{\pi}^*(\mathfrak{d}G_M)\mathcal{V}G_P \to \mathfrak{d}G_P \xrightarrow{\omega} \mathcal{V}G_P \to 0.$$

Let us note that, by analogy with associated smooth bundles, one can introduce associated superbundles and superconnections on these superbundles. In particular, every supervector bundle of fibre dimension (r, s) is a superbundle associated with $\widehat{GL}(r|s;\Lambda)$-principal superbundle [20].

9.5 Graded principal bundles

Graded principal bundles and connections on these bundles can be studied similarly to principal superbundles and principal superconnections, though the theory of graded principal bundles preceded that of principal superbundles [5, 183]. Therefore, we will touch on only a few elements of the graded bundle technique (see, e.g. [285] for a detailed exposition).

Let $(Z, \mathcal{A})$ be a graded manifold of dimension (n, m). A useful object in the graded manifold theory, not mentioned above, is the *finite dual* $\mathcal{A}(Z)^\circ$ of the algebra $\mathcal{A}(Z)$ which consists of elements a of the dual $\mathcal{A}(Z)^*$ which vanish on an ideal of $\mathcal{A}(Z)$ of finite codimension. This is a graded commutative coalgebra with the comultiplication

$$(\Delta^\circ(a))(f \otimes f') \stackrel{\text{def}}{=} a(ff'), \qquad \forall f, f' \in \mathcal{A}(Z),$$

and the counit

$$\epsilon^\circ(a) \stackrel{\text{def}}{=} a(1_{\mathcal{A}}).$$

In particular, $\mathcal{A}(Z)^\circ$ includes the evaluation elements δ_z such that

$$\boxed{\delta_z(f) = (\sigma(f))(z).}$$

Given an evaluation element δ_z, elements $u \in \mathcal{A}(Z)^\circ$ are called *primitive elements* with respect to δ_z if they obey the relation

$$\Delta^\circ(v) = u \otimes \delta_z + \delta_z \otimes u. \tag{9.5.1}$$

These elements are derivations of $\mathcal{A}(Z)$ at z, i.e.,

$$u(ff') = (uf)(\delta_z f') + (-1)^{[u][f]}(\delta_z f)(uf').$$

DEFINITION 9.5.1. A *graded Lie group* $(G, \mathcal{G})$ is defined as a graded manifold such that G is an ordinary Lie group, the algebra $\mathcal{G}(G)$ is a graded Hopf algebra (Δ, ϵ, S), and the algebra epimorphism $\sigma : \mathcal{G}(G) \to C^\infty(G)$ is a morphism of graded Hopf algebras. □

One can show that $\mathcal{G}(G)^\circ$ is also equipped with the structure of a Hopf algebra with the multiplication law

$$a \star b \stackrel{\text{def}}{=} (a \otimes b) \circ \Delta, \qquad \forall a, b \in \mathcal{G}(G)^\circ. \tag{9.5.2}$$

With respect to this multiplication, the evaluation elements δ_g, $G \in G$, constitute a group $\delta_g \star \delta_{g'} = \delta_{gg'}$ isomorphic to G. Therefore, they are also called *group-like elements*. It is readily observed that the set of primitive elements of $\mathcal{G}(G)^\circ$ with respect to δ_e, i.e., the tangent space $T_e(G, \mathcal{G})$ is a Lie superalgebra with respect to the multiplication (9.5.2). It is called the *Lie superalgebra* $\mathfrak{g}$ of the graded Lie group $(G, \mathcal{G})$.

One says that a graded Lie group $(G, \mathcal{G})$ acts on a graded manifold $(Z, \mathcal{A})$ on the right if there exists a morphism

$$(\varphi, \Phi) : (Z, \mathcal{A}) \times (G, \mathcal{G}) \to (Z, \mathcal{A})$$

such that the corresponding algebra morphism

$$\Phi : \mathcal{A}(Z) \to \mathcal{A}(Z) \otimes \mathcal{G}(G)$$

defines a structure of a right $\mathcal{G}(G)$-comodule on $\mathcal{A}(Z)$, i.e.,

$$(\mathrm{Id} \otimes \Delta) \circ \Phi = (\Phi \otimes \mathrm{Id}) \circ \Phi, \qquad (\mathrm{Id} \otimes \epsilon) \circ \Phi = \mathrm{Id}.$$

For a right action (φ, Φ) and for each element $a \in \mathcal{G}(G)^\circ$, one can introduce the linear map

$$\Phi_a = (\mathrm{Id} \otimes a) \circ \Phi : \mathcal{A}(Z) \to \mathcal{A}(Z). \tag{9.5.3}$$

In particular, if a is a primitive element with respect to δ_e, then $\Phi_a \in \mathfrak{d}\mathcal{A}(Z)$.

Let us consider a right action of $(G, \mathcal{G})$ on itself. If $\Phi = \Delta$ and $a = \delta_g$ is a group-like element, then Φ_a (9.5.3) is a homogeneous graded algebra isomorphism of degree zero which corresponds to the right translation $G \to Gg$. If $a \in \mathfrak{g}$, then Φ_a is a derivation of $\mathcal{G}(G)$. Given a basis $\{u_i\}$ for $\mathfrak{g}$, the derivations Φ_{u_i} constitute the global basis for $\mathfrak{d}\mathcal{G}(G)$, i.e., $\mathfrak{d}\mathcal{G}(G)$ is a free left $\mathcal{G}(G)$-module. In particular, there is the decomposition

$$\begin{aligned}
&\mathcal{G}(G) = \mathcal{G}'(G) \otimes_R \mathcal{G}''(G), \\
&\mathcal{G}'(G) = \{f \in \mathcal{G}(G) \,:\, \Phi_u(f) = 0, \ \forall u \in \mathfrak{g}_0\}, \\
&\mathcal{G}''(G) = \{f \in \mathcal{G}(G) \,:\, \Phi_u(f) = 0, \ \forall u \in \mathfrak{g}_1\}.
\end{aligned}$$

Since $\mathcal{G}'(G) \cong C^\infty(G)$, one finds that every graded Lie group $(G, \mathcal{G})$ is the sheaf of sections of some trivial exterior bundle $G \times \mathfrak{g}_1^* \to G$ [5, 36, 183].

Turn now to the notion of a graded principal bundle. A right action (φ, Φ) of $(G, \mathcal{G})$ on $(Z, \mathcal{A})$ is called *free* if, for each $z \in Z$, the morphism $\Phi_z : \mathcal{A}(Z) \to \mathcal{G}(G)$ is such that the dual morphism $\Phi_{z*} : \mathcal{G}(G)^\circ \to \mathcal{A}(Z)^\circ$ is injective.

A right action (φ, Φ) of $(G, \mathcal{G})$ on $(Z, \mathcal{A})$ is called *regular* if the morphism

$$(\varphi \times \mathrm{pr}_1) \circ \Delta : (Z, \mathcal{A}) \times (G, \mathcal{G}) \to (Z, \mathcal{A}) \times (Z, \mathcal{A})$$

defines a closed *graded submanifold* of $(Z, \mathcal{A}) \times (Z, \mathcal{A})$.

Remark 9.5.1. Note that $(Z', \mathcal{A}')$ is said to be a graded submanifold of $(Z, \mathcal{A})$ if there exists a morphism $(Z', \mathcal{A}') \to (Z, \mathcal{A})$ such that the corresponding morphism $\mathcal{A}'(Z')^\circ \to \mathcal{A}(Z)^\circ$ is an inclusion. A graded submanifold is called closed if $\dim(Z', \mathcal{A}') < \dim(Z, \mathcal{A})$. •

Then we come to the following variant of the well-known theorem on the quotient of a graded manifold [5, 285].

THEOREM 9.5.2. A right action (φ, Φ) of $(G, \mathcal{G})$ on $(Z, \mathcal{A})$ is regular if and only if the quotient $(Z/G, \mathcal{A}/\mathcal{G})$ is a graded manifold, i.e., there exists an epimorphism of graded manifolds $(Z, \mathcal{A}) \to (Z/G, \mathcal{A}/\mathcal{G})$ compatible with the projection $Z \to Z/G$. □

In view of this Theorem, a *graded principal bundle* $(P, \mathcal{A})$ can be defined as a locally trivial submersion $(P, \mathcal{A}) \to (P/G, \mathcal{A}/\mathcal{G})$ with respect to the right regular free action of $(G, \mathcal{G})$ on $(P, \mathcal{A})$. In an equivalent way, one can say that a graded principal bundle is a graded manifold $(P, \mathcal{A})$ together with a free right action of a graded Lie group $(G, \mathcal{G})$ on $(P, \mathcal{A})$ such that the quotient $(P/G, \mathcal{A}/\mathcal{G})$ is a graded manifold and the natural surjection $(P, \mathcal{A}) \to (P/G, \mathcal{A}/\mathcal{G})$ is a submersion. Obviously, $P \to P/G$ is an ordinary G-principal bundle.

A *graded principal connection* on a graded $(G,\mathcal{G})$-principal bundle $(P,\mathcal{A}) \to (X,\mathcal{B})$ can be introduced similarly to a superconnection on a principal superbundle. This is defined as a $(G,\mathcal{G})$-invariant splitting of the sheaf $\mathfrak{d}\mathcal{A}$, and is represented by a $\mathfrak{g}$-valued graded connection form on $(P,\mathcal{A})$ [285].

Remark 9.5.2. In an alternative way, one can define *graded connections* on a graded bundle $(Z,\mathcal{A}) \to (X,\mathcal{B})$ as sections Γ of the jet graded bundle $J^1(Z/X) \to (Z,\mathcal{A})$ of sections of $(Z,\mathcal{A}) \to (X,\mathcal{B})$ [5], which is also a graded manifold [260]. In the case of a $(G,\mathcal{G})$-principal graded bundle, these sections Γ are required to be $(G,\mathcal{G})$-equivariant (cf. Definition 6.1.1). •

9.6 SUSY-extended field theory

Sections 9.4 and 9.5 provided the general mathematical formalism for field models with supersymmetries (SUSY field models). Here, we show that field theory on a wide class of fibre bundles $Y \to X$ can be extended in a standard way to SUSY field theory which is invariant under the Lie supergroup ISp(2). In comparison with the SUSY field theory in Ref. [42, 60], this extension is formulated in terms of simple graded manifolds, and is the direct generalization of the BRS mechanics of E.Gozzi and M.Reuter [136, 137, 138, 213]. The SUSY-extended field theory is constructed as the BRS-generalization of the vertical extension of field theory on the fibre bundle $VY \to X$ in Section 4.5 [125]. From the physical viewpoint, it may describe odd deviations of physical fields

Given a fibre bundle $Y \to X$ and the vertical tangent bundle $VY \to X$, let us consider the vertical tangent bundle VVY of $VY \to X$ and the simple graded manifold $(VY, \mathcal{A}_{VVY})$ whose body manifold is VY and whose characteristic vector bundle is $VVY \to VY$. Its local basis is $(c^i, \overline{c}^i)$, where $\{c^i, \overline{c}^i\}$ is the fibre basis for V^*VY, dual of the holonomic fibre basis $\{\partial_i, \dot{\partial}_i\}$ for $VVY \to VY$. Graded vector fields and graded exterior 1-forms are introduced on VY as sections of the vector bundles $\mathcal{V}_{VVY}$ and $\mathcal{V}^*_{VVY}$, respectively. Let us complexify these bundles as $\mathbb{C} \underset{X}{\otimes} \mathcal{V}_{VVY}$ and $\mathbb{C} \underset{X}{\otimes} \mathcal{V}^*_{VVY}$. By the *BRS operator* on graded functions on VY is meant the complex graded vector field

$$u_Q = c^i \partial_i + i\dot{y}^i \frac{\partial}{\partial \overline{c}^i}. \tag{9.6.1}$$

It satisfies the nilpotency rule $u_Q^2 = 0$.

The configuration space of the SUSY-extended field theory is the simple graded manifold $(VJ^1Y, \mathcal{A}_{VVJ^1Y})$ whose characteristic vector bundle is the vertical tangent bundle $VVJ^1Y \to VJ^1Y$ of $VJ^1Y \to X$. Its local basis is $(c^i, \overline{c}^i, c^i_\lambda, \overline{c}^i_\lambda)$ which is the fibre basis for V^*VJ^1Y dual of the holonomic fibre basis $\{\partial_i, \dot\partial_i, \partial^\lambda_i, \dot\partial^\lambda_i\}$ for $VVJ^1Y \to VJ^1Y$. The affine fibration $\pi^1_0 : VJ^1Y \to VY$ and the corresponding vertical tangent morphism $V\pi^1_0 : VVJ^1Y \to VVY$ yields the associated morphism of graded manifolds $(VJ^1Y, \mathcal{A}_{VVJ^1Y}) \to (VY, \mathcal{A}_{VVY})$ (9.2.17).

Let us introduce the operator of the total derivative

$$d_\lambda = \partial_\lambda + y^i_\lambda \partial_i + \dot y^i_\lambda \dot\partial_i + c^i_\lambda \frac{\partial}{\partial c^i} + \overline{c}^i_\lambda \frac{\partial}{\partial \overline{c}^i}.$$

With this operator, the coordinate transformation laws of c^i_λ and $\overline{c}^i_\lambda$ read

$$c'^i_\lambda = d_\lambda c'^i, \qquad \overline{c}'^i_\lambda = d_\lambda \overline{c}'^i. \tag{9.6.2}$$

Then one can treat c^i_λ and $\overline{c}^i_\lambda$ as the jets of c^i and $\overline{c}^i$. Note that this is not the notion of jets of graded bundles in [260]. The transformation laws (9.6.2) show that the BRS operator u_Q (9.6.1) on VY can give rise to the complex graded vector field

$$J^u_Q = u_Q + c^i_\lambda \partial^\lambda_i + i\dot y^i_\lambda \frac{\partial}{\partial \overline{c}^i_\lambda} \tag{9.6.3}$$

on the VJ^1Y (cf. (1.3.10)).

In a similar way, the simple graded manifold with the characteristic vector bundle $VVJ^kY \to VJ^kY$ can be defined. Its local basis is $(c^i, \overline{c}^i, c^i_\Lambda, \overline{c}^i_\Lambda)$, $0 < |\Lambda| \leq k$. Let us introduce the operators

$$\partial_c = c^i\partial_i + c^i_\lambda \partial^\lambda_i + c^i_{\lambda\mu}\partial^{\lambda\mu}_i + \cdots, \qquad \partial_{\overline{c}} = \overline{c}^i\partial_i + \overline{c}^i_\lambda \partial^\lambda_i + \overline{c}^i_{\lambda\mu}\partial^{\lambda\mu}_i + \cdots, \tag{9.6.4}$$

$$d_\lambda = \partial_\lambda + y^i_\lambda \partial_i + c^i_\lambda \frac{\partial}{\partial c^i} + \overline{c}^i_\lambda \frac{\partial}{\partial \overline{c}^i} + \cdots. \tag{9.6.5}$$

It is easily verified that

$$d_\lambda \partial_c = \partial_c d_\lambda, \qquad d_\lambda \partial_{\overline{c}} = \partial_{\overline{c}} d_\lambda. \tag{9.6.6}$$

As in the BRS mechanics, the main criterion of the SUSY extension of Lagrangian formalism is its invariance under the BRS transformation (9.6.3). The BRS-invariant extension of the vertical Lagrangian L_V (4.5.1) is the graded n-form

$$L_S = L_V + i\partial_c \partial_{\overline{c}} \mathcal{L}\omega \tag{9.6.7}$$

such that $\mathbf{L}_{Ju_Q} L_S = 0$. The corresponding Euler–Lagrange equations are defined as the kernel of the Euler–Lagrange operator

$$\mathcal{E}_{L_S} = (dy^i \delta_i + d\dot{y}^i \dot{\delta}_i + dc^i \frac{\delta}{\delta c^i} + d\overline{c}^i \frac{\delta}{\delta \overline{c}^i}) \mathcal{L}_S \wedge \omega. \tag{9.6.8}$$

They read

$$\dot{\delta}_i \mathcal{L}_S = \delta_i \mathcal{L} = 0, \tag{9.6.9a}$$

$$\delta_i \mathcal{L}_S = \delta_i \mathcal{L}_V + i\partial_{\overline{c}} \partial_c \delta_i \mathcal{L} = 0, \tag{9.6.9b}$$

$$\frac{\delta}{\delta c^i} \mathcal{L}_S = -i\partial_{\overline{c}} \delta_i \mathcal{L} = 0, \tag{9.6.9c}$$

$$\frac{\delta}{\delta \overline{c}^i} \mathcal{L}_S = i\partial_c \delta_i \mathcal{L} = 0, \tag{9.6.9d}$$

where the relations (9.6.6) are used. The equations (9.6.9a) are the Euler–Lagrange equations for the initial Lagrangian L, while (9.6.9b) - (9.6.9d) can be seen as the equations for a Jacobi field $\delta y^i = \overline{\varepsilon} c^i + \overline{c}^i \varepsilon + i\overline{\varepsilon}\varepsilon \dot{y}^i$ modulo terms of order > 2 in the odd parameters ε and $\overline{\varepsilon}$.

A momentum phase space of the SUSY-extended field theory is the complexified simple graded manifold $(V\Pi, \mathcal{A}_{VV\Pi})$ whose body manifold is $V\Pi$ and whose characteristic vector bundle is $VV\Pi \to V\Pi$. Its local basis is $(c^i, \overline{c}^i, c_i^\lambda, \overline{c}_i^\lambda)$, where such that c_i^λ and $\overline{c}_i^\lambda$ have the same transformation laws as p_i^λ and $\dot{p}_i^\lambda$, respectively. The corresponding graded vector fields and graded 1-forms are introduced on $V\Pi$ as sections of the vector bundles $\mathbb{C} \underset{X}{\otimes} \mathcal{V}_{VV\Pi}$ and $\mathbb{C} \underset{X}{\otimes} \mathcal{V}^*_{VV\Pi}$, respectively.

In accordance with the above mentioned transformation laws of c_i^λ and $\overline{c}_i^\lambda$, the BRS operator u_Q (9.6.1) on VY can give rise to the complex graded vector field

$$\widetilde{u}_Q = \partial_c + i\dot{y}^i \frac{\partial}{\partial \overline{c}^i} + i\dot{p}_i^\lambda \frac{\partial}{\partial \overline{c}_i^\lambda} \tag{9.6.10}$$

on $V\Pi$ (cf. (4.4.1). The BRS-invariant extension of the polysymplectic form Ω_{VY} (4.5.6) on $V\Pi$ is the TX-valued graded form

$$\Omega_S = [d\dot{p}_i^\lambda \wedge dy^i + dp_i^\lambda \wedge d\dot{y}^i + i(d\overline{c}_i^\lambda \wedge dc^i - d\overline{c}^i \wedge dc_i^\lambda)] \wedge \omega \otimes \partial_\lambda,$$

where $(c^i, -i\overline{c}_i^\lambda)$ and $(\overline{c}^i, ic_i^\lambda)$ are the conjugate pairs. Let γ be a Hamiltonian connection for a Hamiltonian form H on Π. Its double vertical prolongation $VV\gamma$

on $VV\Pi \to X$ (see (2.7.18) and (4.5.11)) is a linear morphism over the vertical connection $V\gamma$ on $V\Pi \to X$, and so defines the composite graded connection

$$(VV\gamma)_S = V\gamma + dx^\mu \otimes [\overline{g}^i_\mu \frac{\partial}{\partial \overline{c}^i} + \overline{g}^\lambda_{\mu i} \frac{\partial}{\partial \overline{c}^\lambda_i} + g^i_\mu \frac{\partial}{\partial c^i} + g^\lambda_{\mu i} \frac{\partial}{\partial c^\lambda_i}]$$

(9.2.26) on $V\Pi \to X$, whose components g and $\overline{g}$ are given by the expressions

$$\overline{g}^i_\lambda = \partial_{\overline{c}} \partial^i_\lambda \mathcal{H}, \quad \overline{g}^\lambda_{\lambda i} = -\partial_{\overline{c}} \partial_i \mathcal{H}, \quad g^i_\lambda = \partial_c \partial^i_\lambda \mathcal{H}, \quad g^\lambda_{\lambda i} = -\partial_c \partial_i \mathcal{H},$$
$$\partial_c = c^i \partial_i + c^\lambda_i \partial^i_\lambda, \qquad \partial_{\overline{c}} = \overline{c}^i \partial_i + \overline{c}^\lambda_i \partial^i_\lambda.$$

This composite graded connection satisfies the relation

$$(VV\gamma)_S \rfloor \Omega_S = -dH_S,$$

and can be regarded as a Hamiltonian graded connection for the Hamiltonian graded form

$$H_S = [\dot{p}^\lambda_i dy^i + p^\lambda_i d\dot{y}^i + i(\overline{c}^\lambda_i dc^i + d\overline{c}^i c^\lambda_i)]\omega_\lambda - \mathcal{H}_S \omega, \tag{9.6.11}$$
$$\mathcal{H}_S = (\partial_V + i\partial_{\overline{c}} \partial_c)\mathcal{H},$$

on $V\Pi$. It is readily observed that this graded form is BRS-invariant, i.e., $\mathbf{L}_{\widetilde{u}_Q} H_S = 0$. Thus, it is the desired SUSY extension of the Hamiltonian form H.

In particular, let $\Gamma^i_\lambda = \Gamma_\lambda{}^i{}_j y^j$ be a linear connection on $Y \to X$ and $\widetilde{\Gamma}$ is a Hamiltonian connection (4.1.23) on $\Pi \to X$ for the Hamiltonian form H_Γ (4.1.13). Given the splitting (4.1.14) of the Hamiltonian form H, there is the corresponding splitting of the SUSY-extended Hamiltonian form

$$\mathcal{H}_S = \mathcal{H}_{\Gamma S} + \widetilde{\mathcal{H}}_{S\Gamma} = \dot{p}^\lambda_i \Gamma_\lambda{}^i{}_j y^j + \dot{y}^j p^\lambda_i \Gamma_\lambda{}^i{}_j + i(\overline{c}^\lambda_i \Gamma_\lambda{}^i{}_j c^j + \overline{c}^j \Gamma_\lambda{}^i{}_j c^\lambda_i)$$
$$+(\partial_V + i\partial_{\overline{c}}\partial_c)\widetilde{\mathcal{H}}_\Gamma$$

with respect to the composite graded connection $(VV\widetilde{\Gamma})_S$ (9.2.26) on the fibre bundle $V\Pi \to X$.

The Hamiltonian graded form H_S (9.6.11) defines the corresponding SUSY extension of the Lagrangian L_H (4.1.20) as follows. The fibration $J^1V\Pi \to V\Pi$ yields the pull-back of the Hamiltonian graded form H_S (9.6.11) onto $J^1V\Pi$. Let us consider the graded generalization

$$h_0 : dc^i \mapsto c^i_\mu dx^\mu, \quad dc^\lambda_i \mapsto c^\lambda_{\mu i} dx^\mu$$

of the horizontal projection h_0 (1.3.15). Then the graded horizontal density

$$\begin{aligned} L_{SH} = h_0(H_S) = (L_H)_S = L_{H_V} + i[(\overline{c}_i^\lambda c_\lambda^i + \overline{c}_\lambda^i c_i^\lambda) - \partial_{\overline{c}}\partial_c\mathcal{H}]\omega = \\ L_{H_V} + i[\overline{c}_i^\lambda(c_\lambda^i - \partial_c\partial_\lambda^i\mathcal{H}) + (\overline{c}_\lambda^i - \partial_{\overline{c}}\partial_\lambda^i\mathcal{H})c_i^\lambda + \overline{c}_i^\lambda c_j^\mu \partial_\lambda^i\partial_\mu^j\mathcal{H} - \overline{c}^i c^j \partial_i\partial_j\mathcal{H}]\omega \end{aligned} \tag{9.6.12}$$

on $J^1V\Pi \to X$ is the SUSY extension (9.6.7) of the Lagrangian L_H (4.1.20). The Euler–Lagrange equations for L_{SH} coincide with the Hamilton equations for H_S, and read

$$y_\lambda^i = \dot{\partial}_\lambda^i\mathcal{H}_S = \partial_\lambda^i\mathcal{H}, \qquad p_{\lambda i}^\lambda = -\dot{\partial}_i\mathcal{H}_S = -\partial_i\mathcal{H}, \tag{9.6.13a}$$

$$\dot{y}_\lambda^i = \partial_\lambda^i\mathcal{H} = (\partial_V + i\partial_{\overline{c}}\partial_c)\partial_\lambda^i\mathcal{H}, \quad \dot{p}_{\lambda i}^\lambda = -\partial_i\mathcal{H}_S = -(\partial_V + i\partial_{\overline{c}}\partial_c)\partial_i\mathcal{H}, \tag{9.6.13b}$$

$$c_\lambda^i = i\frac{\partial\mathcal{H}_S}{\partial\overline{c}_i^\lambda} = -\partial_c\partial_\lambda^i\mathcal{H}, \qquad c_{\lambda i}^\lambda = i\frac{\partial\mathcal{H}_S}{\partial\overline{c}^i} = -\partial_c\partial_i\mathcal{H}, \tag{9.6.13c}$$

$$\overline{c}_\lambda^i = -i\frac{\partial\mathcal{H}_S}{\partial c_i^\lambda} = -\partial_{\overline{c}}\partial_\lambda^i\mathcal{H}, \qquad \overline{c}_{\lambda i}^\lambda = -i\frac{\partial\mathcal{H}_S}{\partial c^i} = -\partial_{\overline{c}}\partial_i\mathcal{H}. \tag{9.6.13d}$$

The equations (9.6.13a) are the Hamilton equations for the initial Hamiltonian form H, while (9.6.13b) – (9.6.13d) describe the Jacobi fields

$$\delta y^i = \overline{\varepsilon}c^i + \overline{c}^i\varepsilon + i\overline{\varepsilon}\varepsilon\dot{y}^i, \qquad \delta p_i^\lambda = \overline{\varepsilon}c_i^\lambda + \overline{c}_i^\lambda\varepsilon + i\overline{\varepsilon}\varepsilon\dot{p}_i^\lambda.$$

Let us study the relationship between SUSY-extended Lagrangian and Hamiltonian formalisms. Given a Lagrangian L on J^1Y, the vertical Legendre map $\widehat{L}_V$ (4.5.4) yields the corresponding morphism (9.2.17) of graded manifolds

$$S\widehat{L}_V : (VJ^1Y, \mathcal{A}_{VVJ^1Y}) \to (V\Pi, \mathcal{A}_{VV\Pi})$$

which is given by the relations (4.5.5) and

$$c_i^\lambda = \partial_c\pi_i^\lambda, \qquad \overline{c}_i^\lambda = \partial_{\overline{c}}\pi_i^\lambda.$$

Let H be a Hamiltonian form on Π. The vertical Hamiltonian map $\widehat{H}_V$ (4.5.14) yields the morphism of graded manifolds

$$S\widehat{H}_V : (V\Pi, \mathcal{A}_{VV\Pi}) \to (VJ^1Y, \mathcal{A}_{VVJ^1Y})$$

given by the relations (4.5.15) and

$$c_\lambda^i = \partial_c\partial_\lambda^i\mathcal{H}, \qquad \overline{c}_\lambda^i = \partial_{\overline{c}}\partial_\lambda^i\mathcal{H}.$$

If a Hamiltonian form H is associated with L, a direct computation shows that the Hamiltonian graded form H_S (9.6.11) is weakly associated with the Lagrangian L_S (9.6.7), i.e.,

$$S\widehat{L}_V \circ S\widehat{H}_V \circ S\widehat{L}_V = S\widehat{L}_V,$$
$$L_S \circ S\widehat{H}_V = (p_i^\lambda \partial_\lambda^i + \dot{p}_i^\lambda \dot{\partial}_\lambda^i + c_i^\lambda \frac{\partial}{\partial c_i^\lambda} + \overline{c}_i^\lambda \frac{\partial}{\partial \overline{c}_i^\lambda})\mathcal{H}_S - \mathcal{H}_S,$$

where the second equality takes place at points of the Lagrangian constraint space $\widehat{L}(J^1Y)$.

The BRS invariance of the SUSY field theory can be extended to the above mentioned Lie supergroup $ISp(2)$ if a fibre bundle $Y \to X$ has affine transition functions (it is not necessarily an affine bundle). Almost all field models are of this type. In this case, the vertical tangent bundle admits the vertical splitting (1.1.13) with respect to the holonomic coordinates $\dot{y}^i$ on VY whose transition functions are independent of y^i. As a consequence, the transformation laws of the frames $\{\partial_i\}$ and $\{\dot{\partial}_i\}$ are the same, and so are the transformations laws of the coframes $\{c^i\}$ and $\{\overline{c}^i\}$. Then the graded vector fields

$$u_{\overline{Q}} = \overline{c}^i\partial_i - i\dot{y}^i\frac{\partial}{\partial c^i}, \qquad u_K = c^i\frac{\partial}{\partial \overline{c}^i}, \qquad u_{\overline{K}} = \overline{c}^i\frac{\partial}{\partial c^i},$$
$$u_C = c^i\frac{\partial}{\partial c^i} - \overline{c}^i\frac{\partial}{\partial \overline{c}^i}. \tag{9.6.14}$$

are globally defined on VY. The graded vector fields (9.6.1) and (9.6.14) constitute the above-mentioned Lie superalgebra of the supergroup ISp(2):

$$[u_Q, u_Q] = [u_{\overline{Q}}, u_{\overline{Q}}] = [u_{\overline{Q}}, u_Q] = [u_K, u_Q] = [u_{\overline{K}}, u_{\overline{Q}}] = 0,$$
$$[u_K, u_{\overline{Q}}] = u_Q, \qquad [u_{\overline{K}}, u_Q] = u_{\overline{Q}}, \qquad [u_K, u_{\overline{K}}] = u_C, \tag{9.6.15}$$
$$[u_C, u_K] = 2u_K, \qquad [u_C, u_{\overline{K}}] = -2u_{\overline{K}}.$$

Similarly to (9.6.3), let us consider the jet prolongation of the graded vector fields (9.6.14) onto VJ^1Y. Using the compact notation $u = u^a\partial_a$, we have the formula

$$J^1u = u + d_\lambda u^a \partial_a^\lambda$$

and, as a consequence, obtain

$$J^1u_{\overline{Q}} = u_{\overline{Q}} + \overline{c}_\lambda^i \partial_i^\lambda - i\dot{y}_\lambda^i \frac{\partial}{\partial c_\lambda^i},$$

$$J^1u_K = u_K + c^i_\lambda \frac{\partial}{\partial \overline{c}^i_\lambda}, \qquad J^1u_{\overline{K}} = u_{\overline{K}} + \overline{c}^i_\lambda \frac{\partial}{\partial c^i_\lambda}, \tag{9.6.16}$$
$$J^1u_C = u_C = c^i_\lambda \frac{\partial}{\partial c^i_\lambda} - \overline{c}^i_\lambda \frac{\partial}{\partial \overline{c}^i_\lambda}.$$

It is readily observed that the SUSY-extended Lagrangian L_S (9.6.7) is invariant under the transformations (9.6.16). The graded vector fields (9.6.3) and (9.6.16) make up the Lie superalgebra (9.6.15).

Graded vector fields (9.6.14) can give rise to $V\Pi$ by the formula

$$\widetilde{u} = u - (-1)^{[y^a]([p_b]+[u^b])} \partial_a u^b p^\lambda_b \frac{\partial}{\partial p^\lambda_a}$$

(cf. (4.4.1)). We have

$$\widetilde{u}_{\overline{Q}} = \partial_{\overline{c}} - i\dot{y}^i \frac{\partial}{\partial c^i} - i\dot{p}^\lambda_i \frac{\partial}{\partial c^\lambda_i},$$
$$\widetilde{u}_K = c^i \frac{\partial}{\partial \overline{c}^i} + c^\lambda_i \frac{\partial}{\partial \overline{c}^\lambda_i}, \qquad \widetilde{u}_{\overline{K}} = \overline{c}^i \frac{\partial}{\partial c^i} + \overline{c}^\lambda_i \frac{\partial}{\partial c^\lambda_i}, \tag{9.6.17}$$
$$\widetilde{u}_C = c^i \frac{\partial}{\partial c^i} + c^\lambda_i \frac{\partial}{\partial c^\lambda_i} - \overline{c}^i \frac{\partial}{\partial \overline{c}^i} - \overline{c}^\lambda_i \frac{\partial}{\partial \overline{c}^\lambda_i}.$$

A direct computation shows that the BRS-extended Hamiltonian form H_S (9.6.11) is invariant under the transformations (9.6.17). Accordingly, the Lagrangian L_{SH} (9.6.12) is invariant under the jet prolongation $J^1\widetilde{u}$ of the graded vector fields (9.6.17). The graded vector fields (9.6.10) and (9.6.17) make up the Lie superalgebra (9.6.15).

With the graded vector fields (9.6.10) and (9.6.17), one can construct the corresponding graded currents $\widetilde{\mathfrak{T}}_u = \widetilde{u}\rfloor H_S = u\rfloor H_S$ (4.4.7). These are the graded $(n-1)$-forms

$$\widetilde{\mathfrak{T}}_{u_Q} = (c^i\dot{p}^\lambda_i - \dot{y}^i c^\lambda_i)\omega_\lambda, \quad \widetilde{\mathfrak{T}}_{u_{\overline{Q}}} = (\overline{c}^i\dot{p}^\lambda_i - \dot{y}^i\overline{c}^\lambda_i)\omega_\lambda,$$
$$\widetilde{\mathfrak{T}}_{u_K} = -ic^\lambda_i c^i\omega_\lambda, \quad \widetilde{\mathfrak{T}}_{u_{\overline{K}}} = i\overline{c}^\lambda_i\overline{c}^i\omega_\lambda, \quad \widetilde{\mathfrak{T}}_{u_C} = i(\overline{c}_i c^i - \overline{c}^i c_i)\omega_\lambda$$

on $V\Pi$. They form the Lie superalgebra (9.6.15) with respect to the product

$$[\widetilde{\mathfrak{T}}_u, \widetilde{\mathfrak{T}}_{u'}] = \widetilde{\mathfrak{T}}_{[u,u']}.$$

The following construction is similar to that in the SUSY and BRS mechanics. Given a function F on the Legendre bundle Π, let us consider the operators

$$F_\beta = e^{\beta F} \circ \widetilde{u}_Q \circ e^{-\beta F} = \widetilde{u}_Q - \beta\partial_c F, \qquad \beta > 0,$$
$$\overline{F}_\beta = e^{-\beta F} \circ \widetilde{u}_{\overline{Q}} \circ e^{\beta F} = \widetilde{u}_{\overline{Q}} + \beta\partial_{\overline{c}} F,$$

called the *SUSY charges*, which act on graded functions on $V\Pi$. These operators are nilpotent, i.e.,

$$F_\beta \circ F_\beta = 0, \qquad \overline{F}_\beta \circ \overline{F}_\beta = 0. \tag{9.6.18}$$

By the BRS-invariant extension of a function F is meant the graded function

$$F_S = -\frac{i}{\beta}(\overline{F}_\beta \circ F_\beta + F_\beta \circ \overline{F}_\beta).$$

We have the relations

$$F_\beta \circ F_S - F_S \circ F_\beta = 0, \qquad \overline{F}_\beta \circ F_S - F_S \circ \overline{F}_\beta = 0.$$

These relations together with the relations (9.6.18) provide the operators F_β, $\overline{F}_\beta$, and F_S with the structure of the Lie superalgebra sl(1/1) [56]. In particular, let F be a local Hamiltonian $\mathcal{H}$ in the expression (4.1.7). Then

$$\mathcal{H}_S = -i(\overline{F}_1 \circ F_1 + F_1 \circ \overline{F}_1)$$

is exactly the local function H_S in the expression (9.6.11). The similar splitting of a super-Hamiltonian is the corner stone of the SUSY mechanics [65, 191].

9.7 The Ne'eman–Quillen superconnection

In this Section, we consider the class of superconnections introduced by Y.Ne'eman in the physical literature [233, 234, 235] and by D.Quillen in the mathematical literature [218, 250]. The fibre bundles that they consider belong neither to above studied graded manifolds and bundles nor superbundles. Ne'eman–Quillen superconnections have been applied to computing the Chern character in K-theory (see below), to non-commutative geometry [237], BRST formalism [195] and some particle unification models (see below).

Let X be an N-dimensional smooth manifold and $\wedge T^*X$ the exterior bundle. Let E_0, and E_1 be two vector bundles over X of dimensions n and m, respectively, One constructs the vector bundle

$$Q = \wedge T^*X \underset{X}{\otimes} E = \wedge T^*X \underset{X}{\otimes}(E_0 \underset{X}{\oplus} E_1), \tag{9.7.1}$$

called hereafter the body of an NQ-superbundle (see Definition 9.7.1 below) or simply an NQ-superbundle.

The typical fibre of the NQ-superbundle Q is

$$V = \wedge\mathbb{R}^N \otimes (B_0 \oplus B_1), \tag{9.7.2}$$

where B_0 and B_1 are the typical fibres of the vector bundles E_0 and E_1 respectively. This typical fibre can be provided with the structure of the superspace $B^{n|m}$ over the Grassmann algebra $\Lambda = \wedge\mathbb{R}^N$. This is the graded envelope of the graded vector space $B = B_0 \oplus B_1$, where B_0 and B_1 are regarded as its even and odd subspaces, respectively. The NQ-superbundle Q inherits this gradation since transition functions of E_0 and E_1 are mutually independent, while the transition functions of T^*X preserve the $\mathbb{Z}$-gradation of the Grassmann algebra $\wedge\mathbb{R}^N$.

Nevertheless, the NQ-superbundle (9.7.1) is not a supervector bundle over X since its transition functions are not Λ-module morphisms. Obviously, one can think of the NQ-superbundle Q as the tensor product of the vector bundle E and the characteristic vector bundle $\wedge T^*X$ of the simple graded manifold $(X, \mathfrak{O}^*_X)$, where $\mathfrak{O}^*_X$ is the sheaf of exterior forms on X. The vector bundle $\mathcal{V}_E$ of graded vector fields and the vector bundle $\mathcal{V}^*_E$ of graded 1-forms (see Section 9.2) have a local structure of an NQ-superbundle (see local isomorphisms (9.2.12) and (9.2.28), respectively).

Let us denote by $Q(X)$ and Q_X the space of global sections of the NQ-superbundle Q and the sheaf of its sections, respectively. Of course, $Q(X) = Q_X(X)$. The space $Q(X)$ has the natural structure of a locally free $C^\infty(X)$-module, while Q_X is the locally free sheaf of C^∞_X-modules of rank $2^N(n+m)$. At the same time, bearing in mind that $(X, \mathfrak{O}^*_X)$ is a graded locally ringed space, one can provide $Q(X) = \mathfrak{O}^*(X) \otimes E(X)$ with the structure of the graded locally free $\mathfrak{O}^*(X)$-module, while

$$Q_X = C^\infty_X \otimes \wedge\mathbb{R}^N \otimes B = \mathfrak{O}^*_X \otimes B$$

can be seen as the graded locally free sheaf of $\mathfrak{O}^*_X$-modules of rank $(n+m)$. We will denote $Q(X)$ and Q_X endowed with the above mentioned structures by $\overline{Q}(X)$ and $\overline{Q}_X$, respectively.

DEFINITION 9.7.1. The pair $\widehat{Q} = (Q, \overline{Q}(X))$ (or the pair $(Q, \overline{Q}_X)$) is called an *NQ-superbundle*. □

Given a trivialization domain $U \subset X$ of the vector bundle $E \to X$, let $\{c_A\}$ and $\{c_i\}$ be fibre bases for the vector bundles E_0 and E_1 over U, respectively. Then every element q of $\overline{Q}(X)$ reads

$$q = q^A c_A + q^i c_i,$$

where q^A, q^i are local exterior forms on U. An element q is homogeneous, if its Grassmann degree is

$$[q] = [q^A] = [q^i] + 1, \qquad [q^A] = |q^A| \,\mathrm{mod} 2, \qquad [q^i] = |q^i| \,\mathrm{mod} 2.$$

Given another trivialization domain $U' \subset X$ of E, the corresponding transition functions read

$$q'^A = \rho^A_B q^B, \qquad q'^i = \rho^i_j q^j, \tag{9.7.3}$$

where ρ^A_B, ρ^i_j are local smooth functions on $U \cap U'$. We call the triple $(U; q^A, q^i)$ together with the transition functions (9.7.3) a *splitting domain* of the NQ-superbundle $\widehat{Q}$.

A connection on the the NQ-superbundle $(Q, \overline{Q}(X))$ can be defined in accordance with Definition (8.2.2). It is easily seen that, in the case of the ring $\mathfrak{O}^*(X)$, the derivation d^1 in Proposition 8.1.6, is exactly the exterior differential d.

DEFINITION 9.7.2. A connection on the NQ-superbundle $(Q, \overline{Q}(X))$, called an *NQ-superconnection*, is defined as a morphism $\nabla : \overline{Q} \to \overline{Q}$ which obeys the Leibniz rule

$$\boxed{\nabla(\phi q) = (d\phi)q + (-1)^{[f]} f \nabla(q),} \quad \forall q \in \overline{Q}(X), \quad \forall \phi \in \mathfrak{O}^*(X). \tag{9.7.4}$$

□

It should be emphasized that an NQ-superconnection is defined as a connection on $\mathfrak{O}^*(X)$-module $Q(X)$, but not on the $C^\infty(X)$-module $Q(X)$. Therefore, it is not an ordinary connection on the smooth vector bundle $Q \to X$.

The Leibniz rule (9.7.4) implies that an NQ-superconnection is an odd morphism. For instance, if $E \to X$ is a trivial bundle, we have the trivial NQ-superconnection $\nabla = d$. Let Γ_0 and Γ_1 be linear connections on the vector bundles E_0 and E_1, respectively, and $\Gamma_0 \oplus \Gamma_1$ a linear connection on $E \to X$. Then the covariant differential ∇^Γ (2.2.8) relative to the connection Γ is an NQ-superconnection

$$\nabla^\Gamma(q) = dx^\lambda \wedge (\partial_\lambda - \Gamma)(q). \tag{9.7.5}$$

Given a splitting domain U, this superconnection reads

$$\boxed{\nabla^\Gamma(q) = (dq^A - dx^\lambda \wedge \Gamma_{0\lambda}{}^A{}_B q^B)c_A + (dq^i - dx^\lambda \wedge \Gamma_{1\lambda}{}^i{}_j q^j)c_i,} \tag{9.7.6}$$

where $\Gamma_{0\lambda}{}^A{}_B$, $\Gamma_{1\lambda}{}^i{}_j$ are local functions on U with the familiar transformation laws under the transition morphisms ρ^A_B and ρ^i_j of the vector bundles E_0 and E_1, respectively.

As it follows directly from the Leibniz rule (9.7.4), the NQ-superconnections on an NQ-superbundle constitute an affine space modelled over the $\mathfrak{O}^*(X)_0$-module $\mathrm{End}(\overline{Q}(X))_1$ of odd degree endomorphisms of $\overline{Q}(X)$, i.e.,

$$\nabla' = \nabla + L, \tag{9.7.7}$$

where L is an odd element of $\mathrm{End}(\overline{Q}(X))$.

It is easy to see that the $\mathfrak{O}^*(X)$-module $\mathrm{End}(\overline{Q}(X))$ is a $C^\infty(X)$-module of sections of the vector bundle

$$\wedge T^*X \underset{X}{\otimes} E \underset{X}{\otimes} E^* \to X.$$

Given a splitting domain U of the NQ-superbundle $\widehat{Q}$, every element of $\mathrm{End}(\overline{Q}(X))$ is represented by a supermatrix function (or simply a supermatrix)

$$L = \begin{pmatrix} L_1 & L_2 \\ L_3 & L_4 \end{pmatrix} \tag{9.7.8}$$

whose entries are local exterior forms on U. The transformation law of this supermatrix under the transition morphisms (9.7.3) is

$$L' = \widehat{\rho} L \widehat{\rho}^{-1},$$

where $\widehat{\rho}$ is the $(n+m)\times(n+m)$ matrix

$$\widehat{\rho} = \begin{pmatrix} \rho^A_B & 0 \\ 0 & \rho^i_j \end{pmatrix} \tag{9.7.9}$$

whose entries are the transition functions ρ^A_B and ρ^i_j.

Due to the relation (9.7.7), any NQ-superconnection on a splitting domain of the NQ-superbundle $\widehat{Q}$ can be written in the form

$$\boxed{\nabla = d + \vartheta,} \tag{9.7.10}$$

where ϑ is a local odd supermatrix (9.7.9), i.e., entries of ϑ_1, ϑ_4 are exterior forms of odd degree, while those of ϑ_2, ϑ_3 are exterior forms of even degree. Obviously, the splitting (9.7.10) is not maintained under the transition morphisms (9.7.3), and we have the transformation law

$$\vartheta' = \widehat{\rho}\vartheta\widehat{\rho}^{-1} - d\widehat{\rho}\rho^{-1},$$

where $d\widehat{\rho}$ is the supermatrix whose entries are 1-forms $d\rho^A_B$ and $d\rho^i_j$.

In accordance with Definition 8.2.4, the *curvature* of the NQ-superconnection ∇ is the morphism

$$\boxed{R = \nabla^2 : \overline{Q} \to \overline{Q}.} \tag{9.7.11}$$

For instance, the curvature of the trivial connection $\nabla = d$ is equal to zero. The curvature of the connection ∇^Γ (9.7.6) is

$$R(q) = \begin{pmatrix} \frac{1}{2}R_{\lambda\mu}{}^A{}_B dx^\lambda \wedge dx^\mu \wedge q^B c_A & 0 \\ 0 & \frac{1}{2}R_{\lambda\mu}{}^i{}_j dx^\lambda \wedge dx^\mu \wedge q^j c_i, \end{pmatrix}, \tag{9.7.12}$$

where $R_{\lambda\mu}{}^A{}_B$, $R_{\lambda\mu}{}^i{}_j$ are the curvatures (2.4.2) of the linear connections Γ_0 and Γ_1, respectively. Given the local splitting (9.7.10) of an NQ-superconnection ∇, its curvature (9.7.11) takes the local form

$$\boxed{R = d(\vartheta) + \vartheta^2,} \tag{9.7.13}$$

and has the transformation law $R' = \widehat{\rho}R\widehat{\rho}^{-1}$ under the transition morphisms (9.7.3). It follows that the curvature of an NQ-superconnection is an even endomorphism of the $\mathfrak{O}^*(X)$-module $\overline{Q}(X)$.

Remark 9.7.1. The notions of an NQ-superbundle and an NQ-superconnection are extended in a straightforward manner to the case of a complex vector bundle $E \to X$ and the exterior algebra $\mathbb{C} \otimes \mathfrak{O}^*(X)$ of complex exterior forms on X. •

Let us discuss briefly the following two applications of NQ-superconnections.

The first one is concerned with the unification models in particle physics (see [195, 233, 234, 256]). Let E_0 be a vector bundle with a structure group G treated as a group of internal symmetries in particle physics, e.g., $SU(2)$, while $E_1 \to X$ is a linear bundle. Then the typical fibre of the NQ-superbundle Q (9.7.1) is the superspace $B^{n|1}$. Let A be an associated linear principal connection (6.1.24) on

$E_0 \to X$, i.e., its components A^p_λ are gauge potentials for the group G. Then one considers the NQ-superconnection which is a sum

$$\nabla = \nabla^A + L$$

of the superconnection $\nabla^A(q)$ (9.7.5) and an even endomorphism

$$L = \begin{pmatrix} 0 & i\phi^* \\ i\phi & 0 \end{pmatrix},$$

where ϕ is a scalar field, treated as a Higgs field. The goal is that gauge potentials and a Higgs field are regarded on the same footing as components of the same NQ-superconnection. It should be emphasized that the authors of [195, 233, 234] follow the convention where the supermatrix multiplication reads

$$\begin{pmatrix} A & C \\ D & B \end{pmatrix}\begin{pmatrix} A' & C' \\ D' & B' \end{pmatrix} = \begin{pmatrix} A\wedge A' + (-1)^{|D'|}C\wedge D' & A\wedge C' + (-1)^{|B'|}C\wedge B' \\ (-1)^{|A'|}D\wedge A' + B\wedge D' & (-1)^{|C'|}D\wedge C' + B\wedge B' \end{pmatrix}.$$

This matrix multiplication does not satisfy the relations (9.1.6), (9.1.7).

Another application addresses the computation of the Chern character [202, 218, 250]. Given an NQ-superbundle Q (9.7.1), let us consider an NQ-superconnection

$$\nabla = \nabla^\Gamma + tL, \tag{9.7.14}$$

where ∇^Γ is the superconnection (9.7.6), L is an odd element of $\mathrm{End}(\overline{Q}(X))$, and t is a real parameter. The curvature (9.7.11) of the connection (9.7.14) reads

$$R = t^2L^2 + t[\nabla^\Gamma, L] + (\nabla^\Gamma)^2,$$

where $(\nabla^\Gamma)^2$ is the curvature (9.7.12) of the NQ-superconnection ∇^Γ (9.7.6). Then we have

$$\mathrm{ch}_t = \mathrm{Str}(\exp(R)) = \sum_{k=0} \mathrm{Str}((\nabla^\Gamma + L)^{2k}). \tag{9.7.15}$$

This series converges since R is a section of a bundle of finite-dimensional algebras over X. It is readily observed that

$$\mathrm{ch}_{t=0} = \mathrm{ch}(E_0) - \mathrm{ch}(E_1), \tag{9.7.16}$$

where $\mathrm{ch}(E_0)$, $\mathrm{ch}(E_1)$ are the Chern characters (6.7.18) of the complex vector bundles E_0 and E_1 (with accuracy to the customary factor $i/2\pi$). This is the Chern character of the difference element $E_0 \ominus E_1$ in the K-theory of complex vector bundles on a manifold X [10, 168] (see the relation (9.8.2) below). The key point is that the De Rham cohomology classes of the exterior forms ch_t (9.7.15) and $\mathrm{ch}_{t=0}$ (9.7.16) are the same. This fact issues from the following two assertions.

PROPOSITION 9.7.3. Given an NQ-superconnection ∇ and an odd endomorphism $T \in \mathrm{End}(\overline{Q}(X))$, one has

$$d(\mathrm{Str}(T)) = \mathrm{Str}([\nabla, T]). \tag{9.7.17}$$

□

Proof. Since the relation (9.7.17) is local, one can assume that ∇ is split as in (9.7.10) and T is a sum of supermatrices of the form $\phi_k T_k$ where ϕ_k is are exterior forms, while T_k are constant supermatrices. Bearing in mind the relation (9.1.7), we have

$$\mathrm{Str}([d+\vartheta, \phi_k T_k]) = \mathrm{Str}(d\phi_k T_k) = d(\mathrm{Str}(\phi_k T_k)).$$

QED

Let $T = R^k$, where R is the curvature of the NQ-superconnection ∇. Since $\mathrm{Str}([\nabla, R^k]) = 0$, we obtain from the relation (9.7.17) that the exterior form $\mathrm{Str}(R^k)$ is closed.

PROPOSITION 9.7.4. [250]. Given the curvature R of an NQ-superconnection ∇, the De Rham cohomology class of the exterior form $\mathrm{Str}(R^k)$ is independent of the choice of the NQ-superconnection ∇. □

The coincidence of the De Rham cohomology classes of the forms ch_t (9.7.15) and $\mathrm{ch}_{t=0}$ (9.7.16) enables one to analyse the Chern character under the different choice of the supermatrix L and the parameter t in the expression (9.7.14).

9.8 Appendix. K-Theory

The characteristic classes discussed in Section 6.7 enable one to describe the equivalence classes of real or complex vector bundles of the same dimension. Let $\mathcal{C}(X)$ be the set of all equivalence classes of vector bundles over a manifold X. It is a commutative monoid with respect to the Whitney sum $\oplus$, where the 0-dimensional vector bundle plays the role of a zero element. The goal is to define the operation $\ominus$.

Recall the following algebraic construction. Given a commutative monoid A, let us consider the quotient $K(A)$ of $A \times A$ with respect to the relation $(a,b) \approx (a',b')$ if there exists an element $p \in A$ such that

$$a + b' + p = b + a' + p.$$

Then $K(A)$ is group, called the *Grothendieck group* of the monoid A. There is the homomorphism $k : A \to K(A)$ such that $k(a)$, $a \in A$, is the equivalence class of the pair $(a,0) \in A \times A$. The inverse element $-k(a)$ is the equivalence class of the pair $(0,a)$. Then any element (a,b) of the group $K(A)$ can be represented as the difference $k(a) - k(b)$, $a,b \in A$. It is easily seen that $k(a) = k(b)$ if and only if there exists an element $p \in A$ such that $a + p = b + p$.

Let us construct the Grothendieck group of the monoid $\mathcal{C}(X)$ of equivalence classes of vector bundles over a compact manifold X [10, 168]. Let us denote $k(E)$, $E \in \mathcal{C}(X)$, simply by $[E]$. Then $[E] = [E']$ if an only if there is a vector bundle $F \to X$ such that $E \oplus F \approx E' \oplus F$.

THEOREM 9.8.1. Let E be a vector bundle over a compact manifold X. There exists a vector bundle E' such that the Whitney sum $E \oplus E'$ is a trivial vector bundle. □

COROLLARY 9.8.2. $[E] = [E']$ in $K(X)$ if and only if $E \oplus I^m \approx E' \oplus I^m$ for some trivial vector bundle I^m over X. □

It follows that vector bundles E and E' belong to the same class in $K(X)$ only if they are of the same dimension, but need not be isomorphic. For example, $[TS^2] = [I^2] = 0 \in K(S^2)$, though the tangent bundle $TS^2 \to S^2$ is not trivial. This example shows that the morphism $k : \mathcal{C}(X) \to K(X)$ is not an injection. It is an injection on equivalence classes of real vector bundles of dimension $m > \dim X$ and on the equivalence classes of complex vector bundles of dimension $m > \dim X/2$.

There is another equivalence relation $\{E\}$ on the monoid $\mathcal{C}(X)$ of equivalence classes of vector bundles over X. We put $\{E\} = \{E'\}$ if and only if there exist trivial vector bundles I^k and I^p such that $E \oplus I^k \approx E' \oplus I^p$. These equivalence classes constitute the group $\widetilde{K}(X)$ whose zero element includes all trivial vector bundles. There is an isomorphism $K(X) = \mathbb{Z} \oplus \widetilde{K}(X)$. For instance, if X is a point, $K(X) = \mathbb{Z}$, while $\widetilde{K}(X) = 0$.

THEOREM 9.8.3. Let X and X' be compact manifolds and manifold morphisms $f_1 : X \to X'$ and $f_2 : X \to X'$ are homotopic. Then f_1 and f_2 yield the same morphisms $K(X) \to K(X')$ and $\widetilde{K}(X) \to \widetilde{K}(X')$. □

Let now $K_C(X)$ be the Grothendieck group of complex vector bundles over a compact manifold X. The Chern character (6.7.18) of these vector bundles defines the map

$$\mathrm{ch} : K_C(X) \to \bigoplus_{i \geq 0} H^{2i}(X, \mathbb{Q}) \tag{9.8.1}$$

such that

$$\mathrm{ch}([E] - [E']) = \mathrm{ch}(E) - \mathrm{ch}(E'). \tag{9.8.2}$$

The morphism (9.8.1) leads to the isomorphism

$$\widetilde{K}_C(X) \otimes \mathbb{Q} = \bigoplus_{i > 0} H^{2i}(X, \mathbb{Q})$$

for complex vector bundles and to the isomorphism

$$\widetilde{K}_R(X) \otimes \mathbb{Q} = \bigoplus_{i > 0} H^{4i}(X, \mathbb{Q})$$

for real vector bundles.

Chapter 10

Connections in quantum mechanics

Quantum mechanics is a vast subject which can be studied from many different point of view. This Chapter is devoted to a few main examples of an application of connections to quantization of mechanical systems. These are linear connections on Kähler manifolds in geometric quantization, symplectic connections in the Fedosov deformation quantization, and Berry connections.

With respect to mathematical prerequisites, the reader is expected to be familiar with the basics of differential analysis on finite-dimensional complex manifolds (see, e.g., [177, 303]), geometry of Poisson and symplectic manifolds (see, e.g., [213, 299]), theory of Hilbert spaces and C^*-algebras [82]. By a Hilbert space throughout this Chapter is meant a complex Hilbert space.

10.1 Kähler manifolds modelled on Hilbert spaces

This Section provides a brief exposition of geometry of Kähler manifolds modelled on infinite-dimensional Hilbert spaces [61, 228]. These manifolds generalize the well-known finite-dimensional Kähler manifolds [177, 303], and are particular Banach manifolds modelled on infinite-dimensional Banach spaces.

Let E be a Hilbert space with a Hermitian form $\langle .|. \rangle$. Let $\overline{E}$ be the dual of E and

$$E \ni x \mapsto \overline{x} \in \overline{E}, \qquad \overline{x}(x') = \langle x'|x \rangle, \quad x' \in E,$$

the corresponding antilinear morphism. Let us consider the direct sum $E\oplus\overline{E}$. There exists its isomorphism, called the *complex structure*,

$$J : E\oplus\overline{E} \ni x+\overline{y} \mapsto ix - i\overline{y} = ix + \overline{(iy)} \in E\oplus\overline{E} \tag{10.1.1}$$

such that $J^2 = -\mathrm{Id}$. The spaces E and $\overline{E}$ are eigenspaces of J characterized by the eigenvalues i and $-i$, respectively. There is the relation $\overline{Jx} = J\overline{x}$.

The notion of holomorphic and antiholomorphic derivatives is extended to Hilbert spaces as follows.

Let E and H be Hilbert spaces, $U\subset E$ an open subset, and $f : U \to H$ a differentiable function between real Banach spaces E and H. It means that, given a point $\zeta_0\in U$, there exists a continuous $\mathbb{R}$-linear map $l_{z_0} : E\to H$ such that

$$f(z) = f(z_0) + l_{z_0}(z-z_0) + o(z-z_0), \qquad z\in U.$$

This condition can be reformulated as follows. There exist morphisms

$$\begin{aligned} a : E \to H, &\qquad a(z) = \frac{1}{2}l_{z_0}(z) - \frac{i}{2}l_{z_0}(iz),\\ b : \overline{E} \to H, &\qquad b(\overline{z}) = \frac{1}{2}l_{z_0}(z) + \frac{i}{2}l_{z_0}(iz), \end{aligned}$$

such that

$$f(z) = f(z_0) + a(z-z_0) + b(\overline{z-z_0}) + o(z-z_0). \tag{10.1.2}$$

They define the $\mathbb{C}$-linear maps

$$(\partial_{z_0}f)(x+\overline{y}) = a(x), \qquad (\overline{\partial}_{z_0}f)(x+\overline{y}) = b(\overline{y}) \tag{10.1.3}$$

from $E\oplus\overline{E}$ to H. The $\mathbb{C}$-linear morphism

$$f'_{z_0} : E\oplus\overline{E} \to H, \qquad f'_{z_0} = \partial_{z_0}f + \overline{\partial}_{z_0}f, \tag{10.1.4}$$

is called a derivative of f at a point $z_0\in U$. The higher order derivatives are defined in a similar way. Accordingly, a function f between Hilbert spaces is said to be

- *smooth* if it is indefinitely differentiable,
- *holomorphic* if it is differentiable and $\overline{\partial}_z f = 0$ for all $z\in E$,
- *antiholomorphic* if it is differentiable and $\partial_z f = 0$ for all $z\in E$.

As in the finite-dimensional case, a holomorphic function is smooth and given by the Taylor series.

Let us consider a *Hilbert manifold* $\mathcal{P}$ modelled on a Hilbert space E. It has an atlas of charts (U_ι, ψ_ι), where the morphisms ψ_ι take their values in the Hilbert space E, while the transition functions $\psi_\zeta\psi_\iota^{-1}$ from $\psi_\iota(U_\iota \cap U_\zeta) \subset E$ to $\psi_\zeta(U_\iota \cap U_\zeta) \subset E$ are smooth. Moreover, we will suppose that transition functions are holomorphic and, consequently, $\mathcal{P}$ is a *holomorphic manifold.* Tangent vectors to a holomorphic manifold $\mathcal{P}$ are defined by analogy with those to a finite-dimensional manifold. Given a point $z \in \mathcal{P}$, let us consider the pair $(v; (U_\iota, \psi_\iota))$ of a vector $v \in E$ and a chart $(U_\iota \ni z, \psi_\iota)$ of the holomorphic manifold $\mathcal{P}$. Two such pairs $(v; (U_\iota, \psi_\iota))$ and $(v'; (U_\zeta, \psi_\zeta))$ are said to be equivalent if

$$v' = \partial_{\psi_\iota(z)}(\psi_\zeta\psi_\iota^{-1})v.$$

The equivalence classes of the above mentioned pairs make up the *holomorphic tangent space* $T_z\mathcal{P}$ to the holomorphic manifold $\mathcal{P}$ at a point $z \in \mathcal{P}$. This tangent space is isomorphic to E regarded as a topological vector space. The dual $\overline{T}_z\mathcal{P}$ of $T_z\mathcal{P}$ is called the *antiholomorphic tangent space.* The *complex tangent space* to the holomorphic manifold $\mathcal{P}$ at a point $z \in \mathcal{P}$ is the direct sum

$$\boxed{\mathcal{T}_z\mathcal{P} = T_z\mathcal{P} \oplus \overline{T}_z\mathcal{P}.}$$

The complex tangent space $\mathcal{T}_z$ is provided with the involution operation

$$v + \overline{u} \mapsto \overline{v} + u$$

and the complex structure J_z. Every *complex tangent vector* $\vartheta \in \mathcal{T}_z\mathcal{P}$ is represented uniquely by a sum $\vartheta = v + \overline{u}$ of its holomorphic and antiholomorphic components. A complex tangent vector ϑ is called *real* if $\vartheta = \overline{\vartheta}$. The disjoint union of complex tangent spaces to the holomorphic manifold $\mathcal{P}$ is the *complex tangent bundle* $\mathcal{T}\mathcal{P}$ of $\mathcal{P}$. Its sections are called *complex vector fields* on $\mathcal{P}$. Complex vector fields on $\mathcal{P}$ constitute the locally free module $\mathcal{T}(\mathcal{P})$ over the ring $\mathbb{C}^\infty(\mathcal{P})$ of smooth complex functions on $\mathcal{P}$.

Accordingly, the dual $\mathcal{T}_z^*\mathcal{P}$ of the complex tangent space $\mathcal{T}_z\mathcal{P}$ is called the *complex cotangent space* to the holomorphic manifold $\mathcal{P}$ at a point $z \in \mathcal{P}$. Complex cotangent spaces make up the *complex cotangent bundle* $\mathcal{T}^*\mathcal{P}$ of $\mathcal{P}$. Its sections are *complex 1-forms* which constitute the $\mathbb{C}^\infty(\mathcal{P})$-module $\mathfrak{O}^1(\mathcal{P})$. One can consider tensor products of complex tangent and cotangent bundles over $\mathcal{P}$. In particular, complex exterior forms and the exterior differential d acting on these forms are well defined on the holomorphic manifold $\mathcal{P}$.

Example 10.1.1. The differential df of a smooth function $f : \mathcal{P} \to \mathbb{C}$ is a complex 1-form on $\mathcal{P}$. Let us denote its holomorphic and antiholomorphic parts by ∂f and $\overline{\partial} f$, respectively. Given a complex tangent vector $\vartheta_z = v_z + \overline{u}_z$ at a point $z \in \mathcal{P}$, we have the relations

$$\begin{aligned}
&\langle \vartheta_z | df \rangle = f'_z(\vartheta_z), \\
&\langle \vartheta_z | \partial f \rangle = \langle v_z | df \rangle = (\partial_z f)(v_z), \\
&\langle \vartheta_z | \overline{\partial} f \rangle = \langle \overline{u}_z | df \rangle = (\overline{\partial}_z f)(\overline{u}_z).
\end{aligned}$$

•

By a *Hermitian metric* on a holomorphic manifold $\mathcal{P}$ is meant a complex bilinear form g on $T\mathcal{P}$ which obeys the conditions:

- $g(\vartheta_z, \nu_z) = 0$ if complex tangent vectors $\vartheta_z, \nu_z \in T_z\mathcal{P}$ are simultaneously holomorphic or antiholomorphic;
- $g(\vartheta_z, \overline{\vartheta}_z) > 0$ for any non-vanishing complex tangent vector $\vartheta_z \in T_z\mathcal{P}$;
- the bilinear form $g_z(.,.)$ defines a topology in the complex tangent space $T_z\mathcal{P}$ which is equivalent to its Hilbert space topology.

As an immediate consequence of this definition, we obtain

$$\begin{aligned}
&\overline{g(\vartheta_z, \nu_z)} = g(\overline{\vartheta}_z, \overline{\nu}_z), \\
&g(J\vartheta_z, J\nu_z) = g(\vartheta_z, \nu_z).
\end{aligned}$$

Example 10.1.2. Let $\mathcal{P} = E$ be a Hilbert space. The Hermitian form $\langle .|. \rangle$ on E defines uniquely the following constant Hermitian metric on $E \oplus \overline{E}$:

$$g : (E \oplus \overline{E}) \times (E \oplus \overline{E}) \to \mathbb{C},$$

$$\boxed{g(\vartheta_1, \vartheta_2) = \langle v_1 | u_2 \rangle + \langle v_2 | u_1 \rangle,} \tag{10.1.5}$$

for all complex vectors $\vartheta_1 = v_1 + \overline{u}_1$ and $\vartheta_2 = v_2 + \overline{u}_2$. Conversely, every Hermitian metric on E provides E with a Hermitian form. •

DEFINITION 10.1.1. The pair $(\mathcal{P}, g)$ of a holomorphic manifold $\mathcal{P}$ and a Hermitian metric g on $\mathcal{P}$ is called a *Hermitian manifold.* □

Given a Hermitian manifold $(\mathcal{P}, g)$, let us define a non-degenerate exterior 2-form ω on $\mathcal{P}$, given by the equality

$$\boxed{\omega(\vartheta_z, \nu_z) = g(J\vartheta_z, \nu_z)} \tag{10.1.6}$$

for any point $z \in \mathcal{P}$ and any pair of complex tangent vectors $\vartheta_z, \nu_z \in T_z\mathcal{P}$. The exterior form ω is called the *fundamental form* of the Hermitian metric g. It satisfies the relations

$$\overline{\omega(\vartheta_z, \nu_z)} = \omega(\overline{\vartheta}_z, \overline{\nu}_z),$$
$$\omega(J_z\vartheta_z, J_z\nu_z) = \omega(\vartheta_z, \nu_z).$$

DEFINITION 10.1.2. A Hermitian metric g on the holomorphic manifold $\mathcal{P}$ is called a *Kähler metric* if $dw = 0$, i.e., its fundamental form is a symplectic form. The $(\mathcal{P}, g)$ is said to be a *Kähler manifold.* □

Example 10.1.3. Let $\mathcal{P} = E$ be a Hilbert space in Example 10.1.2 together with the Hermitian metric g (10.1.5). The corresponding fundamental form (10.1.6) on $E \oplus \overline{E}$ reads

$$\boxed{\omega(\vartheta_1, \vartheta_2) = i\langle v_1|u_2\rangle - i\langle v_2|u_1\rangle} \tag{10.1.7}$$

for all complex vectors $\vartheta_1 = v_1 + \overline{u}_1$ and $\vartheta_2 = v_2 + \overline{u}_2$. It is constant on E, and $d\omega = 0$. Therefore, g (10.1.5) is a Kähler metric. •

In accordance with Definition 8.2.2, a *connection* ∇ on a Hermitian manifold $\mathcal{P}$ is defined as a morphism

$$\nabla : \mathcal{T}(\mathcal{P}) \to \mathfrak{O}^1(\mathcal{P}) \otimes \mathcal{T}(\mathcal{P})$$

which obeys the Leibniz rule

$$\nabla(f\vartheta) = df \otimes \vartheta + f\nabla(\vartheta), \qquad f \in \mathbb{C}^\infty(\mathcal{P}), \qquad \vartheta \in \mathcal{T}(\mathcal{P}).$$

Similarly, a connection is introduced on any $\mathbb{C}^\infty(\mathcal{P})$-module, e.g., on the structure module $T(\mathcal{P})$ of a tensor bundle T over the holomorphic manifold $\mathcal{P}$. Let us denote the holomorphic and antiholomorphic parts of ∇ by D and $\overline{D}$, respectively. Given a complex vector field $\vartheta = v + \overline{u}$ on $\mathcal{P}$, we have the relations

$$\nabla_\vartheta = D_\vartheta + \overline{D}_\vartheta,$$
$$D_\vartheta = \nabla_v, \qquad \overline{D}_\vartheta = \nabla_{\overline{u}},$$
$$D_{J\vartheta} = iD_\vartheta, \qquad \overline{D}_{J\vartheta} = -i\overline{D}_\vartheta.$$

If f is a smooth complex function on a holomorphic manifold $\mathcal{P}$, then $\nabla f = df$. There are the equalities

$$(\nabla\nabla f)(\vartheta_z, \nu_z) = (\nabla\nabla f)(\nu_z, \vartheta_z),$$
$$(DDf)(\vartheta_z, \nu_z) = (DDf)(\nu_z, \vartheta_z),$$
$$(\overline{D}\overline{D}f)(\vartheta_z, \nu_z) = (\overline{D}\overline{D}f)(\nu_z, \vartheta_z),$$
$$(\overline{D}Df)(\vartheta_z, \nu_z) = (D\overline{D}f)(\nu_z, \vartheta_z).$$

Given a Hermitian manifold $(\mathcal{P}, g)$, there always exists a *metric connection* on $\mathcal{P}$ such that the covariant differential ∇g vanishes everywhere on $\mathcal{P}$. We have the equality

$$\overline{\nabla_\vartheta \nu} = \nabla_{\overline{\vartheta}}\overline{\nu}$$

for any pair of complex vector fields ϑ and ν on $\mathcal{P}$. If $(\mathcal{P}, g)$ is a Kähler manifold, then $\nabla\omega = 0$ and $\nabla J = 0$, where J is regarded as a section of the tensor bundle $T^*\mathcal{P} \otimes T\mathcal{P}$.

Example 10.1.4. Let a Hilbert space E be regarded as a Kähler manifold with the Kähler metric g (10.1.5). The metric connection on E is trivial, and $D = \partial$, $\overline{D} = \overline{\partial}$. •

A Kähler metric g and its fundamental form ω on a holomorphic manifold $\mathcal{P}$ define the isomorphisms

$$g^\flat : T\mathcal{P} \ni v \mapsto v \rfloor g \in T^*\mathcal{P}, \tag{10.1.8}$$
$$\omega^\flat : T\mathcal{P} \ni v \mapsto v \rfloor \omega \in T^*\mathcal{P}. \tag{10.1.9}$$

Let us denote by $g^\sharp$ and $\omega^\sharp$ the inverse isomorphisms $T^*\mathcal{P} \to T\mathcal{P}$. They have the properties

$$g^\sharp(\phi)\rfloor\sigma = g^\sharp(\sigma)\rfloor\phi, \qquad \phi, \sigma \in T^*\mathcal{P},$$
$$\omega^\sharp(\phi)\rfloor\sigma = -\omega^\sharp(\sigma)\rfloor\phi,$$
$$g^\sharp = J\omega^\sharp, \qquad \omega^\sharp = -Jg^\sharp.$$

If ∇ is a metric connection on $\mathcal{P}$, then

$$\nabla g^\sharp = 0, \qquad \nabla \omega^\sharp = 0,$$

where $g^\sharp$ and $\omega^\sharp$ are regarded as sections of the tensor bundle $\mathcal{T}^*\mathcal{P} \otimes \mathcal{T}\mathcal{P}$.

In particular, every smooth complex function f on the Kähler manifold $(\mathcal{P}, g)$ defines:

- the complex vector field

$$g^\sharp(df) = g^\sharp(\overline{\partial} f) + g^\sharp(\partial f), \tag{10.1.10}$$

  where $g^\sharp(\overline{\partial} f)$ and $g^\sharp(\partial f)$ are its holomorphic and antiholomorphic parts, respectively;

- *complex Hamiltonian vector field*

$$\boxed{\omega^\sharp(df) = -J(g^\sharp(df)) = -ig^\sharp(\overline{\partial} f) + ig^\sharp(\partial f).} \tag{10.1.11}$$

In conclusion, let us consider the important example of a *projective Hilbert space* PE made up by the complex 1-dimensional subspaces of a Hilbert space E. This is a holomorphic manifold which possesses the following standard atlas. Given a non-zero element $x \in E$, let us denote by x a point of PE such that $x \in \mathrm{x}$. Then any normalized element $h \in E$, $||h|| = 1$, defines a chart (U_h, ψ_h) of the projective Hilbert space PE such that

$$U_h = \{\mathrm{x} \in PE \ : \ \langle x|h\rangle \neq 0\}, \qquad \psi_h(\mathrm{x}) = \frac{x}{\langle x|h\rangle} - h. \tag{10.1.12}$$

The image of U_h in the Hilbert space E is the subspace

$$E_h = \{z \in E \ : \ \langle z|h\rangle = 0\}. \tag{10.1.13}$$

Conversely, the inverse image $\psi_h^{-1}(z)$ of any element $z \in E_h$ is an element in U_h such that $z + h \in \psi_h^{-1}(z)$. The set of the charts $\{(U_h, \psi_h)\}$ is a holomorphic atlas of the projective Hilbert space PE. In particular, given a point $\rho \in PE$, one can choose the centred chart E_h, $h \in \mathrm{x}$, such that $\psi_h(\mathrm{x}) = 0$.

The projective Hilbert space PE admits a unique Hermitian metric g such that the corresponding distance function on PE is

$$\rho(\mathrm{x}, \mathrm{x}') = \sqrt{2}\arccos(|\ \langle x|x'\rangle\ |), \tag{10.1.14}$$

where x, x' are normalized elements of E. This is a Kähler metric, called the *Fubini–Studi metric.* Given a local chart (U_h, ψ_h), this metric reads

$$g(\vartheta_1, \vartheta_2) = \frac{\langle v_1|u_2\rangle + \langle v_2|u_1\rangle}{1 + \|z\|^2} - \frac{\langle z|u_2\rangle\langle v_1|z\rangle + \langle z|u_1\rangle\langle v_2|z\rangle}{(1+\|z\|^2)^2}, \qquad z \in E_h, \tag{10.1.15}$$

for any complex tangent vectors $\vartheta_1 = v_1 + \overline{u}_1$ and $\vartheta_2 = v_2 + \overline{u}_2$ in T_zPE. The corresponding fundamental form is given by the expression

$$\omega(\vartheta_1, \vartheta_2) = i\frac{\langle v_1|u_2\rangle - \langle v_2|u_1\rangle}{1 + \|z\|^2} - i\frac{\langle z|u_2\rangle\langle v_1|z\rangle - \langle z|u_1\rangle\langle v_2|z\rangle}{(1+\|z\|^2)^2}. \tag{10.1.16}$$

Written in the coordinate chart centred at a point $z = 0$, the expressions (10.1.15) and (10.1.16) come to the expressions (10.1.5) and (10.1.7), respectively.

10.2 Geometric quantization

There are three main approaches to quantization of Poisson and symplectic systems of classical mechanics. These are Berezin's quantization, geometric quantization and deformation quantization.

Let $(Z, \{.,.\})$ be a finite-dimensional Poisson manifold. The ring $C^\infty(Z)$ of real smooth functions on Z is provided with the structure of a *Poisson algebra* $\mathcal{A}(Z)$ with respect to the Poisson bracket

$$(f, f') \mapsto \{f, f'\}, \qquad f, f' \in C^\infty(Z). \tag{10.2.1}$$

Recall that, in local canonical coordinates (q^i, p_i) on Z, the Poisson bracket reads

$$\{f, f'\} = \partial^i f \partial'_i - \partial_i f \partial^i f'.$$

The Poisson algebra $\mathcal{A}(Z)$ describes a classical Poisson system. Its quantization implies an assignment of a Hermitian operator $\widehat{f}$ to each element $f \in \mathcal{A}$ such that *Dirac's condition*

$$[\widehat{f}, \widehat{f'}] = -i\hbar\widehat{\{f, f'\}} \tag{10.2.2}$$

holds. The above mentioned Berezin's, geometric and deformation quantizations suggest different variants of such an assignment.

In our book, we are concerned with geometric quantization and deformation quantization (see the next Section) which involve different types of connections.

We refer the reader to [182, 282, 309] for the basics of the geometric quantization technique. Our exposition is concentrated to the role of connections in this technique. We will start from the following particular model.

Let M be an m-dimensional manifold. Its cotangent bundle T^*M is endowed with the canonical symplectic form

$$\Omega = dp_i \wedge dq^i, \tag{10.2.3}$$

written with respect to holonomic coordinates $(q^i, p_i = \dot{q}_i)$ on T^*M. Let us consider the trivial bundle

$$T^*M \times \mathbb{C} \to T^*M. \tag{10.2.4}$$

It can be provided with the linear connection

$$\Gamma = dp_j \otimes \partial^j + dq^j \otimes (\partial_j - 2\pi i p_j c \partial_c), \tag{10.2.5}$$

where c is a coordinate of $\mathbb{C}$. The curvature form (2.4.2) of this connection is

$$R = -2\pi i c \Omega, \tag{10.2.6}$$

where we omit ∂_c for the sake of simplicity. Given a function f in the Poisson algebra $\mathcal{A}(T^*M)$, let us consider the following first order differential operator on sections s of the fibre bundle (10.2.4):

$$\widehat{f}(s) = (\nabla_{\vartheta_f} - 2\pi i f)s = (\partial^j f(\partial_j + 2\pi i p_j) - \partial_j f \partial^j - 2\pi i f)s, \tag{10.2.7}$$

where ∇_{ϑ_f} is the covariant derivative relative to the connection Γ (10.2.5) along the Hamiltonian vector field

$$\vartheta_f \rfloor \Omega = -df,$$
$$\vartheta_f = \partial^i f \partial_i - \partial_i f \partial^i,$$

for the function f. Then we obtain Dirac's condition

$$\boxed{\widehat{f} \circ \widehat{g} - \widehat{g} \circ \widehat{f} = \widehat{\{f, g\}}} \tag{10.2.8}$$

for all $f, g \in \mathcal{A}(T^*M)$. This equality is the corollary of the particular form (10.2.6) of the curvature of the connection Γ.

Remark 10.2.1. Using standard geometric terms, we will follow Dirac's relation (10.2.8). To restart Dirac's relation (10.2.2) with physical coefficients, one should choose the connection

$$\Gamma = dp_j \otimes \partial^j + dq^j \otimes (\partial_j + \frac{i}{\hbar} p_j c \partial_c)$$

and write the operator

$$\widehat{f}(s) = -i\hbar(\nabla_{\vartheta_f} + \frac{i}{\hbar} f)s.$$

•

Now we generalize the above construction to an arbitrary symplectic manifold (Z, Ω). Let us consider a complex line bundle $\zeta : C \to Z$ coordinated by (z^μ, c). This is a fibre bundle with the structure group $U(1)$ (see Section 6.7). We denote by C_0 its subbundle with the typical fibre $\mathbb{C}_0 = \mathbb{C} \setminus \{0\}$ and with the same structure group. Let

$$\Gamma = dz^\mu \otimes (\partial_\mu + \Gamma_\mu c \partial_c) \tag{10.2.9}$$

be a linear connection on the fibre bundles $C \to Z$ and $C_0 \to Z$. The corresponding covariant differential on sections s of these fibre bundles reads

$$\nabla s = \nabla_\mu s dz^\mu = (\partial_\mu s - \Gamma_\mu s) dz^\mu.$$

Let us consider the complex 1-form

$$\alpha = \frac{1}{2\pi i}\left(\frac{dc}{c} - \Gamma_\mu dz^\mu\right) \tag{10.2.10}$$

on the fibre bundle $C_0 \to Z$. It is readily observed that

$$d\alpha = -\frac{1}{\pi i c} R,$$

where

$$R = \frac{1}{2} R_{\nu\mu} dz^\nu \wedge dz^\mu,$$
$$R_{\nu\mu} = (\partial_\nu \Gamma_\mu - \partial_\mu \Gamma_\nu) c,$$

is the curvature 2-form of the connection Γ, and

$$\nabla_u s = 2\pi i (u \rfloor (s^* \alpha)) s$$

for an arbitrary vector field u on Z and any section s of $C_0 \to Z$. It follows that the symplectic manifold Z admits a complex 2-form ω such that the pull-back form $\zeta^*\omega$ on C_0 coincides with $d\alpha$, i.e.,

$$\boxed{d\alpha = \zeta^*\omega.} \tag{10.2.11}$$

This form ω is given by the coordinate expression

$$\omega = -\frac{1}{2\pi i}\partial_\nu\Gamma_\mu dz^\nu \wedge dz^\mu.$$

It is closed, but not necessarily exact because $\Gamma_\mu dz^\mu$ is not a 1-form on Z in general.

Let g be a Hermitian fibre metric in fibre bundles C and C_0. This metric is said to be *invariant* with respect to a linear connection Γ (10.2.9) if

$$u\rfloor d(g(s,s')) = g(\nabla_u s, s') + g(s, \nabla_u s')$$

for an arbitrary vector field u on Z and any sections s, s' of the complex line bundle $C \to Z$.

PROPOSITION 10.2.1. Given a linear connection Γ on the complex line bundle C, this fibre bundle admits a Γ-invariant Hermitian fibre metric g if and only if the exterior form $2\pi i(\alpha - \overline{\alpha})$ on C_0 is exact. Then the metric g is given by the relation

$$\boxed{2\pi i(\alpha - \overline{\alpha}) = d\ln(g(c,c)).} \tag{10.2.12}$$

□

If a connection Γ on the complex line bundle C obeys the conditions of Proposition 10.2.1, then the form $(\Gamma + \overline{\Gamma})/c$ is exact, and the 2-form ω in the expression (10.2.11) is real, i.e.,

$$\zeta^*(\omega - \overline{\omega}) = d(\alpha - \overline{\alpha}) = 0.$$

It means that ω is a presymplectic form on Z.

Let us assume that the complex line bundle C admits a linear connection Γ (10.2.9), called an *admissible connection*, such that $\omega = \Omega$, i.e., the curvature R of this connection satisfies the relation (10.2.6). Then, by virtue of Proposition 10.2.1, there exists a Γ-invariant Hermitian fibre metric in C. For instance, if Γ is the connection (10.2.5), the corresponding Γ-invariant fibre metric reads

$$\boxed{g(c,c) = c\overline{c}.} \tag{10.2.13}$$

Let Γ be an admissible connection. Following the expression (10.2.7), one can assign to any function $f \in \mathcal{A}(Z)$ the differential operator $\widehat{f}$ on sections s of the complex line bundle C in accordance with the formula

$$\boxed{\widehat{f}(s) \stackrel{\text{def}}{=} (\nabla_{\vartheta_f} - 2\pi i f)s,} \tag{10.2.14}$$

called the *Kostant–Souriau formula*. The operators (10.2.14) obey Dirac's condition (10.2.8) for all elements f, g of the Poisson algebra $\mathcal{A}(Z)$.

A criteria of the existence of an admissible connection is based on the fact that the Chern form c_1 (6.7.11) of this connection is

$$c_1 = \frac{i}{2\pi}F = \frac{i}{2\pi c}R = \Omega. \tag{10.2.15}$$

Since, this is a representative of an integral cohomology class in the De Rham cohomology group $H^2(Z)$, the complex line bundle $C \to Z$ over a symplectic manifold (Z, Ω) has an admissible connection if and only if the symplectic form Ω belongs to an integral De Rham cohomology class. For instance, the canonical symplectic form (10.2.3) on T^*M is exact, i.e., has the zero De Rham cohomological class.

DEFINITION 10.2.2. A complex line bundle $C \to Z$ over a symplectic manifold (Z, Ω) is called a *prequantization bundle* if its Chern form c_1 coincides with the symplectic form Ω. □

It should be emphasized that the procedure of constructing the prequantization bundle is only the first step of geometric quantization. The operators $\widehat{f}$ (10.2.14) act in the subspace of smooth functions of the Hilbert space $L^2(Z, \mu_{\rm L})$ of complex functions on Z, which are square-integrable with respect to the Liouville measure

$$d\mu_{\rm L} = \frac{1}{(2\pi)^m}\prod_{i=1}^m dp_i \wedge dq^i, \qquad 2m = \dim Z. \tag{10.2.16}$$

However, this representation of the Poisson algebra $\mathcal{A}(Z)$ fails to be satisfactory (cf. Proposition 10.2.3 below). For instance, we have a non-conventional operator

$$\widehat{q^j} = -\partial^j + 2\pi i q^j$$

(10.2.14) assigned to the local function $f = q^j$.

The second step of geometric quantization consists in the following. Let (Z, Ω) be a symplectic manifold. By a *polarization* is meant a maximal involutive distribution $\mathbf{T} \subset TZ$ such that

$$\Omega(\vartheta, \upsilon) = 0, \qquad \forall \vartheta, \upsilon \in \mathbf{T}_z, \qquad z \in Z.$$

This term also stands for the algebra $\mathcal{T}_\Omega$ of sections of the distribution $\mathbf{T}$. We denote by $\mathcal{A}_\mathcal{T}$ the subalgebra of the Poisson algebra $\mathcal{A}(Z)$ which consists of the functions f such that

$$[\vartheta_f, \mathcal{T}_\Omega] \subset \mathcal{T}_\Omega.$$

Elements of this subalgebra only are quantized.

Let the symplectic form Ω on a manifold Z belongs to an integer cohomology class of the De Rham cohomology group $H^2(Z)$. Let $C \to Z$ be the complex line bundle whose Chern class c_1 coincides with the cohomology class of Ω. By $\mathcal{D}$ we denote the complex vector bundle of *half-densities* ρ over Z whose transformation law is

$$\rho' = |\det(\frac{\partial z'^\mu}{\partial z^\nu})|^{1/2}\rho$$

(see [309] for a detailed exposition). Given an admissible connection Γ and the corresponding Γ-invariant Hermitian metric g in C, the prequantization formula (10.2.14) is extended to sections $s \otimes \rho$ of the fibre bundle $C \otimes \mathcal{D} \to Z$ as follows:

$$\widehat{f}(s \otimes \rho) = (\nabla_{\vartheta_f} - 2\pi i f)(s \otimes \rho) = (\widehat{f}s) \otimes \rho + s \otimes \mathbf{L}_{\vartheta_f}\rho, \tag{10.2.17}$$

$$\nabla_{\vartheta_f}(s \otimes \rho) = (\nabla_{\vartheta_f}s) \otimes \rho + s \otimes \mathbf{L}_{\vartheta_f}\rho,$$

where

$$\mathbf{L}_{\vartheta_f}\rho = \frac{1}{2}\partial_\mu(\vartheta^\mu_f\rho)$$

is the Lie derivative. It is readily observed that the operators (10.2.17) obey Dirac's condition (10.2.8). Let us denote by $\mathcal{H}_Z$ the set of sections q of the fibre bundle $C \otimes \mathcal{D} \to Z$ such that

$$\nabla_\vartheta q = 0, \qquad \forall \vartheta \in \mathcal{T}_\Omega.$$

PROPOSITION 10.2.3. For any function $f \in \mathcal{A}_\mathcal{T}$ and an arbitrary section $q \in \mathcal{H}_Z$, the relation

$$\boxed{\widehat{f}q \in \mathcal{H}_Z} \tag{10.2.18}$$

holds. □

Thus, we have a representation of the algebra $\mathcal{A}_T$ in the space $\mathcal{H}_Z$. Therefore, by quantization of a function $f \in \mathcal{A}_T$ is meant the restriction of the operator $\widehat{f}$ (10.2.17) to $\mathcal{H}_Z$.

It should be emphasized that the space $\mathcal{H}_Z$ does not necessarily exist. If Z is a compact manifold without a boundary, the Hermitian form

$$\langle s_1 \otimes \rho_1 | s_2 \otimes \rho_2 \rangle = \frac{i^{m^2}}{\pi^m} \int\limits_Z g(s_1, s_2) \rho_1 \overline{\rho}_2 \qquad (10.2.19)$$

brings $\mathcal{H}_Z$ into a pre-Hilbert space. Its completion is called a *quantum Hilbert space* H_Z, and the operators $i\widehat{f}$ in this Hilbert space are Hermitian.

Now we will consider geometric quantization of holomorphic manifolds, e.g., a projective Hilbert space. As was shown, a projective Hilbert space admits a standard complex line bundle, and we have a standard procedure of prequantization of this space.

Let Z be an m-dimensional holomorphic manifold and $C \to Z$ a complex line bundle equipped with complex coordinates $(c, z^j, \overline{z}^j)$. It means that the coordinate transition functions on Z obey the condition

$$\frac{\partial z'^i}{\partial \overline{z}^k} = \frac{\partial \overline{z}'^i}{\partial z^k} = 0.$$

Let us assume that the complex line bundle C admits a linear connection

$$\Gamma = dz^i \otimes (\partial_i + \Gamma_i c \partial_c) + d\overline{z}^i \otimes (\overline{\partial}_i - \overline{\Gamma}_i c \partial_c)$$

with the curvature form

$$R = -2\pi i c \Omega,$$

where Ω is a non-degenerate real 2-form on Z. It implies that the manifold Z is provided with a symplectic structure, given by the symplectic form Ω, and with a Γ-invariant Hermitian fibre metric g. With respect to the local *complex canonical coordinates* $(z^j, \overline{z}^j)$, the above mentioned symplectic form reads

$$\boxed{\Omega = 2i \sum_j dz^j \wedge d\overline{z}^j,}$$

while the Hermitian metric g is given by the expression (10.2.13). Therefore, geometric quantization of a holomorphic manifold is similar to that of a real symplectic manifold if complex and real canonical coordinates are connected with each other by the relations

$$z^j = \frac{1}{2}(p_j + iq^j), \qquad \overline{z}^j = \frac{1}{2}(p_j - iq^j).$$

The holomorphic manifold Z has the canonical polarization

$$\mathbf{T} = \{v \in TZ \,:\, Jv = -iv\},$$

whose sections $\vartheta \in \mathcal{T}_\Omega$ are complex vector fields $\vartheta = \overline{\vartheta}^i \overline{\partial}_i$ on Z. We use the notation

$$\partial_i = \frac{\partial}{\partial z^i}, \qquad \overline{\partial}_i = \frac{\partial}{\partial \overline{z}^i},$$
$$d = \partial + \overline{\partial}, \qquad \partial = dz^i \partial_i, \qquad \overline{\partial} = d\overline{z}^i \overline{\partial}_i.$$

As in the case of a real symplectic manifold, let us consider the subalgebra $\mathcal{A}_\mathcal{T}$ of the Poisson algebra $\mathcal{A}(Z)$ which consists of smooth complex functions f on Z such that

$$[\vartheta_f, \mathcal{T}_\Omega] \subset \mathcal{T}_\Omega,$$

where

$$\vartheta_f = \frac{1}{2i} \sum_j (-\overline{\partial}_j f \partial_j + \partial_j f \overline{\partial}_j)$$

is the Hamiltonian vector field for a function f. It is readily observed that holomorphic functions on Z do not belong to $\mathcal{A}_\mathcal{T}$.

To construct the representation space of the algebra $\mathcal{A}_\mathcal{T}$, we take the sections s of the complex line bundle $C \to Z$ such that

$$\nabla_\vartheta s = 0, \qquad \vartheta \in \mathcal{T}_\Omega,$$

i.e., $(\overline{\partial}_i + \Gamma_i)s = 0$, and the holomorphic sections

$$\rho = \rho_{1\ldots m} dz^1 \wedge \cdots \wedge dz^m$$

of the cotangent bundle $T^{*(m,0)} \to Z$. Let

$$q = s \otimes \rho = s\rho_{1\ldots m} dz^1 \wedge \cdots \wedge dz^m \tag{10.2.20}$$

be sections of the tensor product $C \underset{Z}{\otimes} T^{*(m,0)}$. Then the operator

$$\boxed{\widehat{f}(s \otimes \rho) = (\nabla_{\vartheta_f} - 2\pi i f)(s) \otimes \rho + s \otimes \mathbf{L}_{\vartheta_f}\rho} \tag{10.2.21}$$

can be assigned to any function $f \in \mathcal{A}_T$. With respect to local complex canonical coordinates, this operator reads

$$\widehat{f}(s \otimes \rho) = \left(-\frac{1}{2i}\sum_k \overline{\partial}_k f(\partial_k - \Gamma_k - 2\pi i f)s\right) \otimes \rho + s \otimes \left(-\frac{1}{2i}\sum_k \partial_k(\rho\overline{\partial}_k f)\right).$$

The operators (10.2.21) obey Dirac's condition (10.2.8).

If Z is a compact manifold without a boundary, sections (10.2.20) form a pre-Hilbert space $\mathcal{H}_Z$ with respect to the scalar product (10.2.19). The operators $\widehat{f}$ (10.2.21) fulfill the relation (10.2.18), and they are anti-Hermitian in $\mathcal{H}_Z$. If Z is not compact, one chooses the pre-Hilbert space $\mathcal{H}_Z$ of sections q (10.2.20) such that the expression (10.2.19) is finite. However, the operators $\widehat{f}$ (10.2.21) in this space fail to be bounded and symmetrical.

We have a standard procedure of geometric quantization if there is an imbedding of a holomorphic manifold Z to a Hilbert space.

Let Z be an m-dimensional holomorphic manifold provided with a symplectic form Ω, and $C \to Z$ a complex line bundle over Z. As before, by H_Z is meant the completion of the above mentioned space $\mathcal{H}_Z$ of sections q (10.2.20) which provide the finite values of the expression (10.2.19). This is a Hilbert space. Let $\Psi = \{(U_\iota, s_\iota, z^k_\iota)\}$ be an atlas of the fibre bundle $C \otimes T^{*(m,0)}$, where z^k_ι are local complex canonical coordinates on Z and s_ι are holomorphic bases for the complex line bundle $C \to Z$ such that $s_\iota \overline{s}_\iota = 1$. The corresponding fibre bases for $C \otimes T^{*(m,0)}$ are $s_\iota \otimes dz^1_\iota \wedge \cdots \wedge dz^m_\iota$. Let us assume that, for all points $z_1, z_2 \in Z$, there exist sections $q_1, q_2 \in H_Z$:

$$\begin{aligned} q_j(z_1) &= q_{j\iota}(z_1) s_\iota \otimes dz^1_1 \wedge \cdots \wedge dz^m_1, \qquad z_1 \in U_\iota, \qquad j = 1, 2, \\ q_j(z_2) &= q_{j\kappa}(z_2) s_\kappa \otimes dz^1_2 \wedge \cdots \wedge dz^m_2, \qquad z_2 \in U_\kappa, \end{aligned}$$

such that

$$\det \begin{pmatrix} q_{1\iota}(z_1) & q_{1\kappa}(z_2) \\ q_{2\iota}(z_1) & q_{2\kappa}(z_2) \end{pmatrix} \neq 0. \tag{10.2.22}$$

This condition is independent of a choice of the above mentioned atlas Ψ.

THEOREM 10.2.4. There is the functional

$$K_\iota(x,z) = K_{\kappa\iota}(x,z)s_\kappa \otimes dx^1 \wedge \cdots \wedge dx^m \tag{10.2.23}$$

such that

$$q_\iota(z) = \langle q(.)|K_\iota(.,z)\rangle = \frac{i^{m^2}}{\pi^m}\int q(x)\overline{K_\iota(x,z)}.$$

□

The functional (10.2.23) is a *reproducing kernel* with the properties:

- $K_{\iota\iota}(z,z) > 0$,
- $\overline{K_{\kappa\iota}(x,z)} = K_{\iota\kappa}(z,x)$,
- $K_{\kappa\iota}(x,z) = \langle K_\iota(.,z)|K_\kappa(.,x)\rangle$.

It defines the mapping

$$\mathcal{K} : Z \supset U_\iota \ni z \mapsto K_\iota(.,z) \in H_Z. \tag{10.2.24}$$

By the assumption (10.2.22), $K_\iota(.,z_1) \neq \lambda K_\kappa(.,z_2)$, $\lambda \in \mathbb{C}$, whenever $z_1, z_2 \in Z$. It follows that the mapping (10.2.24) is an inclusion of Z into the projective Hilbert space PH_Z. This inclusion is a manifold imbedding if $\mathcal{K}$ is of constant rank. Let us denote

$$\xi_z = \mathcal{K}(z) = K_\iota(.,z) \in H_Z. \tag{10.2.25}$$

Using the properties of the reproducing kernel (10.2.23), one can observe that the vectors ξ_z, $z \in Z$, constitute a reproducing system which provides the decomposition of the identity operator in the Hilbert space H_Z with respect to the Liouville measure (10.2.16), i.e., one can think of the vectors (10.2.25) as being coherent states [4, 242].

Thus, we come to projective Hilbert spaces described in the previous Section. Since a projective Hilbert space admits the canonical complex line bundle (see (10.2.26) below), we have the following standard procedure of its quantization.

Let E be a Hilbert space and PE the corresponding projective Hilbert space. For the sake of simplicity, E is assumed to be a separable Hilbert space. Let us consider the fibre bundle

$$C = \{(z,\mathrm{z}) \in E \times PE \,:\, z \in \mathrm{z}\} \tag{10.2.26}$$

together with the corresponding projections $\zeta : C \to PE$ and $\pi : C \to E$. It is called the *universal line bundle*. There is its subbundle

$$C_0 = E \setminus \{0\} \to PE, \qquad \pi : z \to \mathrm{z},$$

with the typical fibre $\mathbb{C} \setminus \{0\}$. Let (U_h, ψ_h) be a chart (10.1.12) of the projective Hilbert space PE. Recall that its image in E is the subspace E_h (10.1.13). We provide this chart with the following coordinate system. Let us take the orthonormal basis $\{h, e_k\}$, $k = 1, \ldots, \mathrm{dimE}-1$, for the Hilbert space E. Then $\{e_k\}$ is an orthonormal basis for its subspace E_h, while the coordinates z_h^k with respect to this basis are the coordinates on the above mentioned chart of the projective Hilbert space PE. The projective Hilbert space PE is provided with the Fubini–Studi metric (10.1.15) and with the corresponding fundamental form ω (10.1.16) which reads

$$\omega = -i \sum_{j,k} \left(\frac{\delta^{jk}}{1 + \|z\|^2} - \frac{z^j \overline{z}^k}{(1 + \|z\|^2)^2} \right), \qquad \|z\|^2 = \sum_k z^k \overline{z}^k.$$

The universal line bundle (10.2.26) has the connection

$$\Gamma = dz^j \otimes \left(\partial_j + 2\pi \frac{\overline{z}^j}{1 + \|z\|^2} c\partial_c \right) + d\overline{z}^j \otimes \left(\overline{\partial}_j - 2\pi \frac{z^j}{1 + \|z\|^2} c\partial_c \right),$$

which preserves the Hermitian fibre metric (10.2.13) in C and satisfies the condition

$$\boxed{R = -2\pi i c\omega.}$$

Now let Z be a holomorphic manifold and $\mathcal{K}$ its inclusion into the projective Hilbert space PE. Then the above manifested standard procedure of geometric quantization of Z consists in the following. Let us propose that the pull-back form $\mathcal{K}^*\omega$ on Z is non-degenerate, i.e., this is a symplectic form. Then one can consider the pull-back line bundle $\mathcal{K}^*C$ over Z. This line bundle admits the pull-back connection $\mathcal{K}^*\Gamma$ with the curvature form $-2\pi i c\mathcal{K}^*\omega$, i.e., $\mathcal{K}^*C$ is a prequantization bundle.

10.3 Deformation quantization

There are well-known quantum groups and algebras which are deformations of ordinary Lie groups and Lie algebras [171, 176, 209]. The method of deformation of algebraic structures can also be applied to quantization, called deformation quantization, of the Poisson algebra on a symplectic manifold. Symplectic connections play an important role in this quantization procedure.

Let (Z, Ω) be a finite-dimensional symplectic manifold and $C^\infty(Z)$ the ring of smooth functions on Z which is also the Poisson algebra $\mathcal{A}(Z)$ with respect to the Poisson bracket (10.2.1). By deformation quantization of $\mathcal{A}(Z)$ is meant a new operation of multiplication of functions $f * f'$ and a new commutator $\{.,.\}_*$ which depend on a real parameter γ such that, when $\gamma \to 0$, the operation $f * f'$ reduces to the usual multiplication of functions, while $\{.,.\}_*$ comes to the Poisson bracket [26, 84, 148]. One can think of the parameter γ as being Planck's constant.

We will start from the following general notion of deformation.

DEFINITION 10.3.1. Let B a $C^\infty(Z)$-valued $\mathbb{R}$-bilinear form on the $\mathbb{R}$-vector space $C^\infty(Z)$. By its *formal deformation* is meant the formal series

$$B_\gamma(f, f') = \sum_{r=0}^{\infty} \gamma^r C_r(f, f'), \tag{10.3.1}$$

where $\gamma \geq 0$ is a real parameter, $C_0(f, f') = B(f, f')$, and $C_r(f, f')$ are $C^\infty(Z)$-valued bilinear forms on $C^\infty(Z)$. □

We will deal with the following two variants of the bilinear form B:

- the familiar pointwise multiplication of functions $B(f, f') = ff'$, whose deformation

$$B_\gamma(f, f') = f *_\gamma f' = ff' + \sum_{r=1}^{\infty} \gamma^r C_r(f, f') \tag{10.3.2}$$

  is associative, but not necessarily commutative;

- the Poisson bracket $B(f, f') = \{f, f'\}$, whose deformation

$$B_\gamma(f, f') = \{f, f'\}_* = \{f, f'\} + \sum_{r=1}^{\infty} \gamma^r S_r(f, f') \tag{10.3.3}$$

  has a Lie algebra structure.

The deformation (10.3.1) is called *associative deformation* of the $\mathbb{R}$-algebra $C^\infty(Z)$, while the deformation (10.3.3) is said to be *Lie deformation* of the Poisson bracket $\{.,.\}$.

Note that every associative deformation $f * f'$ defines the Lie deformation

$$\{f, f'\}_* = \frac{i}{\gamma}(f * f' - f' * f). \tag{10.3.4}$$

An associative deformation must satisfy the condition

$$(f_1 *_\gamma f_2) *_\gamma f_3 - f_1 *_\gamma (f_2 *_\gamma f_3) = \sum_{k=1}^{\infty} \gamma^k D_k(f_1, f_2, f_3) = 0, \tag{10.3.5}$$

$$D_k(f_1, f_2, f_3) = \sum_{s+r=k,\, s,r\geq 0} C_r(C_s(f_1, f_2), f_3) - C_r(f_1, C_s(f_2, f_3)), \tag{10.3.6}$$

i.e., $D_k(f_1, f_2, f_3) = 0$ for all $k = 1, 2, \ldots$. This condition is phrased in terms of the Hochshild cohomology.

Remark 10.3.1. Recall briefly the notion of the Hochshild cohomology (see, e.g., [299]). Let A be an associative $\mathbb{R}$-algebra. One defines p-cochains as p-linear maps $C : \overset{p}{\otimes} A \to A$, and introduces their homomorphisms

$$(\partial C)(a_0, \ldots, a_p) = a_0 C(a_1, \ldots, a_p) + \tag{10.3.7}$$
$$\sum_{i=0}^{p-1} (-1)^{i+1} C(a_0, \ldots, a_i a_{i+1}, \ldots, a_p) + (-1)^{p+1} C(a_0, \ldots, a_{p+1}) a_p.$$

It is readily observed that $\partial^2 = 0$. The cohomology of the corresponding cochain complex is called the *Hochshild cohomology*. •

Turn to the expression (10.3.6). If $E_k(f_1, f_2, f_3)$ denotes the sum of the terms with indices $s, r \geq 1$ in its right-hand side, this expression takes the form

$$D_k(f_1, f_2, f_3) = E_k(f_1, f_2, f_3) - (\partial C_k)(f_1, f_2, f_3), \tag{10.3.8}$$

where ∂ is the operator (10.3.7). Then one can obtain associative deformation in the framework of the following recurrence procedure. Let us assume that there are 2-cochains $C_i(f, f')$, $i \leq k$, such that $D_i = 0$ for all $i \leq k$. We need a cochain C_{k+1} such that $D_{k+1} = 0$. It should be emphasized that E_{k+1} depends only on $C_{i\leq k}$, and one can show that, if $D_{i\leq k} = 0$, then $\partial E_{k+1} = 0$, i.e., E_{k+1} is a 3-cocycle. If this cocycle is a coboundary, i.e., it belongs to the zero element of the Hochshild cohomology group $H^3_{\rm H}(C^\infty(Z))$, then a desired cochain C_{k+1} can be found. Therefore, if the Hochshild cohomology group $H^3_{\rm H}(C^\infty(Z))$ vanishes, the associative deformation exists. For instance, associative deformation exists on a symplectic manifold [78]. Note that, in order to start the requirency procedure from $D_1 = 0$, one can choose

$$C_1(f, f') = \lambda\{f, f'\},$$

where λ is a constant (see the Moyal product below).

Similarly, one can examine the existence of the Lie deformation (10.3.3), where

$$S_r(f, f') + S_r(f', f) = 0.$$

Any Lie deformation must satisfy the Jacobi identity

$$\{f_1, \{f_2, f_3\}_*\}_* + \{f_2, \{f_3, f_1\}_*\}_* + \{f_3, \{f_1, f_2\}_*\}_* = \sum_{k=1}^{\infty} T_k(f_1, f_2, f_3) = 0, \tag{10.3.9}$$

$$T_k(f_1, f_2, f_3) = \sum_{r+s=k,\, s,r\geq 0} [S_r(f_1, S_s(f_2, f_3)) + S_r(f_2, S_s(f_3, f_1)) + S_r(f_3, S_s(f_1, f_2))]. \tag{10.3.10}$$

If $Q_k(f_1, f_2, f_3)$ denotes the sum of the terms with indices $s, r \geq 1$ in the right-hand side of the expression (10.3.10), this expression takes the form

$$S_k(f_1, f_2, f_3) = Q_k(f_1, f_2, f_3) - (\widetilde{\partial} S_k)(f_1, f_2, f_3), \tag{10.3.11}$$

where

$$(\widetilde{\partial} S_k)(f_1, f_2, f_3) = [\{f_1, S_k(f_2, f_3)\} + \{f_2, S_k(f_3, f_1)\} + \{f_3, S_k(f_1, f_2)\}] + [S_k(f_1, \{f_2, f_3\}) + S_k(f_2, \{f_3, f_1\}) + S_k(f_3, \{f_1, f_2\})]$$

is the coboundary operator ($\widetilde{\partial}^2 = 0$) for the Chevalley–Eilenberg cohomology (see Section 14.2). Therefore, Lie deformation can also be constructed by a recurrence procedure. Let us assume that there are 2-cochains $S_i(f, f')$, $i \leq k$, such that $T_i = 0$ for all $i \leq k$. We need a cochain S_{k+1} such that $T_{k+1} = 0$. The term Q_{k+1} in the right-hand side of the expression (10.3.11) depends only on $S_{i\leq k}$, and one can show that, if $T_{i\leq k} = 0$, then Q_{k+1} is a 3-cocycle. Therefore, if the Chevalley–Eilenberg cohomology group $H^3_{\rm CE}(C^\infty(Z))$ vanishes, then the Lie deformation exists.

In particular, let $Z = \mathbb{R}^{2m}$ be provided with the coordinates (Q^i, P_i) and the corresponding canonical symplectic form. Of course, both the Hochshild cohomology and the Chevalley–Eilenberg cohomology for this space vanish. Let us consider the operator

$$f\overset{\leftrightarrow}{P}f' = \{f,f'\}, \qquad f,f' \in C^\infty(\mathbb{R}^{2m}).$$

We have the associative deformation

$$f * f' = f\exp[\frac{-i\gamma}{2}\overset{\leftrightarrow}{P}]f', \tag{10.3.12}$$

called the *Moyal product* [145, 167]. The Moyal product (10.3.12) defines the corresponding Lie deformation (10.3.3).

The following construction, called the *Fedosov deformation*, generalizes the Moyal product to any symplectic manifold [102, 103]. The main ingredient in this construction is a connection on a symplectic manifold.

Let Z be a $2m$-dimensional symplectic manifold coordinated by (z^λ) and provided with the symplectic form

$$\Omega = \frac{1}{2}\Omega_{\alpha\beta}dz^\alpha \wedge dz^\beta.$$

DEFINITION 10.3.2. A *formal Weyl algebra* $\mathcal{A}_z$ over the tangent space T_zZ is the associative algebra with a unit whose elements are the formal sums

$$a(y,\gamma) = \sum_{2k+r\geq 0} \gamma^k a_{k,\alpha_1\ldots\alpha_r} y^{\alpha_1}\cdots y^{\alpha_r}, \tag{10.3.13}$$

where $y^\mu = \dot z^\mu$ are holonomic coordinates on the tangent space T_zZ. The algebra $\mathcal{A}_z$ is provided with the *Weyl product*

$$a\circ a' = \exp\left[-\frac{i\gamma}{2}\Omega^{\alpha\beta}(z)\frac{\partial}{\partial y^\alpha}\frac{\partial}{\partial y'^\beta}\right] a(y,\gamma)a'(y',\gamma)|_{y=y'} = \tag{10.3.14}$$

$$\sum_{k=0}^\infty \left(-\frac{i\gamma}{2}\right)^k \frac{1}{k!}\Omega^{\alpha_1\beta_1}(z)\cdots\Omega^{\alpha_k\beta_k}(z)\frac{\partial^k a}{\partial^{\alpha_1}\ldots\partial^{\alpha_k}}\frac{\partial^k a'}{\partial^{\beta_1}\ldots\partial^{\beta_k}}.$$

□

Of course, this definition is independent of a coordinate system. The disjoint union of the Weyl algebra $\mathcal{A}_z$, $z\in Z$, is the *Weyl algebra bundle* $A\to Z$ whose sections read

$$a(z,y,\gamma) = \sum_{2k+r\geq 0} \gamma^k a_{k,\alpha_1\ldots\alpha_r}(z) y^{\alpha_1}\cdots y^{\alpha_r}, \tag{10.3.15}$$

where $a_{k,\alpha_1\ldots\alpha_r}(z)$ are sections of the tensor bundles $\overset{r}{\vee} TZ$. The set $\mathcal{A}$ of sections (10.3.15) is also an associative algebra with respect to the fibrewise multiplication (10.3.14). Its unit element is $a(z,y,\gamma)=1$. The centre $\mathcal{Z}$ of the algebra $\mathcal{A}$ consists of the elements

$$a=\sum_{k=0}^{\infty}\gamma^k a_k(z), \tag{10.3.16}$$

independent of coordinates y^λ. There is a filtration $\mathcal{A}\supset\mathcal{A}_1\supset\cdots$ in the algebra $\mathcal{A}$ with respect to the total degree $2k+r$ of the terms of the series (10.3.15).

Let us consider the tensor product $\mathcal{A}\otimes\mathfrak{D}^*(Z)$ whose elements are A-valued exterior forms on the manifold Z:

$$\boxed{a(z,y,dz,\gamma)=\sum\gamma^k a_{k,\alpha_1\ldots\alpha_r,\beta_1\ldots\beta_s}(z)y^{\alpha_1}\cdots y^{\alpha_r}dz^{\beta_1}\wedge\cdots\wedge dz^{\beta_s},} \tag{10.3.17}$$

called simply A-forms. Their multiplication is defined as the exterior product $\wedge$ of exterior forms and the Weyl product $\circ$ (10.3.14) of polynomials in y^α. Let the symbol $\circ$ also stand for this multiplication. With this multiplication law, the algebra $\mathcal{A}\otimes\mathfrak{D}^*(Z)$ has the structure of a graded algebra over the graded commutative ring $\mathfrak{D}^*(Z)$. The corresponding bracket of two A-forms a, a' (10.3.17) is defined as

$$[a,a']=a\circ a'-(-1)^{|a||a'|}a'\circ a, \tag{10.3.18}$$

where $|a|\in\mathcal{A}\otimes\mathfrak{D}^{|a|}(Z)$, $|a'|\in\mathcal{A}\otimes\mathfrak{D}^{|a'|}(Z)$. One says that an element a belongs to the centre of the algebra $\mathcal{A}\otimes\mathfrak{D}^*(Z)$ if its bracket (10.3.18) with any element of this algebra vanishes. This centre is $\mathcal{Z}\otimes\mathfrak{D}^*(Z)$. There are two projections of an A-form $a(z,y,dz,\gamma)$ (10.3.17) to this centre. These are

$$a_0=a(z,0,dz,\gamma), \qquad a_{00}=a(z,0,0,\gamma). \tag{10.3.19}$$

We also have the following two operators acting on A-forms:

$$\delta:\mathcal{A}_r\otimes\mathfrak{D}^s(Z)\to\mathcal{A}_{r-1}\otimes\mathfrak{D}^{s+1}(Z),$$
$$\delta a=dz^\alpha\wedge\frac{\partial}{\partial y^\alpha}a,$$

$$\delta^*:\mathcal{A}_r\otimes\mathfrak{D}^s(Z)\to\mathcal{A}_{r+1}\otimes\mathfrak{D}^{s-1}(Z),$$
$$\delta^* a=y^\alpha\frac{\partial}{\partial y^\alpha}\rfloor a.$$

These operators possess the properties:

- the operator δ is an antiderivative, i.e.,

$$\delta(a \circ a') = (\delta a) \circ a' + (-1)^{|a|} a \circ \delta a',$$

and is represented by the bracket

$$\delta a = -[\frac{i}{\gamma}\Omega_{\alpha\beta} y^\alpha dz^\beta, a];$$

- $\delta^2 = (\delta^*)^2 = 0$;

- for any monomial

$$a = y^{\alpha_1} \cdots y^{\alpha_r} dz^{\beta_1} \wedge \cdots \wedge dz^{\beta_s}, \tag{10.3.20}$$

we have

$$(\delta\delta^* + \delta^*\delta)a = (r+s)a.$$

One also introduces the operator

$$\delta^{-1} : \mathcal{A}_r \otimes \mathfrak{O}^s(Z) \to \mathcal{A}_{r+1} \otimes \mathfrak{O}^{s-1}(Z)$$

which acts on monomials (10.3.20) by the law

$$\delta^{-1} a = \begin{cases} (r+s)^{-1}\delta^* a, & r+s > 0, \\ \delta^{-1} a = 0, & r+s = 0. \end{cases}$$

Then any A-form (10.3.17) has the decomposition

$$\boxed{a = (\delta\delta^{-1} + \delta^{-1}\delta)a + a_{00}.} \tag{10.3.21}$$

Turn now to a connection on the symplectic manifold (Z, Ω). It is given by the tangent-valued form

$$\Gamma = dz^\lambda \otimes \left(\frac{\partial}{\partial z^\lambda} + \Gamma_\lambda{}^\mu{}_\nu y^\nu \frac{\partial}{\partial y^\mu}\right). \tag{10.3.22}$$

The connection Γ (10.3.22) is called a *symplectic connection* if it is torsionless and the covariant differential of the symplectic form relative to this connection vanishes, i.e.,

$$\nabla\Omega_{\alpha\beta} = dz^\lambda\nabla_\lambda\Omega_{\alpha\beta} = dz^\lambda(\frac{\partial}{\partial z^\lambda}\Omega_{\alpha\beta} + \Gamma_\lambda{}^\mu{}_\alpha\Omega_{\mu\beta} + \Gamma_\lambda{}^\mu{}_\beta\Omega_{\alpha\mu}) = 0.$$

With respect to the local canonical coordinates where Ω =const., the coefficients

$$\Gamma_{\lambda\mu\nu} = \Omega_{\mu\alpha}\Gamma_\lambda{}^\alpha{}_\nu \tag{10.3.23}$$

of the symplectic connection are symmetric over all indices. A symplectic connection on a symplectic manifold exists, but is not unique [35, 118, 298]. Coefficients (10.3.23) of different symplectic connections differ from each other in symmetric tensor fields $\sigma_{\lambda\mu\nu}$.

Remark 10.3.2. Let us formulate more general result (see [118]). Let ω be a non-degenerate 2-form on a manifold Z and Γ a linear connection on Z. We denote

$$\Gamma_{\lambda\mu\nu} = \omega_{\mu\alpha}\Gamma_\lambda{}^\alpha{}_\nu, \qquad \tau_{\lambda\mu\nu} = \frac{1}{2}(\Gamma_{\lambda\mu\nu} + \Gamma_{\lambda\nu\mu}).$$

A connection Γ preserving the 2-form ω can be written as the sum

$$\Gamma_{\lambda\mu\nu} = \frac{1}{2}(\partial_\lambda\omega_{\mu\nu} - \partial_\mu\omega_{\nu\lambda} - \partial_\nu\omega_{\lambda\mu}) + (\tau_{\lambda\mu\nu} + \tau_{\nu\mu\lambda} - \tau_{\mu\nu\lambda}).$$

In the case when ω is closed (hence, symplectic), this formula takes the form

$$\Gamma_{\lambda\mu\nu} = \partial_\lambda\omega_{\mu\nu} + (\tau_{\lambda\mu\nu} + \tau_{\nu\mu\lambda} - \tau_{\mu\nu\lambda}).$$

If Γ is a torsionless connection (i.e., a symplectic connection), we have

$$\Gamma_{\lambda\mu\nu} = \partial_\lambda\omega_{\mu\nu} + \tau_{\lambda\mu\nu}.$$

If Γ and Γ' are different symplectic connections, then

$$\Gamma'_{\lambda\mu\nu} - \Gamma_{\lambda\mu\nu} = \tau'_{\lambda\mu\nu} - \tau_{\lambda\mu\nu} = \sigma_{\lambda\mu\nu}$$

is a symmetric tensor field. A torsionless linear connection preserving a non-degenerate 2-form ω on a manifold Z exists if and only if ω is closed, i.e., symplectic.
•

A symplectic connection Γ on the symplectic manifold Z defines a connection on the graded $\mathfrak{O}^*(Z)$-algebra $\mathcal{A} \otimes \mathfrak{O}^*(Z)$ of A-forms (10.3.17) by the rule

$$\nabla a = dz^\lambda \wedge \nabla_\lambda a.$$

In local canonical coordinates (z^α), we have

$$\nabla a = da + \left[\frac{i}{\gamma} G, a\right],$$

where

$$d = dz^\lambda \wedge \frac{\partial}{\partial \mathrm{z}^\lambda}, \qquad G = \frac{1}{2}\Gamma_{\lambda\mu\nu} y^\mu y^\nu dz^\lambda.$$

In accordance with Definition 8.2.20 extended to an algebra over a graded ring, this connection is a graded derivation

$$\nabla(a \circ a') = \nabla a \circ a' + (-1)^{|a|} a \circ \nabla a'$$

which obeys the Leibniz rule

$$\nabla(\phi \wedge a) = d\phi \wedge a + (-1)^k \phi \wedge \nabla a$$

for any exterior form $\phi \in \mathfrak{O}^k(Z)$. Its curvature reads

$$\nabla^2 a = \left[\frac{i}{\gamma} R, a\right],$$
$$R = \frac{1}{4} R_{\lambda\alpha\mu\nu} y^\mu y^\nu dz^\lambda \wedge dz^\alpha,$$
$$R_{\lambda\alpha\mu\nu} = \Omega_{\mu\beta} R_{\lambda\alpha}{}^\beta{}_\nu.$$

We have the relation

$$\boxed{(\nabla\delta + \delta\nabla) a = 0.}$$

Given a symplectic connection Γ on the symplectic manifold (Z, Ω), let us consider more general connections on the algebra $\mathcal{A} \otimes \mathfrak{O}^*(Z)$, namely, connections of the form

$$\widetilde{\nabla} a = \nabla a + [\frac{i}{\gamma}\tau, a] = da + [\frac{i}{\gamma}(G + \tau), a], \tag{10.3.24}$$

where τ is a A-valued 1-form on Z. The A-form τ in the expression (10.3.24) is not defined uniquely. For the uniqueness of τ, we will require that its projection τ_0 (10.3.19) to the centre $\mathcal{Z} \otimes \mathfrak{O}^*(Z)$ vanishes. Then the curvature

$$\widetilde{\nabla}^2 a = \left[\frac{i}{\gamma}\widetilde{R}, a\right]$$

of the connection $\widetilde{\nabla}$ (10.3.24) takes the form

$$\frac{i}{\gamma}\widetilde{R} = \frac{i}{\gamma}(R + \nabla\tau + \tau^2). \tag{10.3.25}$$

The connection (10.3.24) is called an *Abelian connection* if

$$\boxed{\widetilde{\nabla}^2 a = 0}$$

for all elements a of the algebra $\mathcal{A} \otimes \mathfrak{O}^*(Z)$, i.e., the curvature form $\widetilde{R}$ (10.3.25) belongs to the centre of this algebra.

PROPOSITION 10.3.3. [102]. For any symplectic connection Γ on a symplectic manifold (Z, Ω), there exists an Abelian connection on the algebra $\mathcal{A}\otimes\mathfrak{O}^*(Z)$ which takes the form

$$\widetilde{\nabla} = \nabla - \delta + [\frac{i}{\gamma}r, .] = \nabla + [\frac{i}{\gamma}(\Omega_{\alpha\beta}y^\alpha dz^\beta + r), .], \tag{10.3.26}$$

where r is an A-valued 1-form such that $r_0 = 0$. □

The curvature form of the connection $\widetilde{\nabla}$ (10.3.26) reads

$$\widetilde{R} = -\frac{1}{2}\Omega_{\alpha\beta}y^\alpha dz^\beta + R - \delta r + \nabla r + \frac{i}{\gamma}r^2.$$

The connection $\widetilde{\nabla}$ is Abelian if

$$\delta r = R + \nabla r + \frac{i}{\gamma}r^2. \tag{10.3.27}$$

Then $\widetilde{R} = -\Omega$ is a central form.

LEMMA 10.3.4. The equation (10.3.27) has a unique solution r such that

$$\delta^{-1}r = 0. \tag{10.3.28}$$

□

Let $\widetilde{\nabla}$ (10.3.26) be a desired Abelian connection where r obeys the relations (10.3.27) and (10.3.28). Let us consider the subalgebra $\mathcal{A}_\nabla$ of the algebra $\mathcal{A}$ which consists of the elements a, called *flat*, such that $\widetilde{\nabla}a = 0$.

THEOREM 10.3.5. [102]. For any element $b \in \mathcal{Z}$ (10.3.16), there exists a unique flat element $a(z, y, \gamma) \in \mathcal{A}_\nabla$ such that

$$\sigma(a) \stackrel{\text{def}}{=} a_0(z, 0, \gamma) = b.$$

□

Then the associative deformation for elements $a, a' \in \mathcal{Z}$ (10.3.16) on the symplectic manifold Z is defined as

$$\boxed{a * a' = \sigma(\sigma^{-1}(a) \circ \sigma^{-1}(a')).} \tag{10.3.29}$$

In particular, let $Z = \mathbb{R}^{2m}$ and Γ be the zero symplectic connection. The corresponding Abelian connection (10.3.26) takes the form

$$\boxed{\widetilde{\nabla} = -\delta + d.}$$

It is readily observed that, in this case, the associative deformation (10.3.29) restarts the Moyal product (10.3.12).

10.4 Quantum time-dependent evolution

We have seen in Section 5.10 that solutions of the Hamilton equations in classical time-dependent mechanics, by definition, are integral sections of a Hamiltonian connection, i.e., evolution in classical Hamiltonian mechanics is described as a parallel transport along time. Following [9, 160, 244], we will treat an evolution of a quantum time-dependent system as a parallel transport.

It should be emphasized that, in quantum mechanics, a time plays the role of a classical parameter. Indeed, all relations between operators in quantum mechanics are simultaneous, while a computation of a mean value of an operator in a quantum state does not imply an integration over a time. It follows that, at each moment, we have a quantum system, but these quantum systems are different at different instants. Although they may be isomorphic to each other.

Recall that, in the framework of algebraic quantum theory, a quantum system is characterized by a C^*-algebra A and a positive (hence, continuous) form ϕ on A which defines the representation π_ϕ of A in a Hilbert space E_ϕ with a cyclic vector ξ_ϕ such that

$$\phi(a) = \langle \pi_\phi(a)\xi_\phi | \xi_\phi \rangle, \qquad \forall a \in A,$$

(see, e.g., [39, 82]). One says that $\phi(a)$ is a mean value of the operator a in the state ξ_ϕ.

Therefore, to describe quantum evolution, one should assign a C^*-algebra A_t to each point $t \in \mathbb{R}$, and treat A_t as a quantum system at the instant t. Thus, we have a family of instantaneous C^*-algebras A_t, parametrised by the time axis $\mathbb{R}$. Let us suppose that all C^*-algebras A_t are isomorphic to each other and to some unital C^*-algebra A. Moreover, let they make up a locally trivial smooth Banach fibre bundle $P \to \mathbb{R}$ with the typical fibre A, whose transition functions are automorphisms of the C^*-algebra A. Smooth sections α of the C^*-algebra bundle $P \to \mathbb{R}$ constitute an involutive algebra with respect to the fibrewise operations. This is also a module $P(\mathbb{R})$ over the ring $C^\infty(\mathbb{R})$ of real functions on $\mathbb{R}$. In accordance with Definition 8.2.7, a connection ∇ on the $C^\infty(\mathbb{R})$-algebra $P(\mathbb{R})$ assigns to the standard vector field ∂_t on $\mathbb{R}$ a derivation

$$\nabla_t \in \mathfrak{d}(P(\mathbb{R})) \tag{10.4.1}$$

which obeys the Leibniz rule

$$\nabla_t(f\alpha) = \partial_t f\alpha + f\nabla_t\alpha, \qquad \alpha \in P(\mathbb{R}), \qquad f \in C^\infty(\mathbb{R}).$$

The fibre bundle $P \to \mathbb{R}$ is obviously trivial, though it has no canonical trivialization in general. Given its trivialization $P = \mathbb{R} \times A$, the derivation ∇_t (10.4.1) reads

$$\boxed{\nabla_t(\alpha) = [\partial_t - \delta(t)](\alpha),} \tag{10.4.2}$$

where $\delta(t)$ at each $t \in \mathbb{R}$ are derivations of the C^*-algebra A, such that

$$\delta_t(ab) = \delta_t(a)b + a\delta_t(b), \qquad \delta_t(a^*) = \delta_t(a)^*.$$

We say that a section $\alpha(t)$ of the fibre bundle $P \to \mathbb{R}$ is an integral section of the connection (10.4.2) if

$$\boxed{\nabla_t(\alpha) = [\partial_t - \delta(t)](\alpha) = 0.} \tag{10.4.3}$$

One can think of the equation (10.4.3) as being the *Heisenberg equation* describing quantum evolution. An integral section $\alpha(t)$ of the connection ∇ is a solution of this equation. We also say that $\alpha(t)$ is a geodesic curve in A.

In particular, let the derivations $\delta(t) = \delta$ be the same for all $t \in \mathbb{R}$. If δ is a generator of a 1-parameter group g_r of automorphisms of the algebra A, then for any $a \in A$, the curve

$$\boxed{\alpha(t) = g_t(a),} \qquad t \in \mathbb{R}, \tag{10.4.4}$$

in A is a solution of the Heisenberg equation (10.4.3).

There are certain conditions in order that a derivation δ of a unital C^*-algebra to define a 1-parameter group of its automorphisms [38, 249].

Remark 10.4.1. Let V be a Banach space. An operator a in V is said to be *bounded* if there is $\lambda \in \mathbb{R}$ such that

$$||av|| \leq \lambda ||v||, \qquad \forall v \in V.$$

The algebra $B(V)$ of bounded operators in a Banach space V is a Banach algebra with respect to the norm

$$||a|| = \sup_{||v||=1} ||av||.$$

It is provided with the corresponding norm topology, called the *norm operator topology*. Another topology in $B(V)$, referred to in the sequel, is the *strong operator topology*. It is given by the following family of open neighbourhoods of $0 \in B(V)$:

$$U_v^\varepsilon = \{a \in B(V) \,:\, ||av|| < \varepsilon\}, \qquad \forall v \in V, \qquad \forall \varepsilon > 0.$$

One also introduces weak, ultra-strong and ultra-weak topologies in $B(V)$ [82]. Note that the s-topology in [9] is the ultra-strong topology in the terminology of [82]. The algebra $B(V)$ is a topological algebra only with respect to the norm operator topology because the morphism

$$B(V) \times B(V) \ni (a, b) \mapsto ab \in B(V)$$

is continuous only in this topology. The norm operator topology is finer than the other above mentioned operator topologies. •

PROPOSITION 10.4.1. If δ is a bounded derivation of a C^*-algebra A, then it is a generator of a 1-parameter group

$$g_r = \exp[r\delta], \qquad r \in \mathbb{R},$$

of automorphisms of A, and *vice versa*. This group is continuous in a norm topology in $\operatorname{Aut}(A)$. Furthermore, for any representation π of A in a Hilbert space E_π, there exists a self-adjoint bounded operator $\mathcal{H}$ in E_π such that

$$\pi(\delta(a)) = i[\mathcal{H}, \pi(a)], \quad \pi(g_r(a)) = e^{ir\mathcal{H}}\pi(a)e^{-ir\mathcal{H}}, \quad \forall a \in A, \quad r \in \mathbb{R}. \quad (10.4.5)$$

□

Note that, if the domain of definition $D(\delta)$ of a derivation δ coincides with A, this derivation is bounded. It follows that, by definition, the derivations δ_t, $t \in \mathbb{R}$, in the connection ∇_t (10.4.2) are bounded. It follows that, if a quantum system is conservative, i.e., $\delta_t = \delta$ are the same for all $t \in \mathbb{R}$, the Heisenberg equation (10.4.3) has a solution (10.4.4) through any point of $\mathbb{R} \times A$ in accordance with Proposition 10.4.1. Proposition 10.4.1 also states that the description of evolution of a quantum conservative system in terms of the Heisenberg equation and that based on the Shrödinger equation are equivalent.

However, non-trivial bounded derivations of a C^*-algebra do not necessarily exist. Moreover, if a curve g_r is continuous in $\operatorname{Aut}(A)$ with respect to the norm operator topology, it implies that the curve $g_r(a)$ for any $a \in A$ is continuous in the C^*-algebra A, but the converse is not true. At the same time, a curve g_r is continuous in $\operatorname{Aut}(A)$ with respect to the strong operator topology in $\operatorname{Aut}(A)$ if and only if the curve $g_r(a)$ for any $a \in A$ is continuous in A.

By this reason, we are also interested in strong-continuous 1-parameter groups of automorphisms of C^*-algebras. We refer the reader to [38, 249] for the sufficient conditions which a derivation δ should satisfy in order to be a generator of such a group. Note only that δ has a dense domain of definition $D(\delta)$ in A, and it is not bounded on $D(\delta)$. If δ is bounded on $D(\delta)$, then δ is extended uniquely to a bounded derivation of A. It follows that, for a strong-continuous 1-parameter group of automorphisms of A, the connection ∇_t (10.4.2) is not defined on the whole algebra $P(\mathbb{R})$. In this case, we deal with a generalized connection which is given by operators of a parallel transport whose generators are not well-defined. It may also happen that a representation π of the C^*-algebra A does not carry out the representation (10.4.5) of a strong-continuous 1-parameter group g_r of automorphisms of

A by unitary operators. Therefore, quantum evolution given by strong-continuous 1-parameter groups of automorphisms need not be described by the Shrödinger equation in general.

Turn to the time-dependent Heisenberg equation (10.4.3). We require that, for all $t \in \mathbb{R}$, the derivations $\delta(t)$ are generators of strong-continuous 1-parameter groups of automorphisms of a C^*-algebra. Then the operator of a parallel transport in A with respect to the connection ∇_t (10.4.2) over the segment $[0, t]$ can be given by the time-ordered exponent

$$G_t = T\exp\left[\int_0^t \delta(t')dt'\right]. \tag{10.4.6}$$

Hence, for any $a \in A$, we have a solution

$$\boxed{\alpha(t) = G_t(a),} \qquad t \in \mathbb{R}^+,$$

of the Heisenberg equation (10.4.3).

Let now all C^*-algebras A_t of instantaneous quantum systems be isomorphic to the von Neumann algebra $B(E)$ of bounded operators in some Hilbert space E. Then we come to quantum evolution phrased in terms of the Shrödinger equation. Let us consider a locally trivial fibre bundle $\Pi \to \mathbb{R}$ with the typical fibre E and smooth transition functions. Smooth sections of the fibre bundle $\Pi \to \mathbb{R}$ constitute a module $\Pi(\mathbb{R})$ over the ring $C^\infty(\mathbb{R})$ of real functions on $\mathbb{R}$. In accordance with Definition 8.2.6, a connection ∇ on $\Pi(\mathbb{R})$ assigns to the standard vector field ∂_t on $\mathbb{R}$ a first order differential operator

$$\nabla_t \in \mathrm{Diff}\,_1^{\rightarrow}(\Pi(\mathbb{R}), \Pi(\mathbb{R})) \tag{10.4.7}$$

which obeys the Leibniz rule

$$\nabla_t(f\psi) = \partial_t f\psi + f\nabla_t\psi, \qquad \psi \in \Pi(\mathbb{R}), \qquad f \in C^\infty(\mathbb{R}).$$

Let us choose a trivialization $\Pi = \mathbb{R} \times E$. Then the operator ∇_t (10.4.7) reads

$$\boxed{\nabla_t(\psi) = (\partial_t - i\mathcal{H}(t))\psi,} \tag{10.4.8}$$

where $\mathcal{H}(t)$ at all $t \in \mathbb{R}$ are bounded self-adjoint operators in E.

Note that every bounded self-adjoint operator $\mathcal{H}$ in a Hilbert space E defines the bounded derivation

$$\delta(a) = i[\mathcal{H}, a], \qquad a \in B(E), \tag{10.4.9}$$

of the algebra $B(E)$. Conversely, every bounded derivation of $B(E)$ is internal, i.e., takes the form (10.4.9) where $\mathcal{H}$ is a self-adjoint element of $B(E)$. Therefore, the operators $\mathcal{H}(t)$ in the expression (10.4.8) are necessarily bounded and self-adjoint.

We say that a section $\psi(t)$ of the fibre bundle $\Pi \to \mathbb{R}$ is an integral section of the connection ∇_t (10.4.8) if it satisfies the equation

$$\boxed{\nabla_t(\psi) = (\partial_t - i\mathcal{H}(t))\psi = 0.} \tag{10.4.10}$$

One can think of this equation as being the *Shrödinger equation* for the Hamiltonian $\mathcal{H}(t)$.

In particular, let a quantum system be conservative, i.e., the Hamiltonian $\mathcal{H}(t) = \mathcal{H}$ in the Shrödinger equation is independent of time. Then, for any point $y \in E$, we obtain the solution

$$\psi(t) = e^{it\mathcal{H}}y, \qquad t \in \mathbb{R},$$

of the conservative Shrödinger equation. If the Shrödinger equation (10.4.10) is not conservative, the operator of a parallel transport in E with respect to the connection ∇_t (10.4.8) over the segment $[0,t]$ can be given by the time-ordered exponent

$$G_t = T\exp\left[i\int_0^t \mathcal{H}(t')dt'\right]. \tag{10.4.11}$$

Then, for any $y \in E$, we obtain a solution

$$\boxed{\psi(t) = G_t y,} \qquad t \in \mathbb{R}^+, \tag{10.4.12}$$

of the Shrödinger equation (10.4.10).

Note that the operator G_t (10.4.11) is an element of the group $U(E)$ of unitary operators in the Hilbert space E. This is a real infinite-dimensional Lie group with respect to the norm operator topology, whose Lie algebra is the real algebra of all anti-self-adjoint bounded operators $i\mathcal{H}$ in E with respect to the bracket $[i\mathcal{H}, i\mathcal{H}']$. The operator G_t (10.4.11), by construction, obeys the equation

$$\partial_t G_t - i\mathcal{H}G_t = 0. \tag{10.4.13}$$

This equation is invariant under right multiplications $G_t \mapsto G_t g$, $\forall g \in U(E)$. Therefore, G_t can be seen as the operator of a parallel transport in the trivial principal bundle $\mathbb{R} \times U(E)$. Accordingly, the operator (10.4.6) can be regarded as the operator of a parallel transport in the trivial group bundle $\mathbb{R} \times \text{Aut}\,(A)$, where the group $\text{Aut}\,(A)$ acts on itself by the adjoint representation.

In the next Section, the description of quantum evolution as a parallel transport in a principal bundle will be extended to quantum systems depending on a set of classical time-dependent parameters in order to explain the Berry's phase phenomenon.

Note that the 1-parameter group G_t defined by the equation (10.4.13) is continuous with respect to the norm operator topology in $U(E)$. Turn to the case when the curve G_t is continuous with respect to the strong operator topology in $B(E)$. Then the curves $\psi(t) = G_t y$, $y \in E$, are also continuous, but not necessarily differentiable in E. Accordingly, a Hamiltonian $\mathcal{H}(t)$ in the Shrödinger equation (10.4.10) is not bounded. Since the group $U(E)$ is not a topological group with respect to the strong operator topology, the product $\mathbb{R} \times U(\mathbb{R})$ is neither principal nor smooth bundle. Therefore, the conventional notion of a connection is not applied to this fibre bundle. At the same time, one can introduce a generalized connection defined in terms of parallel transport curves and operators, but not their generators [9].

10.5 Berry connections

We refer the reader to [7, 17, 31, 175, 225, 279, 310] and references therein for the geometric and topological analysis of the Berry's phase phenomenon in quantum systems depending on classical time-dependent parameters. In Section 5.12, classical mechanical systems with time-dependent parameters have been described in terms of composite fibre bundles and composite connections. Here, this description is extended to quantum systems.

Let us consider quantum systems depending on a finite number of real classical parameters given by sections of a smooth parameter bundle $\Sigma \to \mathbb{R}$. For the sake of simplicity, we fix a trivialization $\Sigma = \mathbb{R} \times Z$, coordinated by (t, σ^m). Although it may happen that the parameter bundle $\Sigma \to \mathbb{R}$ has no preferable trivialization, e.g., if one of parameters is a velocity of a reference frame.

In the previous Section, we have characterized the time as a classical parameter in quantum mechanics. This characteristic is extended to other classical parameters. Namely, we assign a C^*-algebra A_σ to each point $\sigma \in \Sigma$ of the parameter bundle Σ, and treat A_σ as a quantum system under fixed values (t, σ^m) of the parameters.

Remark 10.5.1. Let us emphasize that one should distinguish classical parameters from coordinates which a wave function can depend on.

Let $\{A_q\}$ be a set of C^*-algebras parameterized by points of a locally compact topological space Q. Let all C^*-algebras A_q are isomorphic to each other and to some C^*-algebra A. We consider a locally trivial topological fibre bundle $P \to Q$ whose typical fibre is the C^*-algebra A, i.e., transition functions of this fibre bundle provide automorphisms of A. The set $P(Q)$ of continuous sections of this fibre bundle is an involutive algebra with respect to fibrewise operations. This involutive algebra exemplifies a locally trivial continuous field of C^*-algebras on the topological space Q [82]. Let us consider a subalgebra $A(Q) \subset P(Q)$ which consists of sections α of $P \to Q$ such that $||\alpha(q)||$ vanishes at infinity of Q. For $\alpha \in A(Q)$, put

$$||\alpha|| = \sup_{q \in Q} ||\alpha(q)|| < \infty. \tag{10.5.1}$$

With this norm, $A(Q)$ is a C^*-algebra [82]. It is called a C^*-algebra defined by a continuous field of C^*-algebras. For example, every liminal C^*-algebra with a Hausdorff spectrum $\widehat{A}$ is defined by a continuous field of C^*-algebras on $\widehat{A}$. Recall that by the *spectrum* of a C^*-algebra is meant the set of its irreducible representations provided with a certain topology. As is well known, any commutative C^*-algebra is isomorphic to the algebra of complex continuous functions on its spectrum, which vanish at infinity (see Section 14.1).

Turn to the C^*-algebra $A(Q)$. One can consider a quantum system characterized by this C^*-algebra. In this case, elements of the set Q are not classical parameters as follows. Given an element $q \in Q$, the assignment

$$A(Q) \ni \alpha \mapsto \alpha(q) \in A \tag{10.5.2}$$

is a C^*-algebra epimorphism. Let π be a representation of A. Then the assignment (10.5.2) yields a representation $\rho(\pi, q)$ of the C^*-algebra $A(Q)$. If π is an irreducible representation of the C^*-algebra A, then $\rho(\pi, q)$ is an irreducible representation of $A(Q)$. Moreover, the irreducible representations $\rho(\pi, q)$ and $\rho(\pi, q')$ of $A(Q)$ are not equivalent. Indeed, there exists a continuous function f on Q such that $f(q) = 1$ and $f(q') = 0$. Then all elements $f\alpha \in A(Q)$ such that $\pi(\alpha(q)) \neq 0$ belongs to the kernel of the representation of $\rho(\pi, q')$, but not to the kernel of the representation $\rho(\pi, q)$. Thus, there is a bijection (but not a homeomorphism) between the spectrum $\widehat{A(Q)}$ of the C^*-algebra $A(Q)$ and the set $Q \times \widehat{A}$. It follows that one can find representations of the C^*-algebra $A(Q)$ among direct integrals of representations of A with respect to some measure on Q. Let μ be a positive measure of total mass 1 on the locally compact space Q , and let ϕ be a positive form on A. Then the function $q \mapsto \phi(\alpha(q))$, $\forall \alpha \in A(Q)$, is a μ-measurable, while the integral

$$\phi(\alpha) = \int \phi(\alpha(q))\mu(q)$$

provides a positive form on the C^*-algebra $A(Q)$. Roughly speaking, a computation of a mean value of an operator $\alpha \in A(Q)$ implies an integration with respect to some measure on Q in general. This is not the case of quantum systems depending on classical parameters $q \in Q$. •

We will simplify repeatedly our consideration in order to single out a desired Berry's phase phenomenon. Let us assume that all algebras A_σ are isomorphic to the von Neumann algebra $B(E)$ of bounded operators in some Hilbert space E, and consider a locally trivial Hilbert space bundle $\Pi \to \Sigma$ with the typical fibre E and smooth transition functions. Smooth sections of $\Pi \to \Sigma$ constitute a module $\Pi(\Sigma)$ over the ring $C^\infty(\Sigma)$ of real functions on Σ. In accordance with Definition 8.2.6, a connection $\widetilde{\nabla}$ on $\Pi(\Sigma)$ assigns to each vector field τ on Σ a first order differential operator

$$\widetilde{\nabla}_\tau \in \mathrm{Diff}\,_1^{\to}(\Pi(\Sigma), \Pi(\Sigma)) \tag{10.5.3}$$

which obeys the Leibniz rule

$$\widetilde{\nabla}_\tau(fs) = (\tau\rfloor df)s + f\widetilde{\nabla}_\tau s, \qquad s \in \Pi(\Sigma), \qquad f \in C^\infty(\Sigma).$$

Let τ be a vector field on Σ such that $dt\rfloor\tau = 1$. Given a trivialization chart of the Hilbert space bundle $\Pi \to \Sigma$, the operator $\widetilde{\nabla}_\tau$ (10.5.3) reads

$$\boxed{\widetilde{\nabla}_\tau(s) = (\partial_t - i\mathcal{H}(t,\sigma^k))s + \tau^m(\partial_m - i\widehat{A}_m(t,\sigma^k))s,} \tag{10.5.4}$$

where $\mathcal{H}(t,\sigma^k)$, $\widehat{A}_m(t,\sigma^k)$ for each $\sigma \in \Sigma$ are bounded self-adjoint operators in the Hilbert space E.

Let us consider the composite Hilbert space bundle $\Pi \to \Sigma \to \mathbb{R}$. Similarly to the case of smooth composite fibre bundles (see Proposition 2.7.1), every section $h(t)$ of the fibre bundle $\Sigma \to \mathbb{R}$ defines the subbundle $\Pi_h = h^*\Pi \to \mathbb{R}$ of the Hilbert space bundle $\Pi \to \mathbb{R}$ whose typical fibre is E. Accordingly, the connection $\widetilde{\nabla}$ (10.5.4) on the $C^\infty(\Sigma)$-module $\Pi(\Sigma)$ defines the pull-back connection

$$\nabla_h(\psi) = [\partial_t - i(\widehat{A}_m(t,h^k(t))\partial_t h^m + \mathcal{H}(t,h^k(t))]\psi \tag{10.5.5}$$

on the $C^\infty(\mathbb{R})$-module $\Pi_h(\mathbb{R})$ of sections ψ of the fibre bundle $\Pi_h \to \mathbb{R}$ (cf. (2.7.11), (5.12.5)).

As in the previous Section, we say that a section ψ of the fibre bundle $\Pi_h \to \mathbb{R}$ is an integral section of the connection (10.5.5) if

$$\boxed{\nabla_h(\psi) = [\partial_t - i(\widehat{A}_m(t, h^k(t))\partial_t h^m + \mathcal{H}(t, h^k(t))]\psi = 0.} \tag{10.5.6}$$

One can think of the equation (10.5.6) as being the Shrödinger equation for a quantum system depending on the parameter function $h(t)$. Its solutions take the form (10.4.12) where G_t is the time-ordered exponent

$$G_t = T\exp\left[i\int_0^t (\widehat{A}_m\partial_{t'}h^m + \mathcal{H})dt'\right]. \tag{10.5.7}$$

The term $i\widehat{A}_m(t, h^i(t))\partial_t h^m$ in the Shrödinger equation (10.5.6) is responsible for the Berry's phase phenomenon, while $\mathcal{H}$ is treated as an ordinary Hamiltonian of a quantum system. To show the Berry's phase phenomenon clearly, we will continue to simplify the system under consideration. Given a trivialization of the fibre bundle $\Pi \to \mathbb{R}$ and the above mentioned trivialization $\Sigma = \mathbb{R}\times Z$ of the parameter bundle Σ, let us suppose that the components $\widehat{A}_m$ of the connection $\widetilde{\nabla}$ (10.5.4) are independent of t and that the operators $\mathcal{H}(\sigma)$ commute with the operators $\widehat{A}_m(\sigma')$ at all points of the curve $h(t) \subset \Sigma$. Then the operator G_t (10.5.7) takes the form

$$G_t = T\exp\left[i\int_{h([0,t])} \widehat{A}_m(\sigma^k)d\sigma^m\right]\cdot T\exp\left[i\int_0^t \mathcal{H}(t')dt'\right]. \tag{10.5.8}$$

One can think of the first factor in the right-hand side of the expression (10.5.8) as being the operator of a parallel transport along the curve $h([0,t]) \subset Z$ with respect to the pull-back connection

$$\boxed{\nabla = i^*\widetilde{\nabla} = d\sigma^m \otimes (\partial_m - i\widehat{A}_m(t, \sigma^k))} \tag{10.5.9}$$

on the fibre bundle $\Pi \to Z$, defined by the imbedding

$$i : Z \hookrightarrow \{0\}\times Z \subset \Sigma.$$

Note that, since $\widehat{A}_m$ are independent of time, one can utilize any imbedding of Z to $\{t\}\times Z$. Moreover, the connection ∇ (10.5.9), called the *Berry connection*, can be seen as a connection on some principal fibre bundle $P \to Z$ for the group $U(E)$ of unitary operators in the Hilbert space E. Let the curve $h([0,t])$ be closed, while the holonomy group of the connection ∇ at the point $h(t) = h(0)$ is not trivial. Then the unitary operator

$$T\exp\left[i \int\limits_{h([0,t])} \widehat{A}_m(\sigma^k)d\sigma^m\right] \tag{10.5.10}$$

is not the identity. For example, if

$$i\widehat{A}_m(\sigma^k) = iA_m(\sigma^k)\mathrm{Id}_E \tag{10.5.11}$$

is a $U(1)$-principal connection on Z, then the operator (10.5.10) is the well-known *Berry phase factor*

$$\exp\left[i \int\limits_{h([0,t])} A_m(\sigma^k)d\sigma^m\right].$$

If (10.5.11) is a curvature-free connection, Berry's phase is exactly the Aharonov–Bohm effect on the parameter space Z (see Section 6.6). In this particular case, Proposition 6.6.1 can be applied to the topological analysis of the Berry's phase phenomenon.

The following variant of the Berry's phase phenomenon leads us to a principal bundle for familiar finite-dimensional Lie groups. Let E be a separable Hilbert space which is the Hilbert sum of n-dimensional eigenspaces of the Hamiltonian $\mathcal{H}(\sigma)$, i.e.,

$$E = \bigoplus_{k=1}^{\infty} E_k, \qquad E_k = P_k(E),$$

where P_k are the projection operators, i.e.,

$$H(\sigma)\circ P_k = \lambda_k(\sigma)P_k$$

(in the spirit of the adiabatic hypothesis). Let the operators $\widehat{A}_m(z)$ be time-independent and preserve the eigenspaces E_k of the Hamiltonian $\mathcal{H}$, i.e.,

$$\widehat{A}_m(z) = \sum_k \widehat{A}^k_m(z)P_k, \tag{10.5.12}$$

where $\widehat{A}^k_m(z)$, $z \in Z$, are self-adjoint operators in E_k. It follows that $\widehat{A}_m(\sigma)$ commute with $\mathcal{H}(\sigma)$ at all points of the parameter bundle $\Sigma \to \mathbb{R}$. Then, restricted to the subspace E_k, the parallel transport operator (10.5.10) is a unitary operator in E_k. In this case, the Berry connection (10.5.9) on the $U(E)$-principal bundle $P \to Z$ can be seen as a composite connection on the composite bundle

$$P \to P/U(n) \to Z,$$

which is defined by some principal connection on the $U(n)$-principal bundle $P \to P/U(n)$ and the trivial connection on the fibre bundle $P/U(n) \to Z$. The typical fibre of $P/U(n) \to Z$ is exactly the classifying space $B(U(n)$ (6.7.3). Moreover, one can consider the parallel transport along a curve in the bundle $P/U(n)$. In this case, a state vector $\psi(t)$ acquires a *geometric phase factor* in addition to the dynamical phase factor. In particular, if $\Sigma = \mathbb{R}$ (i.e., classical parameters are absent and Berry's phase has only the geometric origin) we come to the case of a Berry connection on the $U(n)$-principal bundle on the classifying space $B(U(n))$ (see [31]). If $n = 1$, this is the variant of Berry's geometric phase of [7].

Chapter 11

Connections in BRST formalism

There are different approaches to BRST formalism. We will consider only its elements where jet manifolds and connections are utilized. In particular, the BRST operator acting on so-called BRST tensor fields is phrased in terms of generalized or BRST connections. We will start our exposition from the basics of the calculus in the infinite order jets extended to odd ghosts, ghost-for-ghosts and antifields.

11.1 The canonical connection on infinite order jets

As was mentioned above, the tangent and cotangent bundles over a fibre bundle $Y \to X$ admit the canonical horizontal splittings (2.2.3) and (2.2.4), but over the jet manifold J^1Y. Choosing a connection $\Gamma : Y \to J^1Y$, one brings these splittings into a true horizontal splitting over Y, but this splitting fails to be canonical. Similarly, the tangent and cotangent bundles over the k-order jet manifold J^kY have canonical horizontal splittings (see (11.1.21) and (11.1.22) below), but these splittings are over $J^{k+1}Y$. One may hope that, in the case of the infinite order jet space $J^\infty Y$, the corresponding canonical splittings are true horizontal splittings over $J^\infty Y$ corresponding to the canonical connection on $J^\infty Y$. This connection provides the canonical decomposition of exterior forms on jet manifolds and the corresponding decomposition $d = d_H + d_V$ of the exterior differential into horizontal and vertical parts. These decompositions lead to the variational bicomplex and the algebraic approach to the calculus in variations (see Section 11.2). This bicomplex extended to a graded algebra plays an important role in the BRST construction phrased in terms of jets (see Section 11.3).

We start from higher order jet manifolds of sections of a fibre bundle $Y \to X$ [123, 179, 212, 274]. They are a natural generalization of the first and second order jet manifolds. We will follow the notation of Section 1.3. Recall that, given fibred coordinates (x^λ, y^i) on a fibre bundle $Y \to X$, by Λ, $|\Lambda| = r$, is meant a collection of numbers $(\lambda_r ... \lambda_1)$ modulo permutations. By $\Lambda + \Sigma$ we denote the collection

$$\Lambda + \Sigma = (\lambda_r \cdots \lambda_1 \sigma_k \cdots \sigma_1)$$

modulo permutations, while $\Lambda\Sigma$ is the union of collections

$$\Lambda\Sigma = (\lambda_r \cdots \lambda_1 \sigma_k \cdots \sigma_1)$$

where the indices λ_i and σ_j are not permuted. Recall the symbol (1.3.14) of the total derivative

$$\boxed{d_\lambda^{(k)} = \partial_\lambda + \sum_{|\Lambda|=0}^{k} y^i_{\Lambda+\lambda} \partial_i^\Lambda.} \tag{11.1.1}$$

We omit the index (k) in this symbol if there is no danger of confusion. We will use the notation

$$\partial_\Lambda = \partial_{\lambda_r} \circ \cdots \circ \partial_{\lambda_1}, \qquad d_\Lambda = d_{\lambda_r} \circ \cdots \circ d_{\lambda_1}, \qquad \Lambda = (\lambda_r ... \lambda_1).$$

The r-*order jet manifold* J^rY of sections of a fibre bundle $Y \to X$ (or simply the r-order jet manifold of $Y \to X$) is defined as the disjoint union

$$J^rY = \bigcup_{x \in X} j^r_x s \tag{11.1.2}$$

of the equivalence classes $j^r_x s$ of sections s of Y so that different sections s and s' belong to the same equivalence class $j^r_x s$ if and only if

$$\boxed{s^i(x) = s'^i(x), \qquad \partial_\Lambda s^i(x) = \partial_\Lambda s'^i(x),} \qquad 0 < |\Lambda| \leq r.$$

In brief, one can say that sections of $Y \to X$ are identified by the $r+1$ terms of their Taylor series at points of X. The particular choice of a coordinate atlas does not matter for this definition. Given an atlas of fibred coordinates (x^λ, y^i) of a fibre bundle $Y \to X$, the set (11.1.2) is endowed with an atlas of the adapted coordinates

$$\begin{aligned} &(x^\lambda, y^i_\Lambda), \qquad 0 \leq |\Lambda| \leq r, \\ &(x^\lambda, y^i_\Lambda) \circ s = (x^\lambda, \partial_\Lambda s^i(x)), \end{aligned} \tag{11.1.3}$$

together with transition functions

$$y'^i_{\lambda+\Lambda} = \frac{\partial x^\mu}{\partial' x^\lambda} d_\mu y'^i_\Lambda. \tag{11.1.4}$$

The coordinates (11.1.3) bring the set J^rY into a smooth manifold of finite dimension

$$\dim J^rY = n + m \sum_{i=0}^{r} \frac{(n+i-1)!}{i!(n-1)!}.$$

The coordinates (11.1.3) are compatible with the natural surjections

$$\pi^r_l : J^rY \to J^lY, \quad r > l,$$

which form the composite bundle

$$\pi^r : J^rY \xrightarrow{\pi^r_{r-1}} J^{r-1}Y \xrightarrow{\pi^{r-1}_{r-2}} \cdots \xrightarrow{\pi^1_0} Y \xrightarrow{\pi} X,$$

$$\pi^k_h \circ \pi^r_k = \pi^r_h, \qquad \pi^h \circ \pi^r_h = \pi^r.$$

A glance at the transition functions (11.1.4) when $\mid \Lambda \mid = r$ shows that the fibration

$$\pi^r_{r-1} : J^rY \to J^{r-1}Y$$

is an affine bundle modelled over the vector bundle

$$\overset{r}{\vee} T^*X \underset{J^{r-1}Y}{\otimes} VY \to J^{r-1}Y. \tag{11.1.5}$$

Remark 11.1.1. To introduce higher order jet manifolds, one can use the construction of the repeated jet manifolds. Let us consider the r-order jet manifold J^rJ^kY of the jet bundle $J^kY \to X$. It is coordinated by

$$(x^\mu, y^i_{\Sigma\Lambda}), \qquad \mid \Lambda \mid \leq k, \qquad \mid \Sigma \mid \leq r.$$

There is the canonical monomorphism

$$\sigma_{rk} : J^{r+k}Y \hookrightarrow J^rJ^kY$$

given by the coordinate relations

$$y^i_{\Sigma\Lambda} \circ \sigma_{rk} = y^i_{\Sigma+\Lambda}.$$

•

In the calculus in r-order jets, we have the r-order jet prolongation functor such that, given fibre bundles Y and Y' over X, every fibred morphism $\Phi : Y \to Y'$ over a diffeomorphism f of X admits the *r-order jet prolongation* to the morphism

$$J^r\Phi : J^rY \ni j^r_x s \mapsto j^r_{f(x)}(\Phi \circ s \circ f^{-1}) \in J^rY' \tag{11.1.6}$$

of the r-order jet manifolds. The jet prolongation functor is exact. If Φ is an injection [surjection], so is $J^r\Phi$. It also preserves an algebraic structure. In particular, if $Y \to X$ is a vector bundle, so is $J^rY \to X$. If $Y \to X$ is an affine bundle modelled over the vector bundle $\overline{Y} \to X$, then $J^rY \to X$ is an affine bundle modelled over the vector bundle $J^r\overline{Y} \to X$.

Every section s of a fibre bundle $Y \to X$ admits the *r-order jet prolongation* to the holonomic section

$$\boxed{(J^rs)(x) = j^r_x s}$$

of the jet bundle $J^rY \to X$.

Every exterior form ϕ on the jet manifold J^kY gives rise to the pull-back form $\pi^{k+i}_k{}^*\phi$ on the jet manifold $J^{k+i}Y$. Let $\mathfrak{O}^*_k = \mathfrak{O}^*(J^kY)$ be the algebra of exterior forms on the jet manifold J^kY. We have the direct system of $\mathbb{R}$-algebras

$$\boxed{\mathfrak{O}^*(X) \xrightarrow{\pi^*} \mathfrak{O}^*(Y) \xrightarrow{\pi^1_0{}^*} \mathfrak{O}^*_1 \xrightarrow{\pi^2_1{}^*} \cdots \xrightarrow{\pi^r_{r-1}{}^*} \mathfrak{O}^*_r \longrightarrow \cdots.} \tag{11.1.7}$$

Sometimes, it is convenient to denote $\mathfrak{O}^*_{-1} = \mathfrak{O}^*(X)$, $\mathfrak{O}^*_0 = \mathfrak{O}^*(Y)$. The subsystem of (11.1.7) is the direct system

$$\boxed{C^\infty(X) \xrightarrow{\pi^*} C^\infty(Y) \xrightarrow{\pi^1_0{}^*} \mathfrak{O}^0_1 \xrightarrow{\pi^2_1{}^*} \cdots \xrightarrow{\pi^r_{r-1}{}^*} \mathfrak{O}^0_r \longrightarrow \cdots} \tag{11.1.8}$$

of the $\mathbb{R}$-rings of real smooth functions $\mathfrak{O}^0_k = C^\infty(J^kY)$ on the jet manifolds J^kY. Therefore, one can think of (11.1.7) and (11.1.8) as being the direct systems of $C^\infty(X)$-modules.

Given the k-order jet manifold J^kY of $Y \to X$, there exists the canonical fibred morphism

$$r_{(k)} : J^kTY \to TJ^kY$$

over $J^kY \underset{X}{\times} J^kTX \to J^kY \underset{X}{\times} TX$ whose coordinate expression is

$$(x^\lambda, y^i_\Lambda, \dot x^\lambda, \dot y^i_\Lambda) \circ r_{(k)} = (x^\lambda, y^i_\Lambda, \dot x^\lambda, (\dot y^i)_\Lambda - \sum (\dot y^i)_{\mu+\Sigma}(\dot x^\mu)_\Xi), \quad 0 \leq \mid \Lambda \mid \leq k,$$

where the sum is taken over all partitions $\Sigma + \Xi = \Lambda$ and $0 <\mid \Xi \mid$. In particular, we have the canonical isomorphism over J^kY:

$$r_{(k)} : J^kVY \to VJ^kY, \qquad (\dot{y}^i)_\Lambda = \dot{y}^i_\Lambda \circ r_{(k)}. \tag{11.1.9}$$

As a consequence, every projectable vector field $u = u^\mu\partial_\mu + u^i\partial_i$ on a fibre bundle $Y \to X$ has the following *k-order jet prolongation* to the vector field on J^kY:

$$\begin{aligned} &J^k u = r_{(k)} \circ J^k u : J^kY \to TJ^kY, \\ &J^k u = u^\lambda\partial_\lambda + u^i\partial_i + u^i_\Lambda\partial^\Lambda_i, \qquad 0 <\mid \Lambda \mid\leq k, \\ &u^i_{\lambda+\Lambda} = d_\lambda u^i_\Lambda - y^i_{\mu+\Lambda}\partial_\lambda u^\mu, \qquad 0 <\mid \Lambda \mid< k, \end{aligned} \tag{11.1.10}$$

(cf. (1.3.10) for $k = 1$). In particular, the k-order jet lift (11.1.10) of a vertical vector field on $Y \to X$ is a vertical vector field on $J^kY \to X$ due to the isomorphism (11.1.9).

A vector field u_r on an r-order jet manifold J^rY is called *projectable* if, for any $k < r$, there exists a projectable vector field u_k on J^kY such that

$$u_k \circ \pi^r_k = T\pi^r_k \circ u_r.$$

A projectable vector field on J^rY has the coordinate expression

$$u_r = u^\lambda\partial_\lambda + u^i_\Lambda\partial^\Lambda_i, \qquad 0 \leq |\Lambda| \leq r,$$

such that u_λ depends only on the coordinates x^μ and every component u^i_Λ is independent of the coordinates y^i_Ξ, $|\Xi| > |\Lambda|$.

Let us denote by $\mathcal{P}^r$ the vector space of projectable vector fields on the jet manifold J^rY. It is easily seen that $\mathcal{P}^r$ is a Lie algebra over $\mathbb{R}$ and that the morphisms $T\pi^r_k$, $k < r$, constitute the inverse system

$$\boxed{\mathcal{P}^0 \overset{T\pi^1_0}{\longleftarrow} \mathcal{P}^1 \overset{T\pi^2_1}{\longleftarrow} \cdots \overset{T\pi^{r-1}_{r-2}}{\longleftarrow} \mathcal{P}^{r-1} \overset{T\pi^r_{r-1}}{\longleftarrow} \mathcal{P}^r \longleftarrow \cdots} \tag{11.1.11}$$

of these Lie algebras.

PROPOSITION 11.1.1. [25, 290]. The k-order jet lift (11.1.10) is the Lie algebra monomorphism of the Lie algebra $\mathcal{P}^0$ of projectable vector fields on $Y \to X$ to the Lie algebra $\mathcal{P}^k$ of projectable vector fields on J^kY such that

$$T\pi^r_k(J^r u) = J^k u \circ \pi^r_k. \tag{11.1.12}$$

□

The jet lift $J^k u$ (11.1.10) is said to be an *integrable* vector field on J^kY. Every projectable vector field u_k on J^kY is decomposed into the sum

$$u_k = J^k(T\pi_0^k(u_k)) + v_k \tag{11.1.13}$$

of the integrable vector field $J^k(T\pi_0^k(u_k))$ and a projectable vector field v_r which is vertical with respect to some fibration $J^kY \to Y$.

Similarly to the exact sequences (1.1.17a) – (1.1.17b) over $J^0Y = Y$, we have the exact sequences

$$0 \to VJ^kY \hookrightarrow TJ^kY \to TX \underset{X}{\times} J^kY \to 0, \tag{11.1.14}$$

$$0 \to J^kY \underset{X}{\times} T^*X \hookrightarrow TJ^kY \to V^*J^kY \to 0 \tag{11.1.15}$$

of vector bundles over J^kY. They do not admit a canonical splitting. Nevertheless, their pull-backs onto $J^{k+1}Y$ are split canonically due to the following canonical bundle monomorphisms over J^kY:

$$\lambda_{(k)} : J^{k+1}Y \hookrightarrow T^*X \underset{J^kY}{\otimes} TJ^kY,$$

$$\lambda_{(k)} = dx^\lambda \otimes d_\lambda^{(k)}, \tag{11.1.16}$$

$$\theta_{(k)} : J^{k+1}Y \hookrightarrow T^*J^kY \underset{J^kY}{\otimes} VJ^kY,$$

$$\theta_{(k)} = \sum (dy^i_\Lambda - y^i_{\lambda+\Lambda}dx^\lambda) \otimes \partial_i^\Lambda, \tag{11.1.17}$$

where the sum is over all multi-indices Λ, $\mid \Lambda \mid \leq k$ (cf. (1.3.5), (1.3.6) for $k = 1$). The forms

$$\boxed{\theta^i_\Lambda = dy^i_\Lambda - y^i_{\Lambda+\lambda}dx^\lambda} \tag{11.1.18}$$

are also called the contact forms. The monomorphisms (11.1.16) and (11.1.17) yield the fibred monomorphisms over $J^{k+1}Y$

$$\widehat{\lambda}_{(k)} : TX \underset{X}{\times} J^{k+1}Y \hookrightarrow TJ^kY \underset{J^kY}{\times} J^{k+1}Y, \tag{11.1.19}$$

$$\widehat{\theta}_{(k)} : V^*J^kY \underset{J^kY}{\times} \hookrightarrow T^*J^kY \underset{J^kY}{\times} J^{k+1}Y. \tag{11.1.20}$$

These monomorphisms split the exact sequences (11.1.14) and (11.1.15) over $J^{k+1}Y$ and define the *canonical horizontal splittings* of the pull-backs

$$\pi^{k+1*}_k TJ^kY = \widehat{\lambda}_{(k)}(TX \underset{X}{\times} J^{k+1}Y) \underset{J^{k+1}Y}{\oplus} VJ^kY, \qquad (11.1.21)$$

$$\boxed{\dot{x}^\lambda\partial_\lambda + \sum \dot{y}^i_\Lambda \partial^\Lambda_i = \dot{x}^\lambda d^{(k)}_\lambda + \sum(\dot{y}^i_\Lambda - \dot{x}^\lambda y^i_{\lambda+\Lambda})\partial^\Lambda_i,}$$

$$\pi^{k+1*}_k T^*J^kY = T^*X \underset{J^{k+1}Y}{\oplus} \widehat{\theta}_{(k)}(V^*J^kY \underset{J^kY}{\times} J^{k+1}Y), \qquad (11.1.22)$$

$$\boxed{\dot{x}_\lambda dx^\lambda + \sum \dot{y}^\Lambda_i dy^i_\Lambda = (\dot{x}_\lambda + \sum \dot{y}^\Lambda_i y^i_{\lambda+\Lambda})dx^\lambda + \sum \dot{y}^\Lambda_i \theta^i_\Lambda,}$$

where summations are over all multi-indices $\mid \Lambda \mid\leq k$.

In accordance with the canonical horizontal splitting (11.1.21), the pull-back

$$\overline{u}_k : J^{k+1}Y \overset{\pi^{k+1}_k \times \mathrm{Id}}{\longrightarrow} J^kY \times J^{k+1} \overset{u_k\times\mathrm{Id}}{\longrightarrow} TJ^kY \underset{J^kY}{\times} J^{k+1}$$

onto $J^{k+1}Y$ of any vector field u_k on J^kY admits the canonical horizontal splitting

$$\boxed{\overline{u} = u_H + u_V = (u^\lambda d^{(k)}_\lambda + \sum y^i_{\lambda+\Lambda}\partial^\Lambda_i) + \sum(u^i_\Lambda - u^\lambda y^i_{\lambda+\Lambda})\partial^\Lambda_i,} \qquad (11.1.23)$$

where the sums are over all multi-indices $|\Lambda| \leq k$. By virtue of the canonical horizontal splitting (11.1.22), every exterior 1-form ϕ on J^kY admits the canonical splitting of its pull-back

$$\boxed{\pi^{k+1*}_k\phi = h_0\phi + (\phi - h_0(\phi)),} \qquad (11.1.24)$$

where h_0 is the horizontal projection (1.3.15).

As was mentioned above, the canonical horizontal splittings (11.1.21) – (11.1.24) are not true horizontal splitting on J^kY because they are defined for the pull-backs from J^kY onto $J^{k+1}Y$. One may hope to overcome this difficulty in the case of infinite order jets.

The direct system (11.1.7) of $\mathbb{R}$-algebras of exterior forms and the inverse system (11.1.11) of the real Lie algebras of projectable vector fields on jet manifolds are defined for any finite order r. These sequences admit the limits for $r \to \infty$ in the category of modules and that of Lie algebras, respectively. Intuitively, one can think of elements of these limits as being the objects defined on the projective limit of the inverse system

$$\boxed{X \overset{\pi}{\longleftarrow} Y \overset{\pi^1_0}{\longleftarrow} \cdots \longleftarrow J^{r-1}Y \overset{\pi^r_{r-1}}{\longleftarrow} J^rY \longleftarrow \cdots} \qquad (11.1.25)$$

of finite order jet manifolds J^rY.

Remark 11.1.2. Recall that, by a *projective limit* of the inverse system (11.1.25) is meant a set $J^\infty Y$ such that, for any k, there exist surjections

$$\pi^\infty : J^\infty Y \to X, \quad \pi_0^\infty : J^\infty Y \to Y, \qquad \pi_k^\infty : J^\infty Y \to J^k Y \tag{11.1.26}$$

which make up the commutative diagrams

$$\begin{array}{ccc} & J^\infty Y & \\ {}^{\pi_k^\infty}\swarrow & & \searrow{}^{\pi_r^\infty} \\ J^kY & \underset{\pi_r^k}{\longrightarrow} & J^rY \end{array}$$

for any admissible k and $r < k$ [217]. •

The projective limit $J^\infty Y$ of the inverse system (11.1.25) exists. It is called the *infinite order jet space*. This space consists of those elements

$$(\ldots, q_i, \ldots, q_j, \ldots), \qquad q_i \in J^iY, \qquad q_j \in J^jY,$$

of the Cartesian product $\prod_k J^kY$ which satisfy the relations $q_i = \pi_i^j(q_j)$ for all $j > i$. Thus, elements of the infinite order jet space $J^\infty Y$ really represent ∞-jets $j_x^\infty s$ of local sections of $Y \to X$. Different sections belong to the same jet $j_x^\infty s$ if and only if their Taylor series at a point $x \in X$ coincide with each other.

Remark 11.1.3. Note that there is a natural surjection of the sheaf Y_X of smooth sections of a fibre bundle $Y \to X$ onto the infinite order jet space $J^\infty Y$ because all sections s of $Y \to X$ with the same germ s_x at $x \in X$ belong to the same jet $j_x^\infty s$, but a converse is not true. •

Remark 11.1.4. The space $J^\infty Y$ is also the projective limit of the inverse subsystem of (11.1.25) which starts from any finite order J^rY. •

The infinite order jet space $J^\infty Y$ is provided with the weakest topology such that the surjections (11.1.26) are continuous. The base of open sets of this topology in $J^\infty Y$ consists of the inverse images of open subsets of J^kY, $k = 0, \ldots$, under the mappings (11.1.26). This topology is paracompact, and admits a smooth partition of unity. The space $J^\infty Y$ can also be provided with some kind of a manifold structure, but it fails to be a well-behaved manifold [25, 290, 291]. Nevertheless, a wide class of differentiable objects on $J^\infty Y$ can be introduced in the terms of the differential calculus in modules in Section 8.1.

The procedure is the following. At first, real smooth functions on $J^\infty Y$ are defined. A real function $f : J^\infty Y \to \mathbb{R}$ is said to be smooth if, for every $q \in J^\infty Y$, there exists a neighbourhood U of q and a smooth function $f^{(k)}$ on $J^k Y$ for some k such that

$$f\mid_U = f^{(k)} \circ \pi^\infty_k \mid_U .$$

Then the same equality takes place for any $r > k$. In particular, the pull-back $\pi^{\infty *}_r f$ of any smooth function on $J^r Y$ is a smooth function on $J^\infty Y$. Smooth functions on $J^\infty Y$ constitute an $\mathbb{R}$-ring $C^\infty(J^\infty Y)$. Vector fields on $J^\infty Y$ are introduced as derivations of this ring. They make up the locally free left $C^\infty(J^\infty Y)$-module $\mathfrak{d}(C^\infty(J^\infty Y))$. The $C^\infty(J^\infty Y)$-module of exterior 1-forms on $J^\infty Y$, in turn, is defined as the dual $\mathfrak{d}^*(C^\infty(J^\infty Y))$ of the module of vector fields.

However, one usually narrows down the class of studied differentiable objects on $J^\infty Y$ to the algebraic limits of the inverse system (11.1.11) of Lie algebras of projectable vector fields and of the direct system (11.1.7) of modules of exterior forms on finite order jet manifolds $J^r Y$. They constitute subsets of the above mentioned $C^\infty(J^\infty Y)$-modules of vector fields and exterior forms on the infinite order jet space.

Let us start from the direct system (11.1.7) of $\mathbb{R}$-modules $\mathfrak{O}^*_k = \mathfrak{O}^*(J^k Y)$ of exterior forms on finite order jet manifolds $J^k Y$. The limit $\mathfrak{O}^*_\infty$ of this direct system, by definition, obeys the following conditions [217]:

- for any r, there exists an injection $\mathfrak{O}^*_r \to \mathfrak{O}^*_\infty$;
- the diagrams

$$\begin{array}{ccc} \mathfrak{O}^*_k & \xrightarrow{\pi^{r*}_k} & \mathfrak{O}^*_r \\ {\scriptstyle \pi^{\infty*}_k} \searrow & & \swarrow {\scriptstyle \pi^{\infty*}_r} \\ & \mathfrak{O}^*_\infty & \end{array}$$

  are commutative for any r and $k < r$.

Such a *direct limit* exists. This is the $\mathbb{R}$-module which is the quotient of the direct sum $\bigoplus\limits_k \mathfrak{O}^*_k$ with respect to identification of the pull-back forms

$$\pi^{\infty*}_r \phi = \pi^{\infty*}_k \sigma, \qquad \phi \in \mathfrak{O}^*_r, \qquad \sigma \in \mathfrak{O}^*_k,$$

if $\phi = \pi_k^{r*}\sigma$. In other words, $\mathfrak{O}_\infty^*$ consists of all exterior forms on finite order jet manifolds module the pull-back identification. Therefore, we will denote the image of $\mathfrak{O}_r^*$ in $\mathfrak{O}_\infty^*$ by $\mathfrak{O}_r^*$ and the elements $\pi_r^{\infty *}\phi$ of $\mathfrak{O}_\infty^*$ simply by ϕ.

Remark 11.1.5. Obviously, $\mathfrak{O}_\infty^*$ is the direct limit of the direct subsystem of (11.1.7) which starts from any finite order r. •

The $\mathbb{R}$-module $\mathfrak{O}_\infty^*$ possesses the structure of the filtered module as follows [185]. Let us consider the direct system (11.1.8) of the commutative $\mathbb{R}$-rings of smooth functions on the jet manifolds J^rY. Its direct limit $\mathfrak{O}_\infty^0$ consists of functions on finite order jet manifolds modulo pull-back identification. Therefore $\mathfrak{O}_\infty^0$ is a subset of the ring $C^\infty(J^\infty Y)$ of all smooth functions on $J^\infty Y$. This is the $\mathbb{R}$-ring filtered by the $\mathbb{R}$-rings $\mathfrak{O}_k^0 \subset \mathfrak{O}_{k+i}^0$. Then $\mathfrak{O}_\infty^*$ has the *filtered* $\mathfrak{O}^0$*-module* structure given by the $\mathfrak{O}_k^0$-submodules $\mathfrak{O}_k^*$ of $\mathfrak{O}_\infty^*$.

DEFINITION 11.1.2. An endomorphism Δ of $\mathfrak{O}_\infty^*$ is called a *filtered morphism* if there exists $i \in \mathbb{N}$ such that the restriction of Δ to $\mathfrak{O}_k^*$ is the homomorphism of $\mathfrak{O}_k^*$ into $\mathfrak{O}_{k+i}^*$ over the injection $\mathfrak{O}_k^0 \hookrightarrow \mathfrak{O}_{k+i}^0$ for all k. □

THEOREM 11.1.3. [217]. Every *direct system of endomorphisms* $\{\gamma_k\}$ of $\mathfrak{O}_k$ such that

$$\pi_i^{j*} \circ \gamma_i = \gamma_j \circ \pi_i^{j*}$$

for all $j > i$ has the *direct limit* γ_∞ in filtered endomorphisms of $\mathfrak{O}_\infty^*$. If all γ_k are monomorphisms [epimorphisms], then γ_∞ is also a monomorphism [epimorphism]. This result also remains true for the general case of morphism between two different direct systems. □

COROLLARY 11.1.4. [217]. The operation of taking homology groups of chain and cochain complexes commutes with the passage to the direct limit. □

The operation of multiplication

$$\phi \to f\phi, \qquad f \in C^\infty(X), \qquad \phi \in \mathfrak{O}_r^*,$$

has the direct limit, and $\mathfrak{O}_\infty^*$ possesses the structure of $C^\infty(X)$-algebra. The operations of the exterior product $\wedge$ and the exterior differential d also have the direct limits on $\mathfrak{O}^*_\infty$. We will denote them by the same symbols $\wedge$ and d, respectively. They provide $\mathfrak{O}^*_\infty$ with the structure of a $\mathbb{Z}$-graded exterior algebra:

$$\mathfrak{O}^*_\infty = \bigoplus_{m=0}^{\infty} \mathfrak{O}^m_\infty,$$

where $\mathfrak{O}^m_\infty$ are the direct limits of the direct systems

$$\boxed{\mathfrak{O}^m(X) \xrightarrow{\pi^*} \mathfrak{O}^m_0 \xrightarrow{\pi^{1*}_0} \mathfrak{O}^m_1 \longrightarrow \cdots \mathfrak{O}^m_r \xrightarrow{\pi^{r+1}_r{}^*} \mathfrak{O}^m_{r+1} \longrightarrow \cdots}$$

of $\mathbb{R}$-modules $\mathfrak{O}^m_r$ of exterior m-forms on r-order jet manifolds J^rY. Elements of $\mathfrak{O}^m_\infty$ are called the *exterior m-forms on the infinite order jet space*. The familiar relations of an exterior algebra take place:

$$\begin{aligned}
&\mathfrak{O}^i_\infty \wedge \mathfrak{O}^j_\infty \subset \mathfrak{O}^{i+j}_\infty,\\
&d : \mathfrak{O}^i_\infty \to \mathfrak{O}^{i+1}_\infty,\\
&d \circ d = 0.
\end{aligned}$$

As a consequence, we have the following *De Rham complex* of exterior forms on the infinite order jet space

$$0 \longrightarrow \mathbb{R} \longrightarrow \mathfrak{O}^0_\infty \xrightarrow{d} \mathfrak{O}^1_\infty \xrightarrow{d} \cdots . \tag{11.1.27}$$

Let us consider the cohomology group $H^m(\mathfrak{O}^*_\infty)$ of this complex. By virtue of Corollary 11.1.4, this is isomorphic to the direct limit of the direct system of homomorphisms

$$H^m(\mathfrak{O}^*_r) \longrightarrow H^m(\mathfrak{O}^*_{r+1}) \longrightarrow \cdots$$

of the cohomology groups $H^m(\mathfrak{O}^*_r)$ of the cochain complexes

$$0 \longrightarrow \mathbb{R} \longrightarrow \mathfrak{O}^0_r \xrightarrow{d} \mathfrak{O}^1_r \xrightarrow{d} \cdots \xrightarrow{d} \mathfrak{O}^l_r \to 0, \qquad l = \dim J^rY,$$

i.e., of the De Rham cohomology groups $H^m(\mathfrak{O}^*_r) = H^m(J^rY)$ of jet manifolds J^rY. The following assertion completes our consideration of cohomology of the complex (11.1.27).

PROPOSITION 11.1.5. The De Rham cohomology $H^*(J^rY)$ of jet manifolds J^rY coincide with the De Rham cohomology $H^*(Y)$ of the fibre bundle $Y \to X$ [25]. □

The proof is based on the fact that the fibre bundle $J^rY \to J^{r-1}Y$ is affine, and it has the same De Rham cohomology than its base. It follows that the cohomology groups $H^m(\mathfrak{O}^*_\infty)$, $m > 0$, of the cochain complex (11.1.27) coincide with the De Rham cohomology groups $H^m(Y)$ of $Y \to X$.

Though we do not discuss a manifold structure on the infinite order jet space, the elements of the direct limit $\mathfrak{O}^*_\infty$ can be considered in the coordinate form as follows. Let U be the domain of a fibred coordinate chart $(U; x^\lambda, y^i)$ of a fibre bundle $Y \to X$. Let $U_r = (\pi^r_0)^{-1}(U)$ be the domain of the corresponding coordinate chart of the bundle $J^rY \to Y$. One can repeat the above procedure for the modules $\mathfrak{O}^*(U_r)$ of the exterior forms defined on U_r, and obtain their direct limit $\mathfrak{O}^*_\infty(U)$. For every r, we have the $\mathbb{R}$-module homomorphism

$$i_{U_r}{}^* : \mathfrak{O}^*_r \to \mathfrak{O}^*(U_r)$$

which sends every exterior form on J^rY onto its pull-back on U_r. Then there exists the $\mathbb{R}$-module homomorphism

$$i_U{}^* : \mathfrak{O}^*_\infty \to \mathfrak{O}^*_\infty(U)$$

such that the diagram

$$\begin{array}{ccc} \mathfrak{O}^*_r & \xrightarrow{\pi^{\infty *}_r} & \mathfrak{O}^*_\infty \\ {\scriptstyle i_{U_r}{}^*}\downarrow & & \downarrow{\scriptstyle i_U{}^*} \\ \mathfrak{O}^*(U_r) & \xrightarrow{\pi^{\infty}_{U_r}{}^*} & \mathfrak{O}^*_\infty(U) \end{array}$$

commutes for any order r [217]. Elements of $\mathfrak{O}^*_\infty(U)$ can be written in the familiar coordinate form.

Given an atlas $\{(U; x^\lambda, y^i)\}$ of fibred coordinates of $Y \to X$, let us consider the vector space $\mathfrak{O}^*_\infty(U)$ of exterior forms on infinite order jets for every coordinate chart $(U; x^\lambda, y^i)$ of this atlas. Every element ϕ of the space $\mathfrak{O}^*_\infty$ is uniquely defined by the collection of elements $\{\phi_U\}$ of the spaces $\mathfrak{O}^*_\infty(U)$, together with the corresponding coordinate transformation rules. Further on, we will utilize the coordinate expressions for exterior forms on infinite order jets, without specifying the coordinate domain U. One can say that an object given by a coordinate expression as an element of each space $\mathfrak{O}^*_\infty(U)$ is globally defined if its coordinate form is preserved under the corresponding coordinate transformations.

In particular, the basic 1-forms dx^λ and the contact 1-forms θ^i_Λ (11.1.18) constitute the set of local generating elements of the filtered $\mathfrak{O}^0_\infty$-module $\mathfrak{O}^1_\infty$ of 1-forms on

$J^\infty Y$. Moreover, the basic 1-forms dx^λ and the contact 1-forms θ^i_Λ have independent coordinate transformation laws. It follows that there is the canonical splitting

$$\boxed{\mathfrak{O}^1_\infty = \mathfrak{O}^{0,1}_\infty \oplus \mathfrak{O}^{1,0}_\infty} \tag{11.1.28}$$

of $\mathfrak{O}^1_\infty$ in the filtered $\mathfrak{O}^0_\infty$-submodules $\mathfrak{O}^{0,1}_\infty$ and $\mathfrak{O}^{1,0}_\infty$ generated separately by basic and contact forms. One can think of this splitting as being the canonical horizontal splitting. It is similar both to the horizontal splitting (2.1.9) of the cotangent bundle of a fibre bundle by means of a connection and the to canonical horizontal splittings (11.1.24) of 1-forms on finite order jet manifolds. Therefore, one can say that the splitting (11.1.28) defines the *canonical connection* on the infinite order jet space $J^\infty Y$.

The canonical horizontal splitting (11.1.28) provides the corresponding splitting of the space of m-forms

$$\boxed{\mathfrak{O}^m_\infty = \mathfrak{O}^{0,m}_\infty \oplus \mathfrak{O}^{1,m-1}_\infty \oplus \ldots \oplus \mathfrak{O}^{m,0}_\infty,} \tag{11.1.29}$$

where elements of $\mathfrak{O}^{k,s-k}$ are called *k-contact forms*. Let us denote by h_k the *k-contact projection*

$$\boxed{h_k : \mathfrak{O}^m_\infty \to \mathfrak{O}^{k,m-k}_\infty, \qquad k \leq m.} \tag{11.1.30}$$

In particular, we restart the horizontal projection h_0 (1.3.15) as the canonical projection

$$\boxed{h_0 : \mathfrak{O}^m_\infty \to \mathfrak{O}^{0,m}_\infty, \qquad \forall m > 0.}$$

Accordingly, the exterior differential on $\mathfrak{O}^*_\infty$ is decomposed into the sum

$$\boxed{d = d_H + d_V} \tag{11.1.31}$$

of the *horizontal differential* d_H and the *vertical differential* d_V. These are defined as follows:

$$\begin{aligned}
&d : \mathfrak{O}^{k,s}_\infty \to \mathfrak{O}^{k+1,s}_\infty \oplus \mathfrak{O}^{k,s+1}_\infty, \\
&d_H : \mathfrak{O}^{k,s}_\infty \to \mathfrak{O}^{k,s+1}_\infty, \qquad d_H \mid \mathfrak{O}^{k,s}_\infty \stackrel{\text{def}}{=} \mathrm{pr}_2 \circ d \mid \mathfrak{O}^{k,s}_\infty, \\
&d_V : \mathfrak{O}^{k,s}_\infty \to \mathfrak{O}^{k+1,s}_\infty, \qquad d_V \mid \mathfrak{O}^{k,s}_\infty \stackrel{\text{def}}{=} \mathrm{pr}_1 \circ d \mid \mathfrak{O}^{k,s}_\infty,
\end{aligned}$$

for $s < n$ and all k. The operators d_H and d_V obey the familiar relations

$$\begin{aligned}
&d_H(\phi \wedge \sigma) = d_H(\phi) \wedge \sigma + (-1)^{|\phi|}\phi \wedge d_H(\sigma), \qquad \phi, \sigma \in \mathfrak{O}^*_\infty, \\
&d_V(\phi \wedge \sigma) = d_V(\phi) \wedge \sigma + (-1)^{|\phi|}\phi \wedge d_V(\sigma),
\end{aligned}$$

and possess the nilpotency property

$$d_H \circ d_H = 0, \qquad d_V \circ d_V = 0, \qquad d_V \circ d_H + d_H \circ d_V = 0. \tag{11.1.32}$$

Recall also the relation

$$h_0 \circ d = d_H \circ h_0.$$

The horizontal differential can be written in the form

$$d_H\phi = dx^\lambda \wedge d_\lambda(\phi), \tag{11.1.33}$$

where

$$\boxed{d_\lambda = d_\lambda^\infty = \partial_\lambda + \sum_{0\leq|\Lambda|} y^i_{\Lambda+\lambda}\partial_i^\Lambda} \tag{11.1.34}$$

are the total derivatives in infinite order jets. It should be emphasized that, though the sum in the expression (11.1.34) is taken with respect to an infinite number of collections Λ, the operator (11.1.34) is well defined since, given any form $\phi \in \mathfrak{O}^*_\infty$, the expression $d_\lambda(\phi)$ involves only a finite number of the terms ∂_i^Λ. The total derivatives satisfy the relations

$$d_\lambda(\phi\wedge\sigma) = d_\lambda(\phi)\wedge\sigma + \phi\wedge d_\lambda(\sigma),$$
$$d_\lambda(d\phi) = d(d_\lambda(\phi)),$$
$$[d_\lambda, d_\alpha] = 0.$$

In contrast with the partial derivatives ∂_λ, they have the coordinate transformation law

$$d'_\lambda = \frac{\partial x^\mu}{\partial x'^\lambda} d_\mu.$$

The reader is referred to Section 1.3 for the explicit expressions for operators d_λ. The corresponding explicit expressions for the operators d_H and d_V read

$$\begin{array}{lll} d_H f = d_\lambda f dx^\lambda, & d_V f = \partial_i^\Lambda f \theta^i_\Lambda, & f \in \mathfrak{O}^0_\infty, \\ d_H(dx^\mu) = 0, & d_V(dx^i) = 0, & \\ d_H(\theta^i_\Lambda) = dx^\lambda \wedge \theta^i_{\lambda+\Lambda}, & d_V(\theta^i_\Lambda) = 0, & 0 \leq |\Lambda|. \end{array}$$

Turn now to the notion of a canonical connection on the infinite order jet space $J^\infty Y$. Given a vector field τ on X, let us consider the map

$$\nabla_\tau : \mathfrak{O}^0_\infty \ni f \to \tau\rfloor(d_H f) \in \mathfrak{O}^0_\infty. \tag{11.1.35}$$

This is a derivation of the ring $\mathfrak{O}^0_\infty$. Moreover, if $\mathfrak{O}^0_\infty$ is regarded as a $C^\infty(X)$-ring, the map (11.1.35) satisfies the Leibniz rule. Hence, the assignment

$$\boxed{\nabla : \tau \mapsto \tau\rfloor d_H = \tau^\lambda d_\lambda} \tag{11.1.36}$$

is the *canonical connection* on the $C^\infty(X)$-ring $\mathfrak{O}^0_\infty$ in accordance with Definition 8.2.7 [219]. Since a derivation is a local operation and $J^\infty Y$ admits a smooth partition of unity, the derivations (11.1.35) can be extended to the ring $C^\infty(J^\infty Y)$ of smooth functions on the infinite order jet space $J^\infty Y$. Accordingly, the connection ∇ (11.1.36) is extended to the canonical connection on the $C^\infty(X)$-ring $C^\infty(J^\infty Y)$. Extended to $C^\infty(J^\infty Y)$, the derivations (11.1.35), by definition, are vector fields on the infinite order jet space $J^\infty Y$. One can also think of such a vector field ∇_τ as the horizontal lift $\tau^\lambda\partial_\lambda \mapsto \tau^\lambda d_\lambda$ onto $J^\infty Y$ of a vector field τ on X by means of a canonical connection on the (topological) fibre bundle $J^\infty Y \to X$.

The vector fields ∇_τ on $J^\infty Y$ are not projectable, though they projected over vector fields on X. Projectable vector fields on $J^\infty Y$ (their definition is a repetition of that for finite order jet manifolds) are elements of the projective limit $\mathcal{P}^\infty$ of the inverse system (11.1.11). This projective limit exists. Its definition is a repetition of that of $J^\infty Y$. This is a Lie algebra such that the surjections

$$T\pi^\infty_k : \mathcal{P}^\infty \to \mathcal{P}^k$$

are Lie algebra morphisms which constitute the commutative diagrams

$$\begin{array}{ccc} & \mathcal{P}^\infty & \\ {\scriptstyle T\pi^\infty_k} \swarrow & & \searrow {\scriptstyle T\pi^\infty_r} \\ \mathcal{P}^k & \underset{T\pi^k_r}{\longrightarrow} & \mathcal{P}^r \end{array}$$

for any k and $r < k$. In brief, we will say that elements of $\mathcal{P}^\infty$ are *vector fields* on the infinite order jet space $J^\infty Y$.

In particular, let u be a projectable vector field on Y. There exists an element $J^\infty u \in \mathcal{P}^\infty$ such that

$$T\pi^\infty_k(J^\infty u) = J^k u, \qquad \forall k \geq 0.$$

One can think of $J^\infty u$ as being the ∞-order jet prolongation of the vector field u on Y. It is given by the recurrence formula (11.1.10) where $0 \leq |\Lambda|$. Then any element of $\mathcal{P}^\infty$ is decomposed into the sum similar to (11.1.13) where $k = \infty$. Of course, it is not the horizontal decomposition. Given a vector field υ on $J^\infty Y$, projected onto a vector field τ on X, we have its horizontal splitting

$$\boxed{\upsilon = \upsilon_H + \upsilon_V = \tau^\lambda d_\lambda + (\upsilon - \tau^\lambda d_\lambda)}$$

by means of the canonical connection ∇ (11.1.36) (cf. (11.1.23)). Note that the component υ_V of this splitting is not a projectable vector field on $J^\infty Y$, but is a vertical vector field with respect to the fibration $J^\infty Y \to X$.

11.2 The variational bicomplex

The role of the horizontal differential d_H in the BRST construction consists in the following. Given the BRST operator $\mathbf{s}$, one introduces the total BRST operator $\mathbf{s}+d_H$ and considers the BRST-cohomology modulo d_H [18, 19, 37, 152] (see Section 11.4). This Section is devoted to the study of the cohomology of the variational complex created by the horizontal differential d_H (see Theorem 11.2.2 below). As was mentioned above, this cohomology also provides the algebraic approach to the calculus in variations (see formula (11.2.15) below).

Remark 11.2.1. We consider the variational complex in the calculus in infinite order jets [8, 76, 123, 291, 297]. In comparison with the finite order variational sequence [188, 301], the essential simplification is that, if the order of jets is not bounded, there is the decomposition (11.1.29) of exterior forms on jet manifolds into contact and basic forms. •

Using on the nilpotency property (11.1.32) of the horizontal and vertical differentials d_H and d_V, one can construct the commutative diagram:

$$
\begin{array}{ccccccccc}
 & & 0 & & 0 & & & & \\
 & & \downarrow & & \downarrow & & & & \\
 & & \mathbb{R} & & \mathbb{R} & & 0 & & \\
 & & \downarrow & & \downarrow & & \downarrow & & \\
0 & \longrightarrow & \mathfrak{O}^0(X) & \xrightarrow{\pi^{\infty *}} & \mathfrak{O}^{0,0}_\infty & \xrightarrow{d_V}\cdots\xrightarrow{d_V} & \mathfrak{O}^{k,0}_\infty & \xrightarrow{d_V}\cdots & \\
 & & {\scriptstyle d}\downarrow & & {\scriptstyle d_H}\downarrow & & {\scriptstyle (-1)^k d_H}\downarrow & & \\
 & & \vdots & & \vdots & & \vdots & & \\
 & & {\scriptstyle d}\downarrow & & {\scriptstyle d_H}\downarrow & & {\scriptstyle (-1)^k d_H}\downarrow & & \\
0 & \longrightarrow & \mathfrak{O}^{n-1}(X) & \xrightarrow{\pi^{\infty *}} & \mathfrak{O}^{0,n-1}_\infty & \xrightarrow{d_V}\cdots\xrightarrow{d_V} & \mathfrak{O}^{k,n-1}_\infty & \xrightarrow{d_V}\cdots & \\
 & & {\scriptstyle d}\downarrow & & {\scriptstyle d_H}\downarrow & & {\scriptstyle (-1)^k d_H}\downarrow & & \\
0 & \longrightarrow & \mathfrak{O}^{n}(X) & \xrightarrow{\pi^{\infty *}} & \mathfrak{O}^{0,n}_\infty & \xrightarrow{d_V}\cdots\xrightarrow{d_V} & \mathfrak{O}^{k,n}_\infty & \xrightarrow{d_V}\cdots & \\
 & & \downarrow & & {\scriptstyle \tau_0}\downarrow & & {\scriptstyle \tau_k}\downarrow & & \\
 & & 0 & \longrightarrow & E_0 & \longrightarrow\cdots\longrightarrow & E_k & \longrightarrow\cdots & \\
 & & & & \downarrow & & \downarrow & & \\
 & & & & 0 & & 0 & &
\end{array}
\qquad (11.2.1)
$$

where

$$E_k = \tau_k(\mathfrak{O}^{k,n}_\infty) \stackrel{\text{def}}{=} \mathfrak{O}^{k,n}_\infty / d_H(\mathfrak{O}^{k,n-1}_\infty). \qquad (11.2.2)$$

Since all columns and rows of this diagram are complexes, it is a complex of complexes.

Remark 11.2.2. The operators d_H and d_V satisfying the relations (11.1.32) define a *bicomplex*, but they do not commute. The operators $(-1)^k d_H$ and d_V in the diagram (11.2.1) mutually commute, and this diagram is called a *complex of complexes* (see [204] for the terminology) •

LEMMA 11.2.1. [123, 297]. The quotient E_k, $k > 0$, (11.2.2) in the bottom row of the diagram (11.2.1) is isomorphic to the complement $\tau_k(\mathfrak{O}^{k,n}_\infty)$ of the subspace $d_H(\mathfrak{O}^{k,n-1}_\infty) \subset \mathfrak{O}^{k,n_\infty}$. □

If follows that τ_k, $k>0$, are the projection maps which have the properties

$$\tau_k\circ\tau_k=\tau_k,\qquad \tau_k\circ d_H=0.$$

The latter leads to the exact sequence

$$0\longrightarrow \operatorname{Ker} e_k \hookrightarrow \mathfrak{O}^{n+k}_\infty \xrightarrow{e_k} E_k\longrightarrow 0,$$

where $e_k=\tau_k\circ h_k$. It is a simple exact sequence because

$$\mathfrak{O}^{n+k}_\infty=\operatorname{Ker} e_k\oplus E_k.$$

One can show that $d(\operatorname{Ker} e_k)\subset \operatorname{Ker} e_{k+1}$.

In view of the above results, we can replace the entities $\mathfrak{O}^{k,n}_\infty$, d_V and τ_k, $k>0$, in the bottom rows of the diagram (11.2.1) with $\mathfrak{O}^{n+K}_\infty$, d and e_k, respectively, and come to the commutative diagram

$$\begin{array}{ccccccccc}
 & & 0 & & 0 & & 0 & & 0 & \\
 & & \downarrow & & \downarrow & & \downarrow & & \downarrow & \\
\cdots\longrightarrow & & \operatorname{Ker} h_0 & \xrightarrow{d} & \operatorname{Ker} h_0 & \xrightarrow{d} & \operatorname{Ker} e_1 & \xrightarrow{d} & \operatorname{Ker} e_2 & \longrightarrow\cdots\\
 & & \downarrow & & \downarrow & & \downarrow & & \downarrow & \\
\cdots\longrightarrow & & \mathfrak{O}^{n-1}_\infty & \xrightarrow{d} & \mathfrak{O}^{n}_\infty & \xrightarrow{d} & \mathfrak{O}^{n+1}_\infty & \xrightarrow{d} & \mathfrak{O}^{n+2}_\infty & \longrightarrow\cdots\\
 & & {\scriptstyle h_0}\downarrow & & {\scriptstyle h_0}\downarrow & & {\scriptstyle e_1}\downarrow & & {\scriptstyle e_2}\downarrow & \\
\cdots\longrightarrow & & \mathfrak{O}^{0,n-1}_\infty & \xrightarrow{d_H} & \mathfrak{O}^{0,n}_\infty & \xrightarrow{\varepsilon_1} & E_1 & \xrightarrow{\varepsilon_2} & E_2 & \longrightarrow\cdots\\
 & & \downarrow & & \downarrow & & \downarrow & & \downarrow & \\
 & & 0 & & 0 & & 0 & & 0 &
\end{array}\tag{11.2.3}$$

Its first and second rows are the subcomplexes of the De Rham complex. Therefore, the last row

$$\boxed{0\to\mathbb{R}\to\mathfrak{O}^0_\infty\xrightarrow{d_H}\mathfrak{O}^{0,1}_\infty\xrightarrow{d_H}\cdots\xrightarrow{d_H}\mathfrak{O}^{0,n}_\infty\xrightarrow{\varepsilon_1}E_1\xrightarrow{\varepsilon_2}E_2\longrightarrow\cdots}\tag{11.2.4}$$

is also a cochain complex, i.e.,

$$\varepsilon_1\circ d_H=0,\qquad \varepsilon_{k+1}\circ\varepsilon_k=0.\tag{11.2.5}$$

This complex is called the *spectral sequence*. Since $E_k\subset\mathfrak{O}^{k,n}_\infty$, the cochain morphisms ε_k of the complex (11.2.4) take the form $\varepsilon_k=\tau_k\circ d$.

One can obtain the morphisms ε_k in an explicit form [24, 123, 297]:

$$\tau_k = \frac{1}{k}\tau \mid \mathfrak{O}^{k,n}_\infty,$$

where τ is the operator, given by the coordinate expression

$$\boxed{\tau(\phi) = (-1)^{|\Lambda|}\theta^i \wedge [d_\Lambda(\partial^\Lambda_i \rfloor \phi)],} \qquad 0 \leq\mid \Lambda \mid,$$

which acts on contact densities $\phi \in \mathfrak{O}^{*,n}_\infty$. Then we obtain

$$\boxed{\varepsilon_k = \tau_k \circ d = \tfrac{1}{k}\delta,} \tag{11.2.6}$$

where

$$\boxed{\delta = \tau \circ d} \tag{11.2.7}$$

is the *variational map* [24, 123] which possesses the nilpotency property

$$\delta \circ \delta = 0, \qquad \delta \circ d_H = 0. \tag{11.2.8}$$

Since the columns of the diagram (11.2.3) are simple exact sequences, the spectral sequence (11.2.4) can be regarded as a subcomplex of the cochain complex

$$\boxed{0 \to \mathbb{R} \longrightarrow \mathfrak{O}^0_\infty \xrightarrow{d_H} \cdots \mathfrak{O}^{0,n-1}_\infty \xrightarrow{d_H} \mathfrak{O}^{0,n}_\infty \xrightarrow{\delta} \mathfrak{O}^{1,n}_\infty \xrightarrow{\delta} \mathfrak{O}^{2,n}_\infty \longrightarrow \cdots.} \tag{11.2.9}$$

It is called the *variational sequence.*

In particular, let

$$\phi = \phi^\Lambda_i \theta^i_\Lambda \wedge \omega \in \mathfrak{O}^{1,n}_\infty$$

be a 1-contact density. We obtain

$$\boxed{\tau_1(\phi) = (-1)^{|\Lambda|} d_\Lambda(\phi^\Lambda_i)\theta^i \wedge \omega.} \tag{11.2.10}$$

A glance at this expression shows that the subspace $E_1 \subset \mathfrak{O}^{1,n}_\infty$ consists of 1-contact densities $\mathcal{E} = \mathcal{E}_i\theta^i \wedge \omega$ which take their values into the tensor bundle

$$T^*Y \wedge (\overset{n}{\wedge} T^*X). \tag{11.2.11}$$

Let $L = \mathcal{L}\omega \in \mathfrak{O}^{0,n}_\infty$ be a horizontal density. Then the cochain morphisms ε_1 and ε_2 take the explicit form

$$\boxed{\varepsilon_1(\mathcal{L}\omega) = (-1)^{|\Lambda|} d_\Lambda(\partial^\Lambda_i \mathcal{L})\theta^i \wedge \omega,} \tag{11.2.12}$$

$$\boxed{\varepsilon_2(\mathcal{E}_i\theta^i \wedge \omega) = \tfrac{1}{2}[\partial^\Lambda_j \mathcal{E}_i \theta^j_\Lambda \wedge \theta^i + (-1)^{|\Lambda|}\theta^j \wedge d_\Lambda(\partial^\Lambda_j \mathcal{E}_i\theta^i)] \wedge \omega,}$$

where summation is over all multi-indices $0 \leq\mid \Lambda \mid$. They are called the *Euler–Lagrange map* and the *Helmholtz–Sonin map*, respectively. Recall that, in fact, L is a horizontal density on some finite order jet manifold J^rY. Therefore, one can think of L as being an r-order Lagrangian. Then the exterior form

$$\mathcal{E}_L = \varepsilon_1(L) = \delta(L), \tag{11.2.13}$$
$$\boxed{\mathcal{E}_L = (-1)^{|\Lambda|}d_\Lambda(\partial_i^\Lambda \mathcal{L})\theta^i \wedge \omega,} \qquad 0 \leq\mid \Lambda \mid\leq r,$$

is called the *Euler–Lagrange form* associated with the r-order Lagrangian L. This form can be seen as the $2r$-order differential operator

$$\mathcal{E}_L : J^{2r}Y \to T^*Y \wedge (\overset{n}{\wedge} T^*X), \tag{11.2.14}$$

called the *Euler–Lagrange operator* associated with L. In particular, if L is a first order Lagrangian on the jet manifold J^1Y, the operator (11.2.14) is exactly the second order Euler–Lagrange operator (3.2.3).

Furthermore, by virtue of Lemma 11.2.1, we have the canonical decomposition

$$\boxed{dL = \tau_1(dL) + (\mathrm{Id} - \tau_1)(dL) = \delta(L) + d_H(\phi), \qquad \phi \in \mathfrak{O}_\infty^{1,n_1},} \tag{11.2.15}$$

which is the *first variational formula* for higher order Lagrangians. In particular, if $r = 1$, it recovers the first variational formula (3.2.2) in the case of vertical vector fields u on $Y \to X$.

Using the spectral sequence (11.2.4), one comes to the following variant of the well-known inverse problem of the calculus of variations. Differential operators which take their values into the tensor bundle (11.2.11) are called *Euler–Lagrange-type operators*. These are elements of the subspace $E_1 \subset \mathfrak{O}_\infty^{1,n}$. An Euler–Lagrange-type operator $\mathcal{E}$ is said to be a *locally variational operator* if

$$\varepsilon_2(\mathcal{E}) = \frac{1}{2}\delta(\mathcal{E}) = 0.$$

In accordance with the relations (11.2.5), any d_H-exact Lagrangian is variationally trivial and every Euler–Lagrange operator is locally variational.

The obstruction for a locally variational operator to be an Euler–Lagrange one lies in the non-zero cohomology group

$$H^{n+1} = \mathrm{Ker}\,\varepsilon_2/\mathrm{Im}\,\varepsilon_1$$

of the complex (11.2.4) at the element E_1. Since columns of the diagram (11.2.3) are exact sequences, we have the following exact sequence of the cohomology groups of its rows denoted by r_1, r_2 and r_3 [204]:

$$\cdots \longrightarrow H^k(r_3) \longrightarrow H^k(r_2) \longrightarrow H^k(r_1) \longrightarrow H^{k+1}(r_3) \longrightarrow \cdots.$$

THEOREM 11.2.2. [76, 297]. If $Y = \mathbb{R}^{l+n} \to \mathbb{R}^n$, the spectral sequence (11.2.4) is exact, i.e.,

$$\boxed{\operatorname{Ker} d_H = \operatorname{Im} d_H, \qquad \operatorname{Ker} \varepsilon_1 = \operatorname{Im} d_H, \qquad \operatorname{Im} \varepsilon_k = \operatorname{Ker} \varepsilon_{k+1}.}$$

□

In view of this theorem, one can say that the complex (11.2.4) and, consequently, the complex (11.2.9) are locally exact.

COROLLARY 11.2.3. Since the variational sequence (11.2.9) is locally exact, a horizontal density $L \in \mathfrak{O}^*_\infty$ is variationally trivial if and only if it is d_H-exact. It follows that there is one-to-one correspondence between the equivalence classes of local horizontal densities L modulo d_H-exact forms and the elements of the set $\operatorname{Im} \varepsilon_1 = \operatorname{Ker} \varepsilon_1$. □

COROLLARY 11.2.4. Let us consider the cochain complex

$$\boxed{0 \to \mathbb{R} \to \mathfrak{O}^0_\infty \xrightarrow{d_H} \mathfrak{O}^{0,1}_\infty \xrightarrow{d_H} \cdots \xrightarrow{d_H} \mathfrak{O}^{0,n}_\infty \xrightarrow{d_H} 0.} \qquad (11.2.16)$$

It is locally exact at all terms, except the last one. Its cohomology group at this terms is

$$\boxed{\mathfrak{O}^{0,n}_\infty / d_H(\mathfrak{O}^{0,n}_\infty) = E_1.} \qquad (11.2.17)$$

□

11.3 Jets of ghosts and antifields

Besides physical fields, the BRST construction involves ghosts, ghost-for-ghosts and antifields which are even and odd algebraic quantities. In order to apply the above jet formalism to the BRST construction, we should provide a geometric description of odd quantities, e.g., define their jets. Note that, in the framework of the jet formulation of the BRST theory, antifields can be introduced on the same footing as fields (see Remark 11.3.2 below).

Let us start from ghosts. In Lagrangian BRST formalism (see, e.g., [130] for a survey), ghosts are associated with parameters of gauge transformations. We will restrict our consideration to the case of bosonic fields and even gauge transformations with a finite number of generators. Then ghosts are odd. This is also the case of Hamiltonian BRST formalism where ghosts are related to constraints (see, e.g., [153]).

Different geometric models of ghosts have been suggested in order to provide a desired BRST transformation law. For instance, ghosts in the Yang–Mills gauge theory on a G-principal bundle can be described as forms on the gauge group manifold $\mathrm{Gau}(P)$ (see, e.g., [32, 163, 275]). This description however is not extended to other gauge theories (see [130] for several examples of gauge models).

Example 11.3.1. Let us consider the above-mentioned gauge theory on a principal bundle $P \to X$ over a compact manifold X whose structure group G is a compact semisimple matrix Lie group. A suitable Sobolev completion makes the gauge group $\mathrm{Gau}(P)$ a Banach Lie group (see Section 12.1). Generators of 1-parameter groups of gauge transformations are G-invariant vector fields on P represented by sections ξ of the Lie algebra bundle V_GP. The Sobolev completion of the set of sections of $V_GP \to X$ is the Lie algebra of the gauge group $\mathrm{Gau}(P)$. The typical fibre of the dual V_G^*P of V_GP is the Lie coalgebra $\mathfrak{g}^*$. Given a generator of gauge transformations $\xi = \xi^r(x)e_r$ (6.3.7), the corresponding generator of gauge transformations of the bundle of principal connections $C \to X$ is ξ_C (6.2.13). Then, if A is a section of $C \to X$, the generator ξ_C acts on A by the law

$$\xi : A = A^q_\lambda dx^\lambda \otimes e_q \mapsto dx^\lambda \otimes (\partial_\lambda \xi^r + c^r_{pq} A^p_\lambda \xi^q)e_r = d\xi + [A, \xi] = \nabla^A \xi \quad (11.3.1)$$

(see (6.1.26), (6.1.29)). Accordingly, gauge parameters themselves are transformed by the coadjoint representation (6.3.8). One can obtain the classical BRST transformations in a naive way by the replacement of gauge parameters in the transformations laws (6.3.8) and (11.3.1) with the ghosts. Namely, let us take the formal generator

$$\boxed{C = C^r e_r} \tag{11.3.2}$$

of gauge transformations and substitute it in the above mentioned expressions. We obtain

$$\boxed{\mathbf{s}A = dC + [A, C], \qquad \mathbf{s}C = -\tfrac{1}{2}[C, C],} \tag{11.3.3}$$

$$\mathbf{s}A^r_\lambda = \partial_\lambda C^r + c^r_{pq} A^p_\lambda C^q, \qquad \mathbf{s}C^p = -\frac{1}{2} c^p_{rq} C^r C^q. \tag{11.3.4}$$

This is exactly the desired classical BRST operator in the Yang–Mills gauge theory. For instance, let $\{C^r\}$ be the local fibre basis for $V^*_G P$ which is dual of e_r, then C (11.3.2) is the canonical section of the tensor bundle $V^*_G P \otimes V_G P$ which defines the identity automorphisms of $V_G P$ and $V^*_G P$ (see Remark 1.1.1). It also coincides with the ghost field η introduced as the Maurer–Cartan form on the gauge group $\mathrm{Gau}(P)$ such that $\eta(\xi) = \xi$, $\xi \in V_G P(X)$. Then the BRST operator $\mathbf{s}$ (11.3.3) can be defined as the coboundary operator of the Chevalley–Eilenberg cohomology (see Section 14.2) of the cochain complex whose elements are q-cochains on $V_G P(X)$ which take values in the exterior algebra of the equivariant $\mathfrak{g}_l$-forms on the principal bundle P [32, 275, 293]. •

Point out the following two peculiarities of ghosts. (i) Since ghosts are considered as odd fields on an ordinary smooth manifold X and are characterized by a ghost number 1, they can be seen as generating elements of a graded algebra. (ii) Example 11.3.1 shows that jets of ghosts should be considered. The following geometric construction fulfills these conditions.

Let $E \to X$ be an m-dimensional vector bundle. Its k-order jet manifold $J^k E$ is also a vector bundle over X. As above, we put $J^0 E = E$. Let us consider the simple graded manifold $(X, \mathcal{A}_{J^k E})$ whose structure vector bundle is $J^k E \to X$ (see the notation in Section 9.2). For the sake of simplicity, it will be denoted by $\widehat{J^k E}$. Its local basis is $\{C^r_\Lambda\}$, $0 \leq |\Lambda| \leq k$, with the transition functions

$$\boxed{C'^r_{\lambda+\Lambda} = d_\lambda(\rho^r_q C^q_\Lambda),} \tag{11.3.5}$$

where

$$d_\lambda = \partial_\lambda + C^r_\lambda \frac{\partial}{\partial C^r} + C^r_{\lambda\mu} \frac{\partial}{\partial C^r_\mu} + \cdots$$

are the *graded total derivatives* (cf. (9.6.5)). In view of the transition functions (11.3.5), one can think of C^r_Λ as *jets of ghosts*. Recall again that C^r_Λ are not jets of graded bundles introduced in [260].

Let

$$\boxed{E \overset{\pi^1_0}{\longleftarrow} J^1E \longleftarrow \cdots \longleftarrow J^{r-1}E \overset{\pi^r_{r-1}}{\longleftarrow} J^rE \longleftarrow \cdots} \tag{11.3.6}$$

be the inverse system of jet manifolds, and $J^\infty E$ its projective limit. The natural projection

$$\pi^r_{r-1} : J^rE \to J^{r-1}E$$

yields the corresponding $\mathbb{R}$-algebra monomorphism of the exterior algebras

$$\pi^{r*}_{r-1} : \wedge J^{r-1}E^* \to \wedge J^rE^* \tag{11.3.7}$$

and the epimorphism

$$S\pi^r_{r-1} : \widehat{J^rE} \to \widehat{J^{r-1}E}, \qquad r > 0, \tag{11.3.8}$$

(9.2.17) of graded manifolds. With the morphisms (11.3.7) and (11.3.8), we have the corresponding direct system of exterior algebras

$$\boxed{\wedge E^* \longrightarrow \cdots \overset{\pi^{r*}_{r-1}}{\longrightarrow} \wedge J^rE^* \to \cdots} \tag{11.3.9}$$

and the inverse system of the graded manifolds

$$\boxed{\widehat{E} \longleftarrow \widehat{J^1E} \longleftarrow \cdots \longleftarrow \widehat{J^{r-1}E} \overset{(\pi^r_{r-1}, \pi^r_{r-1*})}{\longleftarrow} \cdots.} \tag{11.3.10}$$

The direct system (11.3.9) has a direct limit $\wedge J^\infty E^*$ in filtered endomorphisms. It consists of the pull-backs of elements of the exterior algebras $\wedge J^kE^*$, $k = 0, 1, \ldots$, onto $J^\infty E$. This direct limit defines the projective limit $\widehat{J^\infty E}$ of the inverse system of graded manifolds (11.3.10). One can think of the pair

$$\widehat{J^\infty E} = (X, \wedge J^\infty E^*(X))$$

as being a graded manifold, while elements of its structure module $\wedge J^\infty E^*(X)$ are graded functions on X. Its coefficients are smooth functions on X.

In order to introduce exterior forms on this graded manifold, let us consider the direct system of the filtered $\wedge J^kE^*(X)$-algebras $\wedge\mathfrak{d}^*(\wedge J^kE^*)(X)$ of graded exterior forms on graded manifolds $\widehat{J^kE}$ with respect to the natural monomorphisms

$$\wedge\mathfrak{d}^*(\wedge E^*)(X) \to \cdots \to \wedge\mathfrak{d}^*(\wedge J^kE^*)(X) \to \cdots.$$

Its direct limit $\wedge\mathfrak{d}^*(\wedge J^\infty E^*)(X)$ consists of graded exterior forms on finite-order graded manifolds modulo these monomorphisms. This direct limit is a locally free filtered $\wedge\mathfrak{d}^*(\wedge J^kE^*)(X)$-algebra generated by the elements

$$(1, dx^\lambda, \theta^r_\Lambda = dC^r_\Lambda - C^r_{\lambda+\Lambda}dx^\lambda), \qquad 0 \leq |\Lambda|,$$

which obey the usual rules for graded exterior forms (see Section 9.2). Similarly to the decomposition (11.1.29), the space $\overset{k}{\wedge}\mathfrak{d}^*(\wedge J^\infty E^*)(X)$ of graded k-forms admits the splitting in the subspaces $\wedge^{k-i,i}\mathfrak{d}^*(\wedge J^\infty E^*)(X)$ of $(k-i)$-contact forms. Accordingly, the graded exterior differential d on the algebra $\overset{k}{\wedge}\mathfrak{d}^*(\wedge J^\infty E^*)(X)$ has the decomposition $d = d_H + d_V$, where the *graded horizontal differential* d_H is

$$d_H(\phi) = dx^\lambda \wedge d_\lambda(\phi), \qquad \phi \in \wedge\mathfrak{d}^*(\wedge J^\infty E^*)(X).$$

The graded differentials d_H and d_V obey the nilpotency property (11.1.32). Since every $\phi \in \wedge\mathfrak{d}^*(\wedge J^\infty E^*)(X)$ is a graded form on some finite order graded manifold, the expression $d_\lambda(\phi)$ contains a finite number of terms. One can consider the horizontal differential d_H as the canonical connection on the algebra $\wedge\mathfrak{d}^*(\wedge J^\infty E^*)(X)$.

The algebra $\wedge\mathfrak{d}^*(\wedge J^\infty E^*)(X)$ provides everything that one needs for the differential calculus in ghosts. Graded forms $\phi \in \wedge\mathfrak{d}^*(\wedge J^\infty E^*)(X)$ are characterized by:

- the $\mathbb{Z}$-graded *ghost number* $\mathrm{gh}(\phi)$ such that

$$\mathrm{gh}(C^i_\Lambda) = 1, \quad \mathrm{gh}(dC^i_\Lambda) = 1, \quad \mathrm{gh}(dx^\lambda) = 0, \quad \mathrm{gh}(f) = 0, \quad f \in C^\infty(X);$$

- the *ghost Grassmann parity* $[\phi] = \mathrm{gh}(\phi)\,\mathrm{mod}2$;
- the usual form degree $|\phi|$ and the form Grassmann parity $|\phi|\,\mathrm{mod}2$;
- the *total ghost number*

$$\mathrm{gh}_T(\phi) = \mathrm{gh}(\phi) + |\phi|.$$

Turn now to cohomology of the algebra $\wedge\mathfrak{d}^*(\wedge J^\infty E^*)(X)$. In accordance with Remark 9.2.6, the graded De Rham cohomology groups of $\wedge\mathfrak{d}^*(\wedge J^k E^*)(X)$ of finite order graded manifolds coincide with the De Rham cohomology groups of the manifold X, and so does their direct limit. By virtue of Theorem 11.1.4, this limit is the graded De Rham cohomology groups of the algebra $\wedge\mathfrak{d}^*(\wedge J^\infty E^*)(X)$. Theorem 11.2.2 can also be extended to this algebra as follows (see [37] and references therein).

THEOREM 11.3.1. If a graded horizontal $(0<k<n)$-form ϕ is locally d_H-closed, i.e., $d_H\phi = 0$, then it is locally d_H-exact, i.e., there exists a graded horizontal $(k-1)$-form σ such that $\phi = d_H\sigma$. A graded horizontal n-form ϕ is locally exact if and only if

$$\delta = \varepsilon_1(\phi) = 0,$$

where ε_1 is the Euler–Lagrange map (11.2.12) extended to the graded exterior algebra. □

Besides physical fields φ^i of vanishing ghost number and ghosts C^r, BRST theory involves ghost-for-ghosts and antifields (see [130] for a survey). In general, an Lth stage reducible theory contains L generations of *ghost-for-ghosts* C^r_l, $l = 1,\ldots,L$, whose ghost numbers and the ghost Grassmann parity are

$$\mathrm{gh}(C^r_l) = \mathrm{gh}(C^r) + l, \qquad [C^r_l] = ([C^r] + l)\,\mathrm{mod}2.$$

The odd ghost-for-ghosts can be introduced in the same manner as ghosts by a choice of the corresponding vector bundle E. Let us denote fields, ghosts, ghost-for-ghosts by the same collective symbol Φ^A, $A = 1,\ldots,N$. *Antifields* Φ^*_A have the following ghost numbers and the ghost Grassmann parity:

$$\mathrm{gh}(\Phi^*_A) = -\mathrm{gh}(\Phi^A) - 1, \qquad [\Phi^*_A] = ([\Phi^A] + 1)\,\mathrm{mod}2.$$

In the jet formalism, antifields Φ^*_A can be introduced on the same footing as fields Φ^A by a choice of the vector bundle $E = Y^*$ dual of the bundle Y for fields Φ^A. Note that gauge potentials are sections of the affine bundle $C \to X$ (6.1.8) modeled over the vector bundle $T^*X \otimes V_G P$. Their odd antifields are modelled on the vector bundle $E = TX \otimes V^*_G P$.

The total system of fields and antifields $\{\zeta^a\}$, called sometimes the *classical basis*, is described by the pointwize exterior product

$$\mathfrak{G}^* = \mathfrak{O}^*(J^\infty E_0) \underset{X}{\wedge} (\wedge \mathfrak{d}^*(\wedge J^\infty E_1^*)(X)) \tag{11.3.11}$$

over X of the $C^\infty(X)$-algebra $\mathfrak{O}^*(J^\infty E_0)$ of even elements of the classical basis and $C^\infty(X)$-algebra $\wedge \mathfrak{d}^*(\wedge J^\infty E_1^*)(X)$ of its odd elements. The exterior algebra $\mathfrak{G}^*$ (11.3.11) is provided with the exterior differential d, which is the sum over X of the exterior differentials on $\mathfrak{O}^*(J^\infty E_0)$ and $\wedge \mathfrak{d}^*(\wedge J^\infty E_1^*)(X)$. The corresponding horizontal differential d_H can be treated as the canonical connection on $\mathfrak{G}^*$. Accordingly, the module $\mathfrak{G}^k$ of k-forms is decomposed into the subspaces $\mathfrak{G}^{k-i,i}$ of $(k-i)$-contact forms. Following the terminology accepted in the physical literature, we will call elements of $\mathfrak{G}^{0,*}$ the *local forms*.

Theorems 11.2.2 and 11.3.1 are true for local forms $f \in \mathfrak{G}^{0,*}$. Hence, the complex

$$\boxed{0 \to \mathbb{R} \to \mathfrak{G}^0 \xrightarrow{d_H} \mathfrak{G}^{0,1} \xrightarrow{d_H} \cdots \xrightarrow{d_H} \mathfrak{G}^{0,n} \xrightarrow{\delta} \mathfrak{G}^{1,n} \xrightarrow{\delta} \cdots} \tag{11.3.12}$$

has the cohomology groups

$$\boxed{H^0(\mathfrak{G}^*) = \mathbb{R}, \qquad H^{0<k<n}(\mathfrak{G}^*) = 0, \qquad H^n(\mathfrak{G}^*) \neq 0.} \tag{11.3.13}$$

Following the usual practice, we use the *right derivatives*

$$\frac{\partial_r f}{\partial \zeta} = (-1)^{[\zeta]([f]+1)} \frac{\partial_l f}{\partial \zeta}, \qquad f \in \mathfrak{G}^0,$$

such that

$$df(\zeta) = d\zeta \frac{\partial_l f}{\partial \zeta} = \frac{\partial_r f}{\partial \zeta} d\zeta.$$

The *left derivatives* $\partial_l/\partial\zeta$ are the derivatives utilized throughout before. By δ_l and δ_r are meant the left and the right variational derivatives given by the coefficients of the Euler–Lagrange map (11.2.12).

Let $f, f' \in \mathfrak{G}^0$ be graded functions. Due to the local isomorphism

$$\mathfrak{G}^0 \ni f \mapsto f\omega \in \mathfrak{G}^{0,n},$$

their *antibracket* is defined as

$$(f, f')_{\rm AB} = \frac{\delta_r f}{\delta \Phi^A} \frac{\delta_l f'}{\delta \Phi^*_A} - \frac{\delta_r f}{\delta \Phi^*_A} \frac{\delta_l f'}{\delta \Phi^A}. \tag{11.3.14}$$

We have

$$\mathrm{gh}((f,f')_{\mathrm{AB}}) = \mathrm{gh}(f) + \mathrm{gh}(f') + 1,$$
$$[(f,f')_{\mathrm{AB}}] = ([f] + [f'] + 1)\,\mathrm{mod}2.$$

Remark 11.3.2. It is readily observed that, in fact, the antibracket (11.3.14) is defined on the elements of the cohomology group $H^n(\mathfrak{G}^*)$ (11.3.13) of the complex (11.3.12). They correspond to local functionals $\int f$ up to surface integrals. At the same time, the antibracket on local functionals implies more intricate geometric interpretation of antifields [173, 308]. •

The antibracket (11.3.14) possesses the properties of the graded Poisson bracket

$$(f,f')_{\mathrm{AB}} = -(-1)^{([f]+1)([f']+1)}(f',f)_{\mathrm{AB}},$$
$$(f,(f',f'')_{\mathrm{AB}})_{\mathrm{AB}} + (-1)^{([f]+1)([f']+[f''])}((f',f'')_{\mathrm{AB}},f)_{\mathrm{AB}} +$$
$$(-1)^{([f'']+1)([f]+[f'])}((f'',f)_{\mathrm{AB}},f')_{\mathrm{AB}} = 0,$$

where the grading degree of f, f' and f'' is their Grassmann parity plus 1. In particular $(f,f)_{\mathrm{AB}} = 0$ if f is odd, and

$$(f,f)_{\mathrm{AB}} = 2\frac{\delta_r f}{\delta\Phi^A}\frac{\delta_l f}{\delta\Phi^*_A}$$

if f is even. One can write

$$(f,f')_{\mathrm{AB}} = \frac{\delta_r f}{\delta\zeta^a}w^{ab}\frac{\delta_l f'}{\delta\zeta^b}, \tag{11.3.15}$$

where w^{ab} is the graded Poisson bivector. It is readily observed that the basis $\{\Phi^A, \Phi^*_A\}$ is canonical for the Poisson structure (11.3.14). With respect to this basis, the Poisson bivector w reads

$$w = \begin{pmatrix} 0 & \delta^A_B \\ -\delta^A_B & 0 \end{pmatrix}.$$

Let $S \in \mathfrak{G}^{0,n}$ be an even local density of vanishing ghost number. The equation

$$(S,S)_{\mathrm{AB}} = 2\frac{\delta_r S}{\delta\Phi^A}\frac{\delta_l S}{\delta\Phi^*_A} = 0 \tag{11.3.16}$$

is called the *classical master equation*. Regarding S as a Lagrangian of the classical basis $\{\zeta^a\}$ of field and antifields, one can think of

$$\frac{\delta_r S}{\delta\zeta^a} = 0 \tag{11.3.17}$$

as being the equations of motion. The equations (11.3.17) are not independent, but obey the relations

$$\frac{\delta_r S}{\delta\zeta^a}\mathcal{R}^a_b = 0, \qquad \mathcal{R}^a_b = w^{ac}\frac{\delta_l\delta_r S}{\delta\zeta^c\delta\zeta^b}. \tag{11.3.18}$$

It follows that a solution S of the master equation (11.3.16) possesses a gauge freedom.

A solution S of the master equation (11.3.16) is called a *proper solution* if the rank of the Hessian

$$\frac{\partial_l\partial_r S}{\partial\zeta^a\partial\zeta^b}$$

at stationary points of $\mathfrak{G}^0$, where (11.3.17) holds, is equal to N. The reason is simple. If S is a proper solution, one can use the above-mentioned gauge freedom in order to remove all antifields.

We refer the reader to [105, 106, 130] and references therein for the problem of existence and uniqueness of a proper solution. Point out only its two properties.

(i) A proper solution S can be expanded in a power series of antifields such that

$$S|_{\Phi^*=0} = L_{\rm cl}$$

is a Lagrangian of physical fields φ^i. This expansion can be seen as an expansion with respect to the *antighost number* defined according to

$$\text{antigh}(\Phi^*_A) = -\text{gh}(\Phi^*_A) = \text{gh}(\Phi^A) + 1, \qquad \text{antigh}(\Phi^A) = 0.$$

(ii) Let S be a proper solution for the classical basis (Φ^A, Φ^*_A). Let Ψ_1 and Ψ_2 be two new fields with the ghost numbers $\text{gh}(\Psi_2) = \text{gh}(\Psi_1) + 1$, and let Ψ^*_1 and Ψ^*_2 be the corresponding antifields. Then $S + \Psi^*_1\Psi_2$ is a proper solution for the classical basis $(\Phi^A, \Psi_1, \Psi_2, \Phi^*_A, \Psi^*_1, \Psi^*_2)$. One calls (Ψ_1, Ψ_2) a *trivial variable pair*. Trivial variable pairs can be be added to the classical basis, while maintaining the classical master equation and its properties. They appear when the gauge-fixing and path integral quantization is considered.

Let S be a proper solution of the master equation (11.3.16). The *BRST operator* is defined as

$$\boxed{\mathrm{s}f = (f, S)_{\mathrm{AB}}.} \tag{11.3.19}$$

One obtains easily from the properties of the antibracket that:

- s is a nilpotent operator ($\mathrm{s}^2 = 0$),
- it is an antiderivation $\mathrm{s}(ff') = f\mathrm{s}f' + (-1)^{[f']}(\mathrm{s}f)f'$,
- $\mathrm{gh}(\mathrm{s}f) = \mathrm{gh}(f) + 1$.

Example 11.3.3. Let us consider the Yang–Mills gauge theory on the bundle of principal connections C (6.1.8) coordinated by (x^λ, a^m_λ). It is irreducible. Therefore its classical basis consists of gauge potentials and ghosts fields $\Phi^A = (a^m_\lambda, C^m)$ together with antifields $\Phi^*_A = (a^{\lambda*}_m, C^*_m)$. The proper solution of the master equation is

$$S = L_{\mathrm{YM}} + a^{\lambda*}_r(C^r_\lambda + c^r_{pq}a^p_\lambda C^q) + \frac{1}{2}C^*_p c^p_{rq}C^r C^q,$$

where L_{YM} is the Yang–Mills Lagrangian (6.3.18). The corresponding BRST operator on A^r_λ and C^p takes the form

$$\boxed{\mathrm{s}a^r_\lambda = C^r_\lambda + c^r_{pq}a^p_\lambda C^q, \qquad \mathrm{s}C^p = -\tfrac{1}{2}c^p_{rq}C^r C^q,} \tag{11.3.20}$$
$$\mathrm{s}a^{\lambda*}_r = \delta^\lambda_r(L_{\mathrm{YM}}) + a^{\lambda*}_p c^p_{rq}C^q, \qquad \mathrm{s}C^*_r = a^{\lambda*}_{\lambda r} - c^p_{qr}a^q_\lambda a^{\lambda*}_p + c^p_{rq}C^*_p C^q,$$

where $\delta^\lambda_r(L_{\mathrm{YM}})$ is the variational derivative of the Yang–Mills Lagrangian L_{YM}. •

Remark 11.3.4. There is another convention where the BRST operator is defined as

$$\mathrm{s}f = (-1)^{[f]}(f, S)_{\mathrm{AB}}. \tag{11.3.21}$$

In contrast with s (11.3.19) it is a derivation

$$\mathrm{s}(ff') = (\mathrm{s}f)f' + (-1)^{[f]}f\mathrm{s}f'.$$

•

11.4 The BRST connection

To make the expression (11.3.19) for the BRST operator $\mathbf{s}$ complete, we should define this operator on the jets of elements of the classical basis. The definition of $\mathbf{s}$ implies that $\mathbf{s}(x^\lambda) = 0$. Therefore, put

$$\boxed{\mathbf{s}\zeta^a_\Lambda = d_\Lambda(\mathbf{s}\zeta^a).}$$

By means of this rule, the BRST operator $\mathbf{s}$ is extended to the subalgebra of local forms $\mathfrak{G}^{0,*}$ such that $\mathbf{s}d_H = -d_H\mathbf{s}$, i.e.,

$$\mathbf{s}(\frac{1}{r!}\phi_{\lambda_1\ldots\lambda_r}dx^{\lambda_1}\wedge\cdots\wedge dx^{\lambda_r}) = \frac{1}{r!}(-1)^r(\mathbf{s}\phi_{\lambda_1\ldots\lambda_r})dx^{\lambda_1}\wedge\cdots\wedge dx^{\lambda_r}. \tag{11.4.1}$$

In particular, with the operator $\mathbf{s}$ (11.4.1), the formula (11.3.20) is rewritten as

$$\mathbf{s}a = -d_HC - [a,C], \qquad \mathbf{s}C = -\frac{1}{2}[C,C], \tag{11.4.2}$$

where $a = a^r_\lambda dx^\lambda \otimes e_r$.

The operators $\mathbf{s}$ and d_H define a bicomplex on the algebra of local forms. This bicomplex is graded by the form degree and the ghost number such that

$$\phi\wedge\phi' = (-1)^{|\phi||\phi'|+[\phi][\phi']}\phi'\wedge\phi, \qquad \phi,\phi'\in\mathfrak{G}^{0,*}. \tag{11.4.3}$$

Following the standard rule [204], one can construct a complex from this bicomplex which is characterized by the cochain operator

$$\boxed{\widetilde{\mathbf{s}} = \mathbf{s} + d_H,} \tag{11.4.4}$$

and is graded by the total ghost number. The operator (11.4.4) is nilpotent ($\widetilde{\mathbf{s}}^2 = 0$) and raises the total ghost number by 1, i.e.,

$$\mathrm{gh}_T(\widetilde{\mathbf{s}}\phi) = \mathrm{gh}_T(\phi) + 1.$$

It is called the *total BRST operator*

Let us study the cohomology of the total BRST operator (11.4.4). In the case of the Yang–Mills gauge theory, this is the so-called local cohomology of the cochain complex mentioned in Remark 11.3.1 [32, 211] which are phrased in the jet terms [18, 37]. Let $\phi_n \in \mathfrak{G}^{0,n}$ be a local density, i.e., a Lagrangian. It is called *locally BRST-closed* if $\mathbf{s}\phi_n$ is a d_H-exact form (see Remark 3.4.5), i.e., satisfies the equality

$$\mathbf{s}\phi_n + d_H\phi_{n-1} = 0, \tag{11.4.5}$$

where $\phi_{n-1} \in \mathfrak{G}^{0,n-1}$ is a local $(n-1)$-form. It is readily observed that ϕ_n is locally BRST-closed if the action functional $\int \phi_n$ is BRST-invariant modulo surface integrals, i.e., ϕ_n is BRST-invariant in the sense of Remark 3.4.5. A local density ϕ_n is said to be *locally BRST-exact* if

$$\phi_n = \mathbf{s}\sigma_n + d_H\sigma_{n-1}.$$

The set $H(\mathbf{s}|d_H)$ of classes of locally BRST-closed local densities modulo locally BRST-exact local densities is called the local BRST cohomology [18, 19, 37, 152].

Let us relate the local BRST cohomology to the cohomology of the total BRST operator $\widetilde{\mathbf{s}}$. Applying $\mathbf{s}$ to the equality (11.4.5) results in $d_H(\mathbf{s}\phi_{n-1}) = 0$. Hence, $\mathbf{s}\phi_{n-1}$ is d_H-closed and, consequently, d_H-exact in accordance with Theorem 11.3.1. Therefore, there is a (possibly vanishing) local $(n-2)$-form ϕ_{n-2} satisfying the equation

$$\mathbf{s}\phi_{n-1} + d_H\phi_{n-2} = 0.$$

Iterating the arguments, one conclude the existence of a set of local forms ϕ_k, $k = k_0, \ldots, n$, satisfying the relations

$$d_H\phi_n = 0, \tag{11.4.6a}$$

$$\mathbf{s}\phi_k + d_H\phi_{k-1} = 0, \qquad n \geq k > k_0, \tag{11.4.6b}$$

$$\mathbf{s}\phi_{k_0} = 0 \tag{11.4.6c}$$

for some k_0. These equations are called the *descent equations* [37]. If $k_0 = 0$, the equation (11.4.6c) takes the general form $\mathbf{s}\phi_0 =$const. The descent equations (11.4.6a) – (11.4.6c) can be compactly written in the form

$$\boxed{\widetilde{\mathbf{s}}\widetilde{\phi} = 0,} \qquad \widetilde{\phi} = \sum_{k=k_0}^{n} \phi_k. \tag{11.4.7}$$

It follows that any solution of the equation (11.4.5) corresponds to an $\widetilde{\mathbf{s}}$-closed form, and *vice versa*. In particular, a solution ϕ_n of the equation (11.4.5) is locally BRST-exact if and only if $\widetilde{\phi} = \widetilde{\mathbf{s}}\widetilde{\sigma}+$const., i.e., every locally BRST-exact local form corresponds to an $\widetilde{\mathbf{s}}$-exact local form modulo constant functions. It states the following.

THEOREM 11.4.1. The local BRST cohomology of local densities ϕ_n of a fixed total ghost number $\mathrm{gh}_T(\phi) = \mathrm{gh}(\phi) + n$ are isomorphic to the $\widetilde{\mathbf{s}}$-cohomology of local forms $\widetilde{\phi}$ of the same total ghost number $\mathrm{gh}_T(\phi)$. □

Since a proper solution S is expanded in a power series of antifields, the BRST operator can be decomposed into the sum

$$\mathbf{s} = \delta + \gamma + \sum_{k\geq 1} s_k \tag{11.4.8}$$

where δ, γ and s_k are the operators of antighost numbers -1, 0 and k, respectively. Since the horizontal differential d_H has the vanishing antighost number, the corresponding decomposition of the total BRST operator reads

$$\widetilde{\mathbf{s}} = \delta + \widetilde{\gamma} + \sum_{k\geq 1} s_k, \qquad \widetilde{\gamma} = \gamma + d_H.$$

The operator δ, called the *Koszul–Tate differential*, is nilpotent. It is non-vanishing only on antifields:

$$\delta\Phi^A = 0, \qquad \delta\varphi_i^* = \frac{\delta_r L_{\rm cl}}{\partial\varphi^i}, \qquad \ldots.$$

The operator γ encodes the gauge transformations with parameters replaced by ghosts:

$$\gamma\varphi^i = R^i_r C^r, \qquad R^i_r = \sum_{k\geq 0} R^{i\Lambda}_r d_\Lambda. \tag{11.4.9}$$

The usefulness of the decomposition (11.4.8) is due to the acyclicity of the Koszul–Tate differential δ on local functions at positive antighost numbers, i.e., $\delta\widetilde{\phi}_k = 0$ (antigh$(\widetilde{\phi}_k) = k$) implies $\widetilde{\phi}_k = \delta\widetilde{\sigma}_{k+1}$ (see [105, 106] for details). One concludes from this fact that an $\widetilde{\mathbf{s}}$-non-exact solution $\widetilde{\phi}$ of $\widetilde{\mathbf{s}}\widetilde{\phi} = 0$ contains necessarily an antifield independent part $\widetilde{\phi}_0$ such that

$$\widetilde{\gamma}\widetilde{\phi}_0 \approx 0, \qquad \widetilde{\phi}_0 \not\approx \widetilde{\gamma}\widetilde{\sigma} + \text{const.}, \qquad \text{antigh}(\widetilde{\phi}_0) = 0, \tag{11.4.10}$$

where $\approx$ denotes the weak equality, i.e., $a_0 \approx 0$ (antigh$(a_0) = 0$) if and only if there exists a_1 (antigh$(a_1) = 1$) such that $a_0 = \delta a_1$. Furthermore, any solution $\widetilde{\phi}_0$ of (11.4.10) can be completed to an $\widetilde{\mathbf{s}}$-closed non-exact local form $\widetilde{\phi}$ such that different completions with the same antifield-independent part belong to the same element of the $\widetilde{\mathbf{s}}$-cohomology. Note that

$$\delta\widetilde{\gamma} + \widetilde{\gamma}\delta = 0, \qquad \widetilde{\gamma}^2 = -(\delta s_1 + s_1\delta),$$

i.e., $\widetilde{\gamma}$ is weakly nilpotent. This establishes the following result.

PROPOSITION 11.4.2. [37]. The cohomology of $\widetilde{\mathbf{s}}$ on local forms is isomorphic to the weak cohomology of $\widetilde{\gamma}$ on antifield-independent local forms. □

Therefore, we can restrict our consideration to antifield-independent local forms. Let us assume that there is a locally invertible change of jet coordinates from the antifield-independent set (Φ^A) to the set (U^l, V^l, W^i) of non-negative ghost number such that

$$\widetilde{\gamma}U^l = V^l, \qquad \widetilde{\gamma}W^i = R^i(W).$$

The (U^l, V^l) are called trivial pairs. Without loss of generality, one can also assume that each of U^l, V^l, W^i has a definite total ghost number. Typically, U^l are components of gauge fields and their jets, while $V^l = \widetilde{\gamma}U^l$ (see Examples 11.4.1 and 11.4.2 below).

PROPOSITION 11.4.3. [37]. If $\widetilde{\gamma}\widetilde{\phi}_0(U, V, W) \approx 0$, then

$$\widetilde{\gamma}\widetilde{\phi}_0(U, V, W) \approx \phi(W) + \widetilde{\gamma}\widetilde{\sigma}(U, V, W),$$

i.e., trivial pairs can be eliminated from the weak $\widetilde{\gamma}$-cohomology. □

Let us denote by T^ι, $\widetilde{C}^L$ and Q^{L_k} those W^i which are of total ghost numbers 0, 1 and $k > 1$, respectively. The T^ι, called *BRST tensor fields*, are 0-forms, whereas $\widetilde{C}^L$ and $\widetilde{Q}^{L_k}$ decompose in general into a sum of local forms with definite form degrees

$$\widetilde{C}^L = \widehat{C}^L + \mathcal{A}^L, \tag{11.4.11}$$
$$\widehat{C}^L \in \mathfrak{G}^0, \ \mathrm{gh}(\widehat{C}^L) = 1, \qquad \mathcal{A}^L \in \mathfrak{G}^{0,1}, \ \mathrm{gh}(\mathcal{A}^L) = 0,$$
$$\widetilde{Q}^{L_k} = \sum_{r=0}^{k} \widehat{Q}_r^{L_k}, \tag{11.4.12}$$
$$\widehat{Q}_r^{L_k} \in \mathfrak{G}^{0,r}, \qquad \mathrm{gh}(\widehat{Q}_r^{L_k}) = k - r.$$

Since $\widetilde{\gamma}$ raises the total ghost number by 1, we can write

$$\widetilde{\gamma}T^\iota = \widetilde{C}^L \Delta_L T^\iota, \qquad \Delta_L = R_L^\iota \frac{\partial}{\partial T^\iota}, \tag{11.4.13}$$

$$\widetilde{\gamma}\widetilde{C}^L = \frac{(-1)^{[\widetilde{C}^N]}}{2} \widetilde{C}^N \widetilde{C}^M f_{MN}^L(T) + \widetilde{Q}^{M_2} Z_{M_2}^L \tag{11.4.14}$$

for some functions R, f and Z of tensor fields T.

The equality (11.4.13) decomposes due to $\widetilde{\gamma} = \gamma + d_H$ and (11.4.11) into

$$\gamma T^\iota = \widehat{C}^L \Delta_L T^\iota, \tag{11.4.15}$$
$$d_\lambda T^\iota = \mathcal{A}^L_\lambda \Delta_L T^\iota, \qquad \mathcal{A}^L = dx^\lambda \mathcal{A}^L_\lambda. \tag{11.4.16}$$

Since the relation (11.4.16) holds identically, it implies the splitting

$$\boxed{d_\lambda = v^m_\lambda V^\mu_m (d_\mu - \mathcal{A}^r_\mu \Delta_r) + \mathcal{A}^r_\lambda \Delta_r,} \tag{11.4.17}$$

and the corresponding splittings of the collections of quantities $\{\mathcal{A}^L_\lambda\} = \{v^m_\lambda; \mathcal{A}^r_\lambda\}$ and $\{\Delta_L\} = \{D_m; \Delta_r\}$, where $v^m_\lambda V^\mu_m = \delta^\mu_\lambda$ and

$$D_m = V^\mu_m (d_\mu - \mathcal{A}^r_\mu \Delta_r).$$

Then the equality (11.4.15) is brought into the form

$$\gamma T^\iota = [\widehat{C}^r \Delta_r + \widehat{C}^m V^\mu_m (d_\mu - \mathcal{A}^r_\mu \Delta_r)] T^\iota. \tag{11.4.18}$$

Accordingly, the operator (11.4.13) takes the form

$$\boxed{\widetilde{\gamma} T^\iota = (\widetilde{C}^m D_m + \widetilde{C}^r \Delta_r) T^\iota.} \tag{11.4.19}$$

In view of this form, the quantities $\{\widetilde{C}^N\}$ are called generalized connections or *BRST connections* [37, 211].

Th following two examples show that this terminology is connected with the usual notion of connections used in physical models.

Example 11.4.1. Let us consider again the Yang–Mills theory on a G-principal bundle. The trivial pairs are given by

$$\{U^l\} = \{a^r_{\Lambda+\lambda},\ 0 \leq |\Lambda|\}, \qquad \{V^l\} = \{\widetilde{\gamma} U^l\}.$$

The BRST connections are

$$\{\widetilde{C}^L\} = \{dx^\lambda;\ \widetilde{C}^r = C^r + dx^\lambda a^r_\lambda\}.$$

The operators Δ (11.4.13) corresponding to these BRST connections read

$$\{\Delta_L\} = \{D_\lambda = d_\lambda - a^r_\lambda \epsilon_r;\ \epsilon_r\}$$

where $\{\epsilon_r\}$ are the basis elements for the Lie algebra $\mathfrak{g}_l$ of the group G. A complete set of BRST tensor fields T^ι consists of algebraically independent components of the strength $\mathcal{F}^r_{\lambda\mu}$ (6.2.19) and its covariant differentials $D_\Lambda \mathcal{F}^r_{\lambda\mu}$. •

Example 11.4.2. In the metric gravitation theory, the BRST transformations of a metric $g_{\mu\nu}$ are general covariant transformations whose parameters are ghost fields ξ^μ:

$$\mathbf{s}g_{\mu\nu} = \xi^\lambda d_\lambda g_{\mu\nu} + (d_\mu \xi^\lambda) g_{\lambda\nu} + (d_\nu \xi^\lambda) g_{\mu\lambda}.$$

The BRST transformations of ghost fields read

$$\mathbf{s}\xi^\mu = \xi^\nu d_\nu \xi^\mu.$$

The trivial pairs $\{U^l, V^l\}$ consist of the jets $d_\Lambda \{_\lambda{}^\nu{}_\mu\}$ of the Christoffel symbols and $\widetilde{\gamma} d_\Lambda \{_\lambda{}^\nu{}_\mu\}$. The BRST connections are

$$\{\widetilde{C}^L\} = \{\widetilde{\xi}^\lambda = \xi^\lambda + dx^\lambda;\ \widetilde{C}^\nu_\lambda = d_\lambda \xi^\nu + \{_\lambda{}^\nu{}_\mu\} \widetilde{\xi}^\mu\}.$$

The operators Δ corresponding to these BRST connections read

$$\{\Delta_L\} = \{D_\lambda = d_\lambda - \{_\lambda{}^\mu{}_\nu\} \Delta^\nu_\mu;\ \Delta^\nu_\mu\},$$

where Δ^ν_μ are generators of the group $GL(n, \mathbb{R})$ acting on world indices according to

$$\Delta^\nu_\mu T_\lambda = \delta^\nu_\lambda T_\mu, \qquad \Delta^\nu_\mu T^\lambda = -\delta^\lambda_\mu T^\nu.$$

The set of BRST tensor fields contains the metric $g_{\mu\nu}$ and the algebraically independent components of the curvature $R_{\lambda\mu}{}^\alpha{}_\beta$ and its covariant differentials $D_\Lambda R_{\lambda\mu}{}^\alpha{}_\beta$. The BRST transformation of a BRST tensor field reads

$$\gamma T_\mu = \xi^\nu D_\nu T_\mu + (d_\mu \xi^\nu + \{_\mu{}^\nu{}_\alpha\} \xi^\alpha) T_\nu = \xi^\nu d_\nu T_\mu + (d_\mu \xi^\nu) T_\nu,$$

i.e., this a general covariant transformation of T_μ whose parameters are ghosts ξ^ν. •

Chapter 12

Topological field theories

By the topological field theory is meant usually:

- a collection of Grassmann (ghost number) graded fields on a Riemannian manifold X,
- a nilpotent odd BRST operator Q,
- physical states defined to be Q-homological classes,
- a metric energy-momentum tensor which is Q-exact

(see [30] and references therein for a survey). One characterizes topological field theories as being either of Witten or Schwartz type. The former is exemplified by the Donaldson theory, while the latter includes models whose action functionals are independent of a metric on X, e.g., the Chern–Simons theory.

Here we will concentrate our consideration to the surprising fact that formulas of the curvature of a connection on the space of principal connections are identical to the BRST transformations of the geometric sector of the above mentioned Donaldson theory. As a consequence, the Donaldson invariants play the role of observables in topological field theory.

12.1 The space of principle connections

As was discussed in Section 6.1, principal connections on a G-principal bundle $P \to X$ are represented by global sections of the affine bundle $C \to X$ (6.1.8). They make up an affine *space of principal connections* **A** modelled on the vector space of sections of the vector bundle $\overline{C} = T^*X \otimes V_G P$ (6.1.9). In gauge theory, principal connections are treated as gauge potentials. Two gauge potentials are believed to be physically equivalent if they differ from each other in a vertical automorphism of the principal bundle P. Therefore, the configuration space of quantum gauge theory is the quotient $\mathbf{A}/\mathcal{G}$, where $\mathcal{G} = \mathrm{Gau}(P)$ is the gauge group in Section 6.3. To provide this configuration space with a smooth structure, its Sobolev completion is considered.

Let us recall briefly the notion of a Sobolev space [2, 214, 232]. Given a domain $U \subset \mathbb{R}^n$, let $L^p(U)$, $1 \leq p < \infty$, be the vector space of all measurable real functions on U such that

$$\int_U |f(x)|^p d^n x < \infty.$$

It is a Banach space with respect to the norm

$$\|f\|_p = \left\{ \int_U |f(x)|^p d^n x \right\}^{1/p}.$$

Of course, functions are identified in the space if they are equal almost everywhere in U. Sobolev spaces are defined over an arbitrary domain $U \subset \mathbb{R}$, and are vector subspaces of various spaces $L^p(U)$.

We will follow the standard notation

$$\partial^\alpha = \frac{\partial^{|\alpha|}}{\partial x_1^{\alpha_1} \cdots \partial x_n^{\alpha_n}},$$

where α is an ordered r-tuple of non-negative integers $(\alpha_1, \ldots, \alpha_r)$ and $|\alpha| = \alpha_1 + \cdots + \alpha_r$. Let us define the functional

$$\|f\|_{k,p} = \left\{ \sum_{|\alpha| \leq k} \|\partial^\alpha f\|_p^p \right\}^{1/p}, \qquad k = 0, 1, \ldots, \tag{12.1.1}$$

for any real function f on X for which the right side of (12.1.1) makes sense. The functional (12.1.1) is a norm on any vector space of such functions. The *Sobolev space* $\mathcal{H}^{k,p}(U)$ is defined as the completion of the set

$$\{f \in C^k(U) \,:\, \|f\|_{k,p} < \infty\}$$

with respect to the norm (12.1.1). A Sobolev space is also defined as another space $W^{k,p}(U)$ which is proved to be isomorphic to $\mathcal{H}^{k,p}(U)$ (see [2] for details). It is important that the set of smooth functions

$$\{f \in C^\infty(U) \, : \, \|f\|_{k,p} < \infty\}$$

is dense in $\mathcal{H}^{k,p}(U)$. Moreover, by a Sobolev space is also meant the closure $W_0^{k,p}(U)$ of a set of smooth functions of compact support in $\mathcal{H}^{k,p}(U)$.

The notion of a Sobolev space can be extended to complex functions and to an arbitrary real k. The Sobolev space $\mathcal{H}^{k,p}(U)$, $k \in \mathbb{R}$, consists of those complex functions and Schwartz distributions f for which the norm

$$\|f\|_{k,p} = \left\{ \int |\widehat{f}(\xi)(1+\xi^2)^{k/2}|^p d\xi \right\}^{1/p}, \tag{12.1.2}$$

where $\widehat{f}$ is the Fourier transform of f, is finite. If k is a non-negative integer, this definition is equivalent to the previous one generalized to complex functions [232]. We will deal with the case $p = 2$ and denote $\mathcal{H}^{k,2}(\mathbb{R}^n) = \mathcal{H}^k$. It is a Hilbert space. In particular, one can demonstrate that H^{-k}, $k > 0$, is the dual of H^k so that the elements of $\mathcal{H}^{-k}$ are natural distributions. Indeed, let $\mathcal{D}$ be the space of smooth complex functions of compact support on $\mathbb{R}^n$. It is provided with the topology determines by the seminorms

$$p_{\{\phi_a\}}(f) = \sup_{x\in Q} | \sum_a \phi_a(x) \partial^\alpha f(x)|,$$

where $\{\phi_a\}$ is a collection of smooth functions such that, on any compact subset $Q \subset \mathbb{R}^n$, only a finite number of these functions differ from zero. The space $\mathcal{D}$ is known as the space of *test functions*. Its (topological) dual $\mathcal{D}'$ is the space of *Schwartz distributions*. Let $\mathcal{S}$ be the nuclear *Schwartz space* of smooth functions rapidly decreasing at infinity. Its dual $\mathcal{S}'$ is the space of *tempered distributions* (also called *generalized functions*). There are the inclusions

$$\mathcal{D} \subset \mathcal{S} \subset \cdots \mathcal{H}^k \subset \cdots \mathcal{H}^0 = L^0(\mathbb{R}^n) \subset \mathcal{H}^{-1} \subset \cdots \mathcal{H}^{-k} \subset \cdots \mathcal{S}' \subset \mathcal{D}'.$$

One can define Sobolev spaces of functions on an arbitrary paracompact manifold X by taking a partition of unity and its associated open covering. The following *Sobolev imbedding theorem* takes place [232].

THEOREM 12.1.1. Let $X = \mathbb{R}^n$ or X be compact. Then $\mathcal{H}^{k,p}(X) \subset C^l(X)$ if $k - n/p > l$. In particular, $\mathcal{H}^k \subset C^n(X)$ if $k > n/2$. □

The notion of a Sobolev space can be generalized to sections of a smooth vector bundle $Y \to X$ if they are of compact support. We will restrict our consideration to the case of vector bundles $Y \to X$ over a compact (closed) oriented n-dimensional Riemannian base X. This assumption is satisfied by most Euclidean gauge theories of physical interest. Given the space $Y(X)$ of global sections of $Y \to X$, one defines its *Sobolev completion* $Y(X)_k$ for a non-negative integer k with respect to the norm

$$\|s\|_k = \left(\sum_{\iota} \int_{U_\iota} d\mathrm{vol} \sum_{|\alpha|\leq k} |\partial^\alpha(\mathrm{pr}_2 \circ \psi_\iota \circ s)|^2_\varrho \right)^{1/2}, \qquad (12.1.3)$$

where $\Psi = \{U_\iota, \psi_\iota\}$ is a bundle atlas of Y over a finite covering $\{U_\iota\}$ of X and $|.|_\varrho$ is a norm with respect to a some fibre metric ϱ in Y. A different choice of an atlas Ψ and a Riemannian metric on X gives the same Sobolev completion. Note also that partial derivatives in the expression (12.1.3) can be replaced with covariant derivatives with respect to a connection on Y.

Now let $P \to X$ be a principal bundle whose structure group G is a compact semisimple matrix Lie group. We start from the Sobolev completion of the gauge group $\mathcal{G}$ of vertical automorphisms of P. Recall that $\mathcal{G}$ is the group of global sections of the group bundle P^G (6.3.6). This group acts on the space of principal connections $\mathbf{A}$ by the law (6.3.3). One defines its normal subgroup which is the stabilizer

$$\mathcal{G}^0 = \{\Phi \in \mathcal{G} \,:\, \Phi(x_0) = \mathbf{1}\}$$

of some point $x_0 \in X$ chosen once for all. $\mathcal{G}^0$ is called the *pointed gauge group*. It acts freely on the space of principal connections $\mathbf{A}$. Note that $\mathcal{G}/\mathcal{G}^0 = G$.

One also introduces the *effective gauge group* $\overline{\mathcal{G}} = \mathcal{G}/\mathcal{Z}$, where $\mathcal{Z}$ is the centre of the gauge group $\mathcal{G}$. The centre $\mathcal{Z}$ coincides with the centre $Z(G)$ of the group G. A principal connection A on P is called *irreducible* if its stabilizer $\mathcal{G}_A$ (i.e., $\Phi(A) = A$, $\forall\Phi \in \mathcal{G}_A$) belongs to $\mathcal{Z}$, and this is the generic case. Let $\overline{\mathbf{A}}$ denote the space of irreducible connections. The effective gauge group $\overline{\mathcal{G}}$ acts freely on $\overline{\mathbf{A}}$.

Remark 12.1.1. Let d_A be the Nijenhuis differential (6.1.30) and $*$ the Hodge duality operator with respect to a Riemannian metric on X, extended to $\mathfrak{g}_r$-valued forms $\phi \in \mathfrak{O}^r(X) \otimes V_G(P)$. Of interest to us later will be the fact that, at an irreducible connection $A \in \overline{\mathbf{A}}$, there exists the Green function

$$G_A = (*d_A * d_A)^{-1}$$

of the *covariant Laplacian*

$$\Delta_A = *d_A * d_A$$

since there is no non-trivial solution of the equation

$$d_A\xi = \nabla^A\xi = 0, \qquad \xi \in V_G P(X)$$

(see (6.1.26)). •

Though $P^G \to X$ is not a vector bundle, the Sobolev completion of the gauge group $\mathcal{G}$ can be constructed as follows [224]. Being a matrix Lie group, G is a subset of the algebra $M(l,\mathbb{C})$ of $l \times l$ complex matrices. We introduce the P-associated fibre bundle

$$P^M = (P \times M(l,\mathbb{C}))/G$$

of $l \times l$ matrices, where G has the adjoint action on $M(l,\mathbb{C})$. This is a vector bundle provided with the fibre norm

$$|L|^2 = \mathrm{Tr}\, LL^*, \qquad L \in M(l,\mathbb{C}).$$

Let $P^M(X)_k$ be the Sobolev completion of the space of sections of P^M. Since $\mathcal{G} \subset P^M(X)$, the Sobolev completion $\mathcal{G}_k$ of $\mathcal{G}$ is defined as the completion of $\mathcal{G}$ in the induced metric. If $k > n/2$, then $\mathcal{G}_k$ is closed in $P^M(X)_k$ and the group operations in $\mathcal{G}_k$ are continuous in accordance with Theorem (12.1.1). Thus $\mathcal{G}_k$ (and also $\mathcal{G}^0_k$, $\overline{\mathcal{G}}_k$) are topological groups. Moreover, they are Lie groups. The Lie algebra of the Lie group $\mathcal{G}_k$ is the Sobolev completion of the space of sections of the gauge algebra bundle $V_G P$ (6.1.6) also seen as a subbundle of P^M [224].

The Sobolev completion $\mathbf{A}_k$ of the space of principal connections is defined as the affine space modelled over the Sobolev completion $\overline{C}(X)_k$ of the space of sections of the vector bundle $\overline{C}$ (6.1.9). Therefore, it is a Hilbert manifold whose tangent space $T_A\mathbf{A}_k$ at a point $A \in \mathbf{A}$ is $\overline{C}(X)_k$. The Sobolev completion $\overline{\mathbf{A}}_k$ of the space of irreducible connections is a dense open subset of $\mathbf{A}_k$.

We will hold $k > 1+n/2$. The crucial point is that the group action $\mathcal{G}_{k+1} \times \mathbf{A}_k \to \mathbf{A}_k$ is smooth, and so are the free actions

$$\mathcal{G}^0_{k+1} \times \mathbf{A}_k \to \mathbf{A}_k, \qquad \overline{\mathcal{G}}_{k+1} \times \overline{\mathbf{A}}_k \to \overline{\mathbf{A}}_k.$$

Moreover, the quotient $\mathcal{O}_k = \overline{\mathbf{A}}_k/\overline{\mathcal{G}}_{k+1}$, called the *orbit space*, is a smooth Hilbert manifold, while the canonical surjection $\overline{\mathbf{A}}_k \to \mathcal{O}_k$ is a smooth fibre bundle (i.e., locally trivial). This is a principal fibre bundle with the structure group $\overline{\mathcal{G}}_{k+1}$ [224]. Note that the quotient $\mathbf{A}_k/\mathcal{G}^0_{k+1}$ is also a smooth Hilbert manifold, while the topological space $\mathbf{A}_k/\mathcal{G}_{k+1}$ admits a stratification into smooth Hilbert manifolds [180] (see also [113, 115, 151]).

In quantum gauge theory, the orbit space $\mathcal{O}$ is well-known to play the role of a configuration space. However, the mathematical nature of this space is essentially unknown. In particular, an important problem is the existence of a global section of the principal bundle $\overline{\mathbf{A}}_k \to \mathcal{O}_k$. If a global section s exists, integration over $\mathcal{O}$ can be replaced with that over $s(\mathcal{O}) \subset \overline{\mathbf{A}}$ with a suitable weight factor such as the Faddeev–Popov determinant. One can think of s as being a *global gauge*. Its non-existence is referred to as the *Gribov ambiguity*. The Gribov ambiguity is proved to take place in a number of gauge models where the principal bundle $\overline{\mathbf{A}}_k \to \mathcal{O}_k$ is non-trivial (see, e.g., [58, 141, 231, 281]).

12.2 Connections on the space of connections

From now on, we believe that all objects requiring Sobolev completions have been completed in appropriate norms, and omit the index k.

The Hilbert manifold $\mathcal{O}$ is modelled on the Hilbert space isomorphic to $\mathrm{Ker}{*}d_A{*}$ for any $A \in \overline{\mathbf{A}}$, where $*d_A*$ acts on $\overline{C}(X)$. Given the fibre bundle

$$\boxed{\overline{\mathbf{A}} \to \mathcal{O},} \tag{12.2.1}$$

we have the canonical splitting of the tangent spaces

$$\boxed{T_A\overline{\mathbf{A}} = V_A\overline{\mathbf{A}} \oplus \mathrm{Ker} * d_A *,} \tag{12.2.2}$$
$$\sigma = d_A G_A * d_A * \sigma + (\sigma - d_A G_A * d_A * \sigma).$$

This splitting defines the canonical connection $\overline{A}$ on the fibre bundle (12.2.1).

Let us consider the product $P \times \overline{\mathbf{A}}$. It is a Hilbert manifold whose tangent space at a point (p, A) is $T_pP \times \overline{C}(X)$. There is a natural action of the effective gauge group $\overline{\mathcal{G}}$ on $P \times \overline{\mathbf{A}}$ which has no fixed points. Therefore

$$\boxed{P \times \overline{\mathbf{A}} \to (P \times \overline{\mathbf{A}})/\overline{\mathcal{G}} = \mathcal{Q}} \tag{12.2.3}$$

is a $\overline{\mathcal{G}}$-principal bundle. Since the action of G on $P \times \overline{\mathbf{A}}$ commutes with that of $\overline{\mathcal{G}}$, the group G acts on $\mathcal{Q}$, and we also have the G-principal bundle

$$\boxed{\mathcal{Q} \to \mathcal{Q}/G = X \times \mathcal{O},} \tag{12.2.4}$$

called the *universal bundle* [13, 30, 284].

The fibre bundles (12.2.1), (12.2.3) and (12.2.4) lead to the commutative diagram of fibrations

$$\begin{array}{ccc} P\times\overline{\mathbf{A}} & \longrightarrow & \mathcal{Q} \\ \downarrow & & \downarrow \\ X\times\overline{\mathbf{A}} & \longrightarrow & X\times\mathcal{O} \end{array} \tag{12.2.5}$$

The left column

$$\boxed{P\times\overline{\mathbf{A}}\to X\times\overline{\mathbf{A}}} \tag{12.2.6}$$

of this diagram is a G-principal bundle (similar to the principal bundle P_C (6.2.5)). It is provided with the canonical connection $\widehat{A}$ given by the splitting

$$\boxed{\dot{x}^\mu\partial_\mu+\sigma+v^qe_q=\dot{x}^\mu(\partial_\mu+A^q_\mu(x)e_q)+\sigma+(v^q-\dot{x}^\mu A^q_\mu(x))e_q} \tag{12.2.7}$$

of the exact sequence

$$0\to V_G(P\times\overline{\mathbf{A}})\hookrightarrow T_G(P\times\overline{\mathbf{A}})\to(P\times\overline{\mathbf{A}})\underset{X\times\overline{\mathbf{A}}}{\times}T(X\times\overline{\mathbf{A}})$$

at each point $(x,A)\in X\times\overline{\mathbf{A}}$ (cf. the universal connection (6.2.4)). Another connection $\widehat{A}$ on the fibre bundle (12.2.6) is given by the splitting

$$\begin{aligned}\dot{x}^\mu\partial_\mu+\sigma+v^qe_q&=\dot{x}^\mu(\partial_\mu+A^q_\mu(x)e_q)+(\sigma-(*d_A*\sigma)(x))+\\&(v^q-\dot{x}^\mu A^q_\mu(x))e_q+(*d_A*\sigma)(x).\end{aligned} \tag{12.2.8}$$

Combining the connections $\overline{A}$ on (12.2.1) and $\widehat{A}$ on (12.2.6) defines a composite connection $\widehat{A}\circ(\mathrm{Id}\,X,\overline{A})$ on the composite fibre bundle

$$P\times\overline{\mathbf{A}}\to X\times\overline{\mathbf{A}}\to X\times\mathcal{O}. \tag{12.2.9}$$

In particular, connections $\widehat{A}$ (12.2.7) and (12.2.8) lead to the same composite connection on (12.2.9).

Let us write a connection $\widehat{A}$ on the G-principal bundle (12.2.6) as the $T_G(P\times\overline{\mathbf{A}})$-valued form $\widehat{A}=A+c$ where A and c are forms on X and $\overline{\mathbf{A}}$, respectively. We will say that they are (1,0)- and (0,1)-forms. Likewise one splits the exterior differential $\widehat{d}$ on $X\times\overline{\mathbf{A}}$ as

$$\boxed{\widehat{d}=d+\delta,} \tag{12.2.10}$$
$$\delta^2=0,\qquad d\circ\delta+\delta\circ d=0.$$

Then the strength of the connection $\widehat{A}$ (see (6.1.20), (6.1.31)) reads

$$\widehat{F} = \frac{1}{2} d_{\widehat{A}} \widehat{A} = \widehat{F}_{(2,0)} + \widehat{F}_{(1,1)} + \widehat{F}_{(0,2)}, \tag{12.2.11}$$

where

$$\widehat{F}_{(2,0)} = \frac{1}{2} d_A A = F_A, \qquad \widehat{F}_{(1,1)} = \frac{1}{2}(\delta_c A + d_A c), \qquad \widehat{F}_{(0,2)} = \frac{1}{2}\delta_c c.$$

The corresponding local expressions are

$$\widehat{F}_{(1,1)} = \delta A + d_A c, \tag{12.2.12}$$

$$\widehat{F}_{(0,2)} = \delta c + \frac{1}{2}[c, c], \tag{12.2.13}$$

where A and c are local connection forms. Since $\delta^2 = 0$, (12.2.12) and (12.2.13) also imply

$$\delta \widehat{F}_{(1,1)} = -[c, \widehat{F}_{(1,1)}] - d_A \widehat{F}_{(0,2)}, \tag{12.2.14}$$

$$\delta \widehat{F}_{(0,2)} = -[c, \widehat{F}_{(0,2)}]. \tag{12.2.15}$$

Put $\psi = \widehat{F}_{(1,1)}$ and $\phi = \widehat{F}_{(0,2)}$. Then the relations (12.2.12) – (12.2.15) are formally identical to the BRST transformations

$$\delta A = \psi - d_A c, \tag{12.2.16}$$

$$\delta c = \phi - \frac{1}{2}[c, c], \tag{12.2.17}$$

$$\delta \psi = -[c, \psi] - d_A \phi, \tag{12.2.18}$$

$$\delta \phi = -[c, \phi] \tag{12.2.19}$$

of fields (A, c, ψ, ϕ) in the geometric sector of the Donaldson theory. These fields are characterized by the ghost numbers 0, 1, 1 and 2, respectively, which coincide with their $(0, k)$ form degrees [30].

If the strength components $\widehat{F}_{(1,1)}$ and $\widehat{F}_{(2,2)}$ vanish, the equations (12.2.16) – (12.2.19) reduce to

$$\delta A = -d_A c, \qquad \delta c = -\frac{1}{2}[c, c]. \tag{12.2.20}$$

These equations look like the BRST transformations (11.4.2) of the Yang–Mills theory (see Section 11.4). If $\widehat{A}$ is the canonical connection (12.2.7) where $c = 0$, the equations (12.2.16) – (12.2.19) take the form

$$\boxed{\delta A = \psi, \qquad \delta \psi = -d_A \phi, \qquad \delta \phi = 0.}$$

They are identical to the BRST transformations used by Witten [307].

The equations (12.2.16) – (12.2.19) imply

$$\widehat{d}\mathrm{Tr}(\widehat{F}\wedge\widehat{F}) = (d+\delta)\mathrm{Tr}(\overset{2}{\wedge}(F_A+\psi+\phi) = 0) \tag{12.2.21}$$

as a consequence of the Bianchi identity

$$\widehat{d}_{\widehat{A}}\widehat{F} = (d+\delta)(F_A+\psi+\phi) + [A+c, F_A+\psi+\phi] = 0. \tag{12.2.22}$$

From the geometric viewpoint, the equality (12.2.21) illustrates the fact that gauge invariant polynomials are closed forms (see Sections 6.7 and 13.1). Let us write

$$\frac{1}{2}\mathrm{Tr}(\overset{2}{\wedge}(F_A+\psi+\phi)) = \sum_{i=0}^{4} w_i, \tag{12.2.23}$$

where w_i are i-forms on X with the ghost number $4-i$ given by

$$w_0 = \frac{1}{2}\mathrm{Tr}(\phi\wedge\phi), \qquad w_1 = \mathrm{Tr}(\psi\wedge\phi), \qquad w_2 = \mathrm{Tr}(F_A\wedge\phi + \frac{1}{2}\psi\wedge\psi),$$
$$w_3 = \mathrm{Tr}(F_A\wedge\psi), \qquad w_4 = \frac{1}{2}\mathrm{Tr}(F_A\wedge F_A).$$

Then the equality (12.2.21) can be expanded in terms of certain ghost number and form degree as the *descent equations*

$$dw_4 = 0, \tag{12.2.24a}$$
$$\delta w_k + dw_{k-1} = 0, \quad k = 1,\ldots,4, \tag{12.2.24b}$$
$$\delta w_0 = 0. \tag{12.2.24c}$$

These equations are similar to the descent equations (11.4.6a) – (11.4.6c)), while (13.2.13) is the analog of (11.4.7). One can say that w_k, $k = 1,\ldots,4$, are locally BRST-closed.

Given a k-cycle γ in X, one can construct the exterior $(4-k)$-form

$$w_k(\gamma) = \int_\gamma w_k, \qquad k = 1,\ldots,4, \tag{12.2.25}$$

on the space of reducible connections $\overline{\mathbf{A}}$. This form is metric independent and gauge invariant. Due to the equality (12.2.24b), it is *BRST-closed*

$$\delta w_k(\gamma) = -\int_\gamma dw_{k-1} = -\int_{\partial\gamma} w_{k-1} = 0$$

and, therefore, an observable in the topological field theory. Moreover, $w_k(\gamma)$ are *BRST-exact* since $\overline{\mathbf{A}}$ is contractible. We have locally

$$\mathrm{Tr} \overset{2}{\wedge} \widehat{F} = \widehat{d}\,\mathrm{Tr}(\widehat{A} \wedge \widehat{F} - \frac{1}{3} \overset{3}{\wedge} \widehat{A}),$$

where $\widehat{A}$ is the local connection form (6.1.17). It follows that the forms $w_k(\gamma)$ are homological in the sense that they only depend on the homology class of γ because

$$w_k(\partial\lambda) = \int_\lambda dw_k = -\delta \int_\lambda w_{k+1} = 0.$$

Let $(\gamma_1, \ldots, \gamma_r)$ be a collection of k_i-cycles in $\overline{\mathbf{A}}$, and let M be a compact submanifold of $\overline{\mathbf{A}}$ of dimension

$$m = \sum_{i=1}^{r}(4 - k_i).$$

Then we obtain the invariant

$$Z : H_{k_1}(X;\mathbb{Z}) \times \cdots \times H_{k_r}(X;\mathbb{Z}) \to \mathbb{R},$$
$$Z(\gamma_1, \ldots, \gamma_r) = \int_M w_{k_1}(\gamma_1) \wedge \cdots \wedge w_{k_r}(\gamma_r).$$

The problem lies in the fact that there is no finite-dimensional compact submanifold $M \subset \overline{\mathbf{A}}$ of physical interest. Therefore, let us turn to the orbit space $\mathcal{O}$.

Since the connections (12.2.2) and (12.2.7) are principal and the groups G and $\overline{\mathcal{G}}$ acting on $P \times \overline{\mathbf{A}}$ commute with each other, the tangent morphism

$$T(P \times \overline{\mathbf{A}}) \to T\mathcal{Q}$$

to the fibration (12.2.3) yields the splitting of $T\mathcal{Q}$ which is a principal connection $\widetilde{A}$ on the universal bundle $\mathcal{Q} \to X \times \mathcal{O}$ (12.2.4). This connection is characterized by the following property. Let

$$s : \mathcal{O} \supset \mathcal{U} \to \overline{\mathbf{A}}$$

be a local gauge. Then the restriction $i^*_{\mathcal{U}}\mathcal{Q}$ of the fibre bundle $\mathcal{Q}$ to $\mathcal{U}$ is isomorphic to the pull-back bundle $(\mathrm{Id}\, X, s)^*(P \times \overline{\mathbf{A}})$, while the connection $\widetilde{A}$ coincides locally with the pull-back connection $(\mathrm{Id}\, X, s)^*\widehat{A}$.

A natural gauge is a background gauge fixing

$$\boxed{d_{A_0} * (A - A_0) = 0.} \tag{12.2.26}$$

This gauge and the condition

$$\boxed{d_A * \psi = 0} \tag{12.2.27}$$

project the topological field theory down from $\overline{\mathbf{A}}$ to $\mathcal{O}$. The exterior differential δ of the equation (12.2.26) gives the equation

$$d_{A_0} * (\psi - d_A c) = 0$$

whose solution is the connection

$$c = (*d_{A_0} * d_A)^{-1} d_{A_0}\psi.$$

The exterior differential δ of the equation (12.2.27) is

$$[\psi, *\psi] + d_A * d_A\phi = 0$$

which implies that

$$\boxed{\phi = -G_A[\psi, *\psi].}$$

This looks like the strength of the *Atiyah–Singer connection* on the universal bundle (12.2.4) [13].

In the topological field theory, one also introduces the condition

$$\boxed{\mathfrak{F}(A) = 0,} \tag{12.2.28}$$

where $\mathfrak{F}$ is a gauge-invariant differential operator on A. This condition singles out a *moduli subspace* $\mathcal{M} \subset \mathcal{O}$ of the orbit space $\mathcal{O}$. Let X be a 4-dimensional compact manifold. Some standard choices for $\mathfrak{F}$ are

$$\mathfrak{F}(A) = F_A^+ = \frac{1}{2}(F_A - *F_A), \tag{12.2.29}$$

$$\mathfrak{F}(A) = F_A, \tag{12.2.30}$$

$$\mathfrak{F}(A) = d_A * F_A. \tag{12.2.31}$$

The corresponding moduli spaces are the moduli spaces of instantones, flat connections and solutions of the Yang–Mills equations, respectively.

12.3 Donaldson invariants

Throughout this Section, X is a 4-dimensional compact oriented smooth or topological manifold, and $P \to X$ is a $SU(2)$-principal bundle.

Remark 12.3.1. Donaldson invariants are differential, but not topological invariants of X [87]. Therefore, let us summarize the well-known peculiarities of 4-dimensional manifolds [88, 110].

- Every topological manifold of less than four dimension has a unique smooth structure.
- In more than four dimension, the homotopy type and the Pontryagin classes of a manifold determine the smooth structure (if it exists) up to a finite ambiguity.
- There are closed topological 4-dimensional manifolds with a countably infinite number of distinct smooth structures.
- There is an uncountable family of distinct smooth structures on $\mathbb{R}^4$, while $\mathbb{R}^{n\neq 4}$ has a unique smooth structure.
- There are rational cohomology invariants (exemplified by the Donaldson polynomials below) which distinguish inequivalent smooth structure, in contrast with the rational Pontryagin classes which are homotopic invariants (see Remark 6.7.1).

The classification problem of smooth structures of 4-dimensional manifolds is non-algorithmic because of the fundamental group $\pi_1(X)$ [108]. Therefore, interest is mainly centred around simply connected manifolds where $\pi_1(X) = 0$. The fundamental invariant of a simply connected 4-dimensional topological manifold X is the *intersection form* ω_X. It is a symmetric bilinear form on the cohomology group $H^2(X;\mathbb{Z})$ defined by

$$\omega_X : H^2(X;\mathbb{Z}) \times H^2(X;\mathbb{Z}) \to \mathbb{Z},$$
$$\boxed{\omega_X : ([a],[b]) \mapsto ([a] \cup [b])[X],} \tag{12.3.1}$$

where $[a] \cup [b]$ is the cup-product and $([a] \cup [b])[X]$ denotes its evaluation on the *fundamental cycle* of X which is the generating element of the homology group

$H_n(X;\mathbb{Z}) = \mathbb{Z}$ of an oriented manifold X. If X is a smooth manifold, the cup product is given by (6.8.5), and the intersection form (12.3.1) reads

$$\omega_X([a],[b]) = \int_X a \wedge b.$$

The intersection form (12.3.1) is non-degenerate and unimodular. It is characterized by the rank $\dim H^2(X;\mathbb{Z})$, which coincides with the second *Betti number* $b_2(X)$ of X, and by the signature

$$\tau(\omega_X) = b_2^+ - b_2^-$$

which is the number of positive minus the number of negative eigenvalues of ω_X. The intersection form ω_X is called *even* if all its diagonal entries $\omega_X([a],[a])$ are even. If X is a smooth manifold, *Hirzebruch's signature theorem* [157] expresses $\tau(\omega_X)$ in terms of the Pontryagin class of X as

$$\tau(\omega_X) = \frac{1}{3}p_1(X).$$

It is called the *signature* $\tau(X)$ of the manifold X. Intersection forms are topological, not only homotopic invariants. Namely, the homeomorphism type of a topological manifold X is uniquely determined by ω_X if ω_X is even, while there are precisely two non-homeomorphic topological manifolds for a given odd ω_X [109]. In particular, there are intersection forms of a 4-dimensional topological manifold which cannot arise as an intersection form of a smooth manifold. By virtue of well-known *Donaldson's theorem* [86], if the intersection form ω_X of a smooth compact (not necessarily simply connected) manifold X is negative definite, then

$$\omega_X \simeq (-1) \oplus \cdots \oplus (-1).$$

•

The main ingredient in the construction of Donaldson invariants is the map

$$\boxed{\mu : H_i(X;\mathbb{Z}) \to H^{4-i}(\mathcal{O}).} \tag{12.3.2}$$

It can be described as follows. Given the universal bundle $\mathcal{Q} \to X \times \mathcal{O}$ (12.2.4), let E be the associated $SU(2)$-bundle and $c_2(E) \in H^4(\mathcal{O})$ its second Chern class. In accordance with the Künneth formula (6.8.4), $c_2(E)$ is decomposed into terms

$$c_{i,4-i} \in H^i(X) \otimes H^{4-i}(\mathcal{O}), \qquad i = 0, \ldots, 4.$$

These terms define the map (12.3.2) as the composition of the homomorphism (6.8.8) and the De Rham duality (6.8.9). Let $\widetilde{A}$ be a principal connection on the universal bundle (12.2.4), $\widetilde{F}$ its curvature and $c_2(\widetilde{F})$ the second Chern form (6.7.13). The above mentioned terms $c_{i,4-i}$ are the De Rham cohomology classes of the $(i,4-i)$-forms $\widetilde{w}_i$ on $X\times\mathcal{O}$ which make up the decomposition of $c_2(\widetilde{F})$ similar to the decomposition (12.2.23). Then the map (12.3.2) is given by the integration

$$\widetilde{w}_i(\gamma)=\int_\gamma \widetilde{w}_i,$$

where γ are i-cycles in X. Note that the rational cohomology classes of $\mathcal{O}$ lie in even dimensions, and are generated by cohomology classes in two and four dimensions. Therefore, we restrict our consideration to the map (12.3.2) where $i=0,2$, and can extend it to the map

$$\boxed{\mu:\overset{m}{\times} H_2(X;\mathbb{Z})\to H^{2m}(\mathcal{O};\mathbb{Q})}$$

via the cup product in $H^{2m}(\mathcal{O};\mathbb{Q})$. This map provides the injection

$$q([\gamma_1],\ldots,[\gamma_m])\mapsto \mu([\gamma_1])\cup\cdots\cup\mu([\gamma_m]) \tag{12.3.3}$$

from the polynomial algebra on $H_2(X;\mathbb{Z})$ into $H^{\rm even}(\mathcal{O};\mathbb{Q})$.

Let $\mathcal{M}$ be the moduli space of irreducible instantones of the instanton number

$$k=\int_X c_2(F_A).$$

Its formal dimension is

$$\dim\mathcal{M}=8k-3(1+b_2^+).$$

It is readily observed that $\dim\mathcal{M}$ is even if and only if b_2^+ is odd. Writing $b_2^+=2p+1$, we have

$$\dim\mathcal{M}=2m,\qquad m=4k-3(1+p).$$

Polynomials (12.3.3) when evaluated on the homology cycle $[\mathcal{M}]\in H_*(\mathcal{O},\mathbb{Q})$ in $\mathcal{O}$ are called the *Donaldson polynomials.* They are expressed into the strength $\widetilde{F}$ of a connection $\widetilde{A}$ on the orbit space $\mathcal{O}$. Treating $\overline{A}$ locally as the pull-back of a connection on $\overline{\mathbf{A}}$ and using the relations (12.2.26) and (12.2.27), we obtain the

Donaldson invariants in the topological field theory, though they are very hard to compute explicitly in general.

Remark 12.3.2. Note that the instanton moduli space $\mathcal{M}$ is usually non-compact. This difficulty can be overcome for a suitably generic metrics which provide that the lower dimensional strata of the compactified moduli space $\overline{\mathcal{M}}$, $k > 1$, are of high enough codimension so as not contribute to the evaluation of compact supported cohomology classes of $\mathcal{O}$ on $\mathcal{M}$ [30]. •

Chapter 13

Anomalies

The anomaly problem lies in violation of conservation laws for quantized fields and the gauge non-invariance of an effective action and a path integral measure in the perturbative quantum field theory (see [29] and references therein for a survey). Here we are only concerned with the geometric origin of anomalies, based on geometry and topology of spaces of principal connections.

13.1 Gauge anomalies

This Section is devoted to anomalies related to the gauge non-invariance of the Chern–Simons form.

Let $P \to X$ be a $GL(N, \mathbb{C})$-principal bundle. Characteristic forms in Section 6.7 exemplify *gauge invariant polynomials* $\mathcal{P}(F)$ in the strength F (6.1.19) of a principal connection on $P \to X$, i.e.,

$$\mathcal{P}(F) = \mathcal{P}(g(F)), \qquad g \in \mathcal{G},$$

where $\mathcal{G}$ is the gauge group of vertical automorphisms of P. Gauge invariant polynomials are complex exterior forms of even degree on X. They possess the properties mentioned in Section 6.7:

- $\mathcal{P}(F)$ is a closed form,
- $\mathcal{P}(F) - \mathcal{P}(F')$ is an exact form, whenever F and F' are strength forms of two distinct principal connections on P.

One can say something more. Any gauge invariant polynomial is a sum of products of gauge invariant polynomials $\mathcal{P}_m(F)$ of definite degree m in F. Then the *transgression formula* [6, 29, 98] takes place

$$\boxed{\mathcal{P}_m(F) - \mathcal{P}_m(F') = dQ_{2m-1}(A, A'),} \tag{13.1.1}$$

$$Q_{2m-1}(A, A') = m \int_0^1 dt P(A - A', F_t), \tag{13.1.2}$$

where F_t is the strength of the *homotopic connection*

$$A_t = A' + t(A - A'), \qquad t \in [0, 1]. \tag{13.1.3}$$

In particular, put $A' = 0$ on a trivialization chart of $P \to X$ (i.e., $A' = \theta_X$ as a $T_G P$-valued form). Then we obtain the *local transgression formula*

$$\boxed{\mathcal{P}_m(F) = dQ_{2m-1}(A, F),} \tag{13.1.4}$$

together with the *Chern–Simons form*

$$Q_{2m-1}(A, F) = m \int_0^1 dt P(A, F_t), \tag{13.1.5}$$

where A is the local connection form (6.1.17) and

$$A_t = tA, \qquad F_t = tF + (t^2 - t)A^2.$$

For the sake of simplicity, we omit the symbol $\wedge$ of the exterior product.

Example 13.1.1. If $\mathcal{P}(F) = c_2(F)$ is the second Chern form (6.7.13) for an $SU(N)$-principle bundle, we have

$$\mathrm{Tr}(F^2) = \frac{1}{8\pi^2} dQ_3,$$

$$\boxed{Q_3 = \mathrm{Tr}(AF - \tfrac{1}{3}A^3) = \mathrm{Tr}(AdA + \tfrac{2}{3}A^3).} \tag{13.1.6}$$

•

Let us obtain the transgression formula (13.1.1) in terms of the homotopy derivation [6, 29, 211]. Given the homotopic connection A_t (13.1.3) and its strength F_t, the *homotopy derivation* is defined as the operator

$$l_t A_t = 0, \qquad l_t F_t = d_t A_t = dt(A - A') \tag{13.1.7}$$

which acts on polynomials $S(A_t, F_t)$ in A_t and F_t and satisfies the antiderivation rule

$$l_t(SS') = l_t(S)S' + (-1)^{|S|}Sl_t(S').$$

One can verify explicitly that

$$l_t \circ d + d \circ l_t = d_t = dt \otimes \partial_t. \tag{13.1.8}$$

Remark 13.1.2. Note that polynomials $S(A_t, F_t)$ fail to be globally defined, unless they are gauge invariant polynomials. Therefore, by their argument A_t is meant the local connection form (6.1.17). •

Let us introduce the operator

$$\mathbf{k} = \int_0^1 l_t. \tag{13.1.9}$$

It is called the *homotopy operator* by analogy with the homotopy operator in Remark 4.1.2. We have the *Cartan homotopy formula*

$$S(A, F) - S(A', F') = (\mathbf{k} \circ d + d \circ \mathbf{k})S(A_t, F_t). \tag{13.1.10}$$

Applied to gauge invariant polynomials $\mathcal{P}_m(F)$, the homotopy operator (13.1.9) reads

$$\mathbf{k}\mathcal{P}_m(F_t) = Q_{2m-1}(A, A')$$

where the polynomial $P(A - A', F_t)$ in the expression (13.1.2) is defined as

$$mP(A - A', F_t) = \mathcal{P}_m(A - A', F_t, \ldots, F_t) + \mathcal{P}_m(F_t, A - A', \ldots, F_t) + \cdots + \mathcal{P}_m(F_t, \ldots, A - A'). \tag{13.1.11}$$

Since gauge invariant polynomials are closed, the Cartan homotopy formula (13.1.10) leads to the transgression formula (13.1.1). Choosing locally $A' = 0$, we obtain the Chern–Simons form (13.1.5) as

$$\boxed{Q_{2m-1}(A, F) = \mathbf{k}\mathcal{P}_m(F_t).} \tag{13.1.12}$$

Let us consider gauge transformations of the Chern–Simons form (13.1.12). In the case of a matrix structure group $G \subset GL(N, \mathbb{C})$, gauge transformations of the local connection form A (6.1.17) and the strength F (6.1.20) read

$$A^g = g^{-1}Ag + g^{-1}dg = g^{-1}(A + \sigma^g)g, \qquad F^g = g^{-1}Fg. \tag{13.1.13}$$

Let us choose the homotopic connection $A_t = tA$, $t \in [0, 1]$. Then

$$A_t^g = (A_t)^g, \qquad F_t^g = (F_t)^g$$

is the homotopy which interpolates continuously between

$$A_{t=0}^g = g^{-1}dg = g^{-1}\sigma^g g, \qquad F_{t=0}^g = 0$$

and

$$A_{t=1}^g = A^g, \qquad F_{t=1}^g = F^g.$$

Applied to the Chern–Simons form containing these homotopies, the Cartan homotopy formula (13.1.10) reads

$$Q_{2m-1}(A^g, F^g) - Q_{2m-1}(g^{-1}dg) = (\mathbf{k} \circ d + d \circ \mathbf{k})Q_{2m-1}(A_t^g, F_t^g) \tag{13.1.14}$$

The gauge transformed Chern–Simons form is

$$Q_{2m-1}(A^g, F^g) = m \int_0^1 dt P(A^g, \widehat{F}_t^g),$$

$$\widehat{F}_t^g = (F^g)_t = g^{-1}\widehat{F}_t g, \qquad \widehat{F}_t = tF + (t^2 - t)(A + \sigma^g)^2.$$

Since $P(A^g, \widehat{F}_t^g)$ is expressed into the gauge invariant polynomials $\mathcal{P}_m$ as (13.1.11), we have

$$P(A^g, \widehat{F}_t^g) = P(A, \widehat{F}_t).$$

It follows that

$$\boxed{Q_{2m-1}(A^g, F^g) = Q_{2m-1}(A + \sigma^g, F)} \tag{13.1.15}$$

and, analogously,

$$Q_{2m-1}(A_t^g, F_t^g) = Q_{2m-1}(A_t + \sigma^g, F_t). \tag{13.1.16}$$

On the other hand, the relations (13.1.4) and (13.1.12) give

$$(\mathbf{k}\circ d)Q_{2m-1}(A^g_t,F^g_t)=\mathbf{k}\mathcal{P}_m(F_t)=Q_{2m-1}(A,F).$$

If we define

$$\alpha_{2m-2}=\mathbf{k}Q_{2m-1}(A^g_t,F^g_t)=\mathbf{k}Q_{2m-1}(A_t+\sigma^g,F_t),$$

the Cartan homotopy formula (13.1.14) takes the form

$$\boxed{Q_{2m-1}(A^g,F^g)=Q_{2m-1}(A,F)+Q_{2m-1}(g^{-1}dg,0)+d\alpha_{2m-2}} \tag{13.1.17}$$

or, by virtue of the relation (13.1.15),

$$\boxed{Q_{2m-1}(A+\sigma^g,F)=Q_{2m-1}(A,F)+Q_{2m-1}(\sigma^g,0)+d\alpha_{2m-2}.} \tag{13.1.18}$$

For instance, let Q_3 be the Chern–Simons form (13.1.6) in Example 13.1.1. For $m=2$, we have

$$\alpha_2=\mathbf{k}Q_3(A_t+\sigma^g,F_t)=\int_0^1 l_t\mathrm{Tr}[(A_t+\sigma^g)F_t-\frac{1}{3}(A_t+\sigma^g)^3]=$$

$$\int_0^1 dt\mathrm{Tr}[-tA^2-\sigma^gA]=-\mathrm{Tr}[\sigma^gA]$$

since $\mathrm{Tr}\,A^2=0$. Then, keeping only the term linear in σ^g, we obtain

$$\boxed{d\alpha_2=\mathrm{Tr}[\sigma^g dA].} \tag{13.1.19}$$

This equation expresses the well-known non-Abelian *anomaly* in two dimensions apart from the normalization [29].

The anomaly (13.1.19) characterizes the non-invariance of the Chern–Simons form (13.1.6) under infinitesimal gauge transformations. Therefore, it can also be calculated as follows. Let us regard the Chern–Simons form Q_3 as a V_GP-valued form on the first order jet manifold J^1C of the bundle of principal connections C (see Section 6.2). It reads

$$Q_3=a^G_{pr}a^p_\alpha(\mathcal{F}^r_{\lambda\mu}-\frac{1}{3}c^r_{lq}a^l_\lambda a^q_\mu)dx^\alpha\wedge dx^\lambda\wedge dx^\mu, \tag{13.1.20}$$

where $a^G_{pr} = \mathrm{Tr}(\varepsilon_p\varepsilon_r)$ is the invariant metric on the Lie algebra $su(N)$. Let ξ_C be the principal vector field (6.3.9) on the fibre bundle C. It is the generator of a 1-parameter group of gauge transformations of C. We have the Lie derivative

$$\boxed{\mathbf{L}_{J^1\xi_C}Q_3 = \mathrm{Tr}(d\xi\wedge d_H a),}$$
$$\xi = \xi^p\varepsilon_p, \qquad a = a^r_\mu dx^\mu\otimes\varepsilon_r,$$

It is readily observed that, if $\sigma^g = d\xi$, then

$$d\alpha_2 = A^*(\mathbf{L}_{J^1\xi_C}Q_3)$$

for any section A of the fibre bundle $C\to X$.

Remark 13.1.3. If $\dim X = 3$, one regards Q_3 (13.1.20) as a Lagrangian of the Chern–Simons topological field model. Being gauge non-invariant, this Lagrangian is not globally defined. Since $\mathbf{L}_{J^1\xi_C}Q_3\neq 0$, neither the Noether current (6.3.21) nor the energy-momentum current (7.4.22) are conserved in the Chern–Simons model [123] •

13.2 Global anomalies

In comparison with gauge anomalies in the previous Section, global anomalies are connected with gauge transformations not in the connected component of the identity.

We start from the notion of the group cohomology [204]. Let G be a multiplicative group, B a right G-module, and B^p, $p = 1,\ldots,$ Abelian groups of morphisms

$$B^p\ni b^p : \overset{p}{\times} G\to B. \tag{13.2.1}$$

One can introduce the coboundary operator

$$\begin{aligned}&\delta^p : B^p\longrightarrow B^{p+1},\\ &(\delta^p b^p)(g_1,\ldots,g_{p+1}) = b^p(g_2,\ldots,g_{p+1})g_1 +\\ &\qquad\sum_{i=1}^p(-1)^i b^p(g_1,\ldots,g_ig_{i+1},\ldots g_{p+1}) + (-1)^{p+1}b^p(g_1,\ldots,g_p).\end{aligned}\tag{13.2.2}$$

With this coboundary operator, we have the cochain complex

$$0\to B\xrightarrow{\delta^0} B^1\xrightarrow{\delta^1} B^2\longrightarrow\cdots. \tag{13.2.3}$$

By B in (13.2.3) is meant the group of constant morphisms

$$B \ni b : G \to b,$$

and

$$(\delta^0 b)(g) = bg - b.$$

It follows that 0-cocycles in B are G-invariant elements of B. We will denote the set of these elements by B^G. The cohomology $H^*(G;B)$ of the complex (13.2.3) is called the *group cohomology* of the group G with coefficients in the module B. For instance, $H^0(G;B) = B^G$. Note that the Abelian group B^p of p-cochains (13.2.1) is a right G-module with respect to the G-action

$$g : b^p(g_1, \ldots, g_p) \mapsto b^p(g^{-1}g_1 g, \ldots, g^{-1}g_p g).$$

This action is trivial on $H^*(G;B)$.

Let H be a normal subgroup of G. One has the exact sequence

$$\begin{aligned} 0 &\longrightarrow H^1(G/H; B^H) \xrightarrow{j} H^1(G;B) \xrightarrow{i} H^1(H;B)^G \longrightarrow \\ &H^2(G/H;B^H) \xrightarrow{j} H^2(G;B). \end{aligned} \tag{13.2.4}$$

The arrow j in (13.2.4) is the composition of the projection $G \to G/H$ with the cocycles of the group G/H, while the arrow i is the restriction of the cocycles of G to H [55].

If G/H is a finite group of order r, there exists a homomorphisms

$$\varrho^* : H^*(H;B) \to H^*(G;B). \tag{13.2.5}$$

For instance, $\varrho^0 : B^H \to B^G$ reads

$$\varrho^0 b = \sum_{c \in G/H} b\overline{c},$$

where $\overline{c}$ is a representative of a coset $c \in G/H$ such that $\overline{c} = e \in G$ if $c = e \in G/H$. We also have

$$(\varrho^1 b^1)(g) = \sum_{c \in G/H} b^1(\overline{c} g \overline{cg}^{-1})\overline{c}.$$

Recall that $\overline{c} g \overline{cg}^{-1} \in H$ for $\forall g \in G$. An important property of the homomorphism (13.2.5) is that $i \circ \varrho$ and $\varrho \circ i$ are both the multiplication by the integer r. It follows that $ra = 0$ for any $a \in H^{k>0}(G/H; B^H)$, i.e., the groups $H^{k>0}(G/H; B^H)$ consist of cyclic elements.

Turn now to anomalies in quantum field theory. Let $P \to X$ be an $SU(N)$-principal bundle over an oriented compact Riemannian manifold X, $\mathcal{G}$ the gauge group and $\mathbf{A}$ the space of principal connections on $P \to X$. As in the previous Chapter, we believe that all objects requiring Sobolev completions have been completed in an appropriate way. One can think of an effective functional of quantum field theory as a complex function $S(A)$ on $\mathbf{A}$, where we omit its dependence on other fields. If $S(A)$ is not gauge invariant, we have

$$S(A^g) = iW^1(A, g) + S(A), \qquad g \in \mathcal{G}, \tag{13.2.6}$$

$$\boxed{W^1(A^g, g') - W^1(A, g'g) + W^1(A, g) = 0,} \tag{13.2.7}$$

(see the notation (13.1.13)).

A standard example of a gauge non-invariant effective action is

$$\mathrm{Tr}\,\ln \widehat{\mathcal{D}}_+ = \ln \det\widehat{\mathcal{D}}_+ = \ln \det\mathcal{D}_+$$

where $\mathcal{D}_+$ is the Weyl operator and

$$\mathcal{D}_+ = i\gamma^\mu[\partial_\mu + \frac{1}{2}A_\mu(\mathbf{1} + \gamma_5)] \tag{13.2.8}$$

is the *Dirac chiral operator* perturbatively equivalent to $\mathcal{D}^+$. A glance at the expression (13.2.7) shows that this is exactly the cocycle condition $\delta^1 W^1 = 0$ of the group cohomology of the gauge group $\mathcal{G}$ with coefficients in the module $\mathbb{C}(\mathbf{A})$ of complex functions on $\mathbf{A}$. The gauge group $\mathcal{G}$ acts on these functions by the law

$$(W^0 g)(A) = W^0(A^g).$$

One can think of $W^1(A, g)$ in (13.2.6) as being a 1-cochain

$$W^1 : \mathcal{G} \ni g \mapsto W^1(A, g) \in C^*(\mathbf{A})$$

of this cohomology. If this cochain is a coboundary

$$W^1(A, g) = W^0(A^g) - W^0(A),$$

then one can add to the effective action $S(A)$ the counterterm $-W^0(A)$ so that the renormalized action $S(A) - W^0(A)$ is gauge invariant. It follows that anomalies in pertubative quantum field theory are characterized by elements of the cohomology group $H^1(\mathcal{G};\mathbb{C}(\mathbf{A}))$, called the *anomaly group*. If the group $\mathcal{G}$ is not connected and $\mathcal{G}_e$ is the connected component of its identity e, one can examine *global anomalies*, i.e., trivial $\mathcal{G}_e$-cocycles that extend non-trivially to $\mathcal{G}$ [55, 281]. The non-trivial and $\mathcal{G}$-invariant $\mathcal{G}_e$-cocycles characterize *local anomalies*.

As was mentioned above, the action of the gauge group $\mathcal{G}$ on the space of principal connections $\mathcal{A}$ is not free. Therefore, one usually considers either the effective gauge group $\overline{\mathcal{G}}$ and the space $\overline{\mathbf{A}}$ of irreducible connections or the pointed gauge group $\mathcal{G}^0$ and the space $\mathbf{A}$. We will restrict our consideration to the first case. There is the composite fibration

$$\overline{\mathbf{A}} \to \overline{\mathbf{A}}/\overline{\mathcal{G}}_e = \overline{\mathbf{A}}_e \to \overline{\mathbf{A}}_e/\pi_0(\overline{\mathcal{G}}) = \mathcal{O}, \qquad \pi_0(\overline{\mathcal{G}}) = \overline{\mathcal{G}}/\overline{\mathcal{G}}_e. \tag{13.2.9}$$

Applying the exact sequence (13.2.4) to the case

$$G = \overline{\mathcal{G}}, \qquad H = \overline{\mathcal{G}}_e, \qquad B = \mathbb{C}(\overline{\mathbf{A}})$$

and observing that $\mathbb{C}(\overline{\mathbf{A}})^{\overline{\mathcal{G}}_e} = \mathbb{C}(\overline{\mathbf{A}}_e)$, we obtain the *anomaly exact sequence*

$$\begin{aligned} 0 &\to H^1(\pi_0(\overline{\mathcal{G}});\mathbb{C}(\overline{\mathbf{A}}_e)) \xrightarrow{j} H^1(\overline{\mathcal{G}};\mathbb{C}(\overline{\mathbf{A}})) \xrightarrow{i} H^1(\overline{\mathcal{G}}_e;\mathbb{C}(\overline{\mathbf{A}}))^{\overline{\mathcal{G}}} \\ &\to H^2(\pi_0(\overline{\mathcal{G}});\mathbb{C}(\overline{\mathbf{A}}_e)). \end{aligned} \tag{13.2.10}$$

Elements of $H^1(\pi_0(\overline{\mathcal{G}});\mathbb{C}(\overline{\mathbf{A}}_e))$ in this exact sequence characterize global anomalies, while elements of $H^1(\overline{\mathcal{G}}_e;\mathbb{C}(\overline{\mathbf{A}}))^{\overline{\mathcal{G}}}$ correspond to local anomalies. The arrow j in the anomaly sequence (13.2.10) is injective. If $H^2(\pi_0(\overline{\mathcal{G}});\mathbb{C}(\overline{\mathbf{A}}_e))$ is trivial, the arrow i in (13.2.10) is surjective. In this case, the anomaly group $H^1(\overline{\mathcal{G}};\mathbb{C}(\overline{\mathbf{A}}))$ is the direct product of groups of global and local anomalies. If the homotopy group $\pi_0(\overline{\mathcal{G}})$ is finite, only cyclic (torsion) elements in the anomaly group can be global anomalies.

To say something more, let us note that the cohomology group $H^1(\overline{\mathcal{G}};\mathbb{C}(\overline{\mathbf{A}}))$ can be seen as the group of $\overline{\mathcal{G}}$-isomorphism classes of the trivial complex line bundle $L(\overline{\mathbf{A}})$ over $\overline{\mathbf{A}}$, called the *determinant line bundle* of the Dirac chiral operator (13.2.8). This geometric interpretation of $H^1(\overline{\mathcal{G}};\mathbb{C}(\overline{\mathbf{A}}))$ applied to the fibration $\overline{\mathbf{A}} \to \mathcal{O}$ gives the exact sequence

$$0 \to H^1(\overline{\mathcal{G}};\mathbb{C}(\overline{\mathbf{A}})) \to H^2(\mathcal{O};\mathbb{Z}) \to H^2(\overline{\mathbf{A}};\mathbb{Z}) \tag{13.2.11}$$

[55]. The second arrow in this exact sequence is the injection of $H^1(\overline{\mathcal{G}};\mathbb{C}(\overline{\mathbf{A}}))$ into the group of equivalence classes of complex line bundles over the orbit space $\mathcal{O}$ (see Section 6.7) due to the identification of elements (A, c) and $(A^g, c + iW^1(A,g))$ for

all $g \in \overline{\mathcal{G}}$. The last arrow in (13.2.11) assign to every complex line bundle over $\mathcal{O}$ its pull-back onto $\overline{\mathbf{A}}$. Since $\overline{\mathbf{A}}$ is an affine space, the cohomology of $\overline{\mathbf{A}}$ is trivial and we find

$$\boxed{H^1(\overline{\mathcal{G}}; \mathbb{C}(\overline{\mathbf{A}})) = H^2(\mathcal{O}; \mathbb{Z}).}$$

Similarly to the exact sequence (13.2.11), one can construct the exact sequences corresponding to the fibrations $\overline{\mathbf{A}} \to \overline{\mathbf{A}}_e$ and $\overline{\mathbf{A}}_e \to \mathcal{O}$ (13.2.9). These exact sequences read

$$0 \to H^1(\overline{\mathcal{G}}_e; \mathbb{C}(\overline{\mathbf{A}})) \to H^2(\overline{\mathbf{A}}_e; \mathbb{Z}) \to H^2(\overline{\mathbf{A}}; \mathbb{Z}), \tag{13.2.12}$$

$$0 \to H^1(\pi_0(\overline{\mathcal{G}}); \mathbb{C}(\overline{\mathbf{A}}_e)) \to H^2(\mathcal{O}; \mathbb{Z}) \to H^2(\overline{\mathbf{A}}_e; \mathbb{Z}). \tag{13.2.13}$$

One can also apply the low-dimensional exact cohomology sequence of the fibre bundle $\overline{\mathbf{A}} \to \overline{\mathbf{A}}_e$. Since $\mathcal{G}_e$ is connected, the Leray spectral sequence [40] gives the exact sequence

$$\begin{aligned} 0 \to H^1(\overline{\mathbf{A}}_e; \mathbb{Z}) \to H^1(\overline{\mathbf{A}}; \mathbb{Z}) \to H^1(\mathcal{G}_e; \mathbb{Z}) \to H^2(\overline{\mathbf{A}}_e; \mathbb{Z}) \\ \to H^2(\overline{\mathbf{A}}; \mathbb{Z}). \end{aligned} \tag{13.2.14}$$

Since $H^2(\overline{\mathbf{A}}; \mathbb{Z}) = 0$, the exact sequences (13.2.12) and (13.2.14) result in

$$\boxed{H^1(\mathcal{G}_e; \mathbb{C}(\overline{\mathbf{A}})) = H^1(\mathcal{G}_e; \mathbb{Z}) = H^2(\overline{\mathbf{A}}_e; \mathbb{Z}).}$$

In particular, the determinant line bundle is trivial and local anomalies are absent if the corresponding complex line bundle over $\overline{\mathbf{A}}_e$ has the vanishing Chern class. This class is computed by the well-known index theorem [29]. We refer the reader to [55] for the derailed analysis of the case of $G = SU(2)$.

13.3 BRST anomalies

In Section 13.1, we have studied the anomalies related to the gauge non-invariance of the Chern–Simons form. This Section is devoted to anomalies caused by the BRST non-invariance of the Chern–Simons form (see, e.g., [29]). We follow the geometric treatment of the Faddeev–Popov ghost field as the local connection $(0,1)$-form c of a principal connection $\widehat{A}$ on the principal bundle (12.2.6) over $X \times \overline{\mathbf{A}}$.

Let $P \to X$ be an $SU(N)$-principal bundle over an oriented compact Riemannian manifold X and $\overline{\mathbf{A}}$ the space of irreducible principal connections on $P \to X$. We consider the $SU(N)$-principal bundle (12.2.6) over $X\times\overline{\mathbf{A}}$ and a connection $\widehat{A}=A+c$ on this principal bundle, where $A(x)$ is a local connection $(1,0)$-form and c is a local connection $(0,1)$-form at a point (x,A) (see Section 12.2). Let us assume that the terms $\widehat{F}_{1,1}$ and $\widehat{F}_{0,2}$ of the strength $\widehat{F}$ (12.2.11)of this connection vanish, i.e.,

$$\boxed{\widehat{F}=(d+\delta)(A+c)+(A+c)^2=F_A.} \tag{13.3.1}$$

Then we have the relations (12.2.20):

$$\boxed{\delta A=-d_Ac, \qquad \delta c=-c^2.} \tag{13.3.2}$$

As was mentioned above, these relations can be treated as the geometric model of the BRST transformations of gauge theory. One can think of δ as being the BRST operator $\mathbf{s}$ (11.4.2) with respect to the ghost field c, while the exterior differential $\widehat{d}$ (12.2.10) is associated with the total BRST operator $\widetilde{s}$ (11.4.4). The BRST operator δ acts on the exterior forms ϕ on $X\times\overline{\mathbf{A}}$, whose $(0,1)$-form degree corresponds to the ghost number. However, the exterior product of these forms, in contrast with (11.4.3), obeys the rule

$$\phi\wedge\phi'=(-1)^{(|\phi|+[\phi])(|\phi'|+[\phi'])}\phi'\wedge\phi,$$

and δ is a derivation on ϕ similarly to the BRST operator (11.3.21), but not (11.3.19). In BRST theory, the equality (13.3.1), being derived from the relations (13.3.2), is called the *Russian formula* [211]. Substituting (13.3.1) in the Bianchi identity (12.2.22), we also obtain

$$\boxed{\delta F_A=[F_A,c].} \tag{13.3.3}$$

As in Section 13.1, let $\mathcal{P}_m$ be a gauge invariant polynomial of degree m in the strength $\widehat{F}$. We have the corresponding local transgression formula (13.1.4), called the *shifted transgression formula*

$$\mathcal{P}_m(\widehat{F})=\widehat{d}Q_{2m-1}(\widehat{A},\widehat{F}_A) \tag{13.3.4}$$

with

$$Q_{2m-1}(\widehat{A},\widehat{F})=m\int_0^1 dtP(\widehat{A},\widehat{F}_t), \tag{13.3.5}$$

$$\widehat{F}_t=t\widehat{F}_A+(t^2-t)\widehat{A}^2.$$

Applying the Russian formula (13.3.1), one can equate the transgression formula (13.1.4) with the shifted one (13.3.4)

$$\boxed{\widehat{d}Q_{2m-1}(A+c,F_A) = dQ_{2m-1}(A,F_A)} \tag{13.3.6}$$

(cf. (13.1.18)). Let us expand the Chern–Simons form $Q_{2m-1}(A+c,F_A)$ in powers of the local connection form c as

$$\begin{aligned} Q_{2m-1}(A+c,F_A) = {} & Q^0_{2m-1}(c,A,F_A) + Q^1_{2m-2}(c,A,F_A) + \\ & \cdots + Q^{2m-1}_0(c), \end{aligned} \tag{13.3.7}$$

where the upper index denotes the $(1,0)$-form degree and the lower index the $(0,1)$-form degree. Substitution of (13.3.7) in (13.3.6) leads to the *descent equations*

$$\mathcal{P}_m(F_A) - dQ^0_{2m-1} = 0, \tag{13.3.8a}$$

$$\delta Q^i_{2m-1-i} + dQ^{i+1}_{2m-2-i} = 0, \qquad i = 0,\ldots,2m-2, \tag{13.3.8b}$$

$$\delta Q^{2m-1}_0 = 0. \tag{13.3.8c}$$

The chain terms Q^i_{2m-1-i} in these descent equations are treated as *BRST anomalies*. We refer the reader to [29] for the list of these anomalies for $i = 0,\ldots,3$. Higher order chain terms Q^i_{2m-1-i}, $i \geq 4$, have no definite physical interpretation.

Note that the descent equations (13.3.8b) – (13.3.8c) are similar both to the descent equations (11.4.6b) – (11.4.6c) written for an arbitrary local form in field-antifield BRST formalism and to the descent equations (12.2.24b) – (12.2.24c) for the Chern form $\mathcal{P}_2(\widehat{F}) = c_2(\widehat{F})$ of an arbitrary principal connection $\widehat{A}$. In particular, it is readily observed that the anomalies Q^i_{2m-1-i} are locally BRST-closed forms, and their local BRST cohomology can be considered. At the same time, the descent equation (13.3.8a) differs from the descent equations (11.4.6a) and (12.2.24a) since, by virtue of the relation (13.3.6), $Q_{2m-1}(A+c,F_A)$ is not a $\widehat{d}$-closed form. It follows that the local BRST cohomology of the anomalies Q^i_{2m-1-i} fails to be connected with the BRST cohomology with respect to the total BRST operator $\widehat{d}$ in general.

It should be emphasized that we have restricted our consideration above to the local transgression formula (13.3.4) valid for a trivial principal bundle $P \to X$. In the case of a non-trivial bundle $P \to X$, one can apply the transgression formula (13.1.1) and put $\delta A' = 0$ [29, 211].

Chapter 14

Connections in non-commutative geometry

There is an extensive literature on non-commutative geometry and its physical applications (see [63, 205, 246] and references therein). In non-commutative geometry, one replaces commutative algebras of smooth functions with associative algebras which are not assumed to be commutative. They are complex involutive algebras as a rule. This Chapter is devoted to the notion of a connection in non-commutative geometry. We follow the algebraic notion of connections in Chapter 8, generalized to modules over non-commutative rings [63, 90, 91]. The problem is that this generalization makes different definitions of algebraic connections non-equivalent.

Note that, in non-commutative geometry over quantum groups [94, 245, 246], connections on quantum principal bundles are defined in terms of pseudotensorial forms similar to connection forms on principal bundles and principal superconnections [93, 95].

14.1 Non-commutative algebraic calculus

In this Section, we recall a few basic facts on modules over associative algebras which are not necessarily commutative.

Let $\mathcal{A}$ be an associative unital algebra over a commutative ring $\mathcal{K}$, i.e., a $\mathcal{A}$ is a $\mathcal{K}$-ring. One considers right [left] $\mathcal{A}$-modules and $\mathcal{A}$-bimodules (or $\mathcal{A}-\mathcal{A}$-bimodules in the terminology of [204]). A bimodule P over an algebra $\mathcal{A}$ is called a *central bimodule* if

$$pa = ap, \qquad \forall p \in P, \qquad \forall a \in \mathcal{Z}(\mathcal{A}), \tag{14.1.1}$$

where $\mathcal{Z}(\mathcal{A})$ is the centre of the algebra $\mathcal{A}$. By a *centre* of a $\mathcal{A}$-bimodile P is called a $\mathcal{K}$-submodule $\mathcal{Z}(P)$ of P such that

$$pa \stackrel{\text{def}}{=} ap, \qquad \forall p \in \mathcal{Z}(P), \qquad \forall a \in \mathcal{A}.$$

Note that, if $\mathcal{A}$ is a commutative algebra, every right [left] module P over $\mathcal{A}$ becomes canonically a central bimodule by putting

$$pa = ap, \qquad \forall p \in P, \qquad \forall a \in \mathcal{A}.$$

At the same time, the bimodule $\text{Diff}_s(P, Q)$ of s-order Q-valued differential operators on a module P in Section 8.1 exemplifies a bimodule which is not central. If $\mathcal{A}$ is a non-commutative algebra, every right [left] $\mathcal{A}$-module P is also a $\mathcal{Z}(\mathcal{A}) - \mathcal{A}$-bimodule [$\mathcal{A} - \mathcal{Z}(\mathcal{A})$-bimodule] such that the equality (14.1.1) takes place, i.e., it is a central $\mathcal{Z}(\mathcal{A})$-bimodule. From now on, by a $\mathcal{Z}(\mathcal{A})$-bimodule is meant a central $\mathcal{Z}(\mathcal{A})$-bimodule. For the sake of brevity, we say that, given an associative algebra $\mathcal{A}$, right and left $\mathcal{A}$-modules, central $\mathcal{A}$-bimodules and $\mathcal{Z}(\mathcal{A})$-modules are A-modules of type $(1,0)$, $(0,1)$, $(1,1)$ and $(0,0)$, respectively, where $A_0 = \mathcal{Z}(\mathcal{A})$ and $A_1 = \mathcal{A}$. Using this notation, let us recall a few basic operations with modules.

- If P and P' are A-modules of the same type (i,j), so is its direct sum $P \oplus P'$.
- Let P and P' be A-modules of types (i,k) and (k,j), respectively. Their tensor product $P \otimes P'$ (see [204]) defines an A-module of type (i,j).
- Given an A-module P of type (i,j), let $P^* = \text{Hom}_{A_i - A_j}(P, \mathcal{A})$ be its $\mathcal{A}$-*dual*. One can show that P^* is the module of type $(i+1, j+1) \bmod 2$ [90]. In particular, P and P^{**} are A-modules of the same type. There is the natural homomorphism $P \to P^{**}$. For instance, if P is a projective module of finite rank, so is its dual P^* and $P \to P^{**}$ is an isomorphism [204].

There are several equivalent definitions of a projective module. One says that a right [left] module P is *projective* if P is a direct summand of a right [left] free module, i.e., there exists a module Q such that $P \oplus Q$ is a free module [204]. Accordingly, a module P is projective if and only if $P = \mathbf{p}S$ where S is a free module and $\mathbf{p}$ is an idempotent, i.e., an endomorphism of S such that $\mathbf{p}^2 = \mathbf{p}$. We have mentioned projective $\mathbb{C}^\infty(X)$-modules of finite rank in connection with the Serre–Swan theorem (see Theorem 14.1.1 below). Recall that a module is said to be of finite rank or simply *finite* if it is a quotient of a finitely generated free module.

Non-commutative geometry deals with unital complex involutive algebras (i.e., unital $*$-algebras) as a rule. Let $\mathcal{A}$ be such an algebra (see [82]). It should be emphasized that one cannot use right or left $\mathcal{A}$-modules, but only modules of type $(1,1)$ and $(0,0)$ since the involution of $\mathcal{A}$ reverses the order of product in $\mathcal{A}$. A central $\mathcal{A}$-bimodule P over $\mathcal{A}$ is said to be a $*$-*module* over a $*$-algebra $\mathcal{A}$ if it is equipped with an antilinear involution $p \mapsto p^*$ such that

$$(apb)^* = b^*p^*a^*, \qquad \forall a,b \in \mathcal{A}, \qquad p \in P.$$

A $*$-module is said to be a finite projective module if it is a finite projective right [left] module.

Non-commutative geometry is developed in main as a generalization of the calculus in commutative rings of smooth functions.

Let X be a locally compact topological space and $\mathcal{A}$ a $*$-algebra $\mathbb{C}^0_0(X)$ of complex continuous functions on X which vanish at infinity of X. Provided with the norm

$$||f|| = \sup_{x\in X} |f|, \qquad f \in \mathcal{A},$$

(cf. (10.5.1)), this algebra is a C^*-algebra [82]. Its spectrum $\widehat{\mathcal{A}}$ is homeomorphic to X. Conversely, any commutative C^*-algebra $\mathcal{A}$ has a locally compact spectrum $\widehat{\mathcal{A}}$ and, in accordance with the well-known *Gelfand–Naĭmark theorem*, it is isomorphic to the algebra $\mathbb{C}^0_0(\widehat{\mathcal{A}})$ of complex continuous functions on $\widehat{\mathcal{A}}$ which vanish at infinity of $\widehat{\mathcal{A}}$ [82]. If $\mathcal{A}$ is a unital commutative C^*-algebra, its spectrum $\widehat{\mathcal{A}}$ is compact. Let now X be a compact manifold. The $*$-algebra $\mathbb{C}^\infty(X)$ of smooth complex functions on X is a dense subalgebra of the unital C^*-algebra $\mathbb{C}^0(X)$ of continuous functions on X. This is not a C^*-algebra, but it is a Fréchet algebra in its natural locally convex topology of compact convergence for all derivatives (see Remark 8.1.7). In non-commutative geometry, one does not use the theory of locally convex algebras (see [221]), but considers dense unital subalgebras of C^*-algebras in a purely algebraic fashion.

The algebra $C^\infty(X)$ of smooth real functions on X is a real subalgebra of $\mathbb{C}^\infty(X)$ which consists of all Hermitian elements of $\mathbb{C}^\infty(X)$. It characterizes the manifold X in accordance with Remark 8.1.8 (see also [300]). In non-commutative geometry, one replaces the algebra of real functions with a Jordan algebra of Hermitian elements of a unital $*$-algebra $\mathcal{A}$.

Turn now to $*$-modules. Let $E \to X$ be a smooth m-dimensional complex vector bundle over a compact manifold X. The module $E(X)$ of its global sections is a $*$-module over the ring $\mathbb{C}^\infty(X)$ of smooth complex functions on X. It is a projective module of finite rank. Indeed, let $(\phi_1, \ldots, \phi_q)$ be a smooth partition of unity such that E is trivial over the sets $U_\zeta \supset \operatorname{supp} \phi_\zeta$, together with the transition functions $\rho_{\zeta\xi}$. Then $p_{\zeta\xi} = \phi_\zeta \rho_{\zeta\xi} \phi_\xi$ are smooth $(m \times m)$-matrix-valued functions on X. They satisfy

$$\sum_\kappa p_{\zeta\kappa} p_{\kappa\xi} = p_{\zeta\xi}, \tag{14.1.2}$$

and so assemble into a $(mq \times mq)$-matrix $\mathbf{p}$ whose entries are smooth complex functions on X. Because of (14.1.2), we obtain $\mathbf{p}^2 = \mathbf{p}$. Then any section s of $E \to X$ is represented by a column $(\phi_\zeta s^i)$ of smooth complex functions on X such that $\mathbf{p}s = s$. It follows that $s \in \mathbf{p}\mathbb{C}(X)^{mq}$, i.e., $E(X)$ is a projective module. The above mentioned *Serre–Swan theorem* [287, 300] provides a converse assertion.

THEOREM 14.1.1. Let P be a finite projective $*$-module over $\mathbb{C}^\infty(X)$. There exists a complex smooth vector bundle E over X such that P is isomorphic to the module $E(X)$ of global sections of E. □

In non-commutative geometry, one therefore thinks of a finite projective $*$-module over a dense unital $*$-subalgebra of a C^*-algebra as being a *non-commutative vector bundle*.

A $*$-module P may be provided with a Hermitiam structure. A right *Hermitian form* on P is a sesquilinear map $(.|.) : P \times P \to \mathcal{A}$ such that [90]:

(i) $(pa|p'b) = a^*(p|p')b$ for $a, b \in \mathcal{A}$, $p, p' \in P$;

(ii) $(p|p') = (p'|p)^*$;

(iii) $(ap|p') = (p|a^*p')$ for $a \in \mathcal{A}$, $p, p' \in P$;

(iv) $(p|p)$, $\forall p \in P$, is a positive element of $\mathcal{A}$ (i.e., $(p|p) = qq^*$ for some element q of $\mathcal{A}$);

(v) $(p|p) = 0$ forces $p = 0$.

This definition of a right Hermitian form, excluding the condition (iii), coincides with the notion of a Hermitian form on a right $\mathcal{A}$-module over a $*$-algebra $\mathcal{A}$ [300].

14.2 Non-commutative differential calculus

One believes that a non-commutative generalization of differential geometry should be given by a $\mathbb{Z}$-graded differential algebra which replaces the exterior algebra of differential forms [210]. This viewpoint is more general than that implicit above where a non-commutative ring replaces a ring of smooth functions. Throughout this Chapter, by a gradation is meant a $\mathbb{Z}$-gradation.

Recall that a *graded algebra* Ω^* over a commutative ring $\mathcal{K}$ is defined as a direct sum

$$\Omega^* = \bigoplus_{k=0} \Omega^k$$

of $\mathcal{K}$-modules Ω^k, provided with the associative multiplication law such that $\alpha \cdot \beta \in \Omega^{|\alpha|+|\beta|}$, where $|\alpha|$ denotes the degree of an element $\alpha \in \Omega^{|\alpha|}$. In particular, Ω^0 is a unital $\mathcal{K}$-algebra $\mathcal{A}$, while $\Omega^{k>0}$ are $\mathcal{A}$-bimodules. A graded algebra Ω^* is called a *graded differential algebra* if it is a cochain complex of $\mathcal{K}$-modules

$$0 \longrightarrow \mathcal{A} \xrightarrow{\delta} \Omega^1 \xrightarrow{\delta} \cdots$$

with respect to a coboundary operator δ such that

$$\delta(\alpha \cdot \beta) = \delta\alpha \cdot \beta + (-1)^{|\alpha|}\alpha \cdot \delta\beta.$$

A graded differential algebra (Ω^*, δ) with $\Omega^0 = \mathcal{A}$ is called the *differential calculus* over $\mathcal{A}$. If $\mathcal{A}$ is a $*$-algebra, we have additional conditions

$$(\alpha \cdot \beta)^* = (-1)^{|\alpha||\beta|}\beta^*\alpha^*,$$
$$(\delta\alpha)^* = \delta(\alpha^*).$$

Remark 14.2.1. The De Rham complex (8.1.42) exemplifies a differential calculus over a commutative ring. To generalize it to a non-commutative ring $\mathcal{A}$, the coboundary operator δ should have the additional properties:

- $\Omega^{k>0}$ are central $\mathcal{A}$-bimodules,

- elements $\delta a_1 \cdots \delta a_k$, $a_i \in \mathcal{Z}(\mathcal{A})$, belong to the centre $\mathcal{Z}(\Omega^k)$ of the module Ω^k. Then, if $\mathcal{A}$ is a commutative ring, the commutativity condition (8.1.27) holds.

•

Let $\Omega^*\mathcal{A}$ be the smallest differential subalgebra of the algebra Ω^* which contains $\mathcal{A}$. As an $\mathcal{A}$-algebra, it is generated by the elements δa, $a \in \mathcal{A}$, and consists of finite linear combinations of monomials of the form

$$\alpha = a_0\delta a_1 \cdots \delta a_k, \qquad a_i \in \mathcal{A}. \tag{14.2.1}$$

The product of monomials (14.2.1) is defined by the rule

$$(a_0\delta a_1)\cdot(b_0\delta b_1) = a_0\delta(a_1 b_0)\cdot\delta b_1 - a_0 a_1\delta b_0\cdot\delta b_1.$$

In particular, $\Omega^1\mathcal{A}$ is a $\mathcal{A}$-bimodule generated by elements δa, $a \in A$. Because of

$$(\delta a)b = \delta(ab) - a\delta b,$$

the bimodule $\Omega^1\mathcal{A}$ can also be seen as a left [right] $\mathcal{A}$-module generated by the elements δa, $a \in \mathcal{A}$. Note that $\delta(\mathbf{1}) = 0$. Accordingly,

$$\Omega^k\mathcal{A} = \underbrace{\Omega^1\mathcal{A}\cdots\Omega^1\mathcal{A}}_{k}$$

are $\mathcal{A}$-bimodules and, simultaneously, left [right] $\mathcal{A}$-modules generated by monomials (14.2.1).

The differential subalgebra $(\Omega^*\mathcal{A}, \delta)$ is a differential calculus over $\mathcal{A}$. It is called the *universal differential calculus* because of the following property [62, 169, 192]. Let (Ω'^*, δ') be another differential calculus over a unital $\mathcal{K}$-algebra $\mathcal{A}'$, and let $\rho : \mathcal{A} \to \mathcal{A}'$ be an algebra morphism. There exists a unique extension of this morphism to a morphism of graded differential algebras

$$\rho^k : \Omega^k\mathcal{A} \to \Omega'^k$$

such that $\rho^{k+1}\circ\delta = \delta'\circ\rho^k$.

Our interest to differential calculi over an algebra $\mathcal{A}$ is caused by the fact that, in commutative geometry, Definition 8.2.2 of an algebraic connection on an $\mathcal{A}$-module requires the module $\mathfrak{O}^1$ (8.1.22). If $\mathcal{A} = C^\infty(X)$, this module is the module of 1-forms on X. To introduce connections in non-commutative geometry, one therefore should construct the non-commutative version of the module $\mathfrak{O}^1$. We may follow the construction of $\mathfrak{O}^1$ in Section 8.1, but not take the quotient by modμ^2 that implies the commutativity condition (8.1.27).

Given a unital $\mathcal{K}$-algebra $\mathcal{A}$, let us consider the tensor product $\mathcal{A}\underset{\mathcal{K}}{\otimes}\mathcal{A}$ of $\mathcal{K}$-modules and the $\mathcal{K}$-module morphism

$$\mu^1 : \mathcal{A}\underset{\mathcal{K}}{\otimes}\mathcal{A} \ni a \otimes b \mapsto ab \in \mathcal{A}.$$

Following (8.1.22), we define the $\mathcal{K}$-module

$$\boxed{\overline{\mathfrak{O}}^1[\mathcal{A}] = \operatorname{Ker}\mu^1.} \tag{14.2.2}$$

There is the $\mathcal{K}$-module morphism

$$d : \mathcal{A} \ni a \mapsto (\mathbf{1}\otimes a - a\otimes\mathbf{1}) \in \overline{\mathfrak{O}}^1[\mathcal{A}] \tag{14.2.3}$$

(cf. (8.1.26)). Moreover, $\overline{\mathfrak{O}}^1[\mathcal{A}]$ is a $\mathcal{A}$-bimodule generated by the elements da, $a \in A$, with the multiplication law

$$b(da)c = b\otimes ac - ba\otimes c, \qquad a,b,c \in \mathcal{A}.$$

The morphism d (14.2.3) possesses the property

$$d(ab) = (\mathbf{1}\otimes ab - ab\otimes\mathbf{1} + a\otimes b - a\otimes b) = (da)b + adb \tag{14.2.4}$$

(cf. (8.1.28)), i.e., d is a $\overline{\mathfrak{O}}^1[\mathcal{A}]$-valued derivation of $\mathcal{A}$. Due to this property, $\overline{\mathfrak{O}}^1[\mathcal{A}]$ can be seen as a left $\mathcal{A}$-module generated by the elements da, $a \in \mathcal{A}$. At the same time, if $\mathcal{A}$ is a commutative ring, the $\mathcal{A}$-bimodule $\overline{\mathfrak{O}}^1[\mathcal{A}]$ does not coincide with the bimodule $\mathfrak{O}^1$ (8.1.22) because $\overline{\mathfrak{O}}^1[\mathcal{A}]$ is not a central bimodule (see Remark 14.2.1).

To overcome this difficulty, let us consider the $\mathcal{Z}(\mathcal{A})$ of derivations of the algebra $\mathcal{A}$. They obey the rule

$$u(ab) = u(a)b + au(b), \qquad \forall a,b \in \mathcal{A}. \tag{14.2.5}$$

It should be emphasized that the derivation rule (14.2.5) differs from that (9.2.9) of a graded algebra and from the derivation rule

$$u(ab) = u(a)b + u(b)a$$

for a general algebra [193]. By virtue of (14.2.5), derivations of an algebra $\mathcal{A}$ constitute a $\mathcal{Z}(\mathcal{A})$-bimodule, but not a left $\mathcal{A}$-module.

The $\mathcal{Z}(\mathcal{A})$-bimodule $\mathfrak{d}\mathcal{A}$ is also a Lie algebra over the commutative ring $\mathcal{K}$ with respect to the Lie bracket

$$[u,u'] = u\circ u' - u'\circ u. \tag{14.2.6}$$

The centre $\mathcal{Z}(\mathcal{A})$ is stable under $\mathfrak{d}\mathcal{A}$, i.e.,

$$u(a)b = bu(a), \quad \forall a \in \mathcal{Z}(\mathcal{A}), \quad b \in \mathcal{A}, \quad u \in \mathfrak{d}\mathcal{A},$$

and one has

$$[u, au'] = u(a)u' + a[u, u'], \quad \forall a \in \mathcal{Z}(\mathcal{A}), \quad u, u' \in \mathfrak{d}\mathcal{A}. \tag{14.2.7}$$

If $\mathcal{A}$ is a unital $*$-algebra, the module $\mathfrak{d}\mathcal{A}$ of derivations of $\mathcal{A}$ is provided with the involution $u \mapsto u^*$ defined by

$$u^*(a) = (u(a^*))^*.$$

Then the Lie bracket (14.2.6) satisfies the reality condition $[u, u']^* = [u^*, u'^*]$.

Let us consider the *Chevalley–Eilenberg cohomology* (see [299]) of the Lie algebra $\mathfrak{d}\mathcal{A}$ with respect to its natural representation in $\mathcal{A}$. The corresponding k-cochain space $\underline{\mathfrak{O}}^k[\mathcal{A}]$, $k = 1, \ldots$, is the $\mathcal{A}$-bimodule of $\mathcal{Z}(\mathcal{A})$-multilinear antisymmetric mappings of $\mathfrak{d}\mathcal{A}^k$ to $\mathcal{A}$. In particular, $\underline{\mathfrak{O}}^1[\mathcal{A}]$ is the $\mathcal{A}$-dual

$$\boxed{\underline{\mathfrak{O}}^1[\mathcal{A}] = \mathfrak{d}\mathcal{A}^*} \tag{14.2.8}$$

of the derivation module $\mathfrak{d}\mathcal{A}$ (cf. (8.2.15)). Put $\underline{\mathfrak{O}}^0[\mathcal{A}] = \mathcal{A}$. The *Chevalley–Eilenberg coboundary operator*

$$d : \underline{\mathfrak{O}}^k[\mathcal{A}] \to \underline{\mathfrak{O}}^{k+1}[\mathcal{A}]$$

is given by

$$(d\phi)(u_0, \ldots, u_k) = \frac{1}{k+1} \sum_{i=0}^{k} (-1)^i u_i(\phi(u_0, \ldots, \widehat{u_i}, \ldots, u_k)) + \tag{14.2.9}$$
$$\frac{1}{k+1} \sum_{0 \le r < s \le k} (-1)^{r+s} \phi([u_r, u_s], u_0, \ldots, \widehat{u_r}, \ldots, \widehat{u_s}, \ldots, u_k),$$

where $\widehat{u_i}$ means omission of u_i. For instance,

$$(da)(u) = u(a), \qquad a \in \mathcal{A}, \tag{14.2.10}$$
$$(d\phi)(u_0, u_1) = \frac{1}{2}(u_0(\phi(u_1)) - u_1(\phi(u_0)) - \phi([u_0, u_1])), \quad \phi \in \underline{\mathfrak{O}}^1[\mathcal{A}]. \tag{14.2.11}$$

It is readily observed that $d^2 = 0$, and we have the *Chevalley–Eilenberg cochain complex* of $\mathcal{K}$-modules

$$0 \longrightarrow \mathcal{A} \xrightarrow{d} \underline{\mathfrak{O}}^k[\mathcal{A}] \xrightarrow{d} \cdots. \tag{14.2.12}$$

Furthermore, the $\mathbb{Z}$-graded space

$$\boxed{\underline{\mathfrak{O}}^*[\mathcal{A}] = \bigoplus_{k=0} \underline{\mathfrak{O}}^k[\mathcal{A}]} \tag{14.2.13}$$

is provided with the structure of a graded algebra with respect to the multiplication $\wedge$ combining the product of $\mathcal{A}$ with antisymmetrization in the arguments. Notice that, if $\mathcal{A}$ is not commutative, there is nothing like graded commutativity of forms, i.e.,

$$\phi \wedge \phi' \neq (-1)^{|\phi||\phi'|}\phi' \wedge \phi$$

in general. If $\mathcal{A}$ is a $*$-algebra, $\underline{\mathfrak{O}}^*[\mathcal{A}]$ is also equipped with the involution

$$\phi^*(u_1, \ldots, u_k) \stackrel{\text{def}}{=} (\phi(u_1^*, \ldots, u_k^*))^*.$$

Thus, $(\underline{\mathfrak{O}}^*[\mathcal{A}], d)$ is a differential calculus over $\mathcal{A}$, called the *Chevalley–Eilenberg differential calculus.*

It is easy to see that, if $\mathcal{A} = \mathbb{C}^\infty(X)$ is the commutative ring of smooth complex functions on a compact manifold X, the graded algebra $\underline{\mathfrak{O}}^*[\mathbb{C}^\infty(X)]$ is exactly the complexified exterior algebra $\mathbb{C} \otimes \mathfrak{O}^*(X)$ of exterior forms on X. In this case, the coboundary operator (14.2.9) coincides with the exterior differential, and (14.2.12) is the De Rham complex of complex exterior forms on a manifold X. In particular, the operations

$$(u\rfloor\phi)(u_1, \ldots, u_{k-1}) = k\phi(u, u_1, \ldots, u_{k-1}), \qquad u \in \mathfrak{d}\mathcal{A},$$
$$\mathbf{L}_u(\phi) = d(u\rfloor\phi) + u\rfloor f(\phi),$$

are the non-commutative generalizations of the contraction and the Lie derivative of differential forms. These facts motivate one to think of elements of $\mathfrak{O}^1[\mathcal{A}]$ as being a non-commutative generalization of differential 1-forms, though this generalization by no means is unique.

Let $\mathfrak{O}^*[\mathcal{A}]$ be the smallest differential subalgebra of the algebra $\underline{\mathfrak{O}}^*[\mathcal{A}]$ which contains $\mathcal{A}$. It is generated by the elements da, $a \in \mathcal{A}$, and consists of finite linear combinations of monomials of the form

$$\phi = a_0 da_1 \wedge \cdots \wedge da_k, \qquad a_i \in \mathcal{A},$$

(cf. (14.2.1)). In particular, $\mathfrak{O}^1[\mathcal{A}]$ is a $\mathcal{A}$-bimodule (14.2.2) generated by da, $a \in \mathcal{A}$. Since the centre $\mathcal{Z}(\mathcal{A})$ of $\mathcal{A}$ is stable under derivations of $\mathcal{A}$, we have

$$bda = (da)b, \qquad adb = (db)a, \qquad a \in \mathcal{A}, \qquad b \in \mathcal{Z}(\mathcal{A}),$$
$$da \wedge db = -db \wedge da, \qquad \forall a \in \mathcal{Z}(\mathcal{A}).$$

Hence, $\mathfrak{O}^1[\mathcal{A}]$ is a central bimodule in contrast with the bimodule $\overline{\mathfrak{O}}^1[\mathcal{A}]$ (14.2.2). By virtue of the relation (14.2.10), we have the isomorphism

$$\boxed{\mathfrak{d}\mathcal{A} = \mathfrak{O}^1[\mathcal{A}]^*} \tag{14.2.14}$$

of the $\mathcal{Z}(\mathcal{A})$-module $\mathfrak{d}\mathcal{A}$ of derivations of $\mathcal{A}$ to the $\mathcal{A}$-dual of the module $\mathfrak{O}^1[\mathcal{A}]$ (cf. (8.1.38)). Combining the duality relations (14.2.8) and (14.2.14) gives the relation

$$\boxed{\underline{\mathfrak{O}}^1[\mathcal{A}] = \mathfrak{O}^1[\mathcal{A}]^{**}.}$$

The differential subalgebra $(\mathfrak{O}^*[\mathcal{A}], d)$ is a universal differential calculus over $\mathcal{A}$. If $\mathcal{A}$ is a commutative ring, then $\mathfrak{O}^*[\mathcal{A}]$ is the De Rham complex (8.1.42).

14.3 Universal connections

Let (Ω^*, δ) be a differential calculus over a unital $\mathcal{K}$-algebra $\mathcal{A}$ and P a left [right] $\mathcal{A}$-module. Similarly to Definition 8.2.2, one can construct the tensor product $\Omega^1 \otimes P$ $[P \otimes \Omega^1]$ and define a connection on P as follows [192, 300].

DEFINITION 14.3.1. A *non-commutative connection* on a left $\mathcal{A}$-bimodule P with respect to the differential calculus (Ω^*, δ) is a $\mathcal{K}$-module morphism

$$\boxed{\nabla : P \to \Omega^1 \otimes P} \tag{14.3.1}$$

which obeys the Leibniz rule

$$\boxed{\nabla(ap) = \delta a \otimes p + a\nabla(p).}$$

□

If $\Omega^* = \Omega^*\mathcal{A}$ is a universal differential calculus, the connection (14.3.1) is called a *universal connection* [192, 300].

The *curvature* of the non-commutative connection (14.3.1) is defined as the $\mathcal{A}$-module morphism

$$\boxed{\nabla^2 : P \to \mathfrak{O}^2[\mathcal{A}] \otimes P}$$

(cf. (8.2.14)) [192]. Note also that the morphism (14.3.1) has a natural extension

$$\nabla : \Omega^k \otimes P \to \Omega^{k+1} \otimes P,$$
$$\nabla(\alpha \otimes p) = \delta\alpha \otimes p + (-1)^{|\alpha|}\alpha \otimes \nabla(p), \qquad \alpha \in \Omega^*,$$

[91, 192].

Similarly, a non-commutative connection on a right $\mathcal{A}$-module is defined. However, a connection on a left [right] module does not necessarily exist. We will refer to the following theorem (see Section 14.6).

THEOREM 14.3.2. A left [right] universal connection on a left [right] module P of finite rank exists if and only if P is projective [72, 192]. □

The problem arises when P is a $\mathcal{A}$-bimodule. If $\mathcal{A}$ is a commutative ring, left and right module structures of an $\mathcal{A}$-bimodule are equivalent, and one deals with either a left or right non-commutative connection on P (see Definition 8.2.2). If P is a $\mathcal{A}$-bimodule over a non-commutative ring, *left* and *right* connections ∇^L and ∇^R on P should be considered simultaneously. However, the pair (∇^L, ∇^R) by no means is a bimodule connection since $\nabla^L(P) \in \Omega^1 \otimes P$, whereas $\nabla^R(P) \in P \otimes \Omega^1$. As a palliative, one assumes that there exists a bimodule isomorphism

$$\varrho : \Omega^1 \otimes P \to P \otimes \Omega^1. \tag{14.3.2}$$

Then a pair (∇^L, ∇^R) of right and left non-commutative connections on P is called a ϱ-compatible if

$$\boxed{\varrho \circ \nabla^L = \nabla^R}$$

[91, 192, 227] (see also [75] for a weaker condition). Nevertheless, this is not a true bimodule connection (see the condition (14.3.6) below).

Remark 14.3.1. If $\mathcal{A}$ is a commutative ring, the isomorphism ϱ (14.1.1) is naturally the permutation

$$\varrho : \alpha \otimes p \mapsto p \otimes \alpha, \qquad \forall \alpha \in \Omega^1, \quad p \in P.$$

•

The above mentioned problem of a bimodule connection is not simplified radically even if $P = \Omega^1$, together with the natural permutations

$$\phi \otimes \phi' \mapsto \phi' \otimes \phi, \qquad \phi, \phi' \in \Omega^1,$$

[90, 227].

Let now $(\mathfrak{O}^*[\mathcal{A}], d)$ be the universal differential calculus over a non-commutative $\mathcal{K}$-ring $\mathcal{A}$. Let

$$\begin{aligned} &\nabla^L : P \to \mathfrak{O}^1[\mathcal{A}] \otimes P, \\ &\nabla^L(ap) = da \otimes p + a\nabla^L(p). \end{aligned} \tag{14.3.3}$$

be a left universal connection on a left $\mathcal{A}$-module P (cf. Definition 8.2.2). Due to the duality relation (14.2.14), there is the $\mathcal{K}$-module endomorphism

$$\nabla^L_u : P \ni p \to u\rfloor\nabla^L(p) \in P \tag{14.3.4}$$

of P for any derivation $u \in \mathfrak{d}\mathcal{A}$. If ∇^R is a right universal connection on a right $\mathcal{A}$-module P, the similar endomorphism

$$\nabla^R_u : P \ni p \to \nabla^L(p)\lfloor u \in P \tag{14.3.5}$$

takes place for any derivation $u \in \mathfrak{d}\mathcal{A}$. Let (∇^L, ∇^R) be a ϱ-compatible pair of left and right universal connections on an $\mathcal{A}$-bimodule P. It seems natural to say that this pair is a bimodule universal connection on P if

$$u\rfloor\nabla^L(p) = \nabla^R(p)\lfloor u \tag{14.3.6}$$

for all $p \in P$ and $u \in \mathfrak{d}\mathcal{A}$. Nevertheless, motivated by the endomorphisms (14.3.4) - (14.3.5), one can suggest another definition of connections on a bimodule, similar to Definition 8.2.6.

14.4 The Dubois-Violette connection

Let $\mathcal{A}$ be $\mathcal{K}$-ring and P an A-module of type (i,j) in accordance with the notation in Section 14.1.

DEFINITION 14.4.1. By analogy with Definition 8.2.6, a *Dubois-Violette connection* on an A-module P of type (i,j) is a $\mathcal{Z}(\mathcal{A})$-bimodule morphism

$$\boxed{\nabla : \mathfrak{d}\mathcal{A} \ni u \mapsto \nabla_u \in \mathrm{Hom}_{\mathcal{K}}(P,P)} \tag{14.4.1}$$

of $\mathfrak{d}\mathcal{A}$ to the $\mathcal{Z}(\mathcal{A})$-bimodule of endomorphisms of the $\mathcal{K}$-module P which obey the Leibniz rule

$$\boxed{\nabla_u(a_ipa_j) = u(a_i)pa_j + a_i\nabla_u(p)a_j + a_ipu(a_j),} \quad \forall p \in P, \quad \forall a_k \in A_k, \tag{14.4.2}$$

[90, 227]. □

By virtue of the duality relation (14.2.14) and the expressions (14.3.4) – (14.3.5), every left [right] universal connection yields a connection (14.4.1) on a left [right] $\mathcal{A}$-module P. From now on, by a connection in non-commutative geometry is meant a Dubois-Violette connection in accordance with Definition (14.4.1).

A glance at the expression (14.4.2) shows that, if connections on an A-module P of type (i,j) exist, they constitute an affine space modelled over the linear space of $\mathcal{Z}(\mathcal{A})$-bimodule morphisms

$$\sigma : \mathfrak{d}\mathcal{A} \ni u \mapsto \sigma_u \in \mathrm{Hom}\,_{A_i-A_j}(P,P)$$

of $\mathfrak{d}\mathcal{A}$ to the $\mathcal{Z}(\mathcal{A})$-bimodule of endomorphisms

$$\sigma_u(a_i p a_j) = a_i\sigma(p)a_j, \quad \forall p \in P, \quad \forall a_k \in A_k,$$

of the A-module P.

Example 14.4.1. If $P = \mathcal{A}$, the morphisms

$$\nabla_u(a) = u(a), \quad \forall u \in \mathfrak{d}\mathcal{A}, \quad \forall a \in \mathcal{A}, \tag{14.4.3}$$

define a canonical connection on $\mathcal{A}$ in accordance with Definition 14.4.1. Then the Leibniz rule (14.4.2) shows that any connection on a central $\mathcal{A}$-bimodule P is also a connection on P seen as a $\mathcal{Z}(\mathcal{A})$–bimodule. •

Example 14.4.2. If P is a $\mathcal{A}$-bimodule and $\mathcal{A}$ has only inner derivations

$$\mathrm{ad}\, b(a) = ba - ab,$$

the morphisms

$$\nabla_{\mathrm{ad}b}(p) = bp - pb, \quad \forall b \in \mathcal{A}, \quad \forall p \in P, \tag{14.4.4}$$

define a canonical connection on P. •

By the *curvature* R of a connection ∇ (14.4.1) on an A-module P is meant the $\mathcal{Z}(\mathcal{A})$-module morphism

$$R : \mathfrak{d}\mathcal{A} \times \mathfrak{d}\mathcal{A} \ni (u,u') \to R_{u,u'} \in \mathrm{Hom}\,_{A_i-A_j}(P,P), \tag{14.4.5}$$

$$\boxed{R_{u,u'}(p) = \nabla_u(\nabla_{u'}(p)) - \nabla_{u'}(\nabla_u(p)) - \nabla_{[u,u']}(p),} \quad p \in P,$$

(cf. (8.2.19)) [90]. We have

$$R_{au,a'u'} = aa'R_{u,u'}, \qquad a, a' \in \mathcal{Z}(\mathcal{A}),$$
$$R_{u,u'}(a_i p b_j) = a_i R_{u,u'}(p) b_j, \quad a_i \in A_j, \quad b_j \in A_j.$$

For instance, the curvature of the connections (14.4.3) and (14.4.4) vanishes.

Let us provide some standard operations with the connections (14.4.1).

(i) Given two modules P and P' of the same type (i,j) and connections ∇ and ∇' on them, there is an obvious connection $\nabla \oplus \nabla'$ on $P \oplus P'$.

(ii) Let P be a module of type (i,j) and P^* its $\mathcal{A}$-dual. For any connection ∇ on P, there is a unique dual connection ∇' on P^* such that

$$u(\langle p,p'\rangle) = \langle \nabla_u(p), p'\rangle + \langle p, \nabla'(p')\rangle, \quad p \in P, \quad p' \in P^*, \quad u \in \mathfrak{d}\mathcal{A}.$$

(iii) Let P_1 and P_2 be A-modules of types (i,k) and (k,j), respectively, and let ∇^1 and ∇^2 be connections on these modules. For any $u \in \mathfrak{d}\mathcal{A}$, let us consider the endomorphism

$$(\nabla^1 \otimes \nabla^2)_u = \nabla^1_u \otimes \mathrm{Id}\, P_1 + \mathrm{Id}\, P_2 \otimes \nabla^2_u \tag{14.4.6}$$

of the tensor product $P_1 \otimes P_2$ of $\mathcal{K}$-modules P_1 and P_2. This endomorphism preserves the subset of $P_1 \otimes P_2$ generated by elements

$$p_1 a \otimes p_2 - p_1 \otimes a p_2,$$

with $p_1 \in P_1$, $p_2 \in P_2$ and $a \in A_k$. Due to this fact, the endomorphisms (14.4.6) define a connection on the tensor product $P_1 \otimes P_2$ of modules P_1 and P_2.

(iv) If $\mathcal{A}$ is a unital $*$-algebra, we have only modules of type $(1,1)$ and $(0,0)$, i.e., $*$-modules and $\mathcal{Z}(\mathcal{A})$-bimodules. Let P be a module of one of these types. If ∇ is a connection on P, there exists a *conjugate connection* ∇^* on P given by the relation

$$\nabla^*_u(p) = (\nabla_{u^*}(p^*))^*. \tag{14.4.7}$$

A connection ∇ on P is said to be *real* if $\nabla = \nabla^*$.

Let a $*$-module P is provided with a Hermitian form $(.|.)$. A connection ∇ on P is called *Hermitian* if

$$d(p|p') = (\nabla_u(p)|p') + (p|\nabla_u(p'), \quad \forall u \in \mathfrak{d}\mathcal{A}, \quad p, p' \in P.$$

Similarly, a Hermitian universal connection on a right $\mathcal{A}$-module P over a $*$-algebra $\mathcal{A}$ is defined. Such a Hermitian connection always exists if P is a projective module of finite rank [300].

Let now $P = \underline{\mathfrak{O}}^1[\mathcal{A}]$. A connection on $\mathcal{A}$-bimodule $\underline{\mathfrak{O}}^1[\mathcal{A}]$ is called a *linear connection* [90, 227]. Note that this is not the term for an arbitrary left [right] connection on $\underline{\mathfrak{O}}^1[\mathcal{A}]$ [91]. If $\underline{\mathfrak{O}}^1[\mathcal{A}]$ is a $*$-module, a linear connection on it is assumed to be real. Given a linear connection ∇ on $\underline{\mathfrak{O}}^1[\mathcal{A}]$, there is a $\mathcal{A}$-bimodule homomorphism, called the *torsion* of the connection ∇,

$$T : \underline{\mathfrak{O}}^1[\mathcal{A}] \to \underline{\mathfrak{O}}^2[\mathcal{A}],$$
$$\boxed{(T\phi)(u,u') = (d\phi)(u,u') - \nabla_u(\phi)(u') + \nabla_{u'}(\phi)(u),} \tag{14.4.8}$$

for all $u, u' \in \mathfrak{d}\mathcal{A}$, $\phi \in \underline{\mathfrak{O}}^1[\mathcal{A}]$.

14.5 Matrix geometry

This Section is devoted to linear connections in matrix geometry when $\mathcal{A} = M_n$ is the algebra of complex $(n \times n)$-matrices [89, 206, 207].

Let $\{\varepsilon_r\}$, $1 \leq r \leq n^2-1$, be an anti-Hermitiam basis of the Lie algebra $su(n)$. Elements ε_r generate M_n as an algebra, while $u_r = \operatorname{ad}\varepsilon_r$ constitute a basis of the right Lie algebra $\mathfrak{d}M_n$ of derivations of the algebra M_n, together with the commutation relations

$$[u_r, u_q] = c^s_{rq} u_s,$$

where c^s_{rq} are structure constants of the Lie algebra $su(n)$. Since the centre $\mathcal{Z}(M_n)$ of M_n consists of matrices $\lambda\mathbf{1}$, $\mathfrak{d}M_n$ is a complex free module of rank $n^2 - 1$.

Let us consider the universal differential calculus $(\mathfrak{O}^*[M_n], d)$ over the algebra M_n, where d is the Chevalley–Eilenberg coboundary operator (14.2.9). There is a convenient system $\{\theta^r\}$ of generators of $\mathfrak{O}^1[M_n]$ seen as a left M_n-module. They are given by the relations

$$\boxed{\theta^r(u_q) = \delta^r_q \mathbf{1}.}$$

Hence, $\mathfrak{O}^1[M_n]$ is a free left M_n-module of rank $n^2 - 1$. It is readily observed that elements θ^r belong to the centre of the M_n-bimodule $\mathfrak{O}^1[M_n]$, i.e.,

$$\boxed{a\theta^r = \theta^r a,} \qquad \forall a \in M_n. \tag{14.5.1}$$

It also follows that

$$\boxed{\theta^r \wedge \theta^q = -\theta^q \wedge \theta^r.} \tag{14.5.2}$$

The morphism $d : M_n \to \mathfrak{O}^1[M_n]$ is given by the formula (14.2.10). It reads

$$d\varepsilon_r(u_q) = \operatorname{ad}\varepsilon_q(\varepsilon_r) = c^s_{qr}\varepsilon_s,$$

that is,

$$\boxed{d\varepsilon_r = c^s_{qr}\varepsilon_s\theta^q.} \tag{14.5.3}$$

The formula (14.2.11) leads to the Maurer–Cartan equations

$$d\theta^r = -\frac{1}{2}c^r_{qs}\theta^q \wedge \theta^s. \tag{14.5.4}$$

If we define $\theta = \varepsilon_r\theta^r$, the equality (14.5.3) can be rewritten as

$$da = a\theta - \theta a, \qquad \forall a \in M_n.$$

It follows that the M_n-bimodule $\mathfrak{O}^1[M_n]$ is generated by the element θ. Since $\mathfrak{d}M_n$ is a finite free module, one can show that the M_n-bimodule $\mathfrak{O}^1[M_n]$ is isomorphic to the M_n-dual $\underline{\mathfrak{O}}^1[M_n]$ of $\mathfrak{d}M_n$.

Turn now to connections on the M_n-bimodule $\mathfrak{O}^1[M_n]$. Such a connection ∇ is given by the relations

$$\begin{aligned} &\nabla_{u=c^r u_r} = c^r\nabla_r,\\ &\nabla_r(\theta^p) = \omega^p_{rq}\theta^q, \qquad \omega^p_{rq} \in M_n. \end{aligned} \tag{14.5.5}$$

Bearing in mind the equalities (14.5.1) – (14.5.2), we obtain from the Leibniz rule (14.4.2) that

$$a\nabla_r(\theta^p) = \nabla_r(\theta^p)a, \qquad \forall a \in M_n.$$

It follows that elements ω^p_{rq} in the expression (14.5.5) are proportional $\mathbf{1} \in M_n$, i.e., complex numbers. Then the relations

$$\boxed{\nabla_r(\theta^p) = \omega^p_{rq}\theta^q,} \qquad \omega^p_{rq} \in \mathbb{C}, \tag{14.5.6}$$

define a linear connection on the M_n-bimodule $\mathfrak{O}^1[M_n]$.

Let us consider two examples of linear connections.

(i) Since all derivations of the algebra M_n are inner, we have the curvature-free connection (14.4.4) given by the relations

$$\nabla_r(\theta^p) = 0.$$

However, this connection is not torsion-free. The expressions (14.4.8) and (14.5.4) result in

$$(T\theta^p)(u_r, u_q) = -c^p_{rq}.$$

(ii) One can show that, in matrix geometry, there is a unique torsion-free linear connection

$$\boxed{\nabla_r(\theta^p) = -c^p_{rq}\theta^q.}$$

14.6 Connes' differential calculus

Connes' differential calculus is based on the notion of a spectral triple [63, 192, 205, 300].

DEFINITION 14.6.1. A *spectral triple* $(\mathcal{A}, \mathcal{H}, D)$ is given by a $*$-algebra $\mathcal{A} \subset B(\mathcal{H})$ of bounded operators on a Hilbert space $\mathcal{H}$, together with an (unbounded) self-adjoint operator $D = D^*$ on $\mathcal{H}$ with the following properties:

- the resolvent $(D - \lambda)^{-1}$, $\lambda \neq \mathbb{R}$, is a compact operator on $\mathcal{H}$,
- $[D, \mathcal{A}] \in B(\mathcal{H})$.

□

The couple $(\mathcal{A}, D)$ is also called a *K-cycle* over $\mathcal{A}$. In many cases, $\mathcal{H}$ is a $\mathbb{Z}_2$-graded Hilbert space equipped with a projector Γ such that

$$\Gamma D + D\Gamma = 0, \qquad [a, \Gamma] = 0, \quad \forall a \in \mathcal{A},$$

i.e., $\mathcal{A}$ acts on $\mathcal{H}$ by even operators, while D is an odd operator. The spectral triple is called *even* if such a grading exists and *odd* otherwise.

Remark 14.6.1. The standard example of a spectral triple is the case of the Dirac operator D on a compact spin-(spin^c)-manifold [64, 111, 112, 165] (see [208] and references therein for the geometry of spin^c-manifolds). •

Given a spectral triple $(\mathcal{A}, \mathcal{H}, D)$, let $(\Omega^*\mathcal{A}, \delta)$ be a universal differential calculus over the algebra $\mathcal{A}$. Let us construct a representation of the graded differential algebra $\Omega^*\mathcal{A}$ by bounded operators on $\mathcal{H}$ when the Chevalley–Eilenberg derivation δ (14.2.9) of $\mathcal{A}$ is replaced with the bracket $[D, a]$, $a \in \mathcal{A}$:

$$\pi : \Omega^*\mathcal{A} \to B(\mathcal{H}),$$
$$\boxed{\pi(a_0\delta a_1 \cdots \delta a_k) \stackrel{\text{def}}{=} a_0[D, a_1] \cdots [D, a_k].} \tag{14.6.1}$$

Since

$$[D, a]^* = -[D, a^*],$$

we have $\pi(\phi)^* = \pi(\phi^*)$, $\phi \in \Omega^*\mathcal{A}$. At the same time, π (14.6.1) fails to be a representation of the graded differential algebra $\Omega^*\mathcal{A}$ because $\pi(\phi) = 0$ does not imply that $\pi(\delta\phi) = 0$. Therefore, one should construct the corresponding quotient in order to obtain a graded differential algebra of operators on $\mathcal{H}$.

Let J_0 be the graded two-sided ideal of $\Omega^*\mathcal{A}$ where

$$J_0^k = \{\phi \in \Omega^k\mathcal{A} \,:\, \pi(\phi) = 0\}.$$

Then it is readily observed that $J = J_0 + \delta J_0$ is a graded differential two-sided ideal of $\Omega^*\mathcal{A}$. By *Connes' differential calculus* is meant the pair $(\Omega^*_D\mathcal{A}, d)$ such that

$$\boxed{\Omega^*_D\mathcal{A} = \Omega^*\mathcal{A}/J,}$$
$$d[\phi] = [\delta\phi],$$

where $[\phi]$ denotes the class of $\phi \in \Omega^*\mathcal{A}$ in $\Omega^*_D\mathcal{A}$. It is a differential calculus over $\Omega^0_D\mathcal{A} = \mathcal{A}$. Its k-cochain submodule $\Omega^*_D\mathcal{A}$ consists of the classes of operators

$$\sum_j a_0^j[D, a_1^j] \cdots [D, a_k^j], \qquad a_i^j \in \mathcal{A},$$

modulo the submodule of operators

$$\{\sum_j [D, b_0^j][D, b_1^j] \cdots [D, b_{k-1}^j] \,:\, \sum_j b_0^j[D, b_1^j] \cdots [D, b_{k-1}^j] = 0\}.$$

Let now P be a right finite projective module over the $*$-algebra $\mathcal{A}$. We aim to study a right connection on P with respect to Connes' differential calculus $(\Omega^*_D\mathcal{A}, d)$. As was mentioned above in Theorem 14.3.2, a right finite projective module has a connection. Let us construct this connection in an explicit form.

Given a generic right finite projective module P over a complex ring $\mathcal{A}$, let

$$\mathbf{p} : \mathbb{C}^N \underset{C}{\otimes} \mathcal{A} \to P,$$
$$i_P : P \to \mathbb{C}^N \underset{C}{\otimes} \mathcal{A},$$

be the corresponding projection and injection, where $\underset{C}{\otimes}$ denotes the tensor product over $\mathbb{C}$. There is the chain of morphisms

$$P \xrightarrow{i_P} \mathbb{C}^N \otimes \mathcal{A} \xrightarrow{\mathrm{Id}\otimes\delta} \mathbb{C}^N \otimes \Omega^1\mathcal{A} \xrightarrow{\mathbf{p}} P \otimes \Omega^1\mathcal{A}, \tag{14.6.2}$$

where the canonical module isomorphism

$$\mathbb{C}^N \underset{C}{\otimes} \Omega^1\mathcal{A} = (\mathbb{C}^N \underset{C}{\otimes} \mathcal{A}) \otimes \Omega^1\mathcal{A}$$

is used. It is readily observed that the composition (14.6.2) denoted briefly as $\mathbf{p}\circ\delta$ is a right universal connection on the module P.

Given the universal connection $\mathbf{p}\circ\delta$ on a right finite projective module P over a $*$-algebra $\mathcal{A}$, let us consider the morphism

$$\boxed{P \xrightarrow{\mathbf{p}\circ\delta} P \otimes \Omega^1\mathcal{A} \xrightarrow{\mathrm{Id}\otimes\pi} P \otimes \Omega^1_D\mathcal{A}.}$$

It is readily observed that this is a right connection ∇_0 on the module P with respect to Connes' differential calculus. Any other right connection ∇ on on P with respect to Connes' differential calculus takes the form

$$\boxed{\nabla = \nabla_0 + \sigma = (\mathrm{Id}\otimes\pi)\circ\mathbf{p}\circ\delta + \sigma} \tag{14.6.3}$$

where σ is an $\mathcal{A}$ module morphism

$$\sigma : P \to P \otimes \Omega^1_D\mathcal{A}.$$

A components σ of the connection ∇ (14.6.3) is called a *non-commutative gauge field*.

Bibliography

[1] E.Abe, *Hopf Algebras*, Cambridge Tracts in Mathematics, **74** (Cambridge Univ. Press, Cambridge, 1980).

[2] R.Adams, *Sobolev Spaces* (Academic Press, N.Y., 1975).

[3] U.Albertin, The diffeomorphism group and flat principal bundles, *J. Math. Phys.* **32** (1991) 1975.

[4] S.T.Ali, J.-P.Antoine, J.-P.Gazeau and U.Mueller, Coherent states and their generalizations: A mathematical overview, *Rev. Math. Phys.* **7** (1995) 1013.

[5] A.Almorox, Supergauge theories in graded manifolds, in *Differential Geometric Methods in Mathematical Physics*, Lect. Notes in Mathematics, **1251** (Springer-Verlag, Berlin, 1987), p.114.

[6] L.Alvarez-Gaumé and P.Ginsparg, The structure of gauge and gravitational anomalies, *Ann. Phys.* **161** (1985) 423.

[7] J.Anandan and Y.Aharonov, Geometric quantum phase and angles, *Phys. Rev. D* **38** (1988) 1863.

[8] I.Anderson, *The Variational Bicomplex* (Academic Press, Boston, 1994).

[9] M.Asorey, J.Cariñena and M.Paramion, Quantum evolution as a parallel transport, *J. Math. Phys.* **23** (1982) 1451.

[10] M.Atiyah, *K-Theory* (Benjamin, N.Y., 1967).

[11] M.Atiyah and I.Macdonald, *Introduction to Commutative Algebra* (Addison-Wesley, London, 1969).

[12] M.Atiyah and R.Bott, The Yang–Mills equations over Riemannian surfaces, *Phil. Trans. R. Soc. Lond. A* **308** (1982) 523.

[13] M.Atiyah and I.Singer, Dirac operators coupled to vector potentials, *Proc. Natl. Acad. Sci. USA* **81** (1984) 2597.

[14] S.Avis and C.Isham, Generalized spin structure on four dimensional space-times, *Commun. Math. Phys.* **72** (1980) 103.

[15] O.Babourova and B.Frolov, Perfect hypermomentum fluid: variational theory and equations of motion, *Int. J. Mod. Phys. A* **13** (1998) 5391.

[16] D.Bak, D.Cangemi and R.Jackiw, Energy-momentum conservation in gravity theories, *Phys. Rev.* **D49** (1994) 5173.

[17] P.Bandyopadhyay, Area preserving diffeomorphism, quantum group and Berry phase, *Int. J. Mod. Phys.* **14** (1998) 409.

[18] G.Barnish, F.Brandt and M.Henneaux, Local BRST cohomology in the antifield formalism. 1. General theorems, *Commun. Math. Phys.* **174** (1995) 57.

[19] G.Barnish and M.Henneaux, Isomorphism between the Batalin–Vilkovisky antibracket and the Poisson bracket, *J. Math. Phys.* **37** (1996) 5273.

[20] C.Bartocci, U.Bruzzo and D.Hernández Ruipérez, *The Geometry of Supermanifolds* (Kluwer Academic Publ., Dordrecht, 1991).

[21] C.Bartocci, U.Bruzzo, D.Hernández Ruipérez and V.Pestov, Foundations of supermanifold theory: the axiomatic approach, *Diff. Geom. Appl.* **3** (1993) 135.

[22] M.Batchelor, The structure of supermanifolds, *Trans. Amer. Math. Soc.* **253** (1979) 329.

[23] M.Batchelor, Two approaches to supermanifolds, *Trans. Amer. Math. Soc.* **258** (1980) 257.

[24] M.Bauderon, Le problème inverse du calcul des variations, *Ann. l'Inst. Henri Poincaré*, **36** (1982) 159.

[25] M.Bauderon, Differential geometry and Lagrangian formalism in the calculus of variations, in *Differential Geometry, Calculus of Variations, and their Applications*, Lecture Notes in Pure and Applied Mathematics, **100** (Marcel Dekker, Inc., N.Y., 1985), p. 67.

[26] F.Bayen, M.Flato, C.Fronsdal, A.Lichnerovicz and D.Sternbeimer, Deformation theory and quantization, *Ann. Phys.* **111** (1978) 61,111.

[27] I.Benn and R.Tucker, *An Introduction to Spinors and Geometry with Applications in Physics* (Adam Hilger, Bristol, 1987).

[28] F.Berezin, A.Kirillov (Ed.), *Introduction to Superanalysis* (Reidel, Dordreht, 1987).

[29] R.Bertlmann, *Anomalies in Quantum Field Theory* (Clarendon Press, Oxford, 1996).

[30] D.Birmingham, M.Blau, M.Rakowski and G.Thompson, Topological field theory, *Phys. Rep.* **209** (1991) 129.

[31] A.Bohm and A.Mostafazadeh, The relation between the Berry and the Anandan–Ahoronov connections for U($\mathcal{N}$) bundles, *J. Math. Phys.* **35** (1994) 1463.

[32] L.Bonora and P.Cotta-Ramusino, Some remarks on BRS transformations, anomalies and the cohomology of the Lie algebra of the group of gauge transformations, *Commun. Math. Phys.* **87** (1983) 589.

[33] A.Borowiec, M.Ferraris, M.Francaviglia and I.Volovich, Energy-momentum complex for nonlinear gravitational Lagrangians in the first-order formalism, *Gen. Rel. Grav.* **26** (1994) 637.

[34] A.Borowiec, M.Ferraris, M.Francaviglia and I.Volovich, Universality of the Einstein equations for Ricci squared Lagrangians, *Class. Quant. Grav.* **15** (1998) 43.

[35] F.Bourgeois and M.Cahen, Variational principle for symplectic connections, *J. Geom. Phys.* **30** (1999) 233.

[36] C.Boyer and O. Sánchez Valenzuela, Lie supergroup action on supermanifolds, *Trans. Amer. Math. Soc.* **323** (1991) 151.

[37] F.Brandt, Local BRST cohomology and covariance, *Commun. Math. Phys.* **190** (1997) 459.

[38] O.Bratelli and D.Robinson, Unbounded derivations of C^*-algebras, *Commun. Math. Phys.* **42** (1975) 253; **46** (1976) 11.

[39] O.Bratelli and D.Robinson, *Operator Algebras and Quantum Statistical Mechanics, Vol.1,* (Springer-Verlag, Berlin, 1979).

[40] G. Bredon, *Sheaf theory* (McGraw-Hill, N.-Y., 1967).

[41] K.Brown, *Cohomology of Groups* (Springer-Verlag, Berlin, 1982).

[42] U.Bruzzo and R.Cianci, Variational calculus on supermanifolds and invariance properties of superspace field theories, *J. Math. Phys.* **28** (1987) 786.

[43] U.Bruzzo, The global Utiyama theorem in Einstein–Cartan theory, *J. Math. Phys.* **28** (1987) 2074.

[44] U.Bruzzo, Supermanifolds, supermanifold cohomology, and super vector bundles, in *Differential Geometric Methods in Theoretical Physics*, ed. K.Bleuler and M.Werner (Kluwer, Dordrecht, 1988), p.417.

[45] U.Bruzzo and V.Pestov, On the structure of DeWitt supermanifolds, *J. Geom. Phys.* **30** (1999) 147.

[46] R.Bryant, S.Chern, R.Gardner, H.Goldschmidt, P.Griffiths, *Exterior Differential Systems* (Springer-Verlag, Berlin, 1991).

[47] P.Budinich and A.Trautman, *The Spinorial Chessboard* (Springer-Verlag, Berlin, 1988).

[48] D.Canarutto, Bundle splittings, connections and locally principle fibred manifolds, *Bull. U.M.I. Algebra e Geometria* Serie VI **V-D** (1986) 18.

[49] A.Carey, D.Crowley and M.Murray, Principal bundles and the Dixmier–Douady class, *Commun. Math. Phys.* **193** (1998) 171.

[50] J.Cariñena, J.Gomis, L.Ibort and N.Román, Canonical transformation theory for presymplectic systems, *J. Math. Phys.* **26** (1985) 1961.

[51] J.Cariñena and M.Rañada, Poisson maps and canonical transformations for time-dependent Hamiltonian systems, *J. Math. Phys.* **30** (1989) 2258.

[52] J.Cariñena, M.Crampin and L.Ibort, On the multisymplectic formalism for first order field theories, *Diff. Geom. Appl.* **1** (1991) 345.

[53] J.Cariñena and J.Fernández-Núñez, Geometric theory of time-dependent singular Lagrangians, *Fortschr. Phys.* **41** (1993) 517.

[54] J.Cariñena and H.Figueroa, Hamiltonian versus Lagrangian formulations of supermechanics, *J. Phys. A* **30** (1997) 2705.

[55] R.Catenacci and A.Lena, A note on global gauge anomalies, *J. Geom. Phys.* **30** (1999) 48.

[56] S.Cecotti and C.Vafa, Topological anti-topological fusion, *Nucl. Phys.* **B367** (1991) 359.

[57] D.Chinea, M.de León, and J.Marrero, The constraint algorithm for time-dependent Lagrangians, *J. Math. Phys.* **35** (1994) 3410.

[58] A.Chodos and V.Moncrief, Geometric gauge conditions in Yang–Mills theory: Some nonexistence results, *J. Math. Phys.* **21** (1980) 364.

[59] R.Cianci, *Introduction to Supermanifolds* (Bibliopolis, Naples, 1990).

[60] R.Cianci, M.Francaviglia and I.Volovich, Variational calculus and Poincaré–Cartan formalism in supermanifolds, *J. Phys. A* **28** (1995) 723.

[61] R.Cirelli, A.Maniá and L.Pizzocchero, A functional representation for non-commutative C^*-algebras, *Rev. Math. Phys.* **6** (1994) 675.

[62] A.Connes, Non-commutative differential geometry, *Publ. I.H.E.S* **62** (1986) 257.

[63] A.Connes, *Noncommutative Geometry* (Academic Press, N.Y., 1994).

[64] A.Connes, Gravity coupled with matter and the foundations of non-commutative geometry, *Commun. Math. Phys.* **182** (1996) 155.

[65] F.Cooper, A.Khare and U.Sukhatme, Supersymmetry and quantum mechanics, *Phys. Rep.* **251** (1995) 267.

[66] L.Cordero, C.Dodson and M de León, *Differential Geometry of Frame Bundles* (Kluwer Acad. Publ., Dordrecht, 1988).

[67] M.Crampin, G.Prince and G.Thompson, A geometrical version of the Helmholtz condition in time-dependent Lagrangian dynamics, *J. Phys. A* **17** (1984) 1437.

[68] M.Crampin, E.Martínez and W.Sarlet, Linear connections for systems of second-order ordinary differential equations, *Ann. Inst. Henri Poincaré* **65** (1996) 223.

[69] J.Crawford, Clifford algebra: Notes on the spinor metric and Lorentz, Poincaré and conformal groups, *J. Math.Phys.* **32** (1991) 576.

[70] F.Croom, *Basic Concepts of Algebraic Topology* (Springer-Verlag, Berlin, 1978).

[71] R.Crowell and R.Fox, *Introduction to Knot Theory* (Springer-Verlag, Berlin, 1963).

[72] J.Cuntz and D.Quillen, Algebra extension and nonsingularity, *J. Amer. Math. Soc.* **8** (1995) 251.

[73] W.Curtis and F.Miller, *Differential Manifolds and Theoretical Physics* (Academic Press, San Diego, 1985).

[74] L.Dabrowski and R.Percacci, Spinors and diffeomorphisms, *Commun. Math. Phys.* **106** (1986) 691.

[75] L.Dabrowski, P.Hajac, G.Lanfi and P.Siniscalco, Metrics and pairs of left and right connections on bimodules, *J. Math. Phys.* **37** (1996) 4635.

[76] P.Dedecker and W.Tulczyjew, Spectral sequences and the inverse problem of the calculus of variations, in *Differential Geometric Methods in Mathematical Physics*, Lect. Notes in Mathematics, **836** (Springer-Verlag, Berlin, 1980), p. 498.

[77] T.Dereli, M.Önder, J.Schray, R.Tucker and C.Yang, Non-Riemannian gravity and the Einstein–Proca system, *Class. Quant. Grav.* **13** (1996) L103.

[78] M.DeWilde and P.B.A.Lecomte, Existence of star-products and of formal deformations of the Poisson-Lie algebra of arbitrary symplectic manifolds, *Lett. Math. Phys.* **7** (1983) 487.

[79] B.DeWitt, *Supermanifolds* (Cambridge Univ. Press, Cambridge, 1984).

[80] R.Dick, Covariant conservation laws from the Palatini formalism, *Int. J. Theor. Phys.* **32** (1993) 109.

[81] W.Dittrich and M.Reuter, *Classical and Quantum Dynamics* (Springer–Verlag, Berlin, 1992).

[82] J.Dixmier, *C^*-Algebras* (North-Holland, Amsterdam, 1977).

[83] C.Dodson, *Categories, Bundles and Spacetime Topology* (Shiva Publishing Limited, Orpington, 1980).

[84] H.-D.Doebner, J.-D.Henning (eds), *Quantum Groups*, Lect. Notes in Physics, **370** (Springer-Verlag, Berlin, 1990).

[85] A.Dold, *Lectures on Algebraic Topology* (Springer-Verlag, Berlin, 1972).

[86] S.Donaldson, The orientation of Yang–Mills moduli space and 4-manifold topology, *J. Diff. Geom.* **26** (1987) 397.

[87] S.Donaldson, Polynomial invariants for smooth four-manifolds, *Topology* **29** (1990) 257.

[88] S.Donaldson and P.Kronheimer, *The Geometry of Four-Manifolds* (Claredon, Oxford, 1990).

[89] M.Dubois-Violette, R.Kerner and J.Madore, Noncommutative differential geometry of matrix algebras, *J. Math. Phys.* **31** (1990) 316.

[90] M.Dubois-Violette and P.Michor, Connections on central bimodules in non-commutative differential geometry, *J. Geom. Phys.* **20** (1996) 218.

[91] M.Dubois-Violette, J.Madore, T.Masson and J.Morad, On curvature in non-commutative geometry, *J. Math. Phys.* **37** (1996) 4089.

[92] B.Dubrovin, A.T.Fomenko and S.Novikov, *Modern Geometry – Methods and Applications: Part III. Introduction to Homology Theory* (Springer-Verlag, Berlin, 1984).

[93] M.Durdević, Geometry of quantum principal bundles, *Commun. Math. Phys.* **175** (1996) 457.

[94] M.Durdević, Quantum principal bundles and corresponding gauge theories, *J. Phys. A* **30** (1997) 2027.

[95] M.Durdević, Geometry of quantum principal bundles, *Rev. Math. Phys.* **9** (1997) 531.

[96] A.Echeverría Enríquez, M.Muñoz Lecanda and N.Román Roy, Geometrical setting of time-dependent regular systems. Alternative models, *Rev. Math. Phys.* **3** (1991) 301.

[97] A.Echeverría Enríquez, M.Muñoz Lecanda and N.Román Roy, Non-standard connections in classical mechanics, *J. Phys. A* **28** (1995) 5553.

[98] T.Eguchi, P.Gilkey, and A.Hanson, Gravitation, gauge theories and differential geometry, *Phys. Rep.* **66** (1980) 213.

[99] L.Fatibene, M.Ferraris and M.Francaviglia, Nöther formalism for conserved quantities in classical gauge field theories, II, *J. Math. Phys.* **35** (1994) 1644.

[100] L.Fatibene, M.Ferraris and M.Francaviglia, Nöther formalism for conserved quantities in classical gauge field theories, *J. Math. Phys.* **38** (1997) 3953.

[101] L.Fatibene, M.Ferraris, M.Francaviglia and M.Godina, Gauge formalism for General Relativity and fermionic matter, *Gen. Rel. Grav.* **30** (1998) 1371.

[102] B.Fedosov, A simple geometrical construction of deformation quantization, *J. Diff. Geom* **40** (1994) 213.

[103] B.Fedosov, *Deformation Quantization and Index Theory* (Akademie Verlag, Berlin, 1996).

[104] M.Ferraris and M.Francaviglia, Energy-momentum tensors and stress tensors in geometric field theories, *J. Math.Phys.* **26** (1985) 1243.

[105] J.Fisch, M.Henneaux, J.Stasheff and S.Teitelboim, Existence, uniqueness and cohomology of the classical BRST charge with ghosts of ghosts, *Commun. Math. Phys.* **120** (1989) 379.

[106] J.Fisch and M.Henneaux, Homological perturbation theory and the algebraic structure of the antifield-antibracket formalism for gauge theories, *Commun. Math. Phys.* **128** (1990) 627.

[107] P.Fiziev, Spinless matter in transposed-equi-affine theory of gravity, *Gen. Rel. Grav.* **30** (1998) 1341.

[108] A.Fomenko, *Differential Geometry and Topology* (Plenum Press, N.Y., 1987).

[109] M.Freedman, The topology of four-dimensional manifold, *J. Diff. Geom.* **17** (1982) 357.

[110] M.Freedman and F.Quinn, *Topology of 4-Manifolds* (Princeton. Univ. Press, Princeton, 1990).

[111] J.Fröhlich, O.Grandjean and A.Recknagel, Sypersymmetric quantum theory and differential geometry, *Commun. Math. Phys.* **193** (1998) 527.

[112] J.Fröhlich, O.Grandjean and A.Recknagel, Supersymmetric quantum theory and non-commutative geometry, *Commun. Math. Phys.* **203** (1999) 117.

[113] J.Fuchs, M.Schmidt and S.Schweigert, On the configuration space of gauge theories, *Nucl. Phys. B* **426** (1994) 107.

[114] R.Fulp, J.Lawson and L.Norris, Geometric prequantization on the spin bundle based on n-symplectic geometry: the Dirac equation, *Int. J. Theor. Phys.* **33** (1994) 1011.

[115] G.Gaeta and P.Morando, Michel theory of symmetry breaking and gauge theories, *Ann. Phys.* **260** (1997) 149.

[116] P.García, Gauge algebras, curvature and symplectic structure, *J. Diff. Geom.* **12** (1977) 209.

[117] K.Gawedski, Supersymmetries-mathematics of supergeometry, *Ann. Inst. Henri Poincaré* **XXVII** (1977) 335.

[118] I.Gelfand, V.Retakh and M.Shubin, Fedosov manifolds, *Advances in Math.* **136** (1998) 104.

[119] R.Geroch, Spinor structure of space-time in general relativity, *J. Math. Phys.* **9** (1968) 1739.

[120] G.Giachetta and L.Mangiarotti, Gauge-invariant and covariant operators in gauge theories, *Int. J. Theor.Phys.* **29** (1990) 789.

[121] G.Giachetta, Jet manifolds in non-holonomic mechanics, *J. Math. Phys.* **33** (1992) 1652.

[122] G.Giachetta and G.Sardanashvily, Stress-energy-momentum of affine-metric gravity. Generalized Komar superportential, *Class. Quant. Grav.* **13** (1996) L67.

[123] G.Giachetta, L.Mangiarotti and G.Sardanashvily, *New Lagrangian and Hamiltonian Methods in Field Theory* (World Scientific, Singapore, 1997).

[124] G.Giachetta, Nonlinear realizations of the diffeomorphism group in metric-affine gauge theory of gravity, *J. Math. Phys.* **40** (1999) 939.

[125] G.Giachetta, L.Mangiarotti and G.Sardanashvily, BRST-extended polysymplectic Hamiltonian formalism for field theory, *Il Nuovo Cimento* **114B** (1999) 939.

[126] G.Giachetta, L.Mangiarotti and G.Sardanashvily, Covariant Hamiltonian equations for field theory, *J. Phys. A* **32** (1999) 6629.

[127] G.Giachetta, L.Mangiarotti and G.Sardanashvily, Nonrelativistic geodesic motion, *Int. J. Theor. Phys.* **38** (1999) N10.

[128] R.Giambò, L.Mangiarotti and G.Sardanashvily, Relativistic and non-relativistic geodesic equations, *Il Nuovo Cimento* **114B** (1999) 749.

[129] W.Goldman, Topological components of spaces of representations, *Invent. Math.* **93** (1988) 557.

[130] J.Gomis, J.Paris and S.Samuel, Antibracket, antifields and gauge theory quantization, *Phy. Rep.* **259** (1995) 1.

[131] F.Gordejuela and J.Masqué, Gauge group and G-structures, *J. Phys. A* **28** (1995) 497.

[132] M.Gotay, On coisotropic imbeddings of presymplectic manifolds, *Proc. Amer. Math. Soc.* **84** (1982) 111.

[133] M.Gotay, A multisymplectic framework for classical field theory and the calculus of variations. I. Covariant Hamiltonian formalism, in *Mechanics, Analysis and Geometry: 200 Years after Lagrange*, ed. M.Francaviglia (North Holland, Amsterdam, 1991), p. 203.

[134] M.Gotay, A multisymplectic framework for classical field theory and the calculus of variations. II. Space + time decomposition, *Diff. Geom. Appl.* **1** (1991) 375.

[135] M.Gotay and J.Marsden, Stress-energy-momentum tensors and the Belinfante-Rosenfeld formula, *Contemp. Mathem.* **132** (1992) 367.

[136] E.Gozzi, M.Reuter and W.Thacker, Hidden BRS invariance in classical mechanics, *Phys. Rev.* **D40** (1989) 3363.

[137] E.Gozzi, M.Reuter and W.Thacker, Symmetries of the classical path integral on a generalized phase-space manifold, *Phys. Rev.* **D46** (1992) 757.

[138] E.Gozzi and M.Reuter, A proposal for a differential calculus in quantum mechanics, *Int. J. Mod. Phys.* **9** (1994) 2191.

[139] M.Greenberg, *Lectures on Algebraic Topology* (W.A.Benjamin, Inc., Menlo Park, 1971).

[140] W.Greub and H.-R. Petry, On the lifting of structure groups, in *Differential Geometric Methods in Mathematical Physics II*, Lect. Notes in Mathematics, **676** (Springer-Verlag, Berlin, 1978), p. 217.

[141] V.Gribov, Quantization of non-abelian gauge theories, *Nucl. Phys. B* **139** (1978) 1.

[142] J.Grifone, Structure preque-tangente et connexions, *Ann. Inst. Fourier* **XXII**, N1 (1972) 287; N3 (1972) 191.

[143] G.Hall, Space-time and holonomy groups, *Gen. Rel. Grav.* **27** (1995) 567.

[144] A.Hamoui and A.Lichnerowicz, Geometry of dynamical systems with time-dependent constraints and time-dependent Hamiltonians: An approach towards quantization, *J. Math. Phys.* **25** (1984) 923.

[145] F.Hansen, Quantum mechanics in phase space, *Rep. Math. Phys.* **19** (1984) 361.

[146] R.Hartshorne, *Algebraic Geometry*, Graduate Texts in Mathematics, **52** (Springer-Verlag, Berlin, 1977).

[147] S.Hawking and G.Ellis, *The Large Scale Structure of a Space-Time* (Cambridge Univ. Press, Cambridge, 1973).

[148] M.Hazewinkel and M.Gerstenhaber (eds), *Deformation Theory of Algebras and Structures and Applications* (Kluwer, Dordrecht, 1988).

[149] F.Hehl, P. von der Heyde, G.Kerlick and J.Nester, General relativity with spin and torsion: Foundations and prospects, *Rev. Mod. Phys.* **48** (1976) 393.

[150] F.Hehl, J.McCrea, E.Mielke and Y.Ne'eman: Metric-affine gauge theory of gravity: field equations, Noether identities, world spinors, and breaking of dilaton invariance, *Phys. Rep.* **258** (1995) 1.

[151] A.Heil, A.Kersch, N.Papadopoulos, B.Reifenhäuser and F.Scheck, Structure of the space of reducible connections for Yang–Mills theories, *J. Geom. Phys.* **7** (1990) 489.

[152] M.Henneaux, Space-time locality of the BRST formalism, *Commun. Math. Phys.* **140** (1991) 1.

[153] M.Henneaux and C.Teitelboim, *Quantization of Gauge Systems* (Prinston Univ. Press, Prinston, 1992).

[154] J.Henninig and A.Jadczyk, Spinors and isomorphisms of Lorentz structures, Preprint of University of Wroclaw, ITP UWr 88/695, 1988.

[155] D.Hernández Ruipérez and J.Muñoz Masqué, Global variational calculus on graded manifolds, *J. Math. Pures Appl.* **63** (1984) 283.

[156] B. van der Heuvel, Energy-momentum conservation in gauge theories, *J. Math. Phys.* **35** (1994) 1668.

[157] F.Hirzebruch, *Topological Methods in Algebraic Geometry* (Springer-Verlag, Berlin, 1966).

[158] N.Hitchin, The self-duality equation on a Riemannian surface, *Proc. London Math. Soc.* **55** (1987) 59.

[159] D.Husemoller, *Fiber Bundles*, Graduate Texts in Mathematics, **20** (Springer-Verlag, Berlin, 1966).

[160] B.Iliev, Quantum mechanics from a geometric-observer's viepoint, *J. Phys. A* **31** (1997) 1297.

[161] D.Ivanenko and G. Sardanashvily, The gauge treatment of gravity, *Phys. Rep.* **94** (1983) 1.

[162] A.Jadczyk and K.Pilch, Superspaces and Supersymmetries, *Commun. Math. Phys.* **78** (1981) 391.

[163] M.Kachkachi, A.Lamine and M.Sarih, Gauge theories: geometry and cohomological invariants, *Int. J. Theor. Phys.* **37** (1998) 1681.

[164] A.Kadić and D.Edelen, *A Gauge Theory of Dislocations and Disclinations* (Springer-Verlag, Berlin, 1983).

[165] W.Kalau and M.Walze, Supersymmetry and noncommutative geometry, *J. Geom. Phys.* **22** (1997) 77.

[166] F.Kamber and P.Tondeur, *Foliated Bundles and Characteristic Classes*, Lect. Notes in Mathematics, **493** (Springer-Verlag, Berlin, 1975).

[167] J.Kammerer, Analysis of the Moyal product in flat space, *J. Math. Phys.* **27** (1986) 529.

[168] M.Karoubi, *K-Theory: An Introduction* (Springer-Verlag, Berlin, 1978).

[169] M.Karoubi, Connexion, courbures et classes caracteristique en K-theorie algebrique, *Can. Nath. Soc. Conf. Proc.* **2** (1982) 19.

[170] M.Karoubi, Homologie cyclique des grupes et algébres, *C.R. Acad. Sci. Paris* **297**, Série 1 (1983) 381.

[171] C.Kassel, *Quantum Groups* (Springer-Verlag, Berlin, 1995).

[172] M.Keyl, About the geometric structure of symmetry breaking, *J. Math. Phys.* **32** (1991) 1065.

[173] O.Khudaverdian, Geometry of superspace with even and odd brackets, *J. Math. Phys.* **32** (1991) 1934.

[174] J.Kijowski and W.Tulczyjew, *A Symplectic Framework for Field Theories* (Springer-Verlag, Berlin, 1979).

[175] E.Kiritsis, A topological investigation of the quantum adiabatic phase, *Commun. Math. Phys.* **111** (1987) 417.

[176] A.Klimyk and K.Schmüdgen, *Quantum Groups and their Representations*, Texts and Monographs in Physics (Springer-Verlag, Berlin, 1997).

[177] S.Kobayashi and K.Nomizu, *Foundations of Differential Geometry, Vol.1,2.* (Interscience Publ., N.Y., 1963,69).

[178] S.Kobayashi, *Transformation Groups in Differential Geometry* (Springer-Verlag, Berlin, 1972).

[179] I.Kolář, P.Michor and J.Slovák, *Natural Operations in Differential Geometry* (Springer-Verlag, Berlin, 1993).

[180] W.Kondracki and P.Sadowski, Geometric structure on the orbit space of gauge connections, *J. Geom. Phys.* **3** (1986) 421.

[181] Y.Kosmann, Dérivées de Lie des spineurs, *Ann. di Matem. Pura ed Appl.* **91** (1972) 317.

[182] B.Kostant, Quantization and unitary representation, in *Lectures in Modern Analysis and Applications III*, Lect. Notes in Mathematics, **170** (Springer-Verlag, Berlin, 1970) p. 87.

[183] B.Kostant, Graded manifolds, graded Lie theory, and prequantization, in *Differential Geometric Methods in Mathematical Physics*, Lect. Notes in Mathematics, **570** (Springer-Verlag, Berlin, 1977) p. 177.

[184] J.Koszul, *Lectures on Fibre Bundles and Differential Geometry* (Tata University, Bombay, 1960).

[185] I.Krasil'shchik, V.Lychagin and A.Vinogradov, *Geometry of Jet Spaces and Nonlinear Partial Differential Equations* (Gordon and Breach, Glasgow, 1985).

[186] D.Krupka, Some geometric aspects of variational problems in fibred manifolds, *Folia Fac. Sci. Nat. UJEP Brunensis* **14** (1973) 1.

[187] D.Krupka and O.Stepánková, On the Hamilton form in second order calculus of variations, in *Proceedings of the Meeting "Geometry and Physics" (Florence, 1982)*, (Pitagora Editrice, Bologna, 1983) p. 85.

[188] D.Krupka, The contact ideal, *Diff. Geom. Appl.* **5** (1995) 257.

[189] D.Krupka, J.Musilova, Trivial Lagrangians in field theory, *Diff. Geom. Appl.* **9** (1998) 293.

[190] O.Krupkova, *The Geometry of Ordinary Variational Equations* (Springer-Verlag, Berlin, 1997).

[191] A.Lahiri, P.K.Roy and B.Bagchi, Supersymmetry in quantum mechanics, *Int. J. Mod. Phys. A* **5** (1990) 1383.

[192] G.Landi, *An Introduction to Noncommutative Spaces and their Geometries*, Lect. Notes in Physics, New series m: Monographs, **51** (Springer-Verlag, Berlin, 1997).

[193] S.Lang, *Algebra* (Addison–Wisley, N.Y., 1993).

[194] H.Lawson and M-L.Michelson, *Spin Geometry* (Princeton Univ. Press, Princeton, 1989).

[195] C.-Y.Lee, D.S.Hwang and Y.Ne'eman, BRST quantization of gauge theory in noncommutative geometry, *J. Math. Phys.* **37** (1996) 3725.

[196] M.de León and P.Rodrigues, *Methods of Differential Geometry in Analytical Mechanics* (North-Holland, Amsterdam, 1989).

[197] M.de León and J. Marrero, Constrained time-dependent Lagrangian systems and Lagrangian submanifolds, *J. Math. Phys.* **34** (1993) 622.

[198] M.de León and D. Martín de Diego, On the geometry of non-holonomic Lagrangian systems, *J. Math. Phys.* **37** (1996) 3389.

[199] M.de León, J. Marrero and D. Martín de Diego, Non-holonomic Lagrangian systems in jet manifolds, *J. Phys. A* **30** (1997) 1167.

[200] R.Littlejohn and M.Reinsch, Internal or shape coordinates in the n-body problem, *Phys. Rev. A* **52** (1995) 2035.

[201] R.Littlejohn and M.Reinsch, Gauge fields in the separation of rotations and internal motions in the n-body problem, *Rev. Mod. Phys.* **69** (1997) 213.

[202] J.Lott, Superconnections and higher index theorem, *Geom. Funct. Anal.* **2** (1992) 421.

[203] R.Maartens and D.Taylor, Lifted transformations on the tangent bundle, and symmetries of particle motion, *Int. J. Theor. Phys.* **32** (1993) 143.

[204] S.Mac Lane, *Homology* (Springer-Verlag, Berlin, 1967).

[205] J.Madore, *An Introduction to Noncommutative Differential Geometry and its Physical Applications* (Cambridge Univ. Press, Cambridge, 1995).

[206] J.Madore, T.Masson and J.Mourad, Linear connections on matrix geometries, *Class. Quant. Grav.* **12** (1995) 1429.

[207] J.Madore, Linear connections on fuzzy manifolds, *Class. Quant. Grav.* **13** (1996) 2109.

[208] S.Maier, Generic metrics and connections on spin- and spin^c-manifolds, *Commun. Math. Phys.* **188** (1997) 407.

[209] S.Majid, *Foundation of Quantum Group Theory* (Cambridge Univ. Press, Cambridge, 1995).

[210] G. Maltsiniotis, Le langage des espaces et des groupes quantiques, *Commun. Math. Phys.* **151** (1993) 275.

[211] J.Mañes, R.Stora and B.Zumino, Algebraic study of chiral anomalies, *Commun. Math. Phys.* **102** (1985) 157.

[212] L.Mangiarotti and M.Modugno, Fibered spaces, jet spaces and connections for field theories, in *Proceedings of the International Meeting on Geometry and Physics (Florence, 1982)*, ed. M. Modugno (Pitagora Editrice, Bologna, 1983), p. 135.

[213] L.Mangiarotti and G.Sardanashvily, *Gauge Mechanics* (World Scientific, Singapore, 1998).

[214] K.Marathe and G.Martucci, *The Mathematical Foundations of Gauge Theories* (North-Holland, Amsterdam, 1992).

[215] A.Masiello, *Variational Methods in Lorentzian Geometry* (Longman Scientific & Technical, Harlow, 1994).

[216] E.Massa and E.Pagani, Jet bundle geometry, dynamical connections and the inverse problem of Lagrangian mechanics, *Ann. Inst. Henri Poincaré* **61** (1994) 17.

[217] W.Massey, *Homology and cohomology theory* (Marcel Dekker, Inc., N.Y., 1978).

[218] V.Mathai and D.Quillen, Superconnections, Thom classes, end equivariant differential forms, *Topology* **25** (1986) 85.

[219] P.McCloud, Jet bundles in quantum field theory: the BRST-BV method, *Class. Quant. Grav.* **11** (1994) 567.

[220] J.McCrea, Irreducible decomposition of non-metricity, torsion, curvature and Bianchi identities in metric-affine space-time, *Class. Quant. Grav.* **9** (1992) 553.

[221] E.Michael, *Locally Multiplicatively Convex Topological Algebras* (Am. Math. Soc., Providence, 1974).

[222] E.Mielke, *Geometrodynamics of Gauge Fields* (Akademie-Verlag, Berlin, 1987).

[223] L.Milnor and J.Stasheff, *Characteristic Classes* (Princeton Univ. Press, Princeton, 1974).

[224] P.Mitter and C.Viallet, On the bundle of connections and the gauge orbit manifold in Yang–Mills theory, *Commun. Math. Phys.* **79** (1981) 457.

[225] R.Montgomery, The connection whose holonomy is the classical adiabatic angles of Hannay and Berry and its generalization to the non-integrable case, *Commun. Math. Phys.* **120** (1988) 269.

[226] G.Morandi, C.Ferrario, G.Lo Vecchio and C.Rubano, The inverse problem in the calculus of variations and the geometry of the tangent bundle, *Phys. Rep.* **188** (1990) 147.

[227] J.Mourad, Linear connections in noncommutative geometry, *Class. Quant. Grav.* **12** (1995) 965.

[228] J.Mujica, *Complex Analysis in Banach Spaces* (North-Holland, Amsterdam, 1986).

[229] M.Muñoz-Lecanda and N.Román-Roy, Lagrangian theory for presymplectic systems, *Ann. Inst. Henri Poincaré* **57** (1992) 27.

[230] G.Murphy, Energy and momentum from the Palatini formalism, *Int. J. Theor. Phys.* **29** (1990) 1003.

[231] M.Narasimhan and T.Ramadas, Geometry of $SU(2)$ gauge fields, *Commun. Math. Phys.* **67** (1979) 121.

[232] C.Nash, *Differential Topology and Quantum Field Theory* (Academic Press, London, 1991).

[233] Y.Ne'eman, Irreducible gauge theory of a consolidated Weinberg–Salam model, *Phys. Lett.* **B81** (1979) 190.

[234] Y.Ne'eman and S.Sternberg, Internal supersymmetry and superconnections, in *Symplectic Geometry and Mathematical Physics*, eds. P.Donato *et al.* (Birkhäuser, Berlin, 1991), p. 326.

[235] Y.Ne'eman, Noncommutative geometry, superconnections and Riemannian gravity as a low-energy theory, *Gen. Rel. Grav.* **31** (1999) 725.

[236] L.Nikolova and V.Rizov, Geometrical approach to the reduction of gauge theories with spontaneous broken symmetries, *Rep. Math. Phys.* **20** (1984) 287.

[237] V.Nistor, Higher McKean–Singer index formula and non-commutative geometry, *Contemp. Mathem.* **145** (1993) 439.

[238] J.Novotný, On the conservation laws in General Relativity, in *Geometrical Methods in Physics, Proceeding of the Conference on Differential Geometry and its Applications (Czechoslovakia 1983)*, ed. D.Krupka (University of J.E.Purkyně, Brno, 1984), p. 207.

[239] J.Novotný, Energy-momentum complex of gravitational field in the Palatini formalism, *Int. J. Theor. Phys.* **32** (1993) 1033.

[240] Yu.Obukhov and S.Solodukhin, Dirac equation and the Ivanenko–Landau–Kähler equation, *Int. J. Theor. Phys.* **33** (1994) 225.

[241] Yu.Obukhov, E.Vlachynsky, W.Esser and F.Hehl, Effective Einstein theory from metric-affine models via irreducible decomposition, *Phys. Rev. D* **56** (1997) 7769.

[242] A.Odzijewicz, Coherent states and geomertic quantization, *Commun. Math. Phys.* **114** (1988) 577.

[243] R.Percacci, *Geometry on Nonlinear Field Theories* (World Scientific, Singapore, 1986).

[244] P.Pereshogin and P.Pronin, Geometrical treatment of nonholonomic phase in quantum mechanics and applications, *Int. J. Theor. Phys.* **32** (1993) 219.

[245] M.Pflaum, Quantum groups on fibre bundles, *Commun. Math. Phys.* **166** (1994) 279.

[246] L.Pittner, *Algebraic Foundations of Non-Commutative Differential Geometry and Quantum Groups* (Springer-Verlag, Berlin, 1996).

[247] J.Pommaret, *Systems of Partial Differential Equations and Lie Pseudogroups* (Gordon and Breach, Glasgow, 1978).

[248] V.Ponomarev and Yu.Obukhov, Generalized Einstein-Maxwell theory, *Gen. Rel. Grav.* **14** (1982) 309.

[249] R.Powers and S.Sakai, Unbounded derivations in operator algebras, *J. Funct. Anal.* **19** (1975) 81.

[250] D.Quillen, Superconnections and the Chern character, *Topology* **24** (1985) 89.

[251] J.Rabin, Supermanifold cohomology and the Wess–Zumino term of the covariant superstring action, *Commun. Math. Phys.* **108** (1987) 375.

[252] B.Reinhart, *Differential Geometry and Foliations* (Springer-Verlag, Berlin, 1983).

[253] F.Riewe, Nonconservative Lagrangian and Hamiltonian mechanics, *Phys. Rev. E* **53** (1996) 1890.

[254] A.Robertson and W.Robertson, *Topological Vector Spaces* (Cambridge Univ. Press., Cambridge, 1973).

[255] W.Rodrigues and Q. de Souza, The Clifford bundle and the nature of the gravitational field, *Found. Phys.* **23** (1993) 1465.

[256] G.Roepstorff, Superconnections and the Higgs field, *J. Math. Phys.* **40** (1999) 2698.

[257] A.Rogers, A global theory of supermanifolds, *J. Math. Phys.* **21** (1980) 1352.

[258] M.Rothstein, The axioms of supermanifolds and a new structure arising from them, *Trans. Amer. Math. Soc.* **297** (1986) 159.

[259] C.Rovelli, Quantum mechanics without time: a model, *Phys. Rev. D* **42** (1990) 2683.

[260] D.Ruipérez and J.Masqué, Global variational calculus on graded manifolds, *J. Math. Pures et Appl.* **63** (1984) 283; **64** (1985) 87.

[261] G.Sardanashvily and M.Gogberashvily, The dislocation treatment of gauge fields of space-time translations, *Mod. Lett. Phys. A* **2** (1987) 609.

[262] G.Sardanashvily, The gauge model of the fifth force, *Acta Phys. Polon.* **B21** (1990) 583.

[263] G.Sardanashvily and O.Zakharov, *Gauge Gravitation Theory* (World Scientific, Singapore, 1992).

[264] G.Sardanashvily, On the geometry of spontaneous symmetry breaking, *J. Math. Phys.* **33** (1992) 1546.

[265] G.Sardanashvily, *Gauge Theory in Jet Manifolds* (Hadronic Press, Palm Harbor, 1993).

[266] G.Sardanashvily and O.Zakharov, On application of the Hamilton formalism in fibred manifolds to field theory, *Diff. Geom. Appl.* **3**, 245 (1993).

[267] G.Sardanashvily, Constraint field systems in multimomentum canonical variables, *J. Math. Phys.* **35** (1994) 6584.

[268] G.Sardanashvily, *Generalized Hamiltonian Formalism for Field Theory. Constraint Systems.* (World Scientific, Singapore, 1995).

[269] G.Sardanashvily, Stress-energy-momentum tensors in constraint field theories, *J. Math. Phys.* **38** (1997) 847.

[270] G.Sardanashvily, Stress-energy-momentum conservation law in gauge gravitation theory, *Class. Quant. Grav.* **14** (1997) 1371.

[271] G.Sardanashvily, Hamiltonian time-dependent mechanics, *J. Math. Phys.* **39** (1998) 2714.

[272] G.Sardanashvily, Covariant spin structure, *J. Math. Phys.* **39** (1998) 4874.

[273] G.Sardanashvily, Universal spin structure, *Int. J. Theor. Phys.* **37** (1998) 1265.

[274] D.Saunders, *The Geometry of Jet Bundles* (Cambridge Univ. Press, Cambridge, 1989).

[275] R.Schmid, Local cohomology in gauge theories, BRST transformations and anomalies, *Diff. Geom. Appl.* **4** (1994) 107.

[276] B.Schmidt, Conditions on a connection to be a metric connection, *Commun. Math. Phys.* 29 (1973) 55.

[277] A.Schwarz, Geometry of Batalin–Vilkovisky quantization, *Commun. Math. Phys.* **155** (1993) 249.

[278] M.Shifman, Anomalies in gauge theories, *Phys. Rep.* **209** (1991) 341.

[279] B.Simon, Holonomy, the quantum adiabatic theorem, and Berry's phase, *Phys. Rev. Lett.* **51** (1983) 2167.

[280] I.Singer and J.Thorpe, *Lecture Notes on Elementary Topology and Differential Geometry* (Springer-Verlag, Berlin, 1967).

[281] I.Singer, Some remarks on the Gribov ambiguity, *Commun. Math. Phys.* **60** (1978) 7.

[282] J.Śniatycki, *Geometric Quantization and Quantum Mechanics* (Springer-Verlag, Berlin, 1980).

[283] M.Socolovsky, Gauge transformations in fiber bundle theory, *J. Math. Phys.* **32** (1991) 2522.

[284] J.Sonnenschein, Topological quantum field theories, moduli spaces and flat gauge connections, *Phys. Rev. D* **42** (1990) 2080.

[285] T.Stavracou, Theory of connections on graded principal bundles, *Rev. Math. Phys.* **10** (1998) 47.

[286] N.Steenrod, *The Topology of Fibre Bundles* (Princeton Univ. Press, Princeton, 1972).

[287] R.Swan, Vector bundles and projective modules, *Trans. Am. Math. Soc.* **105** (1962) 264.

[288] S.Switt, Natural bundles. II. Spin and the diffeomorphism group, *J. Math. Phys.* **34** (1993) 3825.

[289] L.Szabados, On canonical pseudotensors, Sparling's form and Noether currents, *Class. Quant. Grav.* **9** (1992) 2521.

[290] F.Takens, Symmetries, conservation laws and variational principles, in *Geometry and Topology*, Lect. Notes in Mathematics, **597** (Springer-Verlag, Berlin, 1977), p. 581.

[291] F.Takens, A global version of the inverse problem of the calculus of variations, *J. Diff. Geom.* **14** (1979) 543.

[292] B.Tennison, *Sheaf Theory* (Cambridge Univ. Press, Cambridge, 1975)

[293] J.Thierry-Mieg, Geometrical reinterpretation of Faddeev–Popov ghost particles and BRS transformations, *J. Math. Phys.* **21** (1980) 2834.

[294] G.Thompson, Non-uniqueness of metrics compatible with a symmetric connection, *Class. Quant. Grav.* **10** (1993) 2035.

[295] A.Trautman, *Differential Geometry for Physicists* (Bibliopolis, Naples, 1984).

[296] R.Tucker and C.Wang, Black holes with Weyl charge and non-Riemannian waves, *Class. Quant. Grav.* **12** (1995) 2587.

[297] W.Tulczyjew, The Euler–Lagrange resolution, in *Differential Geometric Methods in Mathematical Physics*, Lect. Notes in Mathematics, **836** (Springer-Verlag, Berlin, 1980), p. 22.

[298] I.Vaisman, Symplectic curvature tensors, *Monatshefte für Mathematik* **100** (1985) 299.

[299] I.Vaisman, *Lectures on the Geometry of Poisson Manifolds* (Birkhäuser Verlag, Basel, 1994).

[300] J.Várilly and J.Grasia-Bondia, Connes' noncommutative differential geometry and the Standard Model, *J. Geom. Phys.* **12** (1993) 223.

[301] R.Vitolo, Finite order variational bicomplex, *Math. Proc. Cambridge Philos. Soc.* **125** (1999) 321.

[302] F.Warner, *Foundations of Differential Manifolds and Lie Groups* (Springer-Verlag, Berlin, 1983).

[303] R.O.Jr.Wells, *Differential Analysis on Complex Manifolds*, Graduate Texts in Mathematics, **65** (Springer-Verlag, Berlin, 1980).

[304] G.W.Whitehead, Generalized homology theories, *Trans. Amer. Math. Soc.* **102** (1962) 227.

[305] G.W.Whitehead, *Elements of Homotopy Theory* (Springer-Verlag, Berlin, 1978).

[306] G.Wiston, Topics on space-time topology, *Int. J. Theor. Phys.* **11** (1974) 341; **12** (1975) 225.

[307] E.Witten, Topological quantum field theory, *Commun. Math. Phys.* **117** (1988) 353.

[308] E.Witten, A note on the antibracket formalism, *Mod. Phys. Lett.* **A5** (1990) 487.

[309] N.Woodhouse, *Geometric Quantization* (Clarendon Press, Oxford, 1980) (2nd ed. 1992).

[310] Y.Wu, Classical non-abelian Berry's phase, *J. Math. Phys.* **31** (1990) 294.

[311] O.Zakharov, Hamiltonian formalism for nonregular Lagrangian theories in fibered manifolds, *J. Math. Phys.* **33** (1992) 607.

[312] R.Zulanke, P.Wintgen, *Differentialgeometrie und Faserbündel*, Hochschulbucher fur Mathematik, **75** (VEB Deutscher Verlag der Wissenschaften, Berlin, 1972).

Index